Engineering Management:
Challenges in the
New Millennium

Engineering Management: Challenges in the New Millennium

C. M. Chang

University at Buffalo
The State University of New York

PEARSON

Prentice Hall

Upper Saddle River, New Jersey 07458

Library of Congress Cataloging-in-Publication Data on File

Vice President and Editorial Director, ECS: *Marcia J. Horton*
Vice President and Director of Production and Manufacturing, ESM: *David W. Riccardi*
Executive Managing Editor: *Vince O'Brien*
Managing Editor: *David A. George*
Production Editors: *James Buckley and Wendy Kopf*
Director of Creative Services: *Paul Belfanti*
Creative Director: *Jayne Conte*
Cover Designer: *Bruce Kenselaar*
Art Editor: *Greg Dulles*
Manufacturing Buyer: *Lisa McDowell*
Senior Marketing Manager: *Holly Stark*

 © 2005 Pearson Education, Inc.
Pearson Prentice Hall
Pearson Education, Inc.
Upper Saddle River, New Jersey 07458

Pearson Prentice Hall® is a trademark of Pearson Education, Inc.

The author and publisher of this book have used their best efforts in preparing this book. These efforts include the development, research, and testing of the theories and programs to determine their effectiveness. The author and publisher make no warranty of any kind, expressed or implied, with regard to these programs or the documentation contained in this book. The author and publisher shall not be liable in any event for incidental or consequential damages in connection with, or arising out of, the furnishing, performance, or use of these programs.

Printed in the United States of America

10 9 8 7 6 5 4

ISBN 0-13-144678-9

Pearson Education Ltd., *London*
Pearson Education Australia Pty. Ltd., *Sydney*
Pearson Education Singapore, Pte. Ltd.
Pearson Education North Asia Ltd., *Hong Kong*
Pearson Education Canada Inc., *Toronto*
Pearson Educación de Mexico, S.A. de C.V.
Pearson Education—Japan, *Tokyo*
Pearson Education Malaysia, Pte. Ltd.
Pearson Education, Inc., *Upper Saddle River, New Jersey*

Dedicated to the memory of my parents:
Mr. Wen-Pei Chang and Mrs. Zhen-Lian Chang

Contents

Preface

Engineers with excellent managerial skills and superior business acumen are needed to lead corporate America in the new century. As the economy grows increasingly global, technologies advance at a faster pace, and the marketplace becomes more dynamic. Consequently, countless industrial companies will need technically trained engineers to turn technological innovations into profitability.

The need for engineering management training is obvious from another point of view. The National Science Foundation estimated in 2000 that about 46 percent of American engineers and scientists were actively working in managerial and administrative capacities. This managerial percentage remained more or less constant across the age groups from under 35 to over 55 years old. As the trend continues, almost one out of every two engineers or scientists will be engaged in managing people, projects, teams, technology, and other resources to add value to their companies.

This book prepares engineers to fulfill their managerial responsibilities, acquire useful business perspectives, and take on much-needed leadership roles to meet the challenges in the new millennium.

A number of themes permeate the book. Value addition, customer focus, and business perspectives are emphasized throughout. Also underlined are discussions of leadership attributes, steps to acquire these attributes, the areas in which engineering managers are expected to add value, the Web-based tools that can be aggressively applied to develop and sustain competitive advantages, the vital tasks of e-transformation, the opportunities offered by market expansion into global regions, and the preparations required for engineering managers to become global leaders.

The book is intended for undergraduate seniors majoring in such disciplines as aerospace, biomedical, chemical, computer, electrical, mechanical, industrial, and systems engineering and for first-year graduate engineering students. The book may also be used as a self-study guide by engineering professionals who aspire to become managers. It should be of value to first-time engineering supervisors or managers who are interested in further advancing their careers along the managerial career path.

The book is organized to contain three major parts: (I) "Functions of Engineering Management," (II) "Business Fundamentals for Engineering Managers," and (III) "Engineering Leadership in the New Millennium."

Part I introduces the basic functions of engineering management such as planning, organizing, leading, and controlling. These functions provide engineers and engineering managers with foundation skills to manage themselves, staff, teams, projects, technologies, and global issues of importance.

Best practices are emphasized as pertinent standards for goal setting and performance measurement. Engineering managers solve problems and minimize conflicts to achieve the company's objectives. They make rational decisions and take lawful and ethical actions. They employ Monte Carlo methods to assess projects that involve risks and uncertainties. Engineering managers engage emerging technologies, motivate a professional workforce of diverse backgrounds, advance new generations of products and services in a timely manner, and constantly surpass the best practices in the industry.

Furthermore, in this book, the roles of engineering managers in strategic planning, employee selection, team building, delegating, decision making, and the management of creativity and innovation are explained. The development of managerial competencies is emphasized.

Part II covers the fundamentals of engineering management, including cost accounting, financial accounting and analysis, managerial finance, and marketing management. This part is constructed to enable engineers and engineering managers to acquire a broadened perspective with respect to the business and stakeholders of the company and to facilitate their interaction with peer groups and units.

The book also prepares engineering managers to make decisions related to cost, finance, products, services, and capital budgets. Discounted cash flow and internal rate of return analyses are reviewed. These deliberations are of critical importance, as decisions made during the product-design phase typically determine up to 85 percent of the final costs of products. Additional deliberations are presented regarding activity-based costing (ABC) to define indirect costs related to products and services and economic value added (EVA) to determine the real profitability of an enterprise or project above and beyond the cost of capital deployed.

Also introduced is capital formation through equity and debt financing, along with resource allocation concepts based on adjusted present value (APV) for assets in place and option pricing for capital investment opportunities. By appreciating the project evaluation criteria and the tools of financial analyses, engineers and engineering managers will be in a better position to win project approvals. A critical step to developing technological projects is the acquisition and incorporation of customer feedback. For managers to lead, the foremost challenges are indeed the initiation, development, and implementation of major technological projects that contribute to the long-term profitability of the company.

The important roles and responsibilities of marketing in any profit-seeking enterprise are then clarified, along with the supporting contributions expected of engineering managers. Numerous progressive enterprises are increasingly concentrating on customer relationship management to grow their business. Such a customer orientation is expected to continue to serve as a key driving force for product design, project management, plant operations, manufacturing, customer service, and a variety of other engineering-centered activities.

Part III addresses five major topics: engineers as managers and leaders, ethics in engineering and business management, Web-based enablers for engineering and management, globalization, and engineering management in the new millennium. These discussions provide additional building blocks to enhance the preparation for engineers and engineering managers to assume technology leadership positions and to meet the challenges of the new millennium.

Engineers are known to possess a strong set of skills that enable them to do extraordinarily well in certain types of managerial work. However, some engineers may also exhibit weaknesses that prevent them from becoming effective leaders in engineering organizations or even from being able to survive as engineers in the industry. The expected norms of effective leaders are described. Steps enabling engineering managers to enhance their leadership qualities and attune themselves to the value-centered business acumen are explained. Certain outlined steps should be of great value to those engineering managers who want to become better prepared to build new products and services based on technology, to integrate technology into organizations, and to lead technology-based organizations.

Many tried-and-true rules are included that serve as suitable guidelines for engineering managers to become excellent leaders. Above all, engineering managers are expected to point the way with a vision of how to apply company core competencies to add value, to have insight into how to capture opportunities offered by the emerging technologies, and to be innovative in making products and services better, faster, and cheaper, so that they constantly improve customer satisfaction. Also deliberated are the concepts of value addition, customer focus, time to market, mass customization, supply chains, enterprise resources integration, and others.

Although engineers are known to be ranked high in trustworthiness and integrity (ahead of businessmen, bankers, certified public accountants, lawyers, and others), it is important for all engineers and engineering managers to remain vigilant in observing a code of ethics, to uphold a high standard of honesty and integrity, and to become sensitive to other topics related to ethics.

The changes wrought by the Internet are transforming most aspects of company business, including information dissemination, product distribution, and customer service. As processor design, software programming, and transmission hardware technologies continue to advance, their roles in business will surely grow and affect various functions of engineering management in the future. Progressive engineering managers need to know what Web-based enablers of engineering and management are currently available and which ones can be applied effectually to promote product customization, expedite new products to market, align supply chains, optimize inventory, foster team creativity and innovation, and upgrade customer service. Presented in considerable detail is a comprehensive set of Web-based tools related to product design, manufacturing, project management, procurement, plant operations, knowledge management, and supply-chain management.

Globalization expands the perspectives of engineers and engineering managers further with respect to divergence in culture, business practices, and value. Globalization is a major business trend that will affect innumerable enterprises in the coming decades. Engineers and engineering managers must become sensitized to the issues associated with globalization. They must prepare themselves to contribute to those enterprises wishing to capture new business opportunities offered in the global emerging markets. They need to be aware of the potential effects of job migration due to globalization and to take steps to prepare themselves to meet such challenges. A useful contribution for engineers and engineering managers to make is to create global technical alliances to take advantage of new technological and business opportunities.

Engineering management will face external challenges in the new millennium. What these specific challenges are, how to prepare to meet them, and how to optimally make use of location-specific opportunities to create competitive advantages will be examined. Progressive companies will also change organizational structures, set up supply chains, expedite e-transformation, and apply advanced tools to serve customers better, cheaper, and faster.

Globalization is also expected to constantly evolve. The United Nations has predicted that, by the year 2020, three of the five biggest national economies will be located in Asia. There will be winners and losers as businesses become more and more global. It is important for future engineering managers to explore prudent corporate strategies for engineering enterprises in the pursuit of globalization, while minimizing any detrimental impact on the environment, respecting human rights, and maintaining acceptable work conditions.

How should engineering managers prepare themselves to add value in the new millennium? What are the success factors for engineering managers in the new century? What might be their social responsibilities in the decades ahead? These questions are addressed in the final chapter of the book. Globalization will create ample opportunities for those who know how to properly prepare and equip themselves with the required global mindset, knowledge, and savvy.

To foster the leadership roles of engineering managers, a six dimensional model is proposed that emphasizes the inside, outside, present, future, local, and global dimensions. The management challenges for engineers in these dimensions are discussed.

During two regular academic semesters, this book could be supplemented by in-class discussions of a number of business cases that focus on engineering and technology management. Questions are also included at the end of each chapter to promote in-class exchange among students. Besides passing a final exam, which could be a take-home, comprehensive analysis of a specific business or engineering case, students may also be required to prepare term papers on engineering management topics of their preference, in order for them to keep abreast of new developments in the marketplace.

In preparation for this book, the author surveyed engineering management texts published in the last 5 to 10 years. None of the textbooks surveyed cover topics presented in "Web-based Enablers for Engineering and Management" (Chapter 12) or "Engineering Management in the New Millennium" (Chapter 14). Only a small number of texts addressed some of the issues raised in "Engineers as Managers/Leaders" (Chapter 10) and "Globalization" (Chapter 13). Furthermore, most of these engineering management texts do not cover "Marketing Management for Engineering Managers" (Chapter 9). The author believes that engineering managers will be able to play key leadership roles in their organizations if they optimally apply their combined strengths in engineering and management.

C. M. CHANG, PH.D., P.E., MBA

Department of Industrial Engineering
University at Buffalo
The State University of New York

Acknowledgments

This book is based on course notes used by the author to teach a two-course sequence, "Principles of Engineering Management I & II," at the University at Buffalo, State University of New York (SUNY), Amherst, New York. The author has taught these courses since 1987, with the exception of a three-year leave of absence during which time he took a business management assignment in China for a Fortune 500 company. The courses were originally developed for the new master-of-engineering degree programs that were initiated in 1987 under the leadership of Dr. George Lee (former Dean) and Samuel P. Capen Professor of engineering. Dr. Lee's encouragement and support are sincerely appreciated. Thanks also are due to Dr. Mark Karwan, current dean of the School of Engineering and Applied Science, University at Buffalo, who continues to support the master-of-engineering programs and the regular offering of these two graduate courses within the School of Engineering.

While regularly revising the course notes over a period of many years, the author benefited tremendously from in-class exchanges and interactions with a large number of graduate students who completed these courses. He is also thankful to many other students who completed their master-of-engineering projects under the author's direction at the Department of Industrial Engineering, University at Buffalo, SUNY.

The author is very appreciative for many helpful comments received from reviewers. Specific thanks are extended to Will Lanes at the University of New Orleans, Julie Dziekan at the University of Michigan at Dearborn, Hojjat Adeli at the Ohio State University, Donald R. McNeeley at Northwestern University, William J. Gray at Washington State University, Paul R. McCright at the University of South Florida, Charles S. Elliot at Arizona State University, Andrew Kusiak at the University of Iowa, and Amit Shukla at Miami University. The author is also indebted to the dedicated staff at Prentice Hall for the excellent execution of the book project, notably Marcia Horton, Editorial Director; Dorothy Marrero, Acquisitions Editor; Eric Svendsen, Executive Editor; James Buckley and Wendy Kopf, Production Editors; Ellen Sanders, Copy Editor; Andrea Messineo, Editorial Assistant; Bruce Kenselaar, Cover Designer; and Joe H. Mize, Developmental Editor.

C. M. CHANG

Department of Industrial Engineering
University at Buffalo
The State University of New York

Chapter 1

Introduction to Management Challenges for Engineers

1.1 INTRODUCTION

In our modern-day economy, customers' needs are changing rapidly, competition is becoming global, and technology is advancing at an ever-increasing speed. To maintain competitiveness in such a challenging environment, companies need effective leaders who understand technology and business. Engineers with proper management training have great opportunities to make valuable and lasting contributions (Babcock 1996; Badawy 1995).

In all companies, managers are select employees who are entrusted with the responsibilities of putting communications means to use, taking decisive actions, utilizing resources, and guiding the behavior of internal teams and external business partners to achieve company objectives (Compton 1997; Shainis, Dekom, and McVinney 1995).

The communications means applied by managers may be verbal or written, with or without body language. The actions taken include planning, organizing, leading, and controlling. The resources utilized are people, time, capital, equipment, facilities, technology, know-how, and business relationships. The teams guided by managers are individual employees (teams of one), projects, task forces, quality circles, and others. The external business partners involved are customers, suppliers, networked partners, and joint ventures or otherwise aligned companies (Silverman 1996; Dorf 1999).

This chapter starts with a brief review of the historical development of management theory and practices. Then it discusses the work of engineering managers and practicing engineers, and delineates the differences between these two types of work. Finally, the chapter addresses the challenges engineering managers face in the new millennium.

1.2 HISTORICAL DEVELOPMENT OF MANAGEMENT THEORY AND PRACTICES

From the management practices of building the Great Wall in China and the Pyramids in Egypt to the customer-focused, supply-chain management strategies of today's global companies, we have come a very long way indeed. This section reviews a slew of major milestones in the historical development of management theory and practices (Compton 1997; Babcock 1996). (See Table 1.1.)

George (1972) summarized many of the ancient management thoughts related to major historical projects, such as the Great Wall in China, the Pyramids of Egypt, monoliths on Easter Island, Mayan Temples in Mexico and Central America, Stonehenge in England, and others. The Pyramids of Egypt (2500 B.C.) represent a remarkable engineering feat, but they are equally notable for the management skills that went into their construction. Many parties (customers, subcontractors, and workers) had to be coordinated, controlled, and monitored. It was estimated to have taken 100,000 men from 20 to 30 years to complete one pyramid. Projects were completed on the basis of trial and error and intuition. There was no systematic documentation of the management strategy and methodology used for these great production-centered projects.

Chinese emperors (2350 B.C.) applied the principles of organizing, planning, directing, and controlling (George 1972). The constitution of Chow was the first known organization chart which specified the roles and responsibilities of officials reporting to the Chinese emperors. Around 500 B.C., the Persian Empire developed a logistic system for transmitting messages by creating posting stations separated from one another by the distance of one day's horse ride, so that the 2600-kilometer-long Royal Road could be traversed in only nine days. Sun Tzu wrote the well-known book *The Art of War*, which offered specific guidelines for strategic and tactical planning in waging wars. India (321 B.C.) was also known to have applied concepts relating to government, commerce, and customs. Alexander the Great (336–323 B.C.) developed an informal council whose members were each entrusted with specific responsibilities. In 120 B.C., the Chinese started to select government officials objectively by holding public examinations. This system has been in use ever since.

The Arsenal of Venice was a large industrial plant (1436), the government shipyard, that was designed to make galleys, arms, and equipment; to store equipment; and to assemble and refit ships on reserve. An assembly-line process was practiced to outfit ships. Also implemented were inventory control, personnel policies, standardization of specifications, accounting (double-entry bookkeeping), and cost control.

James Watt developed the steam engine in 1769. The steam engine, along with many other inventions during this era, changed the ways products were manufactured. Factories were built and workers were assembled. The Industrial Revolution destroyed the cottage industry in England. Chaos erupted due to crime, brutality, child labor, and overcrowded living conditions for workers. By 1800, factory layout planning, inventory control, production planning, workflow analysis, and cost analysis were developed in response to this factory chaos. Production management became the driving force to achieve productivity.

In 1790, emigrants from England set up textile mills in the United States. Railroads, steel mills, and canals were built. Industrial development in the United States was started.

TABLE 1.1 Historical Development of Management Theory and Practices

Era	Persons or Events	Accomplishments
Ancient Management Thoughts	The Great Wall in China, Pyramids of Egypt, monoliths on Easter Island, Mayan Temples in Mexico and Central America, Stonehenge in England	Involved management practices of coordination, control, and monitoring of many people over extended periods of time. (No records were available.)
	Chinese emperors (2350 B.C.)	Practiced organizing, directing, and controlling.
	Constitution of Chow (1100 B.C.)	Organization chart for officials and craft specialization.
	Persepolis in Persia (500 B.C.)	Built the 2600-km Royal Road and set up message systems using horse riders.
	Sun Tzu (500 B.C.)	The art of war—planning and directing.
	Alexander the Great (336–332 B.C.)	Practiced informal council with specific roles and responsibilities to its members.
	India (321 B.C.)	Practiced the concepts of government, commerce, and custom.
	China (120 B.C.)	Selected and classified officials by examinations into nine specific grades.
Production Practice (15th Century)	Arsenal of Venice (1436)	Streamlined production process of outfitting ships, inventory control, standardization of specification, double-entry accounting, and cost control.
Industrial Revolution (18th Century)	Steam engine invented by James Watt (1769); other technological inventions	Factories are formed involving equipment and workers; destroyed the cottage industry in England; created problems related to child labor, poor living conditions for workers, crime, and brutality; induced the creation of factory layout planning, inventory control, production planning, work-flow analysis, and cost analysis.
Industrial Development in the United States (19th Century)	Railroads, textile mills, steel mills, and waterways were built	
	Charles Babbage (1792–1871)	Advanced the concepts of division of labor, factory size optimization, profit-sharing scheme, method of observing manufactures, and the time-study method.
	West Point Military Academy (1817) started teaching engineering and management; Norwich University (1819), Rensselear Polytechnic Institute (1823), Union College (1845), Harvard, Yale, and Michigan (1847)	Expansion of engineering and management education in the United States.

(*Continued*)

TABLE 1.1 (Continued)

Era	Persons or Events	Accomplishments
	Morrill Land Grant Act (1862)	Authorized federal land for each state to establish at least one college to teach "scientific and classical studies … agricultural and mechanical arts." The mechanical arts became engineering.
	Formation of several associations: American Society of Engineering Education (1893), American Society of Mechanical Engineers (1880), and American Society of Civil Engineers (1982)	Promoted the exchange of best practices in engineering and management.
Scientific Management (20th Century)	Frederick Taylor (1856–1915)	Pioneered the time-and-motion study to break down a complex job into elementary motions and find the most efficient procedure of doing the job. Taylor's study formed the corner stone of the discipline of industrial engineering.
	Frank Gillbreth (1868–1924) and Lillian M. Gillbreth (1878–1972)	Pioneered the study of human factors in the workplace.
	Gantt (1861–1919)	Developed charts and graphed performance against time for project management.
	Henri Fayol (1841–1925)	Divided the industrial undertaking into six groups: technical (production), commercial (marketing), financial, security, accounting, and administrative activities (planning/forecasting, organization, command, coordination, and control).
		Developed 14 "general principles of administration" which remain valid today.
	Max Weber (1864–1920)	Developed a model for rational and efficient organizations involving position charter, roles and responsibilities, compensation policy, and others.
Human Factors (20th Century)	Douglas M. McGregor (1906–1964)	Developed Theory X and Theory Y.
	William Ouchi (1943–)	Developed Theory Z.
	Elton Mayo (1880–1945) and Fritz J. Roethlisberger (1898–1974)	Conducted extensive studies at Hawthorne Works near Cicero, Illinois, to study the impact of environmental, psychological, group factors, and other factors affecting worker productivity.

4

Babbage (1832) published the first study known that described the use of machinery and the organization of human resources for production purposes. He advanced the ideas of division of labor, factory size determination, profit-sharing schemes, methods of observing manufactures, and time-study methods.

Taylor (1911) studied work methods and shop management. He decomposed a factory job of a mechanical nature into a set of elementary motions, discarded unnecessary motions, and examined the remaining motions (with stopwatch studies) to find the most efficient method and sequence of motion elements. He developed wage rates to pay workers. He came up with a "frequent resting" method to minimize the physical fatigue of workers who carried iron blocks weighing 92 pounds apiece from the factory up an inclined plank to railroad flat cars. The load transferred was subsequently increased from 12.5 tons to 47.5 tons per day per worker.

Taylor was focusing on the production aspects without paying attention to the most important element in the process; namely, the workers. According to Taylor, work is divided into planning/training (a management responsibility) and rote execution (by the uneducated worker of the day). Workers were regarded as passive partners in carrying out production orders. As a result, some U.S. managers have been "Taylorized" into thinking that they do not need the input of workers. A major deficiency of scientific management has indeed been its failure to engage workers as an active part of the overall management system. Only in recent years have empowered teams become accepted as an improved organizational system over the Taylor model.

At about the same time as Taylor, Frank and Lillian Gilbreth started to emphasize the aspects of human factors in the workplace. Meanwhile, behavioral studies at Hawthorne Works of Western Electric Company (near Cicero, Illinois) were initiated in 1927 by Elton Mayo and Fritz Roethlisberger of Harvard Business School to study the environmental, psychological, group, and other factors which impacted the work output.

Fayol (1949) divided the industrial undertakings into six groups: technical (production), commercial (marketing), financial, security, accounting, and administrative activities (planning/forecasting, organization, command, coordination, and control). He also developed 14 general principles of administration which have remained valid to this day.

Weber (1947) developed a model for rational and efficient organizations involving position charter, roles and responsibilities, compensation policy, and other features.

According to Urwick (1972), the study of management started with engineers. It was in the sciences underlying the engineering practice—mathematics, physics, mechanics, and so on—which were first applied by Taylor to analyzing and measuring the tasks assigned to individuals, that the science of management got started.

Management theory and practices were also studied from several other points of view: *quantitative approach* (decision making is the central focus of management—decision theory and group decision making), *operations research* (expressing management problems in terms of mathematical symbols and relationships), and *systems approach* (organizations as interacting systems influenced by the external environment).

McGregor (1957) developed the Theory X and Theory Y of management. Theory X says that workers are passive and they are to be induced by management to contribute to the organizational objectives. Workers may be motivated according to

Maslow's model of need hierarchy. McGregor proposed to optimize the worker's motivation by management. According to Theory Y, workers can be motivated. Management must arrange workplace conditions and methods of operations (participatory management) to influence the worker's motivation.

Ouchi (1981) came up with the Theory Z to place emphasis on the following: the bottom-up process, the senior manager as a facilitator, the middle manager as an initiator and coordinator, decision by consensus, and concern for the employee. Theory Z advances four specific principles:

1. Employees should be offered lifetime employment.
2. Promotions should be based on length of service, as distinct from evaluation of immediate job performance.
3. Individuals should not be specialized but should be moved throughout the company.
4. Decisions should be made through a collective decision-making process.

Today, not all Japanese companies practice Theory Z as recommended. Furthermore, Theory Z aims at satisfying only the first three levels of hierarchical human needs (i.e., physiological, safety, and social). Paying attention to ways of satisfying the next two higher level needs (i.e., esteem and self-actualization) may bring about a management system superior to the current systems in practice.

This is certainly not the end of the story. Greater development of management theory and practice is yet to come, in view of today's workforce mobility and diversity, worker preference for independence, and business globalization. Furthermore, not every knowledge worker has the same high-intensity needs at the ego, peer recognition, and self-actualization levels. The needs of production workers may be quite different from those of professional workers. In addition, the needs of individual workers are expected to change in time due to personal circumstances (such as life stage, family circumstances, and value perceptions), business climate, and other factors involved.

The author believes that an effective worker motivation strategy may need to be individually tailored in order to maximize employee satisfaction and maintain organizational productivity. Such an approach demands that managers and leaders understand the individual workers and motivate them according to their respective unsatisfied needs.

1.3 ENGINEERING MANAGEMENT AND LEADERSHIP

Engineers interested in moving into leadership positions need to know what it takes to be selected as managers by their employers. In order to lead their employers to greater prosperity, they need to prepare themselves to accept leadership roles and responsibilities by honing their people skills and interacting effectively with peer managers in other corporate functions, and become versed in the use of Web-based management and technology tools. They also must nurture a clear vision for the companies they work for, be innovative and creative in product development and other ways of utilizing emerging technologies, and possess the required business savvy with a customer focus and a global orientation (Kossiakoff 2003; Hermone 1998). This book assists engineers in acquiring the skills and attributes deemed essential for assuming leadership roles in the new millennium.

Not all engineers are interested in becoming managers (Mintzberg 1990). Indeed, in 2003, the author conducted a small-scale survey that showed that about one-third of an engineering undergraduate class each responded "yes," "no," and "maybe" to the survey question: "Wanna be a Manager?" This book, *Engineering Management: Challenges in the New Millennium*, is useful to both engineers who want to remain technical contributors and those who elect to become managers.

1.3.1 Making Engineers Effective Technical Contributors

As a technical contributor, every engineer reports to a superior who is typically a supervisor or manager. The superior makes decisions that will have a profound impact on the engineer's contributions to the company and hence on his professional career. Exposure to the functions, concepts, skills, and best practices of engineering management allows the engineer to better align his own work with the needs of his superior and hence that of the company. The engineer is then in a better position to accept the fact that his superior will typically decide whether or not to adopt a new technology, program, or project primarily based on the value it can add to the company. The decision is likely to take into account the resources (e.g., money, people, time, technologies, and business relationships) required for implementation. The decision will not likely be made based on inherent technological sophistication, innovative strength, rational elegance, or theoretical robustness.

Value is said to have been added if the company's profitability is derived from

- **Increased sales revenues**—for example, due to new product features, better customer service, novel logistics in product and service delivery, shorter time to market, etc.
- **Reduced cost to do business**—for example, due to improved engineering and manufacturing productivity, raised operational efficiency, new synergy among aligned business partners, simplified product design, better quality control, etc.

The superior sets project priorities. Priorities are usually set based on whether the anticipated value of a project is large or small, short term or long term, direct or indirect, and certain or uncertain.

Exposure to *Engineering Management: Challenges in the New Millennium* makes engineers more effective in increasing their cumulative value contributions to their employers; such an objective is shared by Covey 1994, Longeneck and Somonetti 2001.

1.3.2 Readying Engineers to Accept Managerial Responsibilities

For those engineers who aspire to become managers, a comprehensive exposure to the topics reviewed in this book enhances their readiness for being selected when such managerial opportunities arise. Knowledge is power. The new knowledge gained from this book can motivate engineers to experiment and to excel. They accumulate experience as they systematically correct their own deficiencies and practice interpersonal skills, decision making, problem solving, delegating, cost accounting, strategic planning, project management, and team coordination. Both the new knowledge gained and the experiences accumulated provide them with a decisive advantage over other management candidates (O'Conner 1994; Matejka and Dunsing 1995).

Engineering managers must be able to lead. This is particularly true in the dynamic marketplace of the new century, which is affected by sophisticated communications tools, Web-based enablers, flexible supply chains, and business operations of global proportions. Web-based enablers are tools that are Internet based. These tools enhance the productivity of product design, project management, plant operations, facility maintenance, innovations, knowledge management, marketing and sales, enterprise resources planning and integration, and procurements. Since programs and projects will become increasingly interdisciplinary and complex in the future, decision making is likely to involve the use of Web-based tools and the participation of team members who have divergent cultural backgrounds, value systems, business priorities, and engineering practices. Also much needed is the push for technological innovations, which many engineers are particularly qualified to provide. Those engineering managers who are innovative and have both technological insights and business savvy will have opportunities to create significant value for and be richly rewarded by their employers in the new millennium (Kales 1998; Uyterhoeven 1989; Noori 1990).

This book is written to prepare both engineers to become better technical contributors and engineering managers to become better leaders in engineering organizations, so that all of them will add substantial value to their employers in the new millennium.

This book shows that certain management principles do not change over time (Shannon 1980, Dhillon 1987; Thamhain 1992; Bennett 1996; Cardulo 1996; Mazda 1997). However, management practices do change in response to changes in customers' needs, employees' attitudes, business models, technologies, organizational structures, resources, and external business relations. Managers must be able to lead and manage these changes.

A good strategy for young engineers is to learn the fundamentals of management (principles, skills, functions, roles and responsibilities, success factors, etc.), and then seek opportunities to actively practice these skills, functions, principles, and management roles. Opportunities to do so may exist in professional societies and volunteer organizations (e.g., the United Way, churches, boy scouts, girl scouts, and so forth). As more management experience is accumulated, proficiency will result and allow the engineer to naturally stand out when management openings become available in the future.

Example 1.1.

Several U.S. universities offer the academic degree program that is concentrated on engineering management. Others have developed the degree program for Management of Technology. Are they fundamentally different from one another?

Answer 1.1.

These two types of degree programs are essentially similar, with minor differences in the course work involved. Both programs are aimed at training managers to point the way in technology.

The Management of Technology degree program was envisaged to correct a deficiency noted in the U.S. educational system (National Research Council 1987). This degree program is designed to address important management issues, such as those enumerated here, which were neglected in the master of business administration program existing at the time:

A. Integrating technology into the overall strategic objectives of the firm
B. Getting into and out of technologies more efficiently
C. Assessing and evaluating technology more effectively
D. Developing better methods for transferring and assimilating new technology
E. Reducing new product development time
F. Managing large, complex, and interdisciplinary or interorganizational projects, programs, and systems
G. Leveraging the effectiveness of technical professionals

Currently, the degree programs in engineering management offered by many U.S. universities address these topics as well.

1.4 DEFINITIONS

It is proper to begin with the language of management. The following are a few management terms that will be used frequently throughout this book (Compton 1997; Babcock 1996):

1.4.1 Management Responsibilities

The management group of a company has the overall responsibility of achieving the company's objectives and meeting the diverse expectations of its stakeholders. The management group is composed of managers at various levels, from CEO (chief executive officer) down to first-line managers (e.g., group leader, section head, manager, etc.).

The stakeholders are groups of people who have a stake in the company's performance. These include shareholders, customers, suppliers, employees, and the community in which the company operates. Typical expectations of these stakeholders include:

A. **Shareholders**—return on investment, dividends, and earnings per share; and increase in stock price over time
B. **Customers**—quality products, acceptable services, and flexibility to accommodate changing needs, efficient delivery, and reasonable prices
C. **Suppliers**—financial stability, market share position, quality production, collaboration efficiency, and on-time payment
D. **Employees**—good working conditions, stable employment, and competitive salary and benefits
E. **Community**—environmentally clean, tax contribution, socially responsible, ethically acceptable reputation, and good corporate citizenship

Over time, company management is responsible for satisfying this diverse set of expectation of all stakeholders.

1.4.2 Type of Work

Work is the task performed to add value to the company. Performing the work involves the use of resources (e.g., time, money, energy, tools, human efforts, technologies, facilities, etc.). There are three types of work:

A. **Management Work**—Plan, organize, lead, and control the efforts of self and others; this requires thinking

B. **Technical Work**—Specialized, nonmanagement work done by engineering managers if others cannot do it for them; this requires doing

C. **Operating Work**—Management and technical work that has been delegated to others; this requires monitoring and controlling

1.4.3 Chain of Command

The chain of command refers to the chain of direct authority relationships between superiors and subordinates. This is derived from military systems.

1.4.4 Principle of Unity of Command

According to this principle, an engineering subordinate reports to a single superior.

1.4.5 Efficiency

Efficiency refers to the accomplishment of a given task with the least amount of effort. Being efficient means not wasting resources (e.g., time, money, equipment, facilities, skills, talents, etc.).

1.4.6 Effectiveness

Effectiveness refers to the accomplishment of tasks with efforts that are commensurate with the value created by these tasks.

The paradigm "All things worth doing are worth doing well" should be replaced by "All things worth doing are worth doing well only to the extent of their contributing value to the company." Engineers and managers need to be value conscious. Perfectionists have no place in a progressive industrial environment.

1.4.7 Strategic and Operational (Tactical) Decisions

Strategic decisions are those which set the direction for the unit, department and company and determine what the right things to do are. Examples include which new markets to pursue, what new products to develop, who should be engaged as supply chain partners, and when the right time is to acquire which new technologies. Operational decisions are those that specify ways to implement a specific task, project, or program and define how to do things correctly.

Engineers with managerial responsibilities are involved in making strategic decisions. Engineers as technical contributors are typically involved with decisions that are operational in nature.

1.5 ENGINEERING MANAGERS

Engineering managers have engineering training and specific technical knowledge and experience, and are accountable for the results of the unit, section, or department they head up.

1.5.1 Prerequisites to Be an Effective Engineering Manager

Generally speaking, to be effective, an engineering manager needs to be motivated to acquire knowledge (e.g., roles, functions, and vocabulary) and skills; prepare mentally (job outlook, management orientation, personality traits, and flexibility); and be determined to diligently practice the principles of engineering management. Many skills of engineering management are learnable.

1.5.2 Characteristics of an Effective Engineering Manager

The effective engineering manager is a rational and organized individual who behaves like a trained professional with regard to ethics, fairness, and honor.

1.5.3 Resources Controlled by an Engineering Manager

The engineering manager has a number of resources at his or her disposal. He or she decides who is to do specific work, what plant or equipment is to be involved, how much money should be spent, which technology is to be applied, and which business relationships and connections should be invoked to achieve predetermined objectives. The technology to be applied may include proprietary innovations, specific know-how, operational procedures, tried-and-true design processes, and others.

1.5.4 Nature of Four-Dimensional Work

The work of engineering managers is four dimensional. Engineering managers need to interface with, and manage the interactions with, subordinates, as well as coordinate their own management actions with those of other managers and peer groups. They manage their own time and efforts. They also attempt to anticipate the requirements of their superiors by making recommendations for future courses of action. Figure 1.1 illustrates this four-dimensional nature of work.

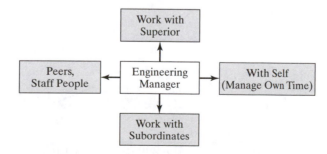

Figure 1.1   Four-dimensional work of engineering managers.

1.5.5 Nature of Management Decisions

Managers decide what should be done (strategic), and technical contributors determine how things should be done (tactical). As an engineering manager rises in an organization, his or her daily decisions will become more strategic. The CEO of a company primarily makes strategic decisions only.

1.5.6 Work of Engineering Managers

Engineering managers plan, organize, lead, and control people, teams, money, technology, facilities, and other resources to achieve the business objectives of the company. To ensure company operations for the short term, they pay attention to problem solving and conflict resolution. As a rule, engineering managers do not perform the technical work themselves. Instead, they work through people. Their job is to decide what the unit, department or company should be doing to advance the objectives of the company and then assign resources to implement their decisions.

An illustration of managerial concern is an issue related to product development. Some companies initiate new product development on a market-driven basis. First, they use market surveys and customer feedback to define product concepts of potential interest to customers. Then they secure resources to develop the product concepts, manufacture the products, and offer customer services to market the products involved. Doing so allows them a high probability of achieving commercial success. Other companies adopt a technology-driven approach. They first invent and develop new technology, and then they incorporate the resulting inventions and innovations in products that they hope to sell to the marketplace. Each of these approaches has advantages and disadvantages. Surveys show that both approaches have yielded successes and failures. Managers decide which approach is the best choice for a company to take.

Another example is the potential difference in opinion between departments when deciding on "buy versus build" options and on setting task priorities. Still another area of potential disagreement is the choice about the level of standardization in product design that reduces cost while allowing a sufficient level of innovation to enhance

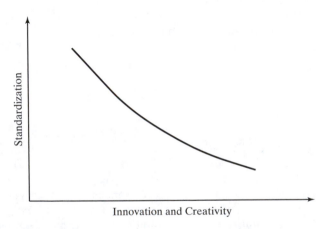

Figure 1.2   Standardization.

competitiveness. In general, enforcing a high level of standardization with strict rules and guidelines tends to impede creativity and innovation, as illustrated by Figure 1.2. Managers are expected to constantly interact and work closely with other managers to resolve such differences.

For those engineers who elect to become managers, there are skills that can be readily learned to make them more efficient and effective. These include time management, work habits, people-related skills (such as team building, communications, and motivation), and use of decision support tools (e.g., Kepner–Tregoe decision methodology, what-if analysis by modeling, risk analysis, Monte Carlo simulation, forecasting, statistics, regression, linear programming, optimization, and office technologies).

Example 1.2.

The company wants to develop a new product to preemptively enter the marketplace. Current information from marketing is sketchy, and the market size cannot be predicted accurately. Indications are that foreign imports are about to foray the market, causing the company to lose the precious opportunity of a preemptive entry.

Should the company initiate a product development program now or wait for more marketing information? Are there other options available to the company?

Answer 1.2.

Yes, there is a third option: The company can act as a distributor and import the foreign product itself, but with its own brand name. This will allow the company to gauge the market acceptance of a low-quality and low-price product. If the results show that customers like the product and the market size is large, then the company can continue importing or develop a low-cost alternative to compete.

Selling a foreign product under the company name requires that the company enter into a private-label production contract with the foreign producer. Typically, such an arrangement includes some of the following elements:

- The contract is good for a predetermined period (e.g., two years) and renewable with mutual consent. The company agrees to pay a unit product cost of x dollars for at least y units per year. The foreign producer agrees to hold the product defect rate at or below z per thousand. The foreign producer remains an exclusive subcontractor to the company for the product types in question during the contract period.

- The company respects all proprietary design and other know-how of the foreign producer. The company is obliged not to use any proprietary design of the foreign producer for the development of its own product or for use by its new production partners.

- The company is responsible for marketing, distributing, selling, and serving the product in the target market (e.g., the United States). The foreign producer agrees to upgrade product design, based on the marketing inputs of the company.

- The company strives to invest in the foreign producer for creating the next generation of products. The foreign producer has the first-refusal rights to accept such investment (i.e., funds and technology).

- Each party can cancel the arrangement after an initial period of collaboration. The foreign producer can go to someone else for marketing the product in the target market. The company may develop its own products or engage another foreign producer as a subcontractor. Thus, selling the foreign product does not preclude the company from selling a similar product after the contract has expired. Companies change subcontractors all the time.

Such a private-label production arrangement is typically a win–win arrangement. The company can preemptively explore the market—that is, test the market and get valuable feedback from customers regarding useful product features—without spending a lot of resources. The foreign producer achieves instant profitability that is assured for the contract period, plus the potential of additional future investment from the company for the next-generation product design and production.

1.6 WORK OF PRACTICING ENGINEERS (TECHNICAL CONTRIBUTORS)

The roles of practicing engineers (technical contributors) are to do things correctly (mostly operational) and to upgrade the ways things are done. When assigned to perform specific work, a practicing engineer is typically told of the specific work objectives as well as any timeline, budgetary, technological, and other constraints that may apply. The engineer is then expected to develop a project plan, jointly define standards with his or her superior, select methods of performing the work, carry out the tasks, and deliver results. A document is prepared to summarize the project; this includes conclusions, the impact of the outcome on the objectives of the group, unit, or company, and possibly suggestions related to potential improvements in methodology and future technical applications of the results.

The performance of practicing engineers is evaluated according to how well they carry out their technical assignments within the time and budget constraints, as well as the value of the project outcome to the company. Young engineers entering an organization are typically assigned to perform tasks of a highly technical content for which their academic training is best utilized.

To perform well as technical contributors, engineers are advised to pay attention to the following five steps:

1. **Demonstrate technical competence and innovative capabilities.** In their early years, engineers need to demonstrate excellent technical skills in performing tasks and projects. They practice fundamental engineering principles correctly, deliver work that is technically free of errors, and are sensitive to time and budget constraints. It is also critically important for the engineers to demonstrate their innovative capabilities in product design, problem solving, and other technological areas.

2. **Practice people skills.** Engineers should ensure that they communicate effectively with others in both verbal and written forms. They need to interact with peers and management in an acceptable manner. Their interactions with management will be strongly enhanced if the engineers become familiar, through self-study or academic courses, with the perspectives of the company managers and with managerial issues and problems. They also need to demonstrate that they are easy to work with and get along well with most people.

3. **Show an unfailing reliability.** Engineers need to show that they are reliable in taking on assignments that add value to their management and are capable of discharging responsibilities delegated to them.

4. **Be proactive.** Engineers should proactively seek team assignments, project coordination, and other roles to practice their managerial skills, foster teamwork, and showcase leadership qualities.

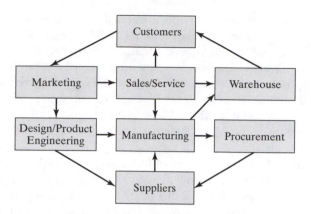

Figure 1.3   Interactions of functional groups in technology companies.

5. **Exhibit a readiness for advancement.** Engineers need to ready themselves for assuming a higher level of technical responsibility so that their employers are able to entrust more challenging responsibilities to them. Engineers should strive to constantly enrich themselves by adding increasingly more value to their employers.

A large, modern-day engineering organization may have many departments, each focusing on a specific function. These functions may include (1) production and manufacturing, (2) construction, (3) design engineering, (3) systems engineering, (4) systems and equipment maintenance, (5) project engineering, (6) program management, (7) process development, (8) product development, (9) technology development, (10) customer service, (11) applied research and development (R&D), and (12) others. Engineers may work in any of these departments. Some engineers may elect to stay in specific functional areas for a long period. Others may prefer to move from one functional area to another in order to attain a well-rounded base of experience.

Figure 1.3 displays the typical interactions between various engineering and nonengineering groups. Engineers are focused on creating value to their employers through their technical work, although they may get involved with various nonengineering functions from time to time.

1.7 ENGINEERING MANAGERS AND LEADERS

This book is aimed at assisting engineering graduates to assume leadership positions in technology-based enterprises of the 21st century. Many of these enterprises are affected by rapid changes in technology and fast-paced development in globalization.

The basic functions of engineering management provide engineers and engineering managers with foundational skills to manage themselves, staff, teams, projects, technologies, and global issues of importance. Engineers innovate to solve problems and minimize conflicts to achieve the company's objectives. They follow the best practices in industry to monitor and control internal operations. They use the Kepner–Tregoe method, among others, to make rational decisions and take lawful and ethical actions.

To facilitate the engineering managers' interaction with peer groups and units, background knowledge in cost accounting, financial analysis, managerial finance, and marketing management should be added. The special cases of activity-based costing, ratio analysis, risk analysis by the Monte Carlo method, and economic value-added accounting are elucidated. These building bocks enable engineers and engineering managers to acquire a broad perspective with respect to the business of the company and its stakeholders.

A critical step to developing technological projects is the acquisition and incorporation of customer feedback. By understanding the project evaluation criteria and the tools of financial analysis, engineers and engineering managers will be in a better position to secure project approvals. For them to lead, a major challenge is indeed the initiation, development, and implementation of major technological projects that contribute to the long-term profitability of the company.

The discussions on ethics, engineers as leaders, Web-based management and engineering enablers, and e-transformation provide additional building blocks to enhance the preparation for engineers and engineering managers to assume technology leadership positions. E-transformation encompasses all activities which electronically transform selected company operations to add value, such as applying enterprise resources planning tools, making procurement more efficient via e-markets, enhancing customer services through Internet portals, improving logistics using global positioning systems, etc. Many tried-and-true rules are included that serve as good guidelines for engineers and engineering managers to become excellent leaders. Above all, so that they constantly augment the satisfaction of customers, engineering managers are expected to lead by their vision of how to utilize company core competencies to add value, by their insights into how to capture opportunities offered by emerging technologies, and by their innovating ability in making products and services better, faster, and cheaper. The concepts of value addition, customer focus, time to market, mass customization, supply chains, enterprise resources integration, and others are also reviewed. E-transformation, as well as Web-based engineering and management enablers, are cited as specific examples of value addition by engineers and engineering managers to "bricks-and-mortar" companies.

Globalization is a major business trend that will affect many enterprises in the next decades. Knowledge of globalization expands the perspectives of engineers and engineering managers with respect to divergences in culture, business practices, and value. Engineers and engineering managers must become sensitive to the issues involved and prepare themselves to contribute to enterprises that wish to capture new business opportunities offered by the high-growth global markets in emerging countries. Engineers and engineering managers need to be aware of the potential effects of job migration due to globalization and take steps to prepare themselves to meet such challenges.

Example 1.3.

John Snyder, the engineering manager, presented to the board of directors a project based on the results generated by Steve Hill, one of his staff members. The board approved the project and praised John for the excellent work done. At that moment, John failed to mention to the board that the work was actually done by Steve. Afterwards, John felt bad about it and recommended to give Steve a bonus.

How would you assess John's handling of this situation?

Answer 1.3.

It would have been more correct for John to initially point out to the board that Steve was the one who did the actual work. However, John's way of handling this situation is acceptable in industry, since he did eventually recognize Steve's contribution by offering a bonus. John should follow through by including Steve in his subsequent monthly progress reports to his boss, the vice president of engineering, to set the records straight and by formally recognizing Steve's work in a staff meeting.

1.8 CHALLENGES IN THE NEW MILLENNIUM

To meet the management challenges of the 21st century, engineering managers need to manage the inside of the company as well as the outside, to lead from present to future, and to act locally and think globally. Table 14.3 contains explanations of this alternative viewpoint.

On the *inside*, engineers and engineering managers practice the basic management functions of planning, organizing, leading, and controlling to implement projects and programs. They manage people, technologies, and other resources to add value to their employers. They enhance the company's core competencies and develop products with features that customers want. They effectively define (by activity-based costing and Monte Carlo simulations), monitor, and control costs. They appraise the company's financial position and seize the right moment to initiate major projects with high-technology contents. These projects are supported by rigorous financial analyses to meet tough corporate evaluation criteria.

On the *outside*, engineers and engineering managers keep abreast of emerging technologies and screen new technologies that might affect the company's products and services. Proactively, to streamline the company's current operations, they define and introduce Web-based tools related to product design, project management, plant operations, facility maintenance, and knowledge management. Engineers and engineering managers identify the best practices in the industry, adopt those practices as standards for evaluating their own in-house practices, and relentlessly strive to surpass these best practices. These engineers look for potential supply-chain partners whose alliances could offer competitive advantages for their employers in production, distribution, product customization, and after-sale service. They are sensitive to the constant need to enhance the management of customer relationships through the use of Web portals and other current technologies. They manage their functions to add value to all stakeholders, namely customers, employees, suppliers, investors, and the communities in which the company operates.

For the *present*, engineers and engineering managers focus on keeping the company operating smoothly by "doing things right." They pay attention to details. They introduce a balanced scorecard to make sure that both financial and nonfinancial metrics are selected to monitor and evaluate the company's performance. They contribute to continuously upgrade current company operations. They take care of tasks (e.g., cost control and waste elimination) needed for the company to achieve profitability in the short term.

For the *future*, engineers and engineering managers seek e-transformation opportunities to generate company profitability in the long term. These opportunities may be created by significantly enhancing the value of the company's products to customers through, for example, distribution, price, service, features, and ordering processes. They develop and introduce new-generation products in a timely manner to ensure a sustainable profitability for the company in the future. Engineers and engineering managers develop a vision for the future, contribute to new company strategies related to technologies, and assist company management in defining what should be done.

At the *local* level, engineers and engineering managers seek the best way to utilize the available resources (skills and business relationships) to achieve the company's objectives. They adjust to local conditions and take lawful, ethical, and proper actions to discharge their daily responsibilities. They maintain their local networks of professional talents and business relationships to enhance the company's productivity. They communicate their experience and preserve lessons learned so that others may benefit at different sites within the company.

At the *global* level, engineers and engineering managers effectuate the optimal use of location-based resources to realize global economies of scale and scope and to derive cost and technology advantages for their employers. They develop global networks of professional talents and business relationships and exploit innovative business opportunities. They acquire a global mindset and become global-business savvy. As companies pursue globalization over time, engineers and engineering managers ready themselves to exercise leadership roles in international settings.

Because many companies are affected by the rapid advancement of technology and the fast-paced development of globalization, the new millennium offers ample opportunities for, and poses new challenges to, engineers and engineering managers. Those engineers and engineering managers who capture these new opportunities and meet these new challenges will be profitably rewarded.

1.9 REFERENCES

Babbage, C. 1832. *On the Economy of Machinery and Manufacturers*. London: Charles Knight. (Reprinted in 1963 by Augustus M. Kelley Publications, New York).

Babcock, D. L. 1996. *Managing Engineering and Technology*. 2d ed. Upper Saddle River, NJ: Prentice Hall.

Badawy, M. 1995. *Developing Managerial Skills in Engineers and Scientist: Succeeding as a Technical Manager*. 2d ed. New York: John Wiley.

Barnard, C. I. 1937. *The Functions of the Executive*. Cambridge, MA: Harvard Business School Press.

Bennett, L. F. 1996. *The Management of Engineering*. New York: John Wiley.

Cardullo, M. W. 1996. *Introduction to Managing Technology*. New York: John Wiley.

Compton, W. D. 1997. *Engineering Management: Creating and Managing World-Class Operations*. Upper Saddle River, NJ: Prentice Hall.

Covey, S. R. 1994. *The Seven Habits of Highly Effective People.* New York: Simon and Schuster.

Dhillon, B. S. 1987. *Engineering Management.* Lancaster, PA: Technomic Publishing Company.

Dorf, R. D. (ed.) 1999. *The Technology Management Handbook.* Boca Raton, FL: CRC Press.

Fayol, H. 1949. *General and Industrial Management.* (Constance Storrs, translator.) London: Sir Isaac Pitman & Sons Ltd.

George, C. S. 1972. *The History of Management Thought.* 2d ed. Englewood Cliffs, NJ: Prentice-Hall.

Hermone, R. H. 1998. *The Management Survival Manual for Engineers.* Boca Raton, FL: CRC Press.

Kales, P. 1998. *Reliability: For Technology, Engineering, and Management.* Upper Saddle River, NJ: Prentice Hall.

Kossiakoff, A. 2003. *Systems Engineering Principles and Practice.* New York: John Wiley.

Longeneck, C. O. and J. L. Somonetti. 2001. *Getting Results: Five Absolutes for High Performance.* San Francisco: Jossey-Bass.

McGregor, D. 1957. *The Professional Manager.* (W. G. Bennis and C. McGregor, editors.). New York: McGraw-Hill.

McGregor, D. 1960. "The Human Side of Enterprise." *Management Review*, Vol. 56.

McGregor, D. 1966. *Leadership and Motivation.* Cambridge, MA: The MIT Press.

Matejka, K. and R. J. Dunsing. 1995. *A Manager's Guide to the Millennium: Today's Strategies for Tomorrow's Success.* New York: AMACOM.

Mazda, F. 1997. *Engineering Management.* Reading, MA: Addison-Wesley.

Mintzberg, H. 1990. "The Manager's Job: Folklore and Fact." *Harvard Business Review*, March–April.

National Research Council. 1987. *Management of Technology: The Hidden Competitive Advantage.* Washington DC: National Academic Press.

Noori, H. 1990. *Managing the Dynamics of New Technology.* Upper Saddle River, NJ: Prentice Hall.

O'Conner, P. D. T. 1994. *The Practice of Engineering Management: A New Approach.* New York: John Wiley.

Ouchi, W. 1981. *Theory Z: How American Business Can Meet the Japanese Challenges.* Reading, MA: Addison Wesley.

Shainis, M. J., A. K. Dekom, and C. R. McVinney. 1995. *Engineering Management: People and Projects.* Columbus, OH: Battelle Press.

Shannon, R. E. 1980. *Engineering Management.* New York: John Wiley.

Silverman, M. 1996. *The Technical Manager's Handbook—A Survival Guide.* New York: Chapman and Hall.

Sloan, A. P. Jr. 1964. *My Years with General Motors.* New York: Doubleday.

Taylor, F. W. 1911. *The Principle of Scientific Management.* New York: Harper & Brothers. (Reissued 1934.)

Thamhain, H. J. 1992. *Engineering Management.* New York: John Wiley.

Urwick, L. 1972. "The Professors and the Professionals." An after-dinner talk at the Oxford Center for Management Studies in Oxford, UK, on October 12.

Uyterhoeven, H. 1989. "General Managers in the Middle." *Harvard Business Review*, September–October.

Weber, M. 1947. *The Theory of Social and Economic Organizations.* (A. M. Henderson and T. Parsons, translators and editors.) New York: The Free Press.

Wren, D. A. 1987. *The Evolution of Management Thought.* 3d ed. New York: John Wiley.

1.10 QUESTIONS

1.1 Tom Taylor, the sales manager, was told by his superior, Carl Bauer, to take an order from a new customer for a batch of products. Both Tom and Carl knew that the products ordered would only partially meet the customer's requirements. But Carl insisted that the order was too valuable to lose. What should Tom do?

1.2 Nancy Bush, the plant manager, needs to decide whether to make or buy a component for the company's core product. She would like the advice of his production supervisors, since they must implement her decision. However, she fears that the supervisors will be biased towards making the component in house, as they tend to favor retaining more work for their people. What should Nancy do?

1.3 Student A, in order to graduate on February 4, works hard to finish her master of engineering report by the due date of January 8. She is planning to return to her home country immediately thereafter and get married. If she graduates on June 10, the next available graduation date, she will have to pay a tuition fee to keep her student status active for one more semester. That would be a substantial financial burden for her.

Her advisor, Professor B, is hesitant to accept the report as presented. The report includes a major marketing activity designed by Student A to promote the new service package of a local company. Because of logistics, this major marketing activity is scheduled to take place on January 20. No customer feedback data, which are required to demonstrate the value brought about by the report, are available before January 8. Professor B cannot bend the rules to pass the report without these data.

Put your innovation hat on and recommend a way to resolve this conflict.

1.4 The engineering manager of Company A proposes to install an automated bar-code scanner costing $4000. He estimates that he can save about 100 hours of labor time per month, as products can now be scanned much faster. He reasons that at the wage rate of $15 per hour, the benefit for using the automated bar-code scanner is $1500 per month, and the scanner can be paid back in 2.67 months.

As the president of the Company A, do you agree or disagree with the way he computes the cost–benefit ratio? Why or why not?

1.5 The new millennium imposes a number of challenges to business managers, who are different from engineering managers and technology managers. Name a few such challenges.

The Functions of Engineering Management

Part I of this book addresses the basic functions of engineering management, such as planning (Chapter 2), organizing (Chapter 3), leading (Chapter 4), and controlling (Chapter 5). These functions provide engineers and engineering managers with foundation skills to manage themselves, staff, teams, projects, technologies, and global issues of importance.

Best practices are emphasized as pertinent standards for goal setting and performance measurement. Engineering managers solve problems and minimize conflicts to achieve the company's objectives. They use the Kepner–Tregoe method, among others, to make rational decisions and take lawful and ethical action. They apply Monte Carlo methods to assess projects involving risks and uncertainties. They engage emerging technologies, motivate a professional workforce of diversified backgrounds, develop new generations of products and services in a timely manner, and constantly surpass the best practices in industry.

The roles of engineering managers in strategic planning, employee selection, team building, delegating, decision making, and managing creativity and innovations are explained. The development of managerial competencies is emphasized.

Chapter 2

Planning

2.1 INTRODUCTION

Planning, a major function of engineering management, is the work done by an engineering manager to predetermine a course of action. Planning defines who will do what, how, where, when, and with which resources. The purpose of planning is to enhance the effectiveness and efficiency of the company by providing focus and direction (Coke 2002; Hamel and Prahalad 1994).

Planning is made necessary by rapid changes in technology (such as Web-based tools, enterprise resource-planning software, broadband communications options, and mobile access), environment (customers, global resources, competition, and marketplace), and organization (such as mergers, acquisitions, supply-chain networks, alliances, and outsourcing).

In this chapter, we will discuss the differences between strategic and operational planning, the planning roles of engineering managers, and the four specific planning activities every engineering manager needs to master.

2.2 TYPES OF PLANNING

There are two types of planning that are engaged in by managers at various levels in a company: strategic planning and operational planning. Both types of planning add value to the company.

2.2.1 Strategic Planning

Strategic planning sets the goals, purpose, and direction of a company. The top-level engineering managers (i.e., chief technology officer and vice president of engineering) are usually involved in strategic planning for the company (Rigsby and Greco 2003; Morrisey 1996A).

Strategic planning focuses on identifying worthwhile future activities. Specifically, strategic planning assures that the company applies its resources—core competencies, skilled manpower resources, business relationships, etc.—effectively to achieve the short- and long-term goals of the company (Corbic 2000). It deals with questions such as the following:

1. What are the company's mission, vision, and value system?

   The mission statement of a company specifies why the company exists in the first place, what entities it serves, and what it will do to serve them.

   The vision statement spells out the aspirations of the company with respect to its asset size, market position, business standing, ranking in industrial sectors, and other factors.

   The value system is the externalization of five or six specific corporate values emphasized by the company. Some typical values favored by U.S. industrial companies include quality, innovation, social responsibility, stability, honesty, quality of life, and empowerment.

2. What business should the company be in?

3. Does the company need to change its product portfolio, market coverage, production system, or service capabilities? If so, why?

4. What specific goals—profitability, market share, sales, technology leadership position, global penetration, etc.—should the company accomplish? By when should these goals be accomplished, with what investment, and by applying which core competencies?

5. What business networks should the company pursue via supplier alliances, comarketing partnerships, production joint ventures, and other forms of collaboration?

6. Which new products should the company offer?

7. What core technologies should the company maintain, develop, acquire, or utilize?

8. Which performance metrics are to be used for monitoring the company's progress?

The horizon of strategic planning is usually spread over five years, although it may be reviewed at more frequent intervals to adjust to changes in the marketplace.

2.2.2 Operational Planning

Managers at both middle levels (managers and directors) and lower levels (supervisors and group leaders) perform operational planning in order to define the specific tactics and action steps needed to accomplish the goals specified by top management (Morrisey 1996B). Managers and directors break down the company goals into short-term objectives. Supervisors and group leaders specify events and tasks that can be implemented with the least amount of resources within the shortest period of time. Operational planning ensures that the company applies its resources efficiently to achieve its stated goals. Questions considered in operational planning include the following:

1. What is the most efficient way of accomplishing a project with known objectives?

2. What is the best way to link up with three top suppliers in the marketplace for needed parts?

3. What are the operational guidelines for performing specific work?

Operational planning involves a process of analysis by which a corporate goal or a set of corporate intentions is broken down into steps. These steps are then formalized for easy implementation. Furthermore, the consequences for the business are articulated at each step. Operational planning focuses on the preservation and rearrangement of established categories (e.g., major strategies defined by upper management, existing products, and organizational structures). Operational planning is essentially a programming task that is aimed at making various given strategies operational.

Operational planning is also called *platform-based planning* because it extrapolates future results from a well-understood, predictable platform of past experience. Results of such planning are predictable because they are based on solid knowledge rather than assumptions.

Compared with strategic planning, operational planning is easier for engineers to accomplish because past experience and examples are usually available as references.

2.3 WHO SHOULD DO THE PLANNING?

Those who have direct knowledge of the specific subject matters involved should take care of the planning.

In the 1960s, strategic planning was accorded emphasis and attention by the top management of American corporations. Company after company set up high-level corporate planning departments made up of full-time planners who would devise business strategies. The approach failed to generate the expected business results. As outlined by Mintzberg (1994), one of the key weaknesses of this approach was that the strategic planners, while being superior analysts of hard business data, were outsiders insofar as the various specific business functions (marketing, production, engineering, and procurements) were concerned. What was not apparent at the time was the fact that planning new strategies for the future required both hard data and intuitive assumptions. The success of the decision to introduce assumptions, and the extent to which these assumptions could be validated, depended very much on the planner's hands-on management experience, intuitive know-how, and in-depth insight of the specific business activities involved. As such, many plans devised by these strategists were poor. Furthermore, business managers in operating departments did not wholeheartedly embrace the plans envisaged by these outsiders. Since the 1960s, many companies have abolished their corporate strategic planning departments altogether and have delegated this important planning function to the business units themselves.

The moral of the story is that the most effective way of creating strategic plans for specific businesses or activities is to entrust such planning to those who are intimately involved with the particular businesses and activities. This paradigm is consistent with the empowerment doctrine whereby decisions are delegated downward to lower level persons who have direct knowledge and in-depth understanding of the subject matters at hand (Barney 2002).

2.4 THE INEXACT NATURE OF STRATEGIC PLANNING

Strategic planning requires an immense amount of strategic thinking (Aaker 2001; Schmetterer 2003). In turn, strategic thinking involves synthesis that likewise requires intuition and creativity. Strategic thinking brings about an integrated perspective of the enterprise, a foresight—albeit not too precisely articulated—of the company's direction that is built upon insights from experience and hard data from market research. Strategic thinking is based on learning by people at various levels involved in conducting specific business activities. Strategic planning invents new categories rather than rearranging existing ones, and synthesizes experience to move the company in a new direction. Managers should encourage others to join in the journey and to shape the company's course thus creating enthusiasm along the way. Broad participation is therefore strongly advisable.

Strategic planners use various kinds of inputs. Study after study has shown that, in addition to hard data, the most effective managers rely on soft information—gossip, hearsay, and various other intangible scraps of information—to develop plans. A key part of strategic planning is to create a vision for the company as to what the company aspires to be. To formulate such a vision, the planners must be able to "see" (Corbic 2000). This is only likely when they are willing to get their hands dirty digging for ideas and extracting the strategic messages from them. Collecting these ideas as building blocks is instrumental to the development of useful strategic plans. Mintzberg (1994) said insightfully, "The big picture is painted with little strokes."

Once information becomes available, strategic thinkers comprehend it, synthesize it, and learn from it. They test ideas and verify the convergence of ideas before they define new strategies. Sometimes strategies must be left in flexible forms, such as broad visions, in order to adapt to a changing environment.

Strategic planning is also called *discovery-driven planning* (McGrath and MacMillian 1995). In situations involving the definition of the company's future direction, most planning inputs are based on assumptions about the future. Because the ratio of assumptions to facts is usually high, the success rate of the resulting plans is typically low. Therefore, strategic planning should be a continuous process and not a single task or event to be taken care of at well-defined milestone dates. It involves constant learning, acquisition, and interpretation of hard data and soft information, as well as staff discussions related to operational decision making, resource allocation, and performance management. Strategic planning requires the discipline of systematically identifying and validating key assumptions introduced in the planning. As more data and knowledge are discovered, more assumptions are validated to form an increasingly solid knowledge base for updating the planning. Strategic planners should engage many participants at various levels to benefit from the relevant corporate expertise available.

The major difficulties of strategic planning can be traced back to three inherent characteristics of such planning:

1. **Prediction of the future.** Certain future events are more predictable than others; for example, seasonal variations of weather and election-year cycles. Other predictions, such as the forecast of discontinuities—technological innovations, price increases, changes of governmental regulations affecting marketplace competition, etc.—are virtually impossible to predict accurately.

2. **Applicable experience and insight.** Strategies cannot be detached from the subject involved. Planners must have in-depth knowledge and relevant hands-on experience of the subject at hand in order to set forth useful strategies.

3. **Random process of strategy making.** The strategy-making process cannot be formalized, as it is not a deductive, but a synthesis process.

Strategic plans often fail due to one or more of the following seven reasons:

1. Not thinking strategically; for example, by limiting the strategy only to short-term needs and processes of the company
2. Failure to identify critical success factors for the company
3. Not having both an internal and external focus
4. Lack of long-term commitment from company management
5. Reluctance of senior management to accept responsibility for tough decisions
6. Not leaving enough flexibility in the plans, causing difficulties in adjusting to the changing environment
7. Failure to properly communicate the plan and thus not securing support and management buy-in

2.5 PLANNING ROLES OF ENGINEERING MANAGERS

Engineering managers at low levels will predominantly devise operational plans to achieve the short-term goals of the unit or department. As the engineering managers move up the corporate ladder, they are expected to participate increasingly in strategic planning, with emphasis placed on technology, product, and production planning. They may find it useful to follow the planning guidelines listed next in order to add value to the company.

2.5.1 Assist Their Own Superiors in Planning

It is important that engineering managers spend time and effort to actively assist their direct superiors in planning. These tasks may include: (a) analyzing hard data (industry, competition, and marketing), (b) offering alternative interpretations to data available, (c) raising insightful questions to challenge conventional assumptions, and (d) communicating the resulting outputs of planning—programs, schedules, and budgets—to help effectuate buy-in from others.

2.5.2 Ask for Support from Subordinates

In order to optimally benefit from the knowledge, expertise, and insights of staff, engineering managers are encouraged to engage their staff and other employees in the planning activities.

2.5.3 Develop Active Plans

It is quite self-evident that engineering managers need to perform a number of planning tasks:

A. **Time management.** All managers need to plan and prioritize their personal daily tasks (such as problem solving, staff meetings, task specification, progress monitoring, performance evaluation), according to the value each task may add, so that high-value tasks are completed before others. This is to maximize the value added by the daily activities (Lane and Wayser 2000).

Also to be included in the daily to-do list are tasks such as networking, continuing education, and scanning emerging technologies, that are deemed important for advancing one's own career.

B. **Projects and programs.** Engineering managers need to plan projects and programs assigned to them by upper management. In planning for projects and programs, engineering managers need to fully understand the applicable project objectives, the relevant performance metrics used to measure success, and the significance of the outcome of such projects and programs to the company. They should carefully select staff members with the relevant skills, expertise and personality to participate in the project and seek their inputs regarding tasks, resource requirements, preferred methodologies, and task duration. They should then integrate all inputs to draft a project plan and distribute the plan among all participants to iteratively finalize the relevant details. These details include budgets, deliverables, and dates of completion. Managers must also secure authorization from upper management before initiating work related to the projects and programs.

C. **Corporate know-how.** The preservation of corporate know-how is of critical importance to the company for maintaining and enhancing its competitiveness in the marketplace. Corporate know-how comes in many types and forms. Certain documentable knowledge, such as patents, published memoranda, operational manuals, and trouble-shooting guides, is easy to retain. Others, such as insights related to procedures or perfected ways of designing specific products and services of the company, may require extra efforts to preserve. Managers should plan to systematically capture, preserve, and widely disseminate such know-how in order to maximize its use within the company.

Certain other cognitive knowledge is typically retained mentally by the experts. Managers need to find effective ways to induce such experts to externalize these skills and insights for use by others in the company.

Problem-solving expertise is yet another type of corporate know-how worth preserving. Typically, engineering managers are busy resolving conflicts that may arise from disagreements in task priorities, personality conflicts, customer complaints, interpretation of data, and other conflicts. Managers need to solve these problems promptly. They should be mentally prepared to jump from one task to another to handle such time-sensitive issues. What should be planned under these circumstances is the preservation of the learning experience garnered from each incident so that the company will become more efficient in solving similar problems in the future.

D. **Proactive tasks.** Engineering managers should plan to devote their efforts to proactively pursue certain other tasks. These tasks include

- Utilizing new technologies to simplify and enhance the products and services of the company
- Looking for business partners to form supply chains

- Offering new or enhanced services to customers (for example, self-service, an information-on-demand system, and a Web-based inquiry center)
- Initiating new programs to promote healthy customer relationships
- Developing new products with upgraded attributes (for example, product customization to serve customers better, cheaper, and faster)
- Reengineering and simplifying specific operational processes to increase efficiency
- Outsourcing specific tasks to augment cost effectiveness

2.6 TOOLS FOR PLANNING

Engineering managers utilize a number of tools to prepare strategic plans. Some of these tools produce hard data, whereas others offer qualitative insights into specific subject areas (Kaufman 2000). The following are examples of some useful planning tools:

2.6.1 Market Research

Market research applies a number of tools to discover the preference of customers with respect to the company's products, services, marketing strategy, product prices, competitive strengths, and brand reputation in the marketplace. Specific tools include polling by questionnaires, product concept testing, focus groups, and pilot testing. The outputs of market research help assess the company's current marketing position and future growth opportunities in the marketplace.

2.6.2 SWOT Analysis

SWOT is the abbreviation for strength, weakness, opportunities, and threats. Each company has strengths and weaknesses in comparison to its competitors. The competitors offer products in direct competition with the products of the company. Because of the company's strengths or core competencies, there may be opportunities offered in the marketplace that the company ought to exploit aggressively. On the other hand, because of the strengths of the competitors and the conditions in the marketplace, the company may be subjected to certain future threats. Such potential threats could be the result of technology advancement, business alliances, marketing partnerships, and other such step changes accomplished by the competition. New governmental regulations and policies may also affect the company's business in the future. A systematic monitoring of publications—patents, technical articles, news release, and financial reports—represents an initial step in conducting a competitive analysis. A well-performed SWOT analysis will bring to the fore the assessment of the company's current position.

As such, the SWOT analysis procreates a road map by which a company can make informed decisions about improving its core competencies to meet its current and future business and operational needs. The analysis answers questions such as the following: (a) What does the company have in place today? (b) In which direction is the company headed in the next three to five years, and (c) What is the company's process of managing changes?

2.6.3 Financial What-If Analysis and Modeling

Spreadsheets are useful in modeling the financial performance of an operation. Financial statements (such as an income statement, balance sheet, and funds flow statement—see Section 7.4) are usually modeled in a spreadsheet program. *What-if* analyses are readily performed to discover the sensitivity of the company's financial performance relative to the changes of specific input variables. In addition, such analyses permit the verification of various assumptions incorporated into the financial models.

Most businesses have inherent risks due to the unpredictable nature of the business climate, the liquidity of financial markets, certain governmental regulations and international trade policies, currency stability, customer preferences, competition in the marketplace, disruptions induced by new technologies, and other factors. What-if analyses and Monte Carlo simulations may be applied to assess the impact of some of these risks on a company's business. (Monte Carlo simulations will be discussed in Section 6.9.)

2.6.4 Performance Benchmarks

Performance benchmarks are those which have been achieved by successful companies in the same industry in which the company operates. When planning, it is important to define these benchmarks to measure corporate progress. Reider (2000) offers an excellent set of broad-based benchmarks, which include:*

A. **Customer-related measures**—product defects, just-in-time delivery, life-cycle product cost, customer satisfaction score, order-processing efficiency, percent sales from new customers, service quality, time taken between orders and product delivery, etc.

B. **Process-related measures**—time to market (i.e., the elapse of time from the initiation of product design and development to product delivery to the marketplace), quality standards, unit product cost, core competence development, labor hours per product, etc.

C. **Financial measures**—gross margin, net income-to-sales ratio, current ratio, sales per employees, return on equity, sales growth rate, market share percentage, inventory turn ratio, etc.

D. **Employee-related measures**—turnover ratio, employee satisfaction score, skill building and development expenses per employee, etc.

E. **Competition-related measures**—market share, cost of innovation, acquisition cost per new customer, number of new products commercialized per year, etc.

A large number of these quantitative metrics are available either from the financial statements of the companies in the same industry or from public sources such as (1) banks that offer loans to companies in a specific industry, (2) financial institutions

---

*Reprinted from R. Reider, *Benchmarking Strategies: A Tool for Project Improvement*, New York: John Wiley, 2000. This material is used by permission of John Wiley & Sons, Inc.

that analyze and compare companies' performance on behalf of investors, and (3) companies that offer credit ratings of companies seeking debt financing (Zagorsky 2003).

These metrics serve well as industrial benchmarks against which to assess the current status of a specific company and to define its new strategic direction.

Example 2.1.

> Quality is usually defined differently by different people in a company. Explain why. Which quality definition is the correct one for the company to adopt?

Answer 2.1.

> Different people in a company may have different interests and perspectives in defining quality. Examples are shown here:
>
> A. **Production**—quality is reject rate and deviation from specifications (view of a production engineer).
> B. **Value based**—quality is defined in terms of price and costs (view of a marketing person).
> C. **User based**—Quality is the degree to which a product satisfies the customer's needs. Customers do not appreciate less or more quality than they need (view of a customer).
> D. **Product based**—Quality is related to the number of attributes offered by a product (view of a product designer).
>
> For the company to succeed in the marketplace, quality is in the eyes of the beholder. The user-based definition is preferred.

2.6.5 Technology Forecasting

Technology forecasting is of critical importance to those companies whose products are composed of high-technology components. Companies must constantly examine, monitor, and apply emerging technologies to enhance business performance.

For example, George Washington University (Halal, Kull, and Leffmann 1997) provides a forecast of emerging technologies in which major events in the fields of energy, environment, information technology, manufacturing, materials, transportation, and others are predicted on the basis of the opinions of a group of selected scholars. The latest forecast findings may be obtained from the GW Forecast website, which is on the Internet at http://www.gwforecast.gwu.edu.

Engineering managers need to understand the value that any of these emerging technologies may have on the products and services offered by their employers and plan accordingly.

Another technological example is the speed of computing. Figure 2.1 shows the improvement in computing speed realized by advanced computing centers in the world.

Currently, the top speed is achieved by Japan, running at 40,9600 gigaflops (GFlops). It is almost a certainty that the computing speed will continue to increase over time. The question for engineering managers is as follows: How can business benefit from such an advancement? Computing speed may be used advantageously in computationally intensive problems, whose solutions of finer granularity provide value

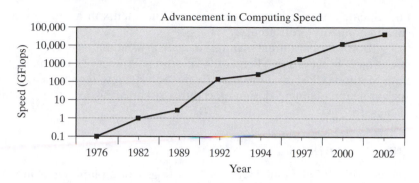

Figure 2.1 Computing speed.

to the business. Here are several examples that illustrate how a refined granularity can help:

1. Data-mining applications related to customer-relations management may be of direct benefit to business. Assuming that customer data (e.g., prices paid, items bought, purchase habits, and payment methods) are collected and available, a detailed analysis could lead to an in-depth understanding of customer behavior, not available heretofore, thus allowing companies to structure customized selling and marketing programs to achieve better customer satisfaction, enhanced brand loyalty, and improved company profitability.

2. Refined modeling of key components in turbomachinery (e.g., compressors, blowers, and pumps) by using computational fluids dynamics programs could raise aerodynamic performance and reduce energy consumption.

3. Plant operations groups typically have encountered a significant amount of data, observations, and experience in maintaining and trouble-shooting equipment and facilities. Such information is dispersed widely, cannot be easily reapplied, and remains useless. A data-mining application could help in getting the information organized and ferreting out valuable knowledge from the wasteful piles of raw data. Application of such knowledge could lead to productivity enhancement in plant operations.

4. To design better, safer, faster, and cheaper cars, sophisticated computer models could be devised to impact-test automobiles, instead of crash-test expensive vehicles at up to only 40 miles per hour of speed.

It is quite certain that engineering managers will be able to envisage many other computationally intensive problems that can be processed to reap business benefits.

2.6.6 Product Life-Cycle Analysis

Every product has a life cycle that moves typically through the stages of initiation, growth, market saturation, and decline. Engineering managers need to examine the life cycles of all products marketed by their employers. Doing so will guide them in introducing new products or product enhancements in a timely manner in order to sustain company profitability.

Tools for operational planning include project management tools: Microsoft® Office Project Standard or Microsoft® Enterprise Project Management Timeline, Critical Path Method (CPM) and Program Evaluation Review Technique (PERT). Other tools, such as action planning, design procedures, risk analysis, and operational guidelines are also useful.

2.7 PLANNING ACTIVITIES

The activities of planning involve forecasting, action planning, issuing policies, and establishing procedures. Strategic planning requires forecasting, action planning, and issuing policies. Operational planning necessitates action planning, issuing policies, and establishing procedures. Some planning activities are proactive, others are reactive in nature.

2.7.1 Forecasting

The objective of forecasting is to estimate and predict future conditions and events. Forecasting activities center on assessing future conditions in technology, products, marketplace, and other factors affecting the business success of the company. The marketplace revolves around customers, competition, economy, global supply chains, human resources, capital, and facilities.

Forecasting helps to define potential obstacles and opportunities and establish the premises for the plan. It sets boundaries for possibilities to help focus on specific future conditions, defines worthwhile objectives, promotes intergroup coordination, provides basis for resources allocation (manpower, budget, facilities, and business relations), and induces innovation through forecasted needs.

Forecasting may be implemented by using the following steps:

1. **Identify**—critical factors that have the most profound effects on the company's profitability.
2. **Determine**—the forecasting horizon as short term (1 year), intermediate term (2 to 5 years) or long term (5 to 10 years).
3. **Select**—forecasting methods such as
    (a) Mechanical projection. The future is projected assuming essentially the same characteristics as in the past.
    (b) Analytical projection. The future is estimated based on an extrapolation of the past (trend analysis). Statistical tools such as linear or nonlinear regression, moving averages, exponential smoothing, time series, and others may be applied.
4. **Forecast**—future eventualities and their likelihood of occurrence.
5. **Prepare**—the forecast, as well as the pertinent database.
6. **Adjust**—forecasts regularly to incorporate pertinent changes related to assumptions and desirable results.
7. **Assure**—understanding and acceptance by all parties affected by the forecast.

Several observations are worth noting. Major economic events such as prices, wages, raw materials, etc. tend to change gradually. The farther an event is projected into the future, the greater the probability of significant deviations between the forecast and reality. Certain future events tend to result from current and past occurrences, as long as there are no disruptive changes in technology or society such as wars, natural disasters, and major incidents. The future may be planned with detailed, factual knowledge of the present and the past under those conditions. It is important to screen ideas by using proper criteria that is consistent with the company's objectives, technical capabilities, financial viability, and marketplace compatibility. Useful inputs may be offered by customers, salespeople, production employees, service clerks, and others who possess intimate knowledge of specific subjects.

Engineering managers are likely to get involved primarily in technology forecasting. As discussed before, forecasting the impact of new technologies on future businesses is particularly difficult. For example, in the past, few companies understood the significance of the Internet to company operations and the marketplace. Questions like those enumerated here did not have clear answers:

1. What will be the impact of broadband technology (cables, optical networks) to communications?
2. How will nanotechnology affect engineering activities such as product design and equipment operation in the future?
3. Will the next wave of new products be smart appliances and intelligent devices?
4. What happens if processors get more powerful and intelligent devices get smaller and more mobile?
5. What about the molecular switching devices that Hewlett Packard is said to be working on that could lead to computational devices about one million times smaller than those we have today?
6. What will be the impact of "pervasive computing" on consumer markets?
7. How quickly will personal computers (PCs) lose their market values, once alternative devices that allow customers to access the Internet, get and send messages, purchase goods and services, activate entertainment programs online, and control home appliances remotely become widely available?
8. How will the new technologies related to intrinsic and extrinsic smart materials, which exhibit sensing and other capabilities, impact on the industrial product design in the future?

To forecast the impact of new technologies, engineering managers must be properly prepared. Some business researchers portend that people with broad perspectives and variable professional experience and exposure may have a better chance of accurately forecasting the impact of emerging technologies, market trends, and other future conditions that require "foresight." Teams whose members have diversified backgrounds in engineering, product design, manufacturing, marketing, service, and sales are said to be better equipped in handling technology forecasting that could benefit from the divergent experience and insights of composite teams.

Engineering managers whose backgrounds are broad based are likely to be more successful in technology forecasting if they are supported by teams that also have diverse experience.

Example 2.2.

The U.S. economy is shaped by a number of factors. The war in Iraq and global anxiety rankle the business environment and influence employment and consumer spending. Correctly reading trends in the economy can make or break a business. Where can an engineering manager find data that could help predict the direction of the economy?

Answer 2.2.

There are leading and trailing indicators for the economy. The 2001 U.S. recession differed from others in its cause, severity, and scope. According a recent assessment (Stock and Watson 2003), many of the commonly used indicators did not forecast well. These indicators include stock prices, unemployment claims, housing starts, orders for new capital equipment, and consumer sentiment.

Gene Sperling, who served as Director of President Clinton's Council of Economic Advisors, offers some advice for business leaders to get ahead of the competition by becoming their own economists (Sperling 2000). Specifically, he suggests that the following set of indicators be used: (1) CEO opinions, (2) temporary jobs, (3) consumer spending, (4) bank loans, (5) semiconductors, (6) commercial structures, and (7) housing markets.

2.7.2 Action Planning

Another important activity related to planning is action planning, the process of establishing specific objectives, action steps, and a schedule and budget related to a predetermined program, task, or project (Kerzner 2003). Action planning helps to focus on critical areas that need attention and action. The identification of critical areas enables the company management to pay attention primarily to planning for deviations that may arise—the principle of management by exception. Furthermore, action planning states specific results to be accomplished. Defining results to be accomplished requires the planner to make judicial selection and exercise judgment. In addition, action planning provides standards as milestones that facilitate control and clarify accountability for results. It also permits an effective delegation of responsibilities (who is responsible for what results), encourages teamwork, and ensures an evaluation of the overall performance of the program, task, or project on a continuous basis.

Action planning mandates engineering managers to take the following specific steps:

1. **Analyze critical needs.** Critical needs are those associated with staff development, staff maintenance, and staff deficiency, as well as those related to special assignments. Managers define these needs by reviewing standards related to position charters, duties, management expectations, and company goals. Short-term needs must be in balance with long-term needs.

2. **Define specific objectives.** Specific objectives need to be defined to satisfy the critical needs. The results statement (who will attain what desirable results by when) must be specific. Establishing objectives predetermines the results to be accomplished.

3. **Define standards.** Standards measure the attainment of the objectives. The standards should preferably be quantitative in terms of performance ratios, percentages, cost figures, resource parameters, and other factors in order to be measurable (Kaplan and Norton 1992).

4. **Define key action steps.** The definition of key action steps establishes the sequence and priority of steps required to attain objectives. Specifically, major steps are lined up in the order in which they are to be performed; this list includes the evaluation of risks for the steps planned, the definition of contingency steps to ensure the expected results, and the specification of who is responsible for each step and who is accountable for achieving the target value associated with each step.

   Action steps must be reasonably implementable. After the expected results are defined, engineering managers should plan these steps with the active participation of involved workers to benefit from their creativity and expertise in the subject.

5. **Create a schedule.** Scheduling establishes both a time sequence for action steps and the interrelationship among the steps, as some might be prerequisites for others. It is advisable to estimate the optimistic (earliest), the pessimistic (latest), and the most likely (most probable) dates of possible completion of each step. Doing so will permit a more realistic modeling of the project schedule.

   Sufficient scheduling flexibility should be included to account for contingency—more for projects related to new development and less for routine design and analysis work. Contingency refers to the slag and cost buffers introduced to account for undefinable, yet generally anticipated, deviations from the plan.

   The most important outcome of the scheduling effort is the definition of the project or program completion date. The engineering manager, as the leader, is accountable for completing the project or program on time.

6. **Develop a budget.** Budgeting allocates resources necessary to accomplish objectives. The planner determines the basic units (man-hours, man-weeks) to accomplish each task, estimates the total resources needed for the project, and adds a contingency to the total amount for potential deviation (e.g., 7 to 10 percent of the total budget, dependent on the customary percentages used in each industry) to arrive at a total budget.

   The budget estimate is typically the basis for seeking management approval for the project or program. The project leader is accountable for completing the project or program within the approved budget.

For complex projects that involve many participants (e.g., peer departments, external suppliers, and outsourcing service organizations) project management tools such as PERT or CPM may be applied. These tools produce timelines, graphically diagram the tasks network to facilitate monitoring and control, and determine the tasks linked along the critical path. Consequently, the shortest time in which the project can be completed can be determined. Managers are then reminded to monitor these critical-path tasks carefully in order to avoid project delays.

A well-developed project plan serves the purpose of promoting communication, monitoring progress, evaluating performance, and managing knowledge.

2.7.3 Issuing Policies

For companies to operate smoothly and consistently, corporate rules and regulations are used to prescribe acceptable practices. Company policies address important issues such as employee hiring and termination, Equal Employment Opportunity (EEO) policies, annual performance appraisals, savings plans, benefits, medical insurance, pension plans, sick leave, safety, contact with representatives of competitors, and other issues. At the departmental level, specific rules may be defined to regulate tasks that are repetitive in nature, such as filing reports after each completed business trip, submitting monthly or quarterly progress reports to summarize achievements, and to outline future work, attending scheduled staff meetings, publishing engineering or scientific articles, participating in professional and technical conferences, and other tasks.

Managers may write policies to offer uniform answers to questions of common concern. In general, policies are continuing directives promulgated to address repetitive issues, tasks, and problems in an organization. Policies are useful for predeciding answers to basic repetitive questions, capturing the distilled experience of the organization, saving management time, and facilitating delegation. Issuing policies is a part of the manager's planning responsibilities.

To be effective, a policy must have certain common characteristics, including the following: (1) applies uniformly to the organization (or specific engineering unit) at large; (2) remains relatively permanent, unless and until repealed; (3) fosters the objectives of the company; (4) frees managers and employees to focus on important matters; (5) encourages effective teamwork by reducing disagreements, conflicts, and differences in interpretation; and (6) is issued by top management or authorized managers with perspective, balance, and objectivity.

2.7.4 Establishing Procedures

Companies perform many important tasks such as product design, plant operation, project management, equipment installation, facility maintenance, manufacturing, system engineering, parts procurement, product delivery, customer service, and others. The specific methods by which these tasks are performed represent the valuable corporate know-how employees have learned to perfect. Over time, companies want to preserve these "tried-and-true" procedures in manuals.

Developing procedures is of critical importance to a company, not only because doing so will preserve the best way to perform repetitive work (to achieve high productivity), but also because doing so will accomplish the following: (1) provide the basis for method improvements; (2) ensure standardized action (such as quality control, resource saving, and work reproducibility); (3) simplify training; and (4) retain corporate memory, such as know-how, knowledge, heuristics, proven safety practices, problem-solving techniques, etc.

Establishing and preserving procedures is part of the planning responsibility of managers. If generated in suitable formats, such procedures could be widely applied within the company and among its business partners to garner competitive advantages in the marketplace. Techniques for developing procedures include:

A. Concentrating on procedures for critical work that is in high demand, repetitive, and time consuming

B. Charting the work required—inputs, workflow, outputs, skills, and resources
C. Reviewing work characteristics carefully in order to decipher (a) why (is the work really necessary?), (b) what (results are to be obtained?), (c) when (is the best time to do it?), (d) where (is the best place—group, station, facility, or equipment—to do it?), and (e) who (is the person with the relevant training to do it?)
D. Proposing procedures in the context of existing objectives, policies, and programs by keeping the procedures to a minimum; this will avoid restricting employee imagination and incentive, as well as assure consistent applications to minimize deviations
E. Defining improvements to procedures and updating them regularly
F. Formulating the procedures in writing
G. Communicating with all affected parties to ensure understanding and acceptance

2.8 SOME SPECIFIC ADVICE ON PLANNING

Good up-front planning is essential for any company to achieve its desired corporate objectives. Managers need to pay sufficient attention to planning activities in order to make sure that certain pivotal factors are sufficiently addressed in the strategic or operational plans they formulate.

2.8.1 Assumptions

Plans are typically built on both hard data and assumptions. Assumptions are usually based on extrapolations of past experience and intuitive projections into the future. It is important for managers to constantly seek and interpret additional resources and insights to verify their assumptions. This is to ascertain that the plans they introduce are built on an increasingly solid foundation.

2.8.2 People

Any plan is worthless unless its objectives are achieved through a successful implementation. Implementation requires dedicated people who are supportive of and ardent about the subject matter involved. Managers need to take into account the suitability of people, including their background, personality, training, mental flexibility, interpersonal skills, collaborative attitudes, adaptability, and emotional attachments to specific ways things are done.

Most plans contain activities related to a change of the current status. Unfortunately, most people resist change, particularly sudden change. Change may induce business instability, technology obsolescence, organizational restructuring, and other unwanted disruptions. People may be more amenable to gradual changes if such changes occur at rate they can understand and accept.

Managing change is a challenging task for managers. Managers need to recognize early that change is coming. They may want to delineate the change in detail and analyze the implications of the change as a way of preparing the staff and allowing them to become gradually accustomed to such a change.

By paying close attention to how changes are being communicated to the staff, managers may be able to minimize the resistance to change and gain support for the implementation of new plans. It is helpful for the managers to isolate and identify areas of threats and opportunity. If needed, they should apply contingency plans for handling threats, but focus on opportunities that will advance the company business.

2.8.3 Benefit versus Cost

When planning, managers need to be guided by the expected value that a given project or program may produce. Low-value projects justify the commitment of low-level efforts, whereas high-value projects justify the allocation of high-level efforts. Efforts applied should be commensurate with the value added by the expected results. Otherwise, corporate resources may be wasted. The saying "things worth doing are worth doing well" is valid only to the extent justifiable by the expected value.

2.8.4 Small but Sure Steps

To be effective in planning, managers should (1) identify clearly the desired end results and the series of small steps required to reach them, (2) allow a timely control and mid-course correction, if needed, and (3) aim at attaining a series of small progressions (or continuous improvements) that are more acceptable in many old-style companies than one large achievement (or a step change) after a long period of time. On the other hand, some start-up companies with an entrepreneurial spirit may be able to exercise patience, take risks, and go for "blow-the-roof-off" breakthrough technologies and step-change products or services. Managers need to adjust accordingly.

2.8.5 Contingency Planning

As discussed before, strategic planning for the future entails considerable risks and uncertainties. Some of the changes in future conditions are unpredictable. Yet strategic planning for the future must be done today. Besides striving for acquiring hard data and soft information to continuously validate the assumptions introduced in the planning, managers should take an additional risk-modulating step: study exhaustively the sensitivity of various assumptions to the company business and incorporate contingency steps, including fallback positions, in order to minimize the adverse impact of questionable assumptions (Childs 2002).

2.8.6 Commitment

Managers need to secure company commitment before any plan can be implemented successfully. Company management must declare their intentions and their readiness to allocate resources needed to achieve the planned objectives. Without a firm company commitment, nothing of value will emerge from the planning efforts.

Example 2.3.

> Joe Engineer took a graduate school course at SUNY-Buffalo where he learned the importance of planning. Joe knows that luck plays a big role in one's life. But he is convinced that proper planning will help him to have an orderly progression in his career. He thinks

that it would be cool to become a CEO of a publicly owned, multinational company at the age of 60 and retire at 65 with a net worth of $5 million. He wants some guidance with career planning. How can you help him?

Answer 2.3.

It is advisable for Joe Engineer to follow a number of planning steps, enumerated here:

1. SET OBJECTIVES AND SPECIFY SUBGOALS

Before starting the planning process, we need to introduce an important assumption. In order for Joe Engineer to be entrusted with a given management position in a publicly held major company, he needs to have acquired and successfully demonstrated certain business management capabilities beforehand. Obviously, this assumption may not be valid for small and medium sized companies that are privately held.

The CEO of a major company must be familiar with many functional areas, such as (1) strategic management, (2) business management, (3) operational management, (4) project or program management, (5) engineering management, (6) production and manufacturing, (7) marketing management, (8) financial control, and (9) globalization. The future CEO must be able to demonstrate sufficiency in various skills, such as

- Public speaking and writing
- Business analysis and planning
- Public relations
- Problem solving and conflict resolution
- Interpersonal skills
- Negotiations
- Business relations development
- Other skills

Therefore, for Joe Engineer to qualify for the CEO job, he must have garnered useful management experience, possibly as a company president a few years back. Future capabilities are, by and large, based on past experience. Applying such a logic in a backward-chaining manner, Joe Engineer could readily establish a set of milestones in his plans:

A. Corporate president at 55

B. Division president at 50

C. Vice president at 45

D. Director at 40

E. Manager at 35

F. Supervisor at 30

G. Group leader at 25

2. DEVELOP ACTION PLANS

A forward chaining plan, which moves from the present to the future, should be considered by Joe Engineer. As examples, the following plan illustrates the qualifications that should be built up when advancing from one stage to another:

A. Preparation (By a Certain Date)

1. Take steps to collect pertinent career development references and acquire perspectives.

2. Talk with experienced engineers to obtain insights related to the costs and benefits of the targeted objectives. The advantages and disadvantages of being a manager are well known: power, prestige, and money versus travel, 50- to 60-hour workweeks, job pressure, office politics, balance between work and home, and related factors.
3. Understand one's own career objectives and the requirements to succeed. What are the "success factors" involved?
4. Be aware of one's own strengths and weaknesses, personality type, value system, personal requirements for happiness.
5. Confirm desirable objectives of moving into the managerial career path.

B. Group Leader

1. Get a master of engineering degree to demonstrate technical competence (by a certain date, say 1/20xx).
2. Become well versed in engineering management concepts and practices (e.g., take courses or training).
3. Practice good interpersonal skills by doing volunteer work.
4. Network inside and outside the company (join technical societies, attend technical conferences, publish technical papers, etc.) and know some professional people well.

C. Supervisor

1. Seek training on supervision and practice teamwork with dedication.
2. Take advanced technical courses, if needed, to help become established as a technical leader.
3. Broaden into marketing, production, and sales through business interactions.
4. Function as a gatekeeper for technology.
5. Demonstrate innovation.
6. Continue networking and become known to many others inside and outside the company.
7. Attain recognizable technical achievements.
8. Demonstrate managerial potential.
9. Become known as a good problem solver.

D. Manager

1. Demonstrate prowess in strategic planning, operation, and all other engineering management skills.
2. Demonstrate capabilities in interacting with sales, marketing, production, service, and customers.
3. Show success in initiating and implementing new technology projects that affect the business success of the company.
4. Achieve organizationwide recognition.
5. Form networks with important people at various levels.

E. Director

1. Become widely known in one's own industry.
2. Participate actively in industrial trade and technical groups.
3. Demonstrate leadership in strategic planning affecting the company.
4. Be recognized for operational efficiency.
5. Make major contributions to direct the company's businesses.
6. Master the technology-marketing interface.

7. Lead the company in applying emerging technology to constantly strengthen competitiveness.
8. Represent the company well to the press.
9. Have real friends in high places.

F. Vice President

Joe Engineer is encouraged to fill in the remainder of this plan as an exercise.

3. BUDGET AND COMMITMENT

A. Invest the proper amount of resources (time, money, and efforts) to ready oneself for the next stages.

B. Make a firm commitment to carry out action steps specified in the plan.

4. REVIEW AND UPDATE

Review the plan and make adjustments regularly to exercise proper control of this career path. Knowing what it takes to move to the next stage, and preparing oneself in time for that big opportunity ahead, represent a good mantra for Joe Engineer to follow.

2.9 CONCLUSION

Both strategic and operational planning are important, because the success of a company depends on creating new paths to the future as well as on implementing short-term operational plans.

Engineering managers are expected to involve themselves in both strategic planning and operational planning. Both types of planning require forecasting and action planning. To regulate work, managers may also participate in issuing policies and establishing procedures. Among these planning activities, forecasting and strategic planning are difficult, as they involve making estimations of the future. The remaining planning activities related to policies and procedures are administrative or operational in nature. These activities are rather straightforward and should appear to be relatively easy to understand and implement. Doing extremely well in these administrative tasks will not necessarily make a manager outstanding, but not doing well in them will project a negative image of the engineering manager.

To demonstrate managerial leadership, engineering managers need to be proficient in technology forecasting and strategic planning. Technology forecasting involves the critical evaluation and adaptation of emerging technologies so that the company's products and services offered to the marketplace become better, cheaper, and faster to deliver. A primary opportunity for engineering managers to add value is to participate actively in creating technology projects affecting the company's future in major ways.

2.10 REFERENCES

Aaker, D. A. 2001. *Developing Business Strategy*. 6th ed. New York: John Wiley.

Barney, J. 2002. *Gaining and Sustaining Competitive Advantages*. Upper Saddle River, NJ: Prentice Hall.

Childs, D. R. 2002. *Contingency Planning and Disaster Recovery: A Small Business Guide*. New York: John Wiley.

Coke, A. 2002. *Seven Steps to a Successful Plan*. New York: AMACOM.

Corbic, C. 2000. *Great Leaders See the Future First: Taking Your Organization to the Top in Five Revolutionary Steps*. New York: Dearborn Trade.

Coveney, M. et al. 2003. *The Strategy Gap: Leveraging Technology to Execute Winning Strategy*. New York: John Wiley.

Halal, W. E., M. D. Kull, and A. Leffmann. 1997. "Emerging Technologies: What's Ahead for 2001–2030." *The Futurist*, November–December.

Hamel, G. and C.K. Prahalad. 1994. "Competing for the Future." *Harvard Business Review*, July–August, pp. 122–128.

Kaufman, R. A. 2000. *Mega Planning: Practical Tools for Organizational Success*. New York: Sage Publications.

Kaplan, R. S. and D. P. Norton. 1992. "The Balance Scoreboard Measures that Drive Performance." *Harvard Business Review*, January–February, pp. 71–79.

Kerzner, H. 2003. *Project Management: A Systems Approach to Planning, Scheduling and Controlling*. 8th ed. New York: Wiley.

Lane, H. L. and C. Wayser. 2000. *Make Every Minute Count: More than 700 Tips and Strategies that Will Revolutionize How You Manage Your Time*. New York: Marlowe & Co.

McGrath, R. G. and Ian C. MacMillian 1995. "Discovery-Driven Planning." *Harvard Business Review*, July–August.

Mintzberg, H. 1994. "The Fall and Rise of Strategic Planning." *Harvard Business Review*, January–February.

Morrisey, G. L. 1996A. *Morrisey on Planning: A Guide to Strategic Thinking: Building Your Planning Foundation*. San Francisco: Jossey-Bass.

Morrisey, G. L. 1996B. *Morrisey on Planning: A Guide to Tactical Planning: Producing Your Short Term Results*. San Francisco: Jossey-Bass.

Reider, R. 2000. *Benchmarking Strategies: A Tool for Project Improvement*. New York: John Wiley.

Rigsby, J. A. and G. Greco. 2003. *Mastering Strategy: Insights from the World's Greatest Leaders and Thinkers*. New York: McGraw-Hill.

Schmetterer, B. 2003. *Leap: A Revolution in Creating Business Strategy*. New York: John Wiley.

Sperling, G. 2003. "The Insider's Guide to Economic Forecasting." *Inc. Magazine*, August.

Stock, J. H. and M. W. Watson. 2003. "How Did Leading Indicator Forecasts Perform During the 2001 Recession?" *Economic Quarterly—Federal Reserve Bank of Richmond*, Summer, Vol. 89, No. 3.

Zagorsky, J. L. 2003. *Business Information: Finding and Using Data in the Digital Age*. New York: McGraw-Hill.

2.11 QUESTIONS

2.1   On the eve of leaving her alma mater, Stacy Engineer remembers the encouraging words of the commencement speaker: "Graduation is the happy beginning of an exciting life ahead." She is, of course, excited about her new master of engineering degree that she received with honor. But she is also a bit confused about what to do now to make her new life exciting and filled with happiness. Apparently, what she needs is a road map into the future. How can you help her?

2.2   The company has always been focused on the high-quality and high-price end of the market. Now, market intelligence indicates that some competitors are planning to enter the low-price and low-quality end of the market. What should the company do?

2.3   Mission and value statements are indicative of the direction in which a company is headed. What are typically included in the statements of mission and values of well-known companies in the United States? Please comment.

2.4   What are included in the typical operational guidelines some industrial companies have developed? Please comment.

2.5   There are always risks (risks of failure) associated with the experimentation of a new manufacturing process or with the entry into a new global market. How should one decide to proceed or not to proceed with a risky venture? What is the proper level of risk for a company to take?

2.6   The marketing director needs to submit a strategic plan for entering a new market. She knows she needs long periods of uninterrupted time. She considers two options: (1) staying at home to do the plan or (2) delegating some parts of the plan to her subordinates. What are the factors the director needs to consider when she chooses the best way to come up with this plan?

2.7   In planning for a project, the Critical Path Method (CPM) is a tool used widely in industry. Elucidate the basic concepts and techniques involved, illustrate by using an example, and review its advantages and disadvantages.

2.8   XYZ Company has been a one-product company focused on developing and marketing a package of innovative ERP software specialized for law firms and operated in computers running on a proprietary operating system software developed by the company. Customers must purchase both the hardware and software as a bundled package from XYZ company. The company also provides around-the-clock services to ensure that the combined hardware and software system performs reliably, as lawyers are known to be typically disinterested in trouble-shooting computer systems. This product-bundling strategy works out well for the company, and the sales revenue of XYZ increases dramatically during its first three years in business.

However, market intelligence shows that new ERP software products are now being introduced by competitors. These new ERP software products are quite capable of performing all of the data processing functions typically required by law firms. Furthermore, these new ERP software products can run on any computers using their existing operating systems, thus eliminating the need for customers to purchase dedicated computers.

The president of XYZ Company recognizes the potential threat imposed by these new ERP software products. He wants to know the best counterstrategy he should plan and implement. Design and explain this counterstrategy.

Chapter 3

Organizing

3.1 INTRODUCTION

Organizing is another important function of engineering management. Organizing means arranging and relating work so that it can be done efficiently by the appropriate people (Galbraith 2002).

Corporate efficiency is usually achieved by a proper partition and distribution of work, as well as by a suitable arrangement of the interrelated groups of people participating in the work that is subject to time constraints, resource limitations, and business priority (O'Reilly and Pfeffer 2000).

Managers are empowered to design the organizational structure—the team, group, department, etc.—and to define the working relationships conducive for attaining the company's objectives. Doing so will:

- Ensure that important work gets done
- Provide continuity
- Form the basis for wage and salary administration
- Aid delegation
- Facilitate communication
- Promote growth and diversification
- Encourage teamwork by minimizing personality conflicts and other problems
- Stimulate creativity

It is generally true that dedicated people can make any organization work. However, dedicated people in well-organized units can get outstanding work done.

This chapter compares several basic forms of organizational structures commonly employed in industry. Special emphasis is placed on teams composed of cross-functional members. Illustrative examples are included for specific organizational structures, which are used to enhance innovation, resolve conflicts at the interface between design and manufacturing, promote collaboration at the interface between research and development

(R&D) and marketing, and foster employee motivation. The critical managerial tasks of assigning responsibilities while maintaining control and establishing work relationships are also discussed.

3.2 DEFINITIONS

Before reviewing the managerial function of organizing, it is useful to introduce a few definitions.

A. **Span of Control.** The span of control refers to the number of people supervised by a manager or supervisor. It may be small (a few people) or large (20 or 30 people). The choice of a small or large span of control depends on workforce diversity, task volume, and complexity of work, as well as on the geographic dispersion of workers. Large span leads to lower costs and greater organizational efficiency, but it also leads to a lower intensity of control. The current trend is moving toward larger span of control, increasing from 7 to 20 or more, due to

- Reduction of middle management levels
- Enhanced communication tools
- Empowered knowledge workers, allowing decision making at lower levels by people with more applicable knowledge
- Improved morale, productivity, and profitability made possible by less detailed supervision, particularly over professional workers

B. **Organization Types.** The *line organization* (e.g., a profit center) performs activities directly related to the company's main goals. Examples include business management, product management, sales and marketing, product design and engineering, production, and customer services.

On the other hand, the *staff organization* (e.g., a cost center) provides advice and comments in support of the line organization's work. Examples include research and development, financial and accounting, information services, procurement, legal affairs, public relations, and facility engineering.

C. **Overlap and Duplication of Responsibility.** This refers to a situation where two or more people do the same work and make the same decisions. Such undesirable situations are to be avoided in any organization, as they represent sources of conflicts.

D. **Specialization.** Specialization refers to the increased degree of skill concentration in narrow technical domains. Specialization of work leads to improved efficiency. However, overspecialization may cause monotony, fatigue, disinterest, and inefficiency on the part of the worker.

E. **Work Arrangement.** Work needs to be arranged in a rational and logical manner. The logical arrangement of work promotes task accomplishments and enhances personal satisfaction for more workers over a longer period of time.

F. **Selected Management Terms.** *Authority* refers to the legal or rightful power of a person, by assignment or by being associated with a position, to command, act, or make decisions—this is the binding force of an organization. *Responsibility* is the duty to perform work assumed by a position holder in an efficient and professional manner. *Accountability* represents an upward-directed obligation to secure the desired results of the assigned work.

3.3 ACTIVITIES OF ORGANIZING

As a function of engineering management, organizing consists of several specific activities described here:

A. **Organizing one's own workplace for productivity**—the organization of one's own office, file systems, and daily routine so that work can be done efficiently (Hemphill 1999)

B. **Developing organizational structure**—the identification and assortment of work so that it can be done efficiently by qualified people in teams, task forces, committees, departments, and other suitable arrangements. (Baguley 2002; Cox 2001)

C. **Delegating**—the entrustment of responsibility and authority to others and the creation of accountability for results. Managers must learn to delegate effectively in order to achieve results by working through people; to distribute the workload while maintaining control to make the best use of available talent in the organization (Burns 2002)

D. **Establishing working relationships**—the creation of conditions necessary for the mutually cooperative efforts of people. Managers must make commitments, set priorities, and provide needed resources (money, physical facilities, skills, and know-how) to foster teamwork and collaboration among people (Straus 2002)

All of these organizing activities exist for the purpose of achieving improved efficiency in performing work.

3.4 ORGANIZING ONE'S OWN WORKPLACE FOR PRODUCTIVITY

How well is the office of a typical engineering manager organized? A simple test is to ask how much time it would take for the engineering manager to find a phone number, a piece of paper, or a file when his or her superior calls. Surveys indicate that an average executive spends about five weeks per year looking for lost items (Von Hoffman 1998).

Engineering managers need to be organized with respect to time, paper, and space. A basic guideline recommended by efficiency experts is as follows: "The less you have, the less you have to sort through." A few rules of thumb that are recommended for the engineering manager to become more efficient are as follows:

1. Use an on-line calendar that indicates time slots blocked out for important tasks. Such a calendar allows others—one's own secretary and peer managers—to schedule meetings conveniently. One should also prepare agendas before holding or attending meetings.

2. Maintain a "to-do" list. Set priority to tasks and separate urgent tasks from others by assigning most urgent tasks to list A, moderately urgent tasks to list B, and least urgent tasks to list C. Consult the lists regularly. If one is computer literate, then use electronic systems.

3. File papers based on "access," or use a logical keyword system under which to find the document later. The file system may be based on categories such as projects, persons, or deadlines. Keep a master copy of the file index nearby and update it often. This master index helps locate a file and safeguards against creating duplicate files. A document should be kept if

   A. The information it contains cannot be easily found elsewhere. (How difficult would it be to obtain or reproduce it again?)

   B. The information it contains helps the engineering manager to reach a goal. (Does this piece of paper require action?)

   C. It has been consolidated as much as possible.

   D. It is up to date. (Is it recent enough to be useful?)

   E. It is really necessary to keep this document. (What is the worst thing that could happen if this document is unavailable?)

   Most professional workers are said to use only about 20 percent of the paperwork they keep. The challenge is, of course, to decide which 80 percent can be thrown away. Question every piece of paper that crosses the desk. Use the wastebaskets frequently. Reserve a time slot during each day (e.g., after work, but before departing for home) to sort, file, and toss unneeded files. Make use of travel time (at the airport, on the plane, and in the hotel) to organize one's own files.

4. Implement a system for keeping track of names and phone numbers (e.g., Rolodex for business cards, address book, Palm Pilots, etc.).

5. Cultivate the use of the phone. Prepare notes before placing calls and make the calls brief (e.g., by standing up).

With practice, every engineering manager can get his or her workspace and daily routine organized for productivity (Hemphill 1999).

Example 3.1.*

David Pope, engineering director, started out the day uptight. His young child had the flu the night before, and he had been up all night to help. Upon arrival at his office, David had to make urgent phone calls to approve a two-week overtime work plan due to a plant fire the night before and to plan for a product committee meeting the next day to counter environmental concerns about a wastewater treatment plant.

Then he spent 30 minutes reviewing the qualifications of new candidates and decided on one. He asked for salary information and wanted to examine the offer before it was sent. He asked for further justification for the budget requested by industrial engineering for a minicomputer. Without reading it, he approved the research proposal from material

*Condensed and adapted from Robert E. Shannon, *Engineering Management*. New York: John Wiley, 1980 p. 32. This material is used by permission of John Wiley & Sons, Inc.

engineering. He rejected an invitation to speak at a regional meeting of the American Society of Plant Engineering by giving an untrue reason.

David made a note for a United Way board meeting coming up soon. At 10:00 A.M., he met with two consultants for one hour and 45 minutes on a formal wage and salary plan and then directed his administrative assistant to work out the details. He promised to inform all department heads and asked for cooperation.

As he walked back to his office after lunch, David noticed several engineers were still playing bridge after 1:30 P.M., and he planned to remind their department heads of this truancy from work.

As soon as he walked into his office after lunch, George Wallace, the general sales manager, called to complain about inadequate responses from engineering to field sales requests. David promised to look into it after receiving specific details. In return, he asked for Wallace's support at the product committee meeting the next day.

David gave a retirement plaque to Glen Sanford in his own office in the presence of the personnel director at 1:45 P.M. Furthermore, he approved the request of two engineers for a week of overtime to design a new, final quality-control station.

At 1:30 P.M., he was asked to attend a three-hour budget meeting at 2:30 P.M. called by the president. In the meeting, guidelines and a timetable for next year's budget requests were discussed. For engineering, he was told there would be an increase of only 10 percent. He then arranged for a meeting with the president and the controller at 2:00 P.M. the next day to request more money.

As he was about to leave for the day around 6:30 P.M., his wife called to say that his child is doing all right, but he has to go to the party of the executive vice president alone.

What do you see are David Pope's problems? How do you suggest improving his day?

Answer 3.1.

David Pope had four major problems: (1) poor time management, resulting in the day being spent responding mostly to others; (2) lack of delegation; (3) inadequate utilization of administrative assistant; and (4) deficient guidelines for handling minor projects. David Pope could improve his day as follows:

A. Review the day's schedule in the morning and call in the administrative assistant to

- Get background information on wage and salary plan for the 10:00 A.M. meeting with the two consultants. Prevent this initial meeting from dragging out to one hour and 45 minutes.
- Request the personnel director to invite peers of Glen Sanford to attend the plaque-awarding ceremony in his own office.
- Collect information on the budget request for the minicomputer from industrial engineering.

B. Return all phone calls.

- Authorize the two-week overtime workplan due to plant fires.
- Send Jamieson (who wrote the report) to the product committee meeting to defend the wastewater treatment plant.
- Approve Oscar Ford to use two engineers for one week to design a new, final quality-control station. Ask the administrative assistant to draft new guidelines for manpower allocation in minor projects.

C. Upon receiving notice for the 2:30 P.M. meeting called by the president, get the administrative assistant to start preparing the engineering positions on budget. Review these positions at 2:00 P.M. Call an urgent meeting for 2:15 P.M. with department heads to finalize the engineering position.

D. Present a plaque to Glen Sanford before many of his peers, express appreciation for his services, and wish him well.

E. Become well-prepared to attend the 2:30 P.M. meeting. If more discussions are needed, request a follow-up meeting with the president. If additional budget preparations are required the next morning, leave a note to the administrative assistant.

F. Go to the party alone, and be happy.

3.5 DEVELOPING ORGANIZATIONAL STRUCTURE

The purpose of developing organizational structure is to help ensure that important work related to the key objectives of the unit or department is performed. By developing the right organizational structure for pursuing specific work, managers hope to eliminate or minimize the overlap and the duplication of responsibilities. Also, by logically grouping work according to positions in the organizational structure, managers will be in a better position to utilize available talents, encourage mutual support among workers, provide technological foci, and facilitate problem solving. Doing so will ensure that management, technical, and operating work are distinguishable so that people can be most efficient in performing such work (Wetlaufer 1999; Dean and Susman 1989; Chesbrough and Teece 2002).

Many industrial organizations adopt one or more of the following structures.

3.5.1 Functional Organization

The functional structure is a very widely used organizational form in industry. Companies that favor this organizational form include (1) manufacturing operations, process industries and other organizations with limited product diversity or high relative stability of workflow; (2) start-up companies; (3) companies with narrow product range, simple marketing pattern, and few production sites; and (4) companies following the lead of their competitors.

Companies that prefer the functional structure establish specific departments responsible for manufacturing, finance, marketing, sales, engineering, design, operation, procurement, and other such functions. (See Figure 3.1.)

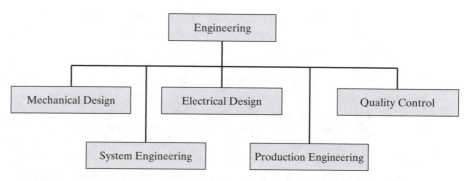

Figure 3.1   Functional organization.

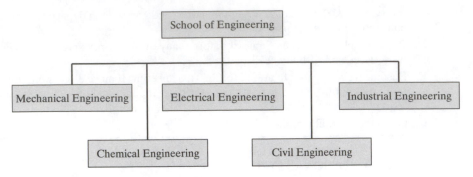

Figure 3.2   Discipline-based organization.

The functional structure has certain advantages, as it (1) permits a hierarchy of skills to be developed and maintained, (2) facilitates specialization in order to achieve high levels of excellence, (3) simplifies coordination as experts in various functional areas are logically grouped together, and (4) allows the use of current technologies and state-of-the-art equipment.

On the other hand, it also has some disadvantages, such as (1) encouraging excessive centralization, (2) delaying decision making due to barriers created by the departmental silos, (3) compounding communication line loss, (4) restricting the development of managerial skills of employees, and (5) limiting employee growth because of constrained exposure to professional experience outside of the departments.

3.5.2 Discipline-Based Organization

Universities, governmental laboratories, and some contract research firms are organized according to disciplines. These organizations contain departments for mechanical engineering, physics, business administration, and other specific disciplines so that specialists may focus on these disciplines in order to excel in research and other activities they pursue. (See Figure 3.2.)

3.5.3 Product- /Region-Based Organization

Large companies may produce and market products of various types to different customers in geographically dispersed locations. More often than not, each of these products may require different production, sales, and business strategies to achieve success in the marketplace. Thus, some companies elect to organize themselves into a product-based structure. (See Figure 3.3.) If the company is marketing products in various geographical regions, each demanding location-specific strategies to penetrate the local markets, and each applying different product-customization strategies according to local needs, a region-based structure may be preferred. (See Figure 3.4.) In either of these cases, a product or regional manager could head up the activities with the overall profit-and-loss (P/L) responsibility for the product or region involved. This manager is further supported by the relevant experts in production, marketing, and other needed functional areas.

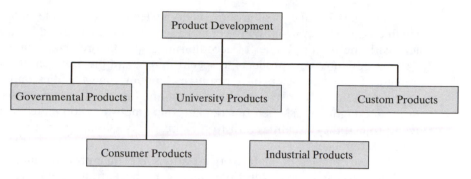

Figure 3.3   Product-based organization.

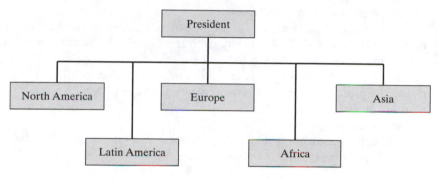

Figure 3.4   Region-based organization.

This type of product- or region-based organization enjoys the following advantages: It (1) focuses on end products or geographical regions for improved local adaptation, (2) facilitates companywide coordination, (3) encourages management development of employees, (4) provides for decentralization, and (5) opens ways for unlimited growth.

The disadvantages for such organizational structure are the following: (1) costs may be high due to layers and autonomous or duplicated facilities; (2) it may require added management talents; (3) specialists may easily become obsolete due to a lack of focus and dedication; and (4) changes are slow to implement because of the complex organizational bureaucracy.

The organizational structure types just described have one thing in common: They all have a hierarchical structure with a clearly defined chain of command (a structure originating from the military).

3.5.4 Matrix Organization

Some companies utilize the matrix organizational structure as a short-term arrangement for specific projects and tasks involving both functional group employees and project managers (Bartlett and Ghoshal 1990).

Managers of functional groups supervise technically capable people who have valuable skills and know-how. Project managers are those entrusted by upper management with the responsibilities of accomplishing specific projects, such as capital projects, the design of new products to specifications, and the creation of business entry strategies. Project managers have resources—money, time, facilities, and management support—and they "borrow" employees from the functional groups to accomplish the work. (See Figure 3.5.) When functional group employees complete their work for a given project, they usually return to their respective home groups to continue their original assignments.

The advantages of matrix organizations are that project managers focus on schedule and cost, whereas functional managers concentrate on quality and expertise. This arrangement offers a balance of workload, and it is excellent for participating employees to achieve wide exposure within the company by interacting with those outside of their special areas of expertise.

The main disadvantage is that this structure requires participating employees to report to two bosses (dual reporting), thus violating the "unity command" principle. When employees are assigned to work on several projects, they may be subjected to marching orders issued by several superiors. In practice, conflicts between the functional and project managers are frequent and severe, mostly with respect to task priority, manpower assignment, interests, quality versus urgency, performance appraisal, employee promotion, and other issues.

Matrix organization demands a delicate balance of power between functional and project managers. Functional managers control manpower, particularly who works

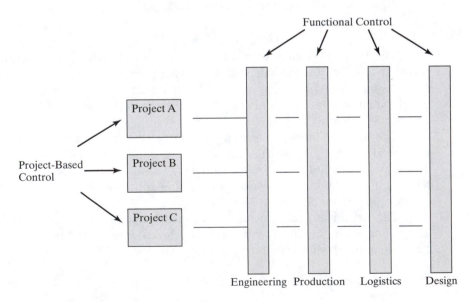

Figure 3.5   Matrix organization.

on what projects, when, and for how long, in addition to controlling knowledge and facilities. On the other hand, project managers have an approved spending budget and the support of upper management. A lack of a balance of power will occur when the functional managers have their own funds to support their own people, thus making them less dependent on project managers. An unbalance of power will also occur when project managers outsource some of the needed work that is not delivered by the functional managers. Under these circumstances, the matrix organization could break down.

Because of the aforementioned built-in conflicts, many companies in industry are moving away from the matrix organizations in favor of teams.

Example 3.2.

Once the functional manager and project manager agree on a project schedule, who is responsible for getting the work performed? Who is accountable for getting the work performed? Why the difference, if any?

Answer 3.2.

Responsibility and accountability are two different management concepts.

In a matrix organization, the project manager delegates tasks to the functional manager, who in turn assigns specific tasks to individual employees in his or her functional group. The functional manager remains responsible for getting the work performed, whereas the project manager is accountable for the results of the work that has been delegated to and done by the functional manager (or his or her people).

The project manager is accountable for achieving specific project objectives. He or she defines the pertinent tasks to be accomplished. If the tasks are defined improperly, causing the objectives to be impossible to attain, the project managers are accountable for such mistakes. The functional manager, on the other hand, is responsible only for supplying the right people with the proper skills and dedication to accomplish the stated tasks. The functional manager is responsible for accomplishing the agreed-upon tasks in an efficient and professional manner.

3.5.5 Team Organization

A team is composed of members who are "on loan" from their respective functional departments and are thus assigned to work full time for the team leader in tackling high-priority, short-duration tasks or projects. (Katzenbach and Smith 2003). Since all team members report to the team leaders only, conflicts arising from dual reporting are eliminated. (See Figure 3.6.) Examples of team organization include product development teams and special task forces.

3.5.6 Network Organization (Weblike Organization)

In response to rapid changes in customers' needs, advancements in technology, marketplace competition, and globalization, some companies have started pursuing certain new business paradigms that are based on thinking globally and acting locally (Ashkenas 2002).

One such paradigm is the inclusion of local suppliers' inherent technical skills and capabilities as a part of corporate strength.

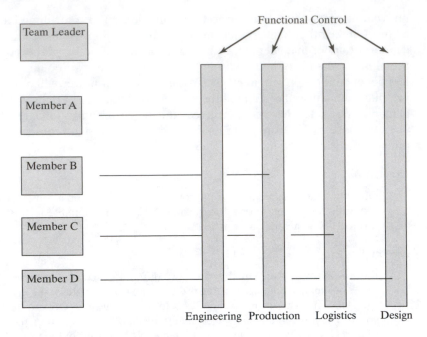

Figure 3.6   Team organization.

Companies form alliances, create business networks, and establish supply chains with regional companies to manufacture, assemble, market, deliver, and service products for specific regional markets. (See Figure 3.7.) At the nodes of such networks are knowledge workers who manage relationships with others (e.g., suppliers, customers, functional groups within an organization, and other such partners).

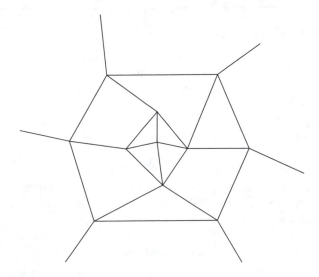

Figure 3.7   Weblike structure.

The number of such business network arrangements is expected to grow with time. Once formed, these networks must be properly maintained. Specific parts of these networks may be activated from time to time for business strategy development, product development, system design, quality control, logistics, customer service, and other such important projects.

Partners linked by these networks may be of different cultural and business backgrounds with divergent value systems and perspectives, and may be dispersed in various regions. Engineering managers serving on such intercompany network organizations may also be challenged by the expected resistance to change, difference in working habits, absence of motivation and control means, and slowness in consensus building and decision making.

Network organizations behave a step closer to organizations in chaos. According to *complexity theory* (sometimes also called *chaos theory*), these organizations exhibit several unique features (Kelly and Allison 1999):

A. All members are independent, flexible, and empowered. They behave responsibly, free of the traditional top-down command and control structures. The Newtonian cause–effect paradigm does not properly describe the organizational behavior that is influenced by the actions of all of its members.

B. Members tend to self-organize themselves by ways of intensive interactions between the members and to form self-directed network organizations.

C. The flexible organizational form fosters creativity and innovation of empowered members.

The complexity theory claims that organizations composed of members with personal autonomy will be better able to operate in economical, political, and social-cultural environments that are turbulent and rapidly varying (Dolan, Garcia, and Auerbach 2003; Hoogerwerf and Poorthuis 2002). Rigid objectives and instructions will no longer be effective in managing such enterprises in the emerging global economy. The following three underlying principles have been formulated for such organizations:

1. **Connectivity.** Members in the network organizations recognize themselves as an integral part of the whole organization and believe that their best interests are served when the interests of the greater whole are served. Members are closely connected with one another.

2. **Indeterminacy.** The turbulent future cannot be readily predicted or planned. Members of the organization need to empower themselves to act with confidence, courage, and integrity.

3. **Consciousness.** Organizations are products of the collective thinking comprising them. The collective consciousness of the organization is defined by all of its members. Everyone is a vital contributor.

Members exercise self-control and are guided by shared essential values, such as honesty, loyalty, integrity, independence, and responsibility. Teamwork is a hallmark of such organizations, wherein individuals are increasingly autonomous, flexible, and dedicated. Product customization is a likely means for such organizations to achieve success in the marketplace.

In practice, research universities behave like semichaotic organizations. As a rule, tenured professors devote 50 percent of their time to academic research, 30 percent to teaching, and 20 percent to services. Each of them is completely independent as far as research is concerned. However, each of them receives instructions as to what courses to teach, which committees to serve on, and how many students to advise in a given year. Although they are all empowered to conduct research with personal autonomy, some of them do more than others. It is not unusual to see school administrators dangling the distinguished professorships, achievement awards, and other inducements to encourage educators to become more active in research.

Another instance of complexity theory in practice is the replacement of a well-known international bridge between the U.S. and Canada. It had been known for some time that the specific bridge under consideration needed to be replaced or expanded due to increased traffic, both commercial and residential. The Canadian provincial government acted decisively by having a truck-processing facility built on their side way ahead of schedule. On the other hand, many groups on the American side have empowered themselves in the decision-making process. Involved are local citizens, politicians, special interest groups, members of Congress, state government officials, business groups, and others. These autonomous groups could not agree on the type of bridge to build (e.g., signature hanging bridge, double-decker, and another design). Some are in favor of speed of construction, whereas others favor a landmark design to add distinction to the region. Lawsuits were initiated. Everyone talked and no one listened. As a result of this complexity theory in action, no agreement was reached as to which bridge to build for a long time.

Network organizations involving independent supply partners could be difficult to manage. It is not likely that much can be accomplished by leaving them alone to direct themselves, as suggested by the complexity theory. Engineering managers need to prepare themselves to effectively lead such network organizations.

Example 3.3.

Company X manufactures automobile jacks, hubcaps, and a variety of fittings. These products are sold as replacement parts through chain auto-supply stores. The business of the company is growing, with production facilities located in rented buildings over various parts of the city. The production staff is expanding constantly. Now the president of the company wants to expand into the brass-fittings business.

However, the president realizes that, after this newest expansion is accomplished, the company should consolidate to make its production operations more efficient.

Which organizational structure should the company adopt now, so that it can best accommodate its current needs of business expansion and also lay the foundation for anticipated consolidation thereafter? What information is needed to set forth such an organizational structure? What are the crucial variables that should be considered in the design of such an organization?

Answer 3.3.

To expand the brass-fittings business, Company X should set up a multidisciplinary team initially. This team will be empowered to come up with a business plan to enter the brass-fittings business on the basis of market research and competitive analysis. The plan should include market-share position, time to market goals, marketing strategies, capital investment, and production or sourcing requirements. It may be useful to retain competent external business consultants for advisement.

The business plan defines the needs of personnel (capabilities, experience, and number), facilities, and other resources required for implementation. A product-centered department is set up to be responsible for the profit and loss of the brass-fittings business. While the distribution of brass fittings may be readily handled by the existing organization in the company, special attention needs to be paid to production, marketing, engineering, and service of brass fittings.

Once the brass-fittings business is fully established, the production facilities of brass fittings could be integrated into the production organization of other company products in order to realize scale of economies advantages.

The organizational design must be flexible to effectively serve the purpose at hand. Crucial variables to be considered in the organizational design include the significance of the brass-fittings business to the company's overall performance (e.g., market size, competition, and profitability), extent of the resources required of the company to enter the brass-fittings business, and the management and technical talents available inside and outside of the company.

3.6 ENHANCING CORPORATE PERFORMANCE BY ORGANIZING—EXAMPLES

Organizing is an efficacious method to achieve the critically important objective of doing things right (or performing tasks in the most efficient ways). Management can put the right people together and keep the wrong people separated so that work efficiency and goal attainment can be greatly enhanced (De Waal 2001). The following are several best practice examples of how company productivity was raised by employing organizing strategies.

3.6.1 Organizing for High Performance by Using a Flexible Structure

Organizational structures are known to have impact on corporate performance. Some companies allow the structure to continuously evolve in order to adapt to changing opportunities in the marketplace.

According to Aufreiter, Lawyer, and Lun (2000) new-style companies achieved great business success by being relentlessly committed and by exercising discipline from the top, as well as practicing three organizational principles:

1. Make it everyone's job to identify new opportunities. Company culture must support such a principle by way of feedback loop and financial incentives.
2. Decide quickly on project priority. Speed and coordination are critically important in implementation. Use technology to support decision making wherever possible, and fill the gaps with fast, centralized, senior–level decision making.
3. Hire people for specific roles such as marketing and technology support. This is needed for implementing ideas to quickly cash in on new opportunities in the marketplace.

The following are some of the best examples of organizational practices in industry:

A. *Starbucks Corporation* sells coffee and related products to consumers. It encourages all employees to suggest new ideas (using a one-page form) and promises a quick response from the senior executive team. The person suggesting a given idea is always invited to be a full-time member of the launching team.

Leaders are critically important for any team to achieve success. Starbucks assembles teams headed by leaders with applicable marketing expertise. Starbucks hires two kinds of people: (1) integrators who are marketers with broad skills and who are capable of carrying out major responsibilities in coordinating the delivery of products and services to the market, and (2) specialists with unique capabilities and expertise for launching projects. All of them need to have an entrepreneurial mindset and be able to thrive in an uncertain and rapidly changing environment. For the "Store of the Future" project, the leader was hired from the outside. Another outsider was hired to spearhead Starbuck's "Lunch Service Concept" project.

Organizational flexibility is a hallmark of new-style companies. When in-house packaging and sales-channel management skills were deemed lacking, Starbucks elected to team up with another company, *Dreyer's Grand Ice Cream*, to launch a new ice cream product. Within four months, its coffee ice cream product became the top-selling brand in the industry. "Frappuccino," a cold coffee drink, was also marketed jointly with *PepsiCo, Inc.*, in 1994. Within one year, its revenue accounted for 11 percent of the total sales of Starbucks in 1995.

Starbucks uses a high-level steering committee for making rapid decisions based on two criteria: (1) the effect on company's revenue growth (the new ideas must have the potential of producing a minimum revenue of about $4 million per year) and (2) the impact on the complexity of the company's retail stores required for implementation.

B. *First USA* is a financial services company. It reconfigures its structure dynamically and makes routine organizational changes. Once new market opportunities worth pursuing are identified, the company determines what specialized skills are required and puts a suitable "dream team" together. The company maintains a pool of managers having special skills that enable them to launch new credit card products. Other people are added from internal resource pools or by hiring from the outside. Within a recent three-year period (1995–1998), First USA issued five times as many credit cards as another well-known financial company organized in traditional ways.

C. *Dell*™, sells computers and other technology products directly to consumers and businesses alike. It focuses on identifying and capturing market opportunities. Managers are encouraged to turn the needs of customers into products and services as quickly as possible. Business units are smaller in size, and more people are empowered to assume the profit and loss (P&L) responsibilities. Successful managers are rewarded accordingly.

D. *Minnesota Mining and Minerals* (officially known as *3M*) is a major industrial company in the United States. It markets a diversified set of products. It is known to organize its employees to create new ideas by allowing them each to spend up to 15 percent of their work time on new initiatives of their own and by supporting such projects with grant money outside of the departmental budgets. To win the rights for funding future new projects, 3M managers must derive at least 30 percent of their sales from ideas brought into being during the four preceding years.

When discussing the merits of an organizational strategy, the basic criterion is, of course, its impact on business results. Aufreiter, Lawyer and Lun (2000) studied the compound annual growth rate (CAGR) of several companies practicing flexible organizational strategies for the period of 1994–1998 and showed that the CAGR of Trilogy, First USA, Dell, Starbucks and Home Depot are significantly higher than the average growth rates of the next three largest competitors in the same industry.

3.6.2 Organizing for Innovation

Some companies are more focused than others on developing and sustaining corporate competitiveness by nurturing innovation. Innovation can be affected by company structure.

According to Chesbrough and Teece (2002), there are two types of innovations: autonomous and systemic inventions. *Autonomous inventions* are those which can be pursued independently of other innovations. For example, inventions related to turbochargers could be pursued independently of inventions related to automobiles, to which turbochargers are applied for boosting power output. Similar relationships exist between filters and air conditioners, motors and compressors, and color-print films and analog cameras. *Systemic inventions*, on the other hand, are those which must be developed in close coordination with others. Examples include the development of product design, supplier management and information technology for lean manufacturing, and films and cameras for instant photography. Because autonomous inventions may be pursued independently of other inventions, they are better suited for virtual organizations. Vertical organizations, on the other hand, are more likely to succeed in pursuing systemic inventions that require close coordination and intensive information sharing.

Organizing a company's structure to be virtual means that all noncore functions are typically outsourced. These virtual companies form supply-chain partnerships and business alliances to develop, manufacture, market, distribute, and support their offerings. They use incentives—signup bonuses, stock options—to attract highly trained independent inventors to generate breakthrough inventions. Because of the self-interests of all participants, the coordination between these partners can be difficult, rendering the resulting inventions of value only to some, but not to others.

Vertically organized companies are those which have rigid hierarchical organizational structures. They maintain control of all functions and typically have well-established procedures for settling conflicts and for planning all activities that promote innovations. Systemic inventions require information sharing and adjustments, which these vertically integrated companies can readily promote and safeguard. However, they cannot offer the high incentives that virtual companies use to attract independent inventors. As a consequence, they may not be able to access top-level talent for creating innovations.

Therefore, when organizing for innovations, one must choose between talents and control. The key is to select the right kind of organizational form to match the type of innovation (autonomous, systemic, or a mix of the two) the company needs. At one extreme, virtual companies are suitable for pursuing autonomous innovations. At the other extreme, vertical companies are excellent for pursuing systemic inventions. As the invention type changes gradually from purely autonomous to purely systemic, the company should consider intermediate forms of organizations such as alliances, joint ventures, and collaborations with autonomous divisions.

Nowadays, few companies can afford to develop all needed technologies internally. A mix of approaches is usually adopted. Some technologies are developed in-house to serve as the core part of the value chain. Other less critical ones are typically purchased outright or acquired through license, partnership, and alliance. Over the long run, however, key value-added advances will need to come from within.

3.6.3 Organizing for Performance at the Design-Manufacturing Interface

Conflicts are known to exist at the interface between product design and manufacturing. These conflicts cause frequent cost overruns and product introduction delays. In many traditional companies, the product-design group signs off on the design, and then they "throw it over the wall." The group responsible for manufacturability takes over and reexamines the design for cost-effective mass production. While product design may have focused on performance and aesthetics, manufacturing looks after production efficiency. Also contributing to these conflicts are other factors such as (1) funding periods for design and manufacturing that do not overlap, (2) differences in education between design and manufacturing staff, and (3) offices that are not at the same location.

Some of these difficulties may be removed by way of organizing. Organizational options for improving the design-manufacturing interface include the following (Dean and Susman 1989):

1. Manufacturing sign-off—manufacturing has veto power over the final product design. Software programs (e.g., *Assembly Evaluation Method* by Hitachi and *Design for Assembly* by Bootheroyd Dewhurst) are available to calculate a producibility score for checking on manufacturability.

2. Appoint an integrator who performs liaison work between design and manufacturing and offers a balanced view.

3. Form a cross-functional team composed of members of design and manufacturing, with the final authority resting with the engineering department. The use of such cross-functional teams is known to have significant benefits such as assuring compatibility between the design and manufacturing processes, saving time, simplifying the design process, and reducing design changes.

4. Combine the manufacturing process and product design into one department.

In general, if the company's culture is conducive to absorbing organizational changes, then the organizational options of the team or combined department are to be preferred. On the other hand, if the products and manufacturing processes are fixed, then the organizational options of the sign-off or integrator tend to make more sense.

3.6.4 Organizing for Heightened Employee Motivation

As a rule, teams are temporary in nature because they are built for specific objectives and will be disbanded after their specific objectives have been achieved. Only in exceptional cases will teams be exhaustively utilized on a permanent basis to achieve business success. This is the case of AES Corporation.

Located in Arlington, Virginia, AES Corporation is the largest global power company, with sales at $9.3 billion (2001) and market capitalization of $1.64 billion (August 30, 2002). In 1999, it operated 90 plants in 13 countries. Seventy-five percent of its business was in contract generation. Figure 3.8 shows its rapid revenue growth for the period 1990–1998.

The company has very few organizational layers and, except for a corporate accounting department, keeps no staff for functional specialties. It is organized into 11 regions, each headed up by a manager. Each region is further organized into 5 to 20 teams, and each team has 5 to 20 members. Teams are created primarily for the combined functions of plant operation and maintenance.

Each team has no more than one of each kind of expert or specialist. As a result, everyone on the team becomes a well-rounded generalist. In-house qualification exams are held to ensure minimum expertise before job rotation requests of employees are approved. Each team owns what it does and is empowered to make decisions with commensurate authority to implement their decisions. The roles for company leaders are limited to advisors, guardians of the company principles, encouragers, and officers accountable to the outside world.

Employees are compensated according to the following formula: (1) 50 percent on financial performance and safety and environmental impact and (2) 50 percent on how well employees follow the four company values—fairness, integrity, social responsibility, and fun. The hiring practice of the company focuses on cultural fit first and technical skills second.

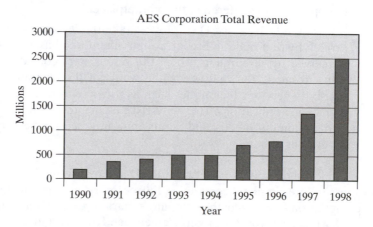

Figure 3.8   Total revenue of AES Corporation.

Company representatives attribute their business success at organizing the company in teams to the heightened level of employee motivation made possible by the team empowerment practice.

It should be noted that the AES Corporation example may indeed apply well to other low-tech operations such as warehouses, distribution centers, supermarkets, hardware stores, and service centers wherein repeat common practices are the norm. Everything you would ever want to know about operating and maintaining a conventional power plant has already been sufficiently preserved in manuals; in-depth technical expertise and innovation are not required (Wetlaufer 1999).

3.6.5 Organizing for High-Tech Marketing

Some companies in industry are high tech, and others are not. "High tech" refers to products and services characterized by (1) their strong scientific–technical bases, (2) the possibility of being quickly obsolete because of new technologies, and (3) the capability to develop or revolutionize markets and demands when built upon new emerging technologies. Examples of high-tech products include semiconductors, microcomputers, and robotics.

High-tech companies need to pay special attention to the interface between R&D and marketing. To achieve business success, a linkage between R&D and marketing must be established. When marketing high-tech products or services, companies typically follow two basic approaches: market driven and innovation driven.

A. **Market Driven.** When pursuing the market-driven strategy, companies use marketing to define the needs of customers and ask R&D to provide the required innovations to satisfy such needs. Customer suggestions are typically good sources of new needs. In this case, marketing uses tools such as *concept testing*, *product prototyping*, and *pilot testing* to define the specific product or service features needed by customers. Marketing efforts precede the R&D efforts. The consequence of practicing a market-driven strategy is that there may be a possible delay in breakthrough innovations, preventing the company from a timely use of the opportunities offered in the marketplace.

B. **Innovation Driven.** R&D employees take the lead in the innovation-driven approach by first making breakthrough inventions based on *preemptive needs* as perceived by researchers. Then, the researchers consider pursuing the inventions to satisfy the real needs and wants of the customers. In such situations, marketing applies techniques such as *focus groups* to verify new product concepts and applications. There are risks associated with this approach, as identified customers' needs and wants may not be satisfied by the breakthrough inventions at hand.

A lack of coordination between R&D and marketing is known to be the cause of business failures in many companies. Furthermore, any new technology advancement by competition can change the company's market position instantly.

Organizing a workable interface between R&D and marketing is a way of avoiding the potential loss of market opportunities due to invention delay and the lack of compatibility of inventions to the actual market needs. Setting up a team of representatives of R&D and marketing to constantly monitor relevant activities and foster communications is a good organizing strategy (Shanklin and Ryans 1984).

3.7 CROSS-FUNCTIONAL TEAMS

Cross-functional teams have become crucial to business success (Parker 2003). In general, teams are set up to (1) generate recommendations, such as a strategy to enter a specific regional market or solve a specific customer-related problem; (2) make or do things—for example, design products, develop new processes, install new assembly lines; and (3) run things—for example, operate plants. The performance of a team is the sum of the performance by individual members plus the work product brought about by the members together. It is the work product delivered by the members together (the team synergism) that is responsible for the superiority of the overall team performance over the sum of the performances of the individuals (Payne 2001). Teams may fail if they are not led properly (Rees 2001; Brounstein 2002).

Organizing multifunctional teams is often the preferred choice to address complex coordination issues at interfaces, in addition to the interface between design and manufacturing discussed before. For example, in a typical functional organization, the development of new products follows a sequential process enumerated here:

1. *Marketing* conducts research to identify the customers' needs and defines product features, such as functionality, reliability, ease of repair, resale value, warranty, etc.

2. *Design engineering* releases specifications, performs functional design, selects material, obtains vendor and supplier inputs, and conducts engineering analyses to incorporate these features into a product.

3. *Manufacturing engineering* reviews and simplifies the product design for manufacturability and reliability considerations.

4. *Service* organization further changes the design to enhance serviceability.

5. *Production* is finally set up to define manufacturing techniques and to mass-produce the product.

Such sequential processes are known to be inherently ineffective with respect to coordination, information sharing, and decision making. In a concurrent engineering team, representatives of all of the functional groups just mentioned, plus those from procurement, finance, vendors and suppliers, product testing, and logistics, are included as members. All aspects related to product development are considered early on and concurrently. The goal is to bring forth an optimum product for the company within the shortest period of time and at the lowest possible cost, while satisfying all constraints and meeting all requirements.

All team members have the full support of their respective departments, functional units and home bases, so that the specific inputs they make on the team, at various stages of product development, are always the best possible inputs (Lencioni 2002). The keys to the success of concurrent engineering teams are the following:

A. Management commitment
B. Ongoing communications that use advanced communication tools such as the intranet, e-mails, and electronic data interchange (EDI)
C. Teamwork training for all members

The value of the concurrent engineering team concept is evident from the following statistics:

- Mercury Computer Systems, Inc., Lowell, Massachusetts—Concurrent engineering team reduced the time to market of its add-on process boards for VME (Versa Module Eurocard) bus from 125 days to 90 days.
- Hewlett-Packard Company, Palo Alto, California—Cut the time to market of its 54600 Oscilloscope by two-thirds.
- Toyota Motor Corporation, Tokyo, Japan—Concurrent engineering decreased its product cost by 61 percent.

In general, concurrent engineering delivered impressive benefits in the order of magnitude shown in Table 3.1 (Yesersky 1993).

TABLE 3.1   Benefits of Concurrent Engineering Teams

Activities	Percentage
Reduction of time of product development	30–70
Reduction of number of engineering changes	65–95
Reduction of time to market	20–90
Improvement of product quality	200–600

Engineering managers need to become proficient in leading and participating in concurrent engineering teams. Teams may be formed to address any important corporate task. For teams to add value, team leaders need to pay attention to team discipline, team learning, and factors affecting team effectiveness.

3.7.1 Team Discipline

For "blow-the-roof-off" performance, a team is often the vehicle of choice. But to excel, the team needs the right ingredients, such as the ability to listen, respond constructively, support one another, share team values, and have a discipline (Katzenback and Smith 2000).

A team is a small number of people (usually between 2 and 25) who are committed to a common purpose and who possess complementary skills, a set of performance goals, and an approach for which they hold themselves mutually accountable.

Mutual accountability refers not only to the team leader, but also to accountability of the team members to each other. Team members are said to have developed mutual accountability if and when they have reached the emotional state of "being in the boat together," based on commitment and a mutual trust that has been established. Team leaders need to strive to organize the team so that team members make commitments and have mutual trust and so that team members hold each other accountable. Without mutual accountability, there can be no team of real value.

3.7.2 Team Learning

One decisive factor that affects the team's responsiveness is its learning capability (Bell and Smith 2003). In corporate settings, teams need to learn new technologies (such as three-dimensional computer-aided design (CAD), visualization software, project management tools, videoconference, Web-based net-meeting tools, and others) or new processes (such as new ways of working and new relationships for collaborative work). How fast a team can learn will affect its overall timely performance in attaining the specific objectives at hand.

A learning team is one that is skilled at creating, acquiring, and transferring knowledge and at modifying its own behavior to reflect new knowledge and insights (Garvin 2003; Garvin 1997). The team needs to have systems and procedures in place to do the following:

A. Solve problems systematically
B. Experiment with new approaches
C. Learn from its own experience and past history
D. Learn from other's experience and best practices
E. Disseminate knowledge effectively throughout the team and organization

Edmondson, Bohmer, and Piscano (2001) studied a large number of cardiovascular surgical teams. They believe that team learning may be speeded up if the team leaders possess both technical and managerial skills. Because learning has both technical and organizational aspects—status, communications—the organizational skills of the team leader affect the team learning. Factors affecting team learning include the following:

A. **Team composition.** When selecting team members, team leaders should give preference to members' technical competence—retention of a mix of skills and expertise, ability to work with others, willingness to deal with ambiguous situations (risk takers), and self-confidence in making suggestions and proposing ideas while not inhibited by other members' ranking and corporate status. The most effective learning takes place during the process. Teams with a mix of expertise and experience tend to be able to draw upon members' relevant past experience and speed up learning.
B. **Team cultures.** Team leaders should build a team culture in which some experimentation is encouraged and failure is acceptable.
C. **Leader's style.** Teams will learn better and faster if, as motivation, the team leaders frame the learning as a challenge for all team members.

Factors that do not affect team learning are said to include (1) educational background, (2) prior experience in applying old technologies, (3) top-management commitment, (4) status of the team leader, and (5) reporting and auditing processes.

Example 3.4.

Some people feel that working as a team, instead of allowing experts to produce more creative outcomes, actually resulted in watered-down compromises and bland solutions. They view teamwork as a series of exercises in "sharing ignorance." Do you agree or disagree, and why? What can be done to advance the technical qualities of the team outcomes?

Answer 3.4.

The concern about the watered-downed outputs of teams is real. Team members of different background and expertise may indeed have different opinions, which often force the team members to compromise. It is quite true that sometimes the views of the domain experts on the team are not shared and accepted by others on the team, who do not and will not want to understand.

One obvious way to ensure the technical quality of the team results is to select people to lead who are technically qualified and able to render technical judgment. Another way is to bring in an impartial outside consultant to comment and advise on the relative technical merits of the options under consideration.

Team consensus is good to have, because it allows the team members to jointly own the team outcome. This ownership represents a strong motivation factor to team members who are then inspired to actively implement the team outcome. A technically superior team outcome adds little value if it is not implemented properly.

3.8 DELEGATING

After a specific form of organization is established—unit, department, team, division, or regional group—the next step is for the engineering manager to delegate the proper responsibility and authority to the selected leaders and workers, and establish the upward-directed accountability needed to achieve the defined organizational objectives. Therefore, delegating is for the purpose of improving the engineering manager's overall efficiency by assigning responsibility and authority and by creating accountability (Burns 2002).

The benefits of delegating are many. To engineering managers, as superiors, delegating is beneficial as it (1) improves the quality and quantity of work performed, (2) relieves the engineering manager for pursuing more important duties or gaining more time for management work, (3) makes the engineering manager knowledgeable of the employee's capabilities, (4) prepares the employee to step in for the engineering manager when needed and hence enabling the engineering manager to be absent from the job occasionally, (5) distributes the workload effectively, (6) develops leadership qualities, (7) eases the engineering manager's job pressure, and (8) reduces costs through more efficient operating decisions.

Delegating is also beneficial to engineers as technical contributors because it (1) makes the job more satisfying, (2) provides encouragement, incentives, and recognition, (3) develops new skills and knowledge, (4) promotes self-confidence, (5) facilitates teamwork, (6) encourages growth and development, and (7) fosters initiative and competence.

It is important for engineering managers to keep in mind what should and should not be delegated. Problems and activities of the following kinds are to be delegated: (1) those that require exploration and recommendation for a decision,

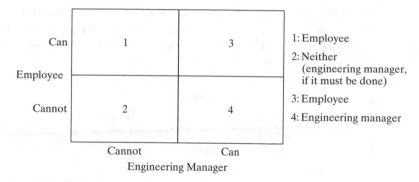

Figure 3.9   Delegation matrix.

(2) those that are within the scope and capabilities of employees involved, (3) those that are needed to achieve company objectives, (4) those that promote the employee's development in technologies, business perspectives, and leadership skills, and (5) those that save the engineering manager's time if done properly by the employees.

For delegating, the following guidelines which is also illustrated in Figure 3.9, may be helpful:

1. What the engineering manager cannot do and the employee can do, the employee does.
2. What both the engineering manager and employee cannot do, the engineering manager does.
3. What both the engineering manager and the employee can do, the employee does.
4. What the engineering manager can do and the employee cannot, the engineering manager does.

Which problems or activities should the engineering manager not delegate? Such problems or activities include (1) planning–creating plans within larger plans and objectives, (2) resolving morale problems in the group or department, (3) reconciling differences and conflicts, (4) coaching and developing employees, (5) reviewing performance of employees, (6) completing assignments given to engineering managers by their superiors, and (7) completing other assignments only engineering managers themselves should handle (such as confidential committee assignments, "pet" projects, and tasks without proper talents to delegate to).

Delegating requires skill and practice. The following are guidelines for efficacious delegation:

A. Explain the importance of the assignment.
B. Check on understanding and confidence.
C. Give the employee leeway in their choice of method, unless the procedure has been specified and developed before.
D. Set a goal, timetable, or deadline. A short-term goal is better than a long-term goal.
E. Be reasonable. Keep the goal within the employee's capabilities.

F. Assign responsibilities that go with the job. Allow commensurate authority of decision making, and let employees accept responsibility for poor as well as good work. (The engineering managers remain accountable for the delegated assignment with respect to their own superiors.)

G. Trust the employee.

H. Give recognition for good work.

I. Share the engineering manager's own worries. Let the employees know your worries about the assignment; explain them openly and fully. Recognize difficulties in the assignment and ask for suggestions on how best to handle the assignment.

J. Make it a project; let the assignment be a challenge.

K. Do not rush in and take over. The employee could use more training if a lack of progress is shown.

L. Do not expect or want perfection.

Certain barriers to delegation do exist. Engineering managers need to beware of these barriers: (1) *Psychological*—engineering managers have fears and worries. If engineering managers let the employee do the work, they may fear their own technological obsolescence, while their employees shine. This fear may be particularly relevant to engineering managers who themselves are technically very strong. (2) *Organizational*—unclear responsibility and relationship and confused understanding of line versus staff positions may hinder effectual delegation.

There are several more noteworthy observations:

- Delegation tends to be limited by the availability of effective controls. If there is no control, there should be no delegation. Engineering managers should delegate safely only to the extent that one can determine if the work can be correctly carried out and decisions can be made in the manner that they should be. Furthermore, engineering managers should make sure that the plans are sound and that controls are in effect.

- Authority must be commensurate with responsibility. Engineering managers should delegate enough authority to allow decision making by the employees related to the work.

- Accountability is demanded of employees who are obligated to their superiors for achieving the expected results. Accountability is achieved by properly discharging responsibility and using authority delegated. Effectiveness of and success in delegation depends on the willingness and ability of the employee to perform the work, make decisions, and achieve results.

- Control must be in place. Engineering managers need to introduce midcourse corrections, if needed. Otherwise, delegation will lead to disaster. Setting up performance metrics and constantly monitoring performance will help.

Many engineers fail to achieve success in managerial ranks, partly because they do not know how to delegate properly. Good delegation is a prerequisite for being a good manager.

3.9 ESTABLISHING WORKING RELATIONSHIPS

Another organizing activity to be performed by engineering managers is the establishment of the proper working relationships between employees and between units (Straus 2002). This is to ensure that people are working together well enough to achieve the company objectives. Specifically, the activity calls for role clarification and conflict resolution.

3.9.1 Role Clarification

In complex organizational settings, clarifying roles addresses the issues of authority and accountability. For a specific project involving personnel of multiple departments or business units, the need for defining roles of all participants is self-evident. Figure 3.10 illustrates an example of role clarification.

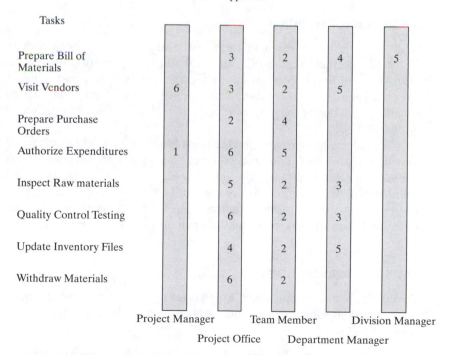

1	General management responsibility
2	Specialized responsibility
3	Must be consulted
4	May be consulted
5	Must be notified
6	Must be approved

Tasks	Project Manager	Project Office	Team Member	Department Manager	Division Manager
Prepare Bill of Materials		3	2	4	5
Visit Vendors	6	3	2	5	
Prepare Purchase Orders		2	4		
Authorize Expenditures	1	6	5		
Inspect Raw materials		5	2	3	
Quality Control Testing		6	2	3	
Update Inventory Files		4	2	5	
Withdraw Materials		6	2		

Figure 3.10   Roles Assignments.

Companies issue organizational charts that describe the roles and responsibilities of major business units. Employees may assume line roles, coordinating roles, and advisory roles.

A. **Line Roles.** Employees with line roles are those in profit centers with monopoly rights within the company to provide products and services to clients and customers. Examples of profit centers include business management, production, and sales. Profit centers are business units empowered to generate profits for the company. Managers of profit centers are accountable for offering quality products and services at competitive prices to ensure that the company makes profits. Profit centers define the services they might need. Managers of profit centers approve the annual budgets of cost centers that provide such needed services.

B. **Coordinating Roles.** Employees in some cost centers have monopoly rights for developing and recommending constraints on the position duties of others. These constraints can take the form of approvals, policies, procedures, or planning objectives—legal, financial control, human resources, etc. They are accountable for achieving higher level organizational objectives such as consistency in work method, integration regarding external contacts, or cost efficiencies.

C. **Advisory Roles.** Employees in other cost centers provide services in support of the profit centers. Examples of such cost centers include R&D, maintenance, investors' relations, financial accounting, procurement, and others.

3.9.2 Conflict Resolution

In the real-world environment, there are conflicts of many types, such as the following: (1) technical, including design, analysis, and interpretation of test results; (2) operational, including procedures to perform specific tasks and assign responsibility; (3) emotional, such as treating bruised egos and hurt personal feelings; and (4) political, such as knowing whom to consult and who has a say on specific projects or issues.

Engineering managers need to learn how to effectively resolve conflicts. Conflicts may be resolved by (1) *dominance*—dictating a solution, (2) *compromise*—negotiation based on a relative power base, and (3) *collaboration* that leads to finding a win–win solution. The key requirement for conflict resolution is openness. By fostering mutual respect and trust, nurturing common interest to achieve project success, and focusing on commitment to task, most conflicts can be successfully resolved.

3.10 INFORMAL ORGANIZATIONS

Beside the formal organization, which is set up to achieve work efficiency, there are informal organizations in every business enterprise. Typical informal organizations are listed as follows:

A. **Social.** People form groups to pursue specific common interests, shared values, and beliefs; for example, beer clubs, bowling clubs, company outings, golf leagues, tennis groups, etc.

B. **Status.** People tend to be drawn toward persons well-known for their technical skills, abilities, special accomplishments, experience, tenure, charisma, interests, peer recognition, and acceptance, and to admire and respect such achievers for their status.

C. **Group.** Coalitions form to advance shared interest. Fitness center on site, day-care centers, toastmasters groups, foreign-language study groups, bridge clubs, etc., are some examples.

D. **Location.** Depending on the flow of vital information, people tend to migrate toward critical locations, such as executive assistants, secretaries, and water coolers.

Engineering managers should be aware that informal organizations encourage additional bonding between employees. Because they contribute toward the smooth operation of an organization and members' job satisfaction, employee participation in them should be encouraged.

3.11 CONCLUSION

Organizing is an important function of engineering management with a direct impact on the manager's ability to get work done efficiently. This function empowers a manager to choose the right organizational forms, be they teams, committees, task forces, functional or matrix arrangements, or other specific organizational structures. Managers assign the right skilled and compatible people to work together, each one having clearly defined roles and responsibilities, along with commensurate authority. Managers assign responsibilities to employees so that work gets done and employees can receive broadened experience. Managers allocate the right resources (such as skills, money, equipment, time, and technology) to accomplish the work efficiently.

Organizing is also important for enhancing the quality of work output. Flexible organizational structures allow companies to better respond to the changes of a dynamic marketplace. Certain organizational forms are superior to others in fostering creativity and inducing innovations. Multifunctional teams are known to be superior in handling conflicts at the interface between design and manufacturing or between R&D and marketing. Teams empowered by management to pursue specific assignments tend to be strongly motivating to the team members.

Engineering managers need to understand the power of organizing and use the function intelligently.

3.12 REFERENCES

Aufreiter, N. A., T. L. Lawyer, and C. D. Lun. 2000. "A New Way to Market." *The McKinsey Quarterly*, No. 2.

Ashkenas, R. 2002. *The Boundaryless Organization: Breaking the Chains of Organizational Structure*. San Francisco: Jossey-Bass.

Baguley, P. 2002. *Teams and Team-Working*. Teach Yourself Books. New York: McGraw-Hill.

Bartlett, C. A. and S. Ghoshal. 1990. "Matrix Management: Not a Structure, a Frame of Mind." *Harvard Business Review*, July–August.

Bell, A. H. and D. M. Smith. 2003. *Learning Team Skills*. Upper Saddle River, NJ: Prentice Hall.

Brounstein, M. 2002. *Managing Teams for Dummies*. New York: John Wiley.

Burns, R. 2002. *Making Delegation Happen: A Simple and Effective Guide to Implementing Successful Delegation*. St. Leonards, New South Wales, Australia: Allen & Unwin.

Chesbrough, H. W. and D. J. Teece. 2002. "Organizing for Innovation: When Is Virtual Virtuous?" *Harvard Business Review*, August.

Cox, T. 2001. *Creating the Multicultural Organization: A Strategy for Capturing the Power of Diversity*. San Francisco: Jossey-Bass.

De Waal, A. 2001. *Power of Performance Management: How Leading Companies Create Sustained Value*. New York: John Wiley.

Dean, J. W. Jr. and G. I. Susman. 1989. "Organizing for Manufacturable Design." *Harvard Business Review*, January–February.

Dolan, S. L., S. Garcia, and A. Auerbach. 2003. "Understanding and Managing Chaos in Organizations." *International Journal of Management*, Vol. 20, No. 1, p. 23.

Edmondson, A., R. Bohmer, and G. Piscano. 2001. "Speeding Up Team Learning." *Harvard Business Review*, October.

Galbraith, J. R. 2002. *Designing Organizations: An Executive Guide to Strategy, Structure and Process*. San Francisco: Jossey-Bass.

Garvin, D. A. 1997. "What Makes for an Authentic Learning Organization: An Interview with David Garvin." *Harvard Business Review*, June 1.

Garvin, D. A. 2003. "Learning in Action: A Guide to Putting the Learning Organization to Work." *Harvard Business Review*, February 13.

Hemphill, B. 1999. "Six Ways to Improve Your Office Organizing Skills." *AFP Exchange*, November–December.

Hoogerwerf, E. C. and A. M. Poothuis. 2002. "The Network Multilogue: A Chaos Approach to Organizational Design." *Journal of Organizational Change Management*, Vol. 15, No. 4, p. 382.

Katzenback, J. R. and D. K. Smith. 2000. "The Discipline of Teams, HBR on Point Article." *Harvard Business Review*, July 14.

Katzenbach, J. R. and D. K. Smith. 2003. *The Wisdom of Teams: Creating the High Performance Organization*. San Francisco: Jossey-Bass.

Kelly, S. and M. A. Allison. 1999. *The Complexity Advantage*. New York: McGraw-Hill.

Lencioni, P. 2002. *The Five Dysfunctions of a Team: A Leadership Fable*. San Francisco: Jossey-Bass.

O'Reilly, C. A. and J. Pfeffer. 2000. *Hidden Value: How Great Companies Achieve Extraordinary Results with Ordinary People*. Cambridge, MA: Harvard Business School Press.

Parker, G. M. 2003. *Cross-Functional Teams: Working with Allies, Enemies, and Other Strangers*. 2d ed. San Francisco: Jossey-Bass.

Payne, V. 2001. *The Team-Building Workshop: A Trainer's Guide*. New York: AMACOM.

Rees, F. 2001. *How to Lead Work Teams: Facilitation Skills*. 2d ed. San Francisco: Jossey-Bass/Pheiffer.

Shanklin, W. L. and J. K. Ryans. 1984. "Organizing for High-Tech Marketing." *Harvard Business Review*, November–December.

Shannon, Robert E. 1990. *Engineering Management*. New York: John Wiley.

Straus, D. 2002. *How to Make Collaboration Work: Powerful Ways to Build Consensus, Solve Problems and Make Decisions*. Berrett-Koehler.

Von Hoffman, C. 1998. "Getting Organized." *Harvard Management Update*, January.

Wetlaufer, S. 1999. "Organizing for Empowerment," *Harvard Business Review*, January–February.

Yesersky, P. T. 1993. "Concurrent Engineering: Your Strategic Weapon in Today's Jungle." *CMA*, Hamilton, July–August.

3.13 QUESTIONS

3.1   What type of organizational structure is best suited for developing a new product that requires a high level of specialization in several functions and for which the time to market represents a critical factor?

3.2   A materials manager suspects that the quality of work within her department has been deteriorating. She wants to introduce a program of change to advance quality. What steps should she take?

3.3   The company has recently concluded a multimillion-dollar contract to supply products to a third-world country. The first elite group of engineers from that country has just completed a two-month training course on maintenance and operations. The company's training manager reports that the level of skill and knowledge of that country's engineers was so low that no amount of training would ever enable them to properly operate and maintain the products in question. "It might be better for that country to buy a less sophisticated product from our competitor," the training manager suggests. What should the company do?

3.4   Six months ago, the company hired an engineer for his expertise in hydraulic drives. The decision to hire him was based on a product development plan that projected a need for such expertise. Market conditions have suddenly changed in favor of more sophisticated electric drives. The new engineer turns out to be very good in his area of specialization, but it is difficult to retrain him for other assignments in the company. Should the company discharge this engineer?

3.5   The company has been making most of its sales to a few large customers. The company president wishes to broaden its customer base. To do so may require changes in the company culture, the product line strategy, marketing and sales programs, and the service organization. How should the president go about making the required changes?

3.6   The company is considering a plan to upgrade its current product line. The cost of upgrading is high. There is a small company that has developed the technology required for this product upgrade. What strategy should the company follow if it wants to continue selling into its current market with the new, upgraded product?

3.7   As the company's sales are coming down unexpectedly, the president asks you to chair a task force with the objective of developing solutions to correct the situation. Who do you want to be on this task force? How should the task force resolve this problem?

3.8   A loyal and high-volume customer has warned the company's marketing department that project X is extremely critical to their needs and that if this project is late, they may be forced to buy elsewhere. The project manager knows that the best estimates available to date from various in-house groups indicate that, at the current rate of progress, project X will be late by about six months. What should the project manager do?

3.9   Sally Lee, the engineering manager, delegates tasks as a good manager should. However, Mark Hayes, the engineering director, has the bad habit of calling up Sally unexpectedly to get detailed reports on various ongoing activities in Sally's department. Sally does not want to hold daily staff meetings in order to satisfy Mark's information needs because Sally is quite certain that asking her professional staff to stand by and make daily reports will definitely be counterproductive, as all of them are known to prefer independence. What should Sally do?

3.10 In an organization offering a dual-ladder career progression system, technically trained people may opt to progress along a technical ladder instead of the traditional managerial ladder. How does this work?

3.11 The organization chart of Company X reveals that different numbers of employees report to its five departments, as shown in Table 3.A1. Why do the numbers differ?

TABLE 3.A1 Distribution of Employees

Department	Number of Employees
A	3
B	7
C	4
D	6
E	9

Chapter 4

Leading

4.1 INTRODUCTION

Leading in engineering management refers to the function of an engineering manager that causes people to take effective action. After deciding what is worth doing, the engineering manager relies on communication and motivation to get people to act. By selecting people who are inclined to collaborate, the engineering manager influences people to take action. In addition, skills training and attitude development may also enable others to take action. In this chapter, five specific leading activities will be discussed: deciding, communicating, motivating, selecting, and developing.

The true measure of the quality of an engineering manager's leadership is his or her demonstrated ability to guide and direct the efforts of others to attain organizational objectives (Bardaracoo 2002). A leader has vision, sets a good personal example, is able to attract and retain good people, motivates people to use their abilities, and induces them to willingly do their best. Managers derive their authority from occupying higher positions within the organization. Leaders, on the other hand, have the power of influence over people. Their power is attained by earning employees' respect and admiration (Pagonis 1992; Zalenznik 1992). However, some leadership skills can be learned. Engineering managers with good leadership qualities are the most valuable to their employers.

In this chapter, specific methods for making decisions are also illustrated. These include the Kepner–Tregoe method, decision making by gut instinct, and decision making in teams. Furthermore, a few special topics on leadership are addressed, including (a) leading changes, (b) advice for new leaders, and (c) guidelines for superior leadership.

4.2 STYLES OF LEADERSHIP

There are five major styles of leadership that are classified according to the attributes of either a *concern for people* or an *emphasis on tasks*. These are defined as follows:

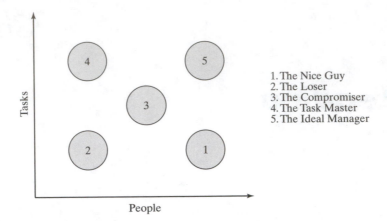

Figure 4.1 Leadership styles.

1. *The Nice Guy*—Places too much value on social acceptance while neglecting technical tasks.
2. *The Loser*—Does not obtain acceptance from others and does not get the job done.
3. *The Compromiser*—Balances both the needs of people and task factors.
4. *The Task Master*—Is interested in getting the job done right without concern for human feelings.
5. *The Ideal Manager*—Gets the job done and at the same time makes everyone happy.

Figure 4.1 illustrates these styles. The principal style of leadership exhibited by a leader is largely determined by his or her personal characteristics derived from traits such as childhood experiences, parental impact, work habits, value systems, and others.

Leadership style is effective or ineffective, flexible or inflexible. Most leaders practice more than one style in their leadership. Different styles are applicable with different people at different times (Cohen 2002). Engineering managers must vary their styles according to the needs of the employees (depending on whether the employees are experienced or novices) and the situation at hand.

4.3 LEADING ACTIVITIES

The managerial function of leading includes performing specific tasks related to leading the engineering unit or department to achieve organizational objectives. The activities involved are outlined here (Kotter 1990):*

---

*Reprinted by permission of *Harvard Business Review*. From Kotter, J. P. 1990. "What Leaders Really Do." May–June, Vol. 68, No. 3. Copyright © 1990 by Harvard Business School Publishing Corporation, all rights reserved.

A. **Deciding.** Arriving at conclusions and judgments with respect to priority, personnel, resources, policies, organizational structures, and strategic directions

B. **Communicating.** Creating understanding by sharing information with others and talking, meeting, or writing to others

C. **Motivating.** Inspiring, encouraging, or impelling others to take required action and creating workplace conditions to ensure work satisfaction

D. **Selecting people.** Choosing the correct employees for positions in the organization or for specific team activities

E. **Developing people.** Helping employees improve their knowledge, attitudes, and skills

Some of these tasks are relatively easy, and others are more difficult. Engineering managers need to practice them in order to become proficient over time in carrying out these activities. Each of these activities will be reviewed in detail in the following sections.

4.4 DECIDING

Making decisions is a key responsibility of engineering managers. The quality of decisions is the hallmark of excellent managers. The purpose of making decisions is to align the choices of project priorities, people, financial resources, technology, and relationships with the attainment of corporate objectives. As the engineering manager gains experience, the overall quality of his or her decisions is expected to increase.

Oftentimes, there is insufficient information available for guidance, or the future business or market conditions are very fluid and fuzzy. Under these circumstances, managers need to make spontaneous decisions based on intuition, gut instinct, and hunch. Otherwise, they make reasonable decisions based on systematic studies and logical analyses of available quantitative data. Engineering managers are typically quite proficient in handling this latter type of decision making, which follows the typical steps enumerated here:

(a) Assessment of facts and evaluation of alternatives

(b) Use of full mental resources

(c) Emphasis on creative aspects of problem solving

(d) Consistency in thinking

(e) Minimization of the probability of errors

Management decisions are usually difficult for the following reasons:

A. They involve problems and issues that are ill defined, as they are wider in scope and affect more people than typical technical problems and issues.

B. Needed data and information may be insufficient or excessive, and there may be no time available to collect or interpret the data.

C. Available information is of poor quality, because it is based on guesswork, rumors, opinions, hunches, or hearsay.

D. Decision making involves human behavior, which is not always predictable.

E. The nature of problems and issues changes continuously.

F. Consequences of management decisions depend on opinions available, and as such, the consequences are also changing.

G. Rarely does a perfect solution exist for management problems, since all options involve compromise, whose validity changes over time.

H. Decision making must consider implementation, which in turn depends on consensus and commitment of the affected people. Oftentimes, political considerations come into play, as well.

I. A critical decision may involve multiple layers of management or peer departments, and thus it requires coordination.

Listed here are several decision-making guidelines useful to engineering managers:

1. Acquire decision-making experience, for example, by recognizing patterns and rules and by studying management cases.

2. Prioritize problems for which decisions must be made. Do nothing with problems that are perceived to have minor significance and impact.

3. Follow a rational process to identify the problem and establish options to remove the root cause of the problem. (See Rational Decision-Making Process, Section 4.4.1.) Asking good questions is also the hallmark of good leadership.

4. Involve others in the decision-making process, especially if the implementation of the pending decision affects them. Group decisions (see Section 4.4.5) are superior from the standpoint of implementation. Such decisions, however, may take longer time to reach, and they usually represent compromises for all involved.

5. Make decisions based on available information and assumptions introduced. Check the validity of all assumptions and update the decisions accordingly. Take necessary risks and avoid becoming paralyzed by stress and uncertainty.

6. Delay making decisions until the last allowable moment, as the problems and available options may continue to change. Above all, meeting all deadlines with a decision is better than having none.

7. Avoid making decisions about issues that are not pertinent at the time or decisions that cannot be practically implemented (or both). Also, do not make a decisions that ought to be made by others.

How can we judge the quality of a given decision? A simple way to find out is to raise the following three questions:

A. Did the decision achieve the stated purpose; did it correct or change the situation that caused the problem in the first place?

B. Is it feasible to implement the decision; is it meaningful with respect to the required resources and the created value?

C. Does the decision generate noticeable adverse consequences or risks to the group or the company?

The decision at hand is regarded as good if the first two questions are answered by a "yes" and the last one by a "no."

As a rule, managers are expected to make decisions. However, there are circumstances in which managers should delegate the decision-making authority to the staff or work alongside with them to come to a decision.

The following are problems or issues that should be handled by the managers only: (1) prioritizing tasks and projects, office assignments, and work group composition; (2) handling personnel assignments, performance evaluations, and job action; (3) dispensing budget allocation; (4) applying administrative policies, procedures, and regulations; and (5) dealing with highly confidential business matters that are explicitly declared to be so by the top management (e.g., compensation, promotion, corporate strategies, and new marketing initiatives).

Managers should include the staff when making decisions on the following problems or issues: (1) considering staff needs for development (e.g., attending professional meetings, technical conferences, seminars, and training courses, as well as committing to study programs at universities), (2) discussing policies and procedures, involving staff interactions with other departments, and (3) determining team membership (e.g., considering personality fit, skills compatibility, working relationships, balancing workload, etc.)

Managers should delegate the decision-making authority to staff members for the following matters: (1) techniques to accomplish assigned tasks or projects; (2) options to continuously improve current operations and work processes; and (3) social events involving staff participation, such as group picnics, golf outings, and Christmas parties.

As the saying goes, "Practice makes perfect." Engineering managers should seek opportunities to constantly acquire experience regarding when, where, and what decisions are to be made. How decisions can be made is deliberated in the next section.

4.4.1 Rational Decision-Making Processes

A rational decision-making process is generally useful in facilitating decision making for numerous problems or issues in engineering when an adequate amount of information is available. It consists of a set of logical steps outlined as follows:

A. Assess the apparent problem based on observed symptoms.
B. Collect the relevant facts. Usually, not all facts are available due to resource, cost, or time constraints. Facts must be related to five decision-making factors:

1. Situation (what, how). The sequence of events leading to the problem and its conditions
2. People (who). Personalities, preferences, personal needs, and egos
3. Place (where). Significance of location
4. Time (when). Pressure to bring forth an immediate solution
5. Cause(why). Why the problem originally occurred, and why it occurred in one situation, but not in another

Past experience indicates that there are several good sources for identifying the relevant facts related to the problems at hand. (See Table 4.1.)

C. Define the real problem. As the most important initial step, the real problem at hand must be defined. It will be helpful to pose the following three questions:

TABLE 4.1   Sources for Facts Related to Problems

Problem Categories	Sources of Useful Facts
Equipment	Plant operations personnel
Technical	Engineers with direct working knowledge
Customer inquiry	Sales people
Customer complaints	Service and sales personnel
Materials and parts	Delivery and inspection personnel
Product quality	Production staff
Customer preference	Marketing personnel
Market competitiveness	Marketing personnel

1. What is the deviation between actual performance and the expected norm?
2. What are the desired measurable results in a problem-solving situation?
3. What represents success based on well-defined metrics and the proper method of measurements?

The answers to these questions are likely to point to the root cause of the problem.

D. Develop alternatives to solve problems. Once the root causes of the problem at hand are defined, it is then useful to come up with options to address them. Engineering managers should (1) freely invite creative suggestions from people who have direct knowledge of the problem at hand, (2) brainstorm in group settings (without criticisms or comments so as not to deter imaginative suggestions), and (3) take into account both short-term and long-term impacts.

E. Select the optimal solution. Engineering managers need to choose among the options to address the root causes. (A useful rational tool for making such a choice is introduced in Section 4.4.2.)

Engineering managers must ensure that the chosen option produces minimum adverse consequences to the company or unit. They also need to plan for contingencies and make midcourse corrections, if required, to secure the greatest probability of achieving the desired results. They should also avoid committing to a final choice prematurely until its implementation becomes feasible.

A decision is nothing but the choice among several available options to solve a problem or address an issue. If there is only one option available, then no decision is needed.

F. Set a course of action to implement the decision. Once a decision is made, engineering managers should devise an applicable action plan to implement the decision. The manager must consider such details as (1) policies that limit possible action, (2) programs (the sequence of action steps), (3) schedules (dates and milestones), (4) procedures (the action steps carried out in an orderly manner), and (5) budgets and expenses for equipment and manpower. Decisions that are not effectively implemented are useless.

4.4.2 Kepner–Tregoe Decision Analysis Tool

The Kepner–Tregoe method is a renowned analysis tool available to support decision making (Kepner and Tregoe 1981). It prescribes the following steps (see Table 4.2) to arrive at a rational decision:

1. Define a set of decision criteria needed for making the decision. The necessary criteria are those that must be met. For example, all entry-level engineering applicants must have undergraduate degrees in engineering to be considered for employment. Some hiring companies may define a grade point average (e.g., 3.5) as the cutoff academic performance level below which an applicant would not be considered. The sufficiency criteria are those that are not necessary, but are good to have. For hiring entry-level engineers, companies may specify these to be summer work experience, internship activities, project work, leadership positions held in student organizations, and others.

2. Rank order the sufficiency criteria by assigning weight factors ranging from 10 (as the most preferable) to 1 (as the least preferable).

3. Evaluate all options against each of the identified as necessary decision criteria. For example, the options that meet the necessary criteria may be designated with words "go."

4. Remove from further consideration those options that fail the necessary criteria.

5. Rank all remaining options relatively, with respect to specific sufficiency criteria. Assign a relative score of 10 to the most satisfactory and 1 to the least satisfactory option.

6. Repeat the scoring process for each of the remaining sufficiency criteria.

7. Compute a weighted score for each option by multiplying its relative score for a specific sufficiency criterion with its corresponding weight factor. Add up the weighted scores for all sufficiency criteria to obtain the overall weighted score for this option. Repeat the computation for each of the remaining options.

8. Compare the overall weighted scores and choose the option with the highest overall weighted score.

The Kepner–Tregoe method forces decision makers to externalize all necessary criteria and to assign weight factors to all sufficiency criteria before making decisions. The chosen criteria must represent a *mutually exclusive and collectively exhaustive* set

TABLE 4.2   Kepner–Tregoe Method

Criteria	Weight Factor	Option A	Option B	Option C
Criteria 1	R	Go	Go	Go
Criteria 2	10	4	8	10
Criteria 3	5	6	10	7
Criteria 4	8	10	6	8
Total Weighted Score		**150**	**178**	**199**

of criteria for the decision at hand. By ranking options against each of the defined criteria, all options are properly evaluated in a rational, equitable, and comprehensive manner (Ragsdale 2000).

The Kepner–Tregoe method is particularly useful in a team environment where members may needlessly argue for specific options without externalizing their decision criteria and the relative ranking they have assigned to the criteria. Oftentimes, the advocates for a specific option make implicit assumptions that remain hidden and unknown to others on the team. In addition, personal biases may influence the relative scores assigned to the options when these options are evaluated against a given decision criteria. Experience has shown that the personal biases tend to become minimized when the relative scores are polled from all teammates during a meeting.

The Kepner–Tregoe method is also effective for decision making on an individual basis. Some engineering managers tend to emotionally overemphasize certain decision criteria and downplay the importance of others. Again, having all decision criteria and their respective weight factors explicitly delineated will facilitate a rational decision.

Example 4.1.

Bill Pickens, manager of the test division, called John Riley, the group head of mechanical testing, into his office and told him that there was a new opening for a manager of product development position in the company. For John, it would be a promotion to a higher managerial rank with an appropriate increase in salary. However, the new position is temporary, in that it may be eliminated in a year. Although Bill hates to lose a very valuable worker like John, he wants to let John himself make the decision. The product development division has specifically requested that this opening be recommended to John. After having given it some thought, John decided to take the new position.

The next day, Bill Pickens and John Riley sit down together again to name a group head successor. Among the three section heads in the group, Dodd is the most experienced. However, Dodd is quiet and does not communicate well. He may have difficulty in selling testing services to others. Yeager is competent, but has made hasty decisions that have been very costly to the group. Bennett is ambitious and aggressive, but has poor interpersonal skills. They concluded that none could be immediately promoted to take over. Finally, they agreed to rotate the acting head job among the three, to test out each of them, since there is an outside chance that John may come back to his old position after one year.

Shortly thereafter, Bill Pickens was promoted out. John decided not to return to take Bill's position. Terry Smith was brought in to take over Bill Pickens's job as test division manager. However, before Bill Pickens left, he indicated to Yeager that Yeager would likely get the job, based on the results of the trial periods.

Terry found significant rivalry and ill feeling between the three section heads. The group had low morale and poor productivity. Under such circumstances, Terry decided to appoint a new employee, Dennis Brown, to the mechanical testing head position instead of one of the three.

Did Terry make the right decision? Apparently, the job rotation idea failed. What would have been the right way for Bill Pickens to handle this problem?

Condensed and adapted from Robert E. Shannon, *Engineering Management*. New York: John Wiley, 1980, p. 203. This material is used by permission of John Wiley & Sons, Inc.

Answer 4.1.

The decision made by Terry was not the right one. The reasons are as follows:

A. The personnel situation was created by Bill Pickens's and John Riley's inability to make a staffing decision by choosing the best one among the three and minimizing the impact of the new head's shortcomings. Rotating the acting head job created chaos due to infighting. This should have been anticipated by experienced managers.

B. Terry's decision negated an implied management promise that one of the three would be promoted after a one-year trial period. This broken promise could be the basis of a future lawsuit.

C. The appointment of a new employee, without consultation with and concurrence by the three section chiefs, reflects a lack of sophistication on the part of Terry. It raises the fairness issue and creates an employee loyalty problem. The three section chiefs are not likely to be motivated to work with the new person.

D. It is not known if Dennis Brown has the necessary technical and managerial skills to be more successful than any one of the three tried entities.

E. The likely results are as follows:

(a) Lost management credibility due to broken promises and a lack of personnel staffing capabilities. (Riley neglected to groom a successor by correcting the perceived shortcomings of his chosen successor during the last five years.)

(b) Management is perceived to be lacking fairness in decision making. (This will result in lower group morale and decreased employee loyalty. Employee turnover may be increased.)

The job rotation idea is very poor. It was selected only because Pickens and Riley were not able to make good staffing decisions. They were looking for a perfect person, and overlooked the possibility that most of the identified shortcomings could be easily compensated for or corrected.

What Pickens and Riley should have done was to make a hard choice in the beginning, either bringing someone in from the outside or promoting one of the three employees. Assuming that no suitable outside candidates were available, then the Kepner–Tregoe decision analysis should have been used to come up with a choice, as shown in Table 4.3.

At the first glance, Dodd appears to be the winner. However, as the candidates weighted scores are rather close, the refinement step shown on Table 4.4 may be taken.

The adjustments are made based on the expected values of improving the relative score of the identified weakness from 5 to 10. The adjusted total score represents the final

TABLE 4.3   Making a Personnel Choice

Criteria	Weight Factor	Dodd	Yeager	Bennett
Minimum technical experience	R	Go	Go	Go
Experience	10	10	8	6
Communications skills	8	5	10	10
Decision-making abilities	6	10	5	10
Human relations skills	6	10	10	5
Total Weighted Score		**260**	**250**	**230**

TABLE 4.4   Making a Refined Personnel Choice

	Rank without improving shortcomings	Probability of correcting shortcomings	Adjusted total score
Dodd	260	80	$292 \ (= 260 + 0.8 \times 8 \times 5)$
Yeager	250	90	$277 \ (= 250 + 0.9 \times 6 \times 5)$
Bennett	230	60	$248 \ (= 230 + 0.6 \times 6 \times 5)$

ranking of these three individuals, after each is allowed to minimize his weaknesses. Dodd remains the winner in this case.

Example 4.2.

Due to global competition, the company faces a tough time in the marketplace, and so it must scale down its workforce. The board has advanced the following options for the employees:

1. *Quit voluntarily.* This could be attractive to several bright young engineers whom the company does not want to lose.
2. *Last in and first out.* This could result in the loss of young and more versatile operators.
3. *Early retirement of those within 10 years of their normal retirement age.* This could cause a loss of engineers with valuable product knowledge.
4. *Reverse ranking in performance records.* This could lead to unfair selection, as the Uniform Performance Appraisal System has been operating in the company only for the last few years.

What methods, or combination of methods, should the company use to reduce employment?

Answer 4.2.

The company should use the Kepner–Tregoe method and assign weight factors to all the criteria. The relative score as displayed in Table 4.5 should be assigned to evaluate all options.

TABLE 4.5   Choosing the Method of Downsizing

Criteria	Weight Factor	Option 1	Option 2	Option 3	Option 4	Option 5 (1&4)	Option 6 (4&1)
Not to lose knowledge and experience	10	5	10	3	8	6	7
Easy for affected employees to find jobs	8	10	10	5	8	9.5	8.5
Easy for company to find replacements	8	5	10	3	8	6	7
Easy for company to avoid legal problems	10	10	9	8	5	8	6
Total Weighted Score		**270**	**350**	**174**	**258**	**264**	**254**

Based on the results obtained from Table 4.5, the method of last in and first out (Option 2) should be chosen to reduce employment.

4.4.3 Additional Support Tools for Decision Making

A number of spreadsheet-based tools are commercially available to engineering managers to support the decision-making process (Nagal 1993, Kirkwood 1997, and Belton et al. 2002). The following tools, among others, are widely used in the industry:

- Forecasting (exponential smoothing, time series, and neural network computing)
- Regression analysis (single variable and multivariable)
- Risk analysis and project management
- What–if solver
- Simulation modeling (Monte Carlo)
- Decision trees
- Optimization (linear programming and integer and dynamic programming)
- Artificial intelligence and pattern recognition tools
- Expert or knowledge-based systems

Engineering managers should familiarize themselves with all of these decision support tools so that they can employ the right tools under the right circumstances (Ragsdale 2004).

Example 4.3.

A certain company makes two products: P_1 and P_2. Each P_1 requires 5 kg of material M and 3 kg of material N. Each P_2 requires 3 kg of M and 3 kg of N. In the warehouse, there are 350 kg of M and 270 kg of N available. The profit is $50 for each P_1 and $40 for each P_2. What mix of P_1 and P_2 products should the company make and sell in order to maximize its total profit?

Answer 4.3.

This is an optimization problem to be solved by linear programming:

$$\text{Maximum Profit} = 50 \times x_1 + 40 \times x_2 \quad \{\text{Objective function}\}$$
$$\text{Subject to} \quad 5x_1 + 3x_2 <= 350 \quad \{\text{Constraints}\}$$
$$3x_1 + 3x_2 <= 270$$
$$x_1 >= 0$$
$$x_2 >= 0$$

Graphically, the solution method can be displayed as shown in Figure 4.2. Because of the four constraints, the solution space is bound by the area $OABCO$: O (0,0), A (90,0), B (40,50), and C (70,0).

As the profit line ($50 \times x_1 + 40 \times x_2 = $ Profit) moves to the right, the profit is maximized when it passes through the point B (40,50), the right-most location it can take, while still satisfying the stated constraints. Thus, the maximum profit is

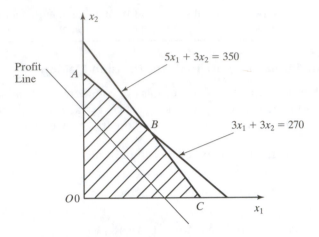

Figure 4.2   Linear Programming Problem.

$4,000 ($=50 \times 40 + 40 \times 50$), and numbers of products P_1 and P_2 to make and sell are 40 and 50, respectively.

The simplex method is programmed to handle linear optimization programs with n independent variables.

Suppose we want to know more about the impact of material constraints on the company's profit. Let us assume that material N can be increased by 30 units, subject to the new constraint

$$3x_1 + 3x_2 <= 300.$$

This new constraint, represented as a straight line $3x_1 + 3x_2 = 300$, intersects the straight line $5x_1 + 3x_2 = 350$ at the point D (25,75), which is not shown in Fig. 4.2. The new optimum solution is then D, producing a new total profit of $4,250 ($=50 \times 25 + 40 \times 75$). This new profit is $250 over the previous total, because of added material N, which has a shadow price of $8.33 ($=$250/30$).

4.4.4 Decision Making by Gut Instinct

Since, up to middle managerial levels, decision making is mostly quantitative, the tools indicated in the preceding section are useful. However, at upper and senior managerial levels, problems and issues get much more complex and ambiguous. When such circumstances defy systematic analyses, decision making is typically based on intuition and gut instinct.

Bob Lutz, president of the Chrysler Corporation, was reported to have made the decision in 1988, by pure instinct while driving alone along a country road, to develop and market a new sports car, the Dodge Viper. This car later turned out to be a great success. Using interviews with several other managers as source material, Hayashi (2001) studied their intuitive decision-making processes.

What is gut instinct? According to Hayashi, our minds process information all the time. Our left brains process conscious, rational, and logical thoughts, whereas our right brains take care of subconscious, intuitive, and emotional thoughts. Some people claim that they can tap into right-brain thinking by jogging, day dreaming, listening to music, or using other meditative techniques. Others have reported that they get innovative ideas while taking long showers or placing themselves in unfamiliar situations.

The theory of intuitive thinking claims that accumulated past experience enables some people to bundle information so that they can easily store and retrieve it. Experts further claim that such information is retrieved from memory by the observation of patterns. Professional judgment can often be reduced to patterns and rules. Accordingly, all other things being equal, people with varied and diverse backgrounds are likely to be more capable of thinking intuitively and learning faster because they recognize more patterns. When using gut instinct, people essentially draw on rules and patterns that reside in their memories. Diverse backgrounds facilitate cross-indexing, allowing one to see similar patterns in disparate fields.

However, instincts can at times be wrong. People who make decisions intuitively are advised to secure constant feedback in order to minimize the impact of incorrect decisions and to learn from these decisions. Over time, the process of learning on the basis of feedbacks has the potential of improving the patterns and rules stored in the peoples' memory, thus enabling them to make better intuitive decisions in the future.

Engineering managers at low- or middle-managerial levels, in order to update and modify their own patterns and rules, should keenly observe how top-level leaders make important decisions and analyze such decision-making processes.

4.4.5 Decision Making in Teams

Typically, individual engineering managers make decisions by using one or more of the methods just described (analytical, rational, or intuitive). However, engineering managers may also elect to make decisions by using the inputs generated by teams. Under such circumstances, additional factors come into play, such as personality clashes, conflicts of interest, and coalitions or alliances among the team members, which affect the resulting decisions. Team leaders need to pay special attention to a set of additional guidelines that foster better decision making in group settings.

Garvin and Roberto (2001)* advanced the idea that the group decision-making process must be managed properly to consider social and organizational aspects. Doing so will secure the needed support for implementation, which ultimately determines the success of any decision. Three factors are important for the team leader to take into account when managing a group decision-making process: conflict, consideration, and closure.

Group decision making requires a set of leadership talents somewhat different from those demanded in other situations. These include (a) active solicitation of divergent

---

*Reprinted by permission of *Harvard Business Review*. From "What You Don't Know about Making Decisions" by D. A. Garvin and M. A. Roberto, Oct. 15, 2001. Copyright © 2001 by Harvard Business School Publishing Corporation, all rights reserved.

viewpoints, (b) acceptance of ambiguity, (c) the wisdom to end a debate, (d) the ability to convince people of the merits of the decision made, and (e) the ability to maintain balance to embrace divergence and unity—divergence in opinion during the debate and the required unity of participants needed to implement the decision.

Engineering managers are encouraged to follow the preceding guidelines when managing team decision-making processes.

4.5 COMMUNICATING

The purpose of communicating is to create understanding and acceptance of the facts, impressions, and feelings being communicated.

When communicating, engineering managers must have a clear purpose in mind and ensure that the message is understood and retained. A proper form of communication needs to be selected, such as a one-on-one meeting, phone conversation, written memo, staff meeting, e-mail, videoconference, Web posting, or net meeting. It is advisable for the engineering managers to keep the communications channels open. They should be straightforward and honest, respect confidential information, welcome suggestions, anticipate resistance to changes, and dispel fears by disclosing full information (Clampitt 2001).

There are four key actions to take to achieve efficacious communication: asking, telling, listening, and understanding. These actions will be discussed next.

4.5.1 Asking

Engineering managers should proactively request information and not wait to be told. A lack of information can prevent understanding. Open-ended questions—those which cannot be answered by "yes" or "no"—should be raised to gain new knowledge. The quality of the questions represents a gauge of the questioner's background, education, and depth of understanding of the issues involved.

4.5.2 Telling

Telling means transmitting information (verbally or in written form, or both) to keep employees informed about matters of concern to them, to inform managers about problems and pertinent development (e.g., to avoid surprises that would trigger spontaneous and low-quality decisions), and to pass information to peers.

Engineering managers need to exercise judgement as to what to tell and what not to tell, as too much information could lead to overload and confusion and too little information could cause employee mistrust and poor productivity. A typical rule of thumb used in industry is that information is dispersed based on *the need to know*. Managers will share information freely if it is needed for performing specific work or has an impact on the individual's work environment.

4.5.3 Listening

Engineering managers need to work on their listening skills to enhance their understanding of both the words (spoken and written) and any possible subtext. They should maintain their concentration by exercising self-discipline and rigorous control of their own urge to talk and interrupt.

4.5.4 Understanding

The ultimate goal of communication is to promote understanding—to hear with the head and to feel with the heart. Engineering managers need to recognize shared meaning (emotional and logical) and to assess the degree of sincerity by observing body language, intonation, and facial expression.

Several communication barriers exist and should be taken into account by engineering managers:

1. **Interpretations of words and terms**—Words are symbols or semantic labels applied to things or concepts. The same words may have different meanings to different people.

2. **Selective seeing**—Some people have the tendency to see only what they want to see and remain blind to other information unfavorable to the position they take.

3. **Selective listening**—Some people hear only what they want to hear by screening out information that may seem threatening to them, thus limiting their ability to appreciate different perspectives and points of view. Others in conflicts may want to understand only that which allows them to pursue their own self-interest.

4. **Emotional barriers**—All people have emotions. Engineering managers need to appreciate the fact that people's feelings are as important as their intellectual knowledge. Sometimes people's attitudes and feelings may be so strong that they impair their understanding of what is being conveyed. Generally, personal biases will distort the understanding of what is being communicated.

The barriers just cited may cause the communications process to fail in creating the desired degree of understanding. Experience has shown that appeals to emotion tend to be understood and accepted much more readily than appeals to reason, analysis, or cold logic.

To communicate efficaciously engineering managers are advised to pay attention to the following guidelines:

A. **Know what to say and say what is meant.** Engineering managers should focus on key messages when communicating. Avoid noise or meaningless sounds, pointless statements, and inconclusive remarks often used by people to impress others, but not to express themselves. Examples of such noise include "The answer is definitely a maybe" and "It is not probable, but still possible."

B. **Understand the audience.** Engineering managers should tailor the communication to the receiver's frame of reference—their beliefs, concerns from the job, background and training, attitudes, experience, and vocabulary.

C. **Secure attention.** Engineering managers should try to appeal to the receiver's interests; anticipate and overcome emotional objections (fear, distrust, and suspicion); talk in the receiver's terms; and lead from the present to the future, the familiar to the unknown, and the agreeable to the disagreeable.

D. **Obtain understanding.** An effective communication technique is to start with agreements and the statement of facts (not conclusions), use simple words (not ponderous, confusing, or abstract terms), and communicate in bursts (avoiding information overflow and knowledge digestion problems).

E. **Ensure retention.** The *rule of four* states the following: (a) Before trying to get an idea across, tell your receivers what you are going to say. (b) Say what you have to say. (c) Tell them what you said. (d) Get them to tell you what they have understood.

Obviously, engineering managers must practice such a rule tactfully when the receiver happens to be the company president instead of a young intern engineer who may have just recently started to work.

F. **Receive feedback.** Engineering managers need to proactively pose questions and learn to listen in order to get feedback from what was communicated.

G. **Get actions to enhance communications.** Engineering managers should have the receivers take action on the just-completed communication as a way to enhance its impact. This could be in the form of a commitment by the receivers to take specific steps by agreed-upon dates.

Creating understanding is what communication is all about. Engineering managers need to practice asking insightful questions, conveying messages clearly, and listening attentively so that understanding is created at each and every communication endeavor.

Example 4.4.

The company decided to move its engineering center to another location, since it was running out of space. The new location was to be modern and had been planned as a showpiece for the company. Management felt certain that the employees would welcome the move. Negotiations were started with several local governmental authorities for suitable accommodations.

To keep the workforce fully informed, it was agreed that the employees would be told that a move was to be made, but that as yet no site had been chosen.

This communication let to wide speculations among the engineers as to the new site, and various rumors circulated. Some engineers with families decided to look for alternative employment elsewhere, fearing that the new location would not be within commuting distance. Morale fell and productivity suffered.

Negotiations took longer than anticipated, and no suitable location had been found after six months. By then morale was so low that the company decided to abandon its relocation plan altogether. To overcome the space problem, the company split the engineering group by putting a smaller team into another factory site nearby.

What went wrong? How would you have handled this case differently?

Answer 4.4.

"To communicate or not to communicate"—that was the question. A well-intended, but premature relocation announcement induced anxiety in the minds of affected engineers. A lack of progress in site negotiations compounded these anxieties, causing low morale and decreased productivity, leading to an eventual abandonment of the plan.

It would have been better for the company management to keep the plan secret initially, negotiate for and decide on a specific site, and then have the company president announce the relocation plan in a town meeting. The announcement should have included:

A. The location of the new site, with emphasis on the advantages in transportation, health care, weather, and historical, cultural, and recreational attractions.

B. A request for the support of all engineers in making the relocation as smooth as possible. The purpose of the relocation is to provide a better facility for everyone. The company is investing x million dollars to support this move, which will allow for further expansion in the near future.

C. The date by which relocation is to be completed.

D. A delineation of the company's plans to fund all relocation costs and offer assistance in selling and buying homes, if the relocation is more than 100 miles away. The company will also assist the affected spouses to find jobs at the new site.

E. The description of a human resources desk that will be set up to answer specific questions.

Example 4.5.

Your department is going to institute a major change. Some members have indicated that they believe the change may be needed. However, in the past, members of the department have tended to resist changes that they did not initiate. The department as a whole has a good performance record. Discuss the advantages of the following alternatives:

(a) Permit the members of the department to determine if the change is needed.

(b) Let the group make recommendations, but see that your objectives are adhered to.

(c) After the group discussion, adjust the goals, if possible, and monitor performance to see that the change is followed.

Which do you think is most appropriate? Are there other strategies that are more appropriate?

Answer 4.5.

People resist changes unless they are convinced that the contemplated changes are necessary. They tolerate changes better if the changes are introduced gradually and they have had some say in making the changes.

Out of the three alternatives given, (b) is the most appropriate. Management must state clearly the objectives of the planned changes and how the attainment of objectives is to be measured. The members of the department should be allowed to participate in deciding how to achieve the stated objectives.

Company management must set the goals, which are not negotiable. Members of the department are not to be empowered to decide if changes are needed or not. Although the department has a good performance record, change may still be needed for achieving significantly better performance in view of the competition in the marketplace. Thus, alternative (a) is not appropriate.

Alternative (c) is also not appropriate because the goals of the department should not be adjusted on the basis of what members of the department would like to do. Performance monitoring is a valid approach to ensure that the stated objectives are met.

In general, staff participation is useful to ensure an active implementation of the decision made. Thus, a combination of alternatives (b) and (c) may be proper. Management specifies the objectives of changes and the ways the attainment of these objectives are to be measured. Members of the department are encouraged to make recommendations regarding the best ways to implement the changes. The actions taken by the department are to be monitored constantly to make sure that the stated objectives are met, even if gradually.

4.6 MOTIVATING

The engineering manager secures results by motivating people. Examples of motivators include opportunities to do challenging, interesting, and important work, leadership, position power, prestige, and compensation.

To motivate is to apply a force that excites and drives an individual to act in preferred ways. In general, emphasis is given to motivational forces that cause the individuals to willingly apply their best efforts.

It is advisable for engineering managers to accept differences in personal preferences, values, and standards, and not try to change people. According to one theory, personality traits are usually firmed up at the early ages of four or five by environmental conditions. Managers should also recognize that every employee has inherent drives to fulfill their own needs, such as self-actualization, recognition, ego, self-esteem, group association, and money.

4.6.1 Methods of Motivation

In general, engineering managers have several methods of motivation at their disposal:

A. **Inspire.** Infuse a spirit of willingness into people to perform most effectually by way of their own personality and leadership qualities, personal examples, and work completed.

B. **Encourage.** Stimulate people to do what has to be done through praise, approval, and help.

C. **Impel.** Force and incite action by any necessary means, including compulsion, coercion, fear, and, if required, punishments.

The first methods are well suited to motivate professionals, and the last category is not. Being assigned challenging work is a useful motivator for professionals.

4.6.2 Specific Techniques to Enhance Motivation

Engineering managers may implement the techniques outlined here to inspire and encourage professionals to act:

A. **Participation.** Invite employees to take part in setting objectives and making decisions. Doing so will ensure emotional ownership and the application of specialized knowledge. Participative management is known to have a positive motivational impact on employees.

B. **Communication.** Set clear standards, relate the importance of the work, keep expectations reasonable, and give answers to suggestions offered by employees.

C. **Recognition.** Give credit where it is due, as sincere praise tends to promote further commitment. Fair appraisals induce employee loyalty and trust.

D. **Delegate Authority.** Trust the employees and do not overcontrol them. Achievers will seek additional responsibilities, and security seekers will not. Delegate what to do and leave how to do it to the individuals. Delegate technically doable work only to those who want it.

E. **Reciprocate Interest.** Show interest in the desired results to motivate employees to achieve these results.

4.6.3 McGregor's Theory of Worker Motivation

Confucius says, "Reciprocity is the foundation of human relations." One very forceful method of employee motivation is indeed to offer help needed by the employees, who will surely be inclined to reciprocate. The key is then to define such needs.

McGregor's theory of worker motivation (McGregor 1957) is built upon the Maslow need hierarchy model (Maslow 1954). According to this model, a person's needs may be grouped into hierarchical levels, as follows:

1. **Physiological needs**—hunger, thirst, and need for clothing and shelter
2. **Safety**—protection from threats and danger
3. **Social**—giving and receiving affection, group membership, and acceptance by peers
4. **Esteem**—ego and self-confidence to achieve recognition
5. **Self-actualization**—continued self-development and realization of one's own potential

A satisfied need no longer dominates the individual's behavior, and the next higher level need takes over. But a higher level need only arises when lower ones are already satisfied. The central premise of the Maslow hierarchy model is that an unsatisfied need acts as a motivator. Accordingly, a need-based motivation strategy suggests that engineering managers should learn to understand the specific needs of their professionals at any given time and find ways to help satisfy these needs.

Experience has shown that the motivation strategies presented here can be helpful in motivating professionals who typically have high-level needs related to self-actualization and esteem:

A. Present a variety of work assignments perceived to be desirable and that offer the opportunity for personal growth.

B. Offer work having a broad enough scope for the employee to develop self-expression and individual creativity.

C. Manage with minimum supervision and control, as professionals favor independence and individuality. Professionals tend to prefer having the freedom to make their own decisions and choose their own work methods for achieving the stated objectives.

D. Provide work that fully utilizes the individual's professional experience, skills, and knowledge.

E. Assign work that enables the employee to receive credit and peers' recognition. Examples include teamwork, publication of technical articles, patents, company awards, and activities in professional and technical societies.

On the other hand, pay and benefits have only a minor impact, as physiological needs do not represent a motivator for most professionals. Because the higher level needs are never completely satisfied, engineering managers have ample opportunities to motivate professionals to act with their best abilities in achieving the corporate objectives.

Example 4.6.

Company X recently installed an incentive system in the production department. Each person receives incentive payments (in addition to hourly wages) for any work done beyond the work standards established for each job. After one month, the production manager noted that there was only a meager increase of 4.5 percent in production. How would you comment on this result?

Answer 4.6.

The incentive program appears to have failed in realizing the projected benefit. This could be due to several reasons:

1. The incentive offered may be too small relative to the base hourly wage that these workers have been earning all along.

2. Management may have made the mistake of not having consulted with workers to understand their specific hierarchy of needs. Additional pay may not be a strong motivator to them in comparison to other nonmonetary factors, such as peer recognition, self-expression, social acceptance among peers, and others. It is known that team participation (such as a quality circle) has been a strong motivating factor for many production workers in the automobile and other industries.

3. If the production workers are unionized, the union leadership may have played a role in discouraging workers to compete against each other for pennies.

It would be worthwhile for the production manager to set the target of a desirable productivity improvement at, for example 10 percent, over the next 12 to 18 months. Then, the production managers should form a team that is empowered to develop recommendations regarding the specific ways to achieve the stated improvement goals in productivity. The team should be made up of workers on the plant floor, union leader, production engineers, and others who have direct knowledge of the production process involved. By having participated in such teamwork, workers on the plant floor become part owners of the resulting action plan. The resulting plan is more likely to be successfully implemented.

4.7 SELECTING ENGINEERING EMPLOYEES

The long-term success of an engineering organization depends on the employees' abilities and the effectual use of these abilities. On the other hand, job satisfaction is known to have a profound impact on employees' willingness to apply their skills to the best of their abilities.

Through employee selection, engineering managers have some control over employees' abilities, their willingness to apply their best efforts, and their job satisfaction. Engineers who are likely to be productive and happy workers in corporate settings are those who are firmly dedicated to their tasks, have excellent interpersonal skills, the team player mentality, sound basic training, and the capability to learn new things quickly.

4.7.1 Selection Process

Typically, the employee selection process includes the following steps:

A. **Define needs.** Specify the needs of the new positions by taking into account the immediate requirements and long-term growth demands of the organization.

B. **Specify jobs.** Compose a job description for each of the open positions to define the roles and responsibilities of the position holders, the position grade levels, and the minimum qualifications of the ideal candidates (i.e., levels of basic training and work experience).

C. **Acquire applicants.** Publicize job openings in newspapers, professional publications, company Web sites, employment agencies, and Internet job sites to solicit candidates.

D. **Review and prescreen.** Select applicants by matching personal objectives with company goals. Check documents and references carefully.

For entry-level candidates who are recent college graduates with no professional experience, many industrial employers place a significant amount of weight on the grade point average (GPA). Some companies have even specified a minimum GPA level as a prescreen criteria.

This overemphasis on the GPA is probably due partially to an ignorance of better, more objective criteria than the GPA in assessing the mastery of basic course subjects. This overemphasis is also due to the notion that the GPA is a composite reflection of the level of personal responsibility demonstrated by an individual in doing his or her principal job of learning during the college years.

E. **Conduct interviews.** Each applicant may be interviewed by several managers. The basis of assessment is typically *studying the past to predict the future*. The quality of past work is a very good predictor for the future, as people are known not to change significantly for the better overnight. A few useful guidelines are enumerated here:

1. Query about the candidate's capabilities pertaining to the new position.

2. Listen carefully to what the candidate says during the interview. Avoid spending too much time selling the job opportunity to the candidate.

3. Prompt the candidate to describe his or her last job. Be cautious of candidates who speak negatively about their past employers.

For example, a recent college graduate may complain loudly about his or her research-centered alma mater's negligence in undergraduate teaching and use it as a reason for the individual's poor GPA records. This individual may be likely to behave as a blame-shifting, finger-pointing, and irresponsible individual in a professional environment.

4. Suggest that candidates tell you something negative that you should know about them. Look for honesty. Determine if the candidate is aware of his or her personal flaws, and what active steps have been taken to correct them.

5. Urge the candidates to explain what they would do if they got in over their heads at work. An employee who turns to a colleague is a team player. An employee who turns to a supervisor behaves like a child. An employee who isolates himself when in trouble can be extremely damaging to the business.

6. Encourage candidates to describe their aspirations. Ask, "Where do you see yourself in five years?" or "What are your future goals?" Companies need employees who can grow and evolve over time. A candidate who does not have goals or ambitions may resist learning new skills or taking on additional responsibilities.

F. **Decide on job candidates.** Match the candidate's personality, technical capabilities, work ethics, values, and other qualities with those of the company.

Generally speaking, the selection process just described, which is widely practiced in industry, has not always yielded desirable results for employers. Typically, four to five managers may interview an engineering applicant during a one-day site visit. For employers, the easy part is to assess the candidate's technical capabilities, as such capabilities are readily supported by documents (e.g., academic records, internship reports, thesis, publications, and reference letters from professors and company executives). The more difficult part is the assessment of the individual's soft skills, as discussed in the next subsection.

4.7.2 Soft Skills

Engineers' future success in a company is strongly affected by their *soft skills* in team work, interpersonal relationships, leadership quality, collaborative attitude, mental flexibility, and adaptability. These soft skills are linked to the engineers' personality traits, psychological profiles, value systems, and deep-rooted beliefs. However, companies generally do not require candidates to undergo specific psychological tests, and most interviewers are not trained to assess candidates for soft skills.

Part of the difficulty in assessing the soft skills of engineers is brought about by the engineers themselves. Nowadays, most engineers, armed with the knowledge of interviewing guides, know quite well how to "talk the talk and walk the walk" in interviews. They have polished responses to almost any type of questions in interviews and are thus proficient in displaying the characteristics they believe many employers are looking for.

Results in literature have indicated over and over again that most professionals who failed in industry—those who have been laid off or voluntarily quit due to personal dissatisfaction—were deficient in soft skills, not in technical capabilities. Future engineering managers need to learn more about how to assess the soft skills of candidates.

Some companies have devoted significant efforts to addressing this issue. Shown here are industrial practices that describe what two progressive companies [Mazda Motor Manufacturing Corporation (U.S.A.), Flat Rock, Michigan, and Diamond Star Motors Corporation, Normal, Illinois] have done to assess the soft skills of their candidates and the selection criteria they used when selecting these blue-collar workers (Hampton 1988):

(a) **Interpersonal skills**—ability to get along with people
(b) **Aptitude for teamwork**—team dedication and participation, focus on the impact on the team and company instead of individual performance
(c) **Flexibility**—learn several jobs, change shifts, and work overtime
(d) **Drive to improve continuously**—make and take constructive criticism

The basic strategy followed by these companies is to *pick the best employees and train them well.* It is noteworthy that "best" is defined by the soft skills of the candidates, not by their hard (technical) skills. These companies select 1300 candidates out of 10,000 applicants at a cost of $13,000 per person, using a multiphase process involving tests, exercises, and role playing in group activities.

4.7.3 Character

In the last few years, the general public has found renewed interest in business ethics, mainly sparked by reported questionable practices by companies such as Enron, Global Crossing, Adelphia Cable, Arthur Andersen, and others.

Chapter 11 provides detailed discussions of various ethical issues. However, it is proper to note here that it serves companies better in the long run to hire employees with character and then train them to acquire the requisite technical skills to become productive.

4.8 DEVELOPING PEOPLE

Developing employees is another important activity of the managerial leader. The objective of developing employees is to shape their knowledge, attitudes, and skills to enhance their contributions to the company and to foster their personal growth. Knowledge is the cognizance of facts, truths, and other information. Attitudes are habitual personal dispositions toward people, things, situations, and information. Skills are the abilities to perform specialized work with recognized competence.

In well-organized companies, managers are evaluated on the basis of several performance metrics, including how they have taken care of the development needs of their employees. To be successful, employees must demonstrate initiative in seeking to continuously improve their own knowledge, attitudes, and skills. Examples of what can be done practically are presented in the next subsections.

4.8.1 Employees

There are several ways in which engineering managers may help develop employees. Employees may be prompted to follow the personal examples of continuous improvement in knowledge, attitude, and skills set by engineering managers. Engineering managers may coach inexperienced employees on the job by demonstrating preferred ways of performing specific tasks. In addition, engineering managers could enrich employees' work experience by institutionalizing job rotation. If the company budget and policy so allow, the engineering managers may send specific employees to attend professional meetings, technical conferences, training seminars, and study programs at universities. Furthermore, team assignments may be used to permit a better utilization of the employees' talents and expertise to other critical projects, while offering the employees an opportunity to become known to a larger circle within the company.

In training employees, engineering managers need to emphasize employee participation, as the goal is to achieve the employee's objectives while simultaneously attaining the company's objectives. Employees should be appraised with respect to their present performance in determining what steps might be needed to qualify them to make greater contributions in the future. If the employee's current performance is deemed to be inadequate, managers need to be positive and forward looking in helping the individual recognize the need for self-improvement. By setting a personal example of continuous improvement, the manager is likely to positively motivate the individual to seek further development.

4.8.2 Successors

Besides training employees, engineering managers are also expected, as a part of their managerial duties, to find suitable candidates within their organizations to succeed themselves one day. This is consistent with career planning programs that some industrial companies are actively implementing in order to promote leaders from within, discourage turnover, and maintain continuity.

4.9 SPECIAL TOPICS ON LEADING

Corporate change needs strong leadership. Leaders promoted into new positions need special strategies to succeed. Successful leaders share certain common attributes.

The next subsection discusses the special topics related to the managerial function of leading.

4.9.1 Leading Changes

In 21st century corporate America, internal changes occur frequently and often in reaction to changes in the external environment. Changes are typically forced upon companies by the market entry of new competitors, the declining market-share position of the company, the emergence of new technologies threatening the company's products or services, financial performance that is worse than expected (e.g., measurements of gross margin, earnings, and other indices), and other factors (Capodagli and Jackson 2001; Murphy and Murphy, 2002; Hesselbein and Johnston 2002).

Changes result in modified ways that the company conducts business. Changes are difficult to introduce because people like to stay in their comfort zones. After changes are introduced, it is important for leaders to see to it that the changes are sustained beyond the transformational period. Corporate changes demand strong leadership. According to Kotter (1995), there are two reasons that many transformation efforts fail:

1. Large-scale corporate transformation takes time. In one specific example, the maximum number of changes in a corporation was reached in the fifth year of

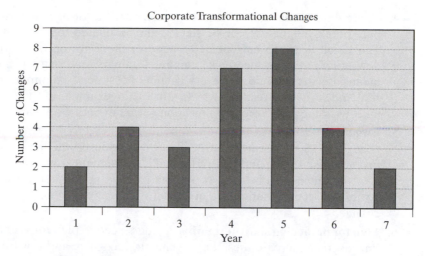

Figure 4.3   Corporate Transformational Changes.

transformation. (See Figure 4.3.) Leaders must be patient in marshalling corporate resources to push forward.

2. Corporate transformation must follow a process of eight consecutive steps to succeed. Every one of these eight steps is critical, as failure in any one will affect the overall transformation performance.

The process of eight consecutive steps delineates essentially the success factors for transformational change. These eight steps are:

1. **Establish a sense of urgency.** Leaders must examine the market and competitive realities and identify and talk about crises, potential crises, or major opportunities available in the global market. The goal is to convince at least 75 percent of the management that remaining in the status quo is more dangerous to the health of the corporation than launching in a new corporate direction.

2. **Form a powerful guiding coalition.** Major renewal programs often start with one or two people. But a leadership coalition must grow over time. In addition to the top leaders, there should be another 5 to 50 people committed to renewal. The group must be powerful in term of titles, reputations, and relationships. Only when there are enough leaders in the senior ranks will the renewal process move forward.

   Leaders should inspire the group to work together as a team. The specific goal is to secure shared commitment to change by the top management and by the most influential people. A line position holder must lead the coalition.

3. **Create a vision.** A coherent and sensible vision is needed to help direct the effort to change. Leaders need to develop strategies to achieve that vision. The vision should be easy to communicate to stockholders, employees, and customers. Ideally, it should be explainable to an audience within five minutes and achieve their understanding and acceptance.

4. **Communicate the vision.** Transformation needs a lot of people to make it happen. Employees need to be persuaded and motivated to help make the changes. The vision needs to be repeated whenever there are opportunities—through a newsletters, review meetings, training seminars, company picnics, and other means. Leaders should use every channel possible to communicate the new vision and strategies, teach new behavior by the example of the guiding coalition, and "walk the talk," as communication occurs in both words and deeds.

5. **Empower others to act on the vision.** Leaders need to do away with obstacles to change, modify systems or structures that seriously undermine the vision, and promote risk-taking and nontraditional ideas, activities, and actions.

    Examples of obstacles to remove include (1) structure (narrow job categories), (2) compensation and appraisal systems, and (3) managers who refuse to change.

6. **Plan for producing short-term wins.** Leaders need to plan for visible performance improvements, create these improvements, recognize and reward employees involved in the improvements, and achieve at least some success within the first one to two years. Otherwise, the renewal effort may lose momentum.

7. **Consolidate improvements and procreate still more change.** Leaders need to use increased credibility to change systems, structures, and policies that do not fit the vision; hire, promote, or develop employees who can implement the vision; and reinvigorate the process with new projects, themes, and change agents.

    Leaders should resist declaring victory too early to avoid killing the momentum.

8. **Institutionalize new approaches.** Leaders should articulate the connections between new behaviors and corporate success, establish the means to ensure leadership development and succession, and anchor the changes in company culture (values, behaviors, and social norms) so that the changes continue into the next generation of top management.*

The eight-step process just described is recommended for top-level engineering leaders who are planning to initiate and implement major corporate transformational changes. With minor modifications, this process applies also to midlevel engineering managers who may be called upon to change the performance of a division or a department.

4.9.2 Advice for Newly Promoted Leaders

If a new engineering manager is hired from the outside to take over a department or division, he or she might need to follow a special strategy during the transition period (e.g., the first six months on the job) in order to be productive. This is because going into an unfamiliar situation is akin to sailing in dense fog and only having visibility for a short distance.

Watkins (2001) points out that new leaders often make a number of common mistakes. These mistakes include being isolated, having "the answer," not strengthening the team, attempting too much, trusting the wrong people, and setting unrealistic expectations. In order for an engineering manager to become effective in a new organization, Watkins offers specific advice consisting of seven rules: leverage time before entry, organize to learn, secure early wins, lay a foundation for major improvements, create a personal vision, build winning coalitions, and manage yourself.

For newly promoted engineering managers, the leadership strategy just outlined serves as a useful guiding light during the initial six-month period of sailing through dense fog. Although the steps recommended above cover all important aspects of a new leadership job—technical, cultural, political, and personal—individual engineering managers may need to further customize these steps to fit their personal style, organizational needs, and the people involved in a given situation.

4.9.3 Guidelines for Superior Leadership

To become superior leaders, engineering managers are advised to focus on the following eight attributes, according to Cohen (2002):

1. **Maintain absolute integrity.** Any doubt about the leader's integrity will be reflected in the trust others place in the leader.

2. **Be knowledgeable.** The leader should be good at what it takes to get the job done.

3. **Declare expectations.** The leader should let people know which direction to go in and what results are expected.

4. **Display unwavering commitment.** The leader must demonstrate his or her clear commitment.

5. **Get out in front.** The leader needs to build, establish, and maintain a strong positive image. The leader should get out of the office see what is going on (for example, to the plant floor, marketplace, customer service center, and technology labs). The leader should also get out in front to be seen, so that others know their manager is committed.

   General MacArthur gave this advice to a young battalion commander during World War II: "Major, when the signal comes to go over the top, if you go first, before your men, your battalion will follow you. Moreover, they will never doubt your leadership or courage in the future."

6. **Expect positive results.** Show self-confidence and work to get favorable results.

7. **Take care of people.** This is the basic reciprocity doctrine of Confucius: "If you take care of people, people will take care of you." Starbucks is said to practice this doctrine by taking care of their employees first, then customers, and finally shareholders.

8. **Put duty before self-interests.** The mission and the employees must be more important than one's own self-interests.

The preceding list of attributes neglects to include the all-important quality of strategic thinking and the leader's capability to create vision. Without vision, a person with the above attributes is merely a hard-working taskmaster who is responsible, goal oriented, and socially assertive.

4.10 CONCLUSION

Leading is another key function of engineering management. It encompasses the specific managerial activities of making decisions and selecting, developing, motivating, and communicating with people. Carrying out these specific activities well will make an engineer into a strong engineering manager.

Decision making plays an important role in the career life of an engineering manager. In the engineering community, the rational decision-making method is regarded as a standard. Engineering managers need to become familiar with a number of other decision support tools so that the right tool can be fittingly applied to specific circumstances.

This chapter offered guidelines for engineering managers to become better prepared for the special cases of (1) introducing major corporate changes, (2) working as a new leader in an engineering management environment, and (3) achieving superior leadership.

Engineering managers are encouraged to practice various guidelines associated with leadership whenever they find opportunities to do so.

4.11 REFERENCES

Bardaracco, J. J. 2002. *Leading Quietly: An Unorthodox Guide to Doing the Right Thing*. Cambridge, MA: Harvard Business School Press.

Belton V. and T. J. Stewart. 2002. *Multiple Criteria Decision Analysis*, Boston: Kluwer Academic Publishers.

Capodagli, B. and L. Jackson. 2001. *Leading at the Speed of Change: Using New Economy Rules to Transform Old Economy Companies*. New York: McGraw-Hill.

Clampitt, P. G. 2001. *Communicating for Managerial Effectiveness*. 2d ed. Thousand Oaks, CA: Sage Publications.

Cohen, W. A. 2002. "The Art of the Successful Leader." *Financial Service Advisor*, July–August.

Garvin, D. A. and M. A. Roberto. 2001. "What You Don't Know about Making Decisions." *Harvard Business Review*, September.

Hampton, W. J. 1988. "How Does Japan Inc. Pick Its American Workers?" *Business Week*, October 3.

Hayashi, A. H. 2001. "When to Trust Your Gut." *Harvard Business Review*, February.

Hesselbein, F. and R. Johnston (editors). 2002. *On Leading Edge Change: A Leader to Leader Guide*. San Francisco: Jossey-Bass.

Ivancevich, J. M. and T. N. Duenig. 2002. *Managing Einstein: Leading High-Tech Workers in the Digital Age*. New York: McGraw-Hill.

Kepner, C. H. and B. B. Tregoe. 1981. *The New Rational Manager*. Princeton, NJ: Princeton Research Press.

Kirkwood, C. W. 1997. *Strategic Decision Making: Multiobjective Decision Analysis with Spreadsheets*. Belmont, CA: Wadsworth Publishing Company.

Kotter, J. P. 1990. "What Leaders Really Do." *Harvard Business Review*, Vol. 68, No. 3, May–June.

Kotter, J. P. 1995. "Leading Change: Why Transformation Efforts Fail." *Harvard Business Review*, Vol. 73, No. 2, March–April.

McGregor, D. 1957. *The Professional Manager*. (C. McGregor and W. G. Bennis, editors.) New York: McGraw-Hill.

Maslow, A. H. 1954. *Motivation and Personality*. New York: Harper & Row.

McKenna, P. and D. H. Maister. 2002. *First among Equals: How to Manage a Group of Professionals*. New York: Free Press.

Murphy, E. C. and M. A. Murphy. 2002. *Leading on the Edge of Chaos: The 10 Critical Elements for Success in Volatile Times*. Upper Saddle River, NJ: Prentice Hall.

Nagal, S. S. 1993. *Computer-Aided Decision Analysis: Theory and Applications*. New York: Quorum Books.

Noyers, R. B. 2001. *The Art of Leading Yourself: Tap the Power of Your Emotional Intelligence*. Fort Bragg, CA: Cypress House.

Pagonis, W. G. 1992. "The Work of the Leader." *Harvard Business Review*, November–December, pp. 118–126.

Ragsdale, C. 2004. *Spreadsheet Modeling and Decision Analysis*. 4th ed. Stamford, CA: Thomson South-Western.

Shannon, R. E. 1980. *Engineering Management*. New York: John Wiley.

Soat, D. M. 1996. *Managing Engineers and Technical Employees: How to Attract, Motivate, and Retain Excellent People*. Norwood, MA: ArtTech House.

Watkins, M. 2001. "Seven Rules for New Leaders." *Harvard Business School Notes*, No. 9-800-288, June.

Zaleznik, A. 1992. "Managers and Leaders: Are They Different?" *Harvard Business Review*, March–April, pp. 126–135.

4.12 APPENDICES

APPENDIX 4.A. FACTORS AFFECTING ONE'S INFLUENCE ON PEOPLE

This section discusses the various factors known to affect a person's influence on people. It is advisable that the engineering manager pay attention to them.

1. **Credibility**

   A person's credibility is based on the following six attributes:

   A. *Composure (ways to handle oneself)*

   Degree of poise, stability, and patience; skills to handle a crisis situation effectively; humor under stress; self-confidence; and ability for public speaking are important attributes. Composure is the most important factor affecting credibility in the short term.

   B. *Character*

   Integrity and honesty (not lying, not cheating, attempting to do things above board and maintaining high-moral standards), cooperative spirit, and professional behavior—return all phones calls and respond to all mail; keep promises, have an open and forthright attitude, and be fair in all situations. Character is the most important factor affecting credibility in the long run (integrity and honesty in professional versus private matters).

   C. *Competence*

   Technical (job-specific skills, experience, and training), managerial (planning, organizing, leading, and controlling), and visionary (capability to envision the future with strategic thinking). Table 4.A1 displays the competence factors that exert an influence on superiors.

TABLE 4.A1 Competence Factors on Influence Exerted Upward (Accuracy +/−10%)

	First-Line Supervisor (percent)	Midmanager (percent)	Executive (percent)
Technical	70	30	5
Managerial	25	40	25
Visionary	5	30	70

D. *Courage*

Commitment to principles; the willingness to stand up for beliefs, challenge others, and admit mistakes; and the ability to make tough decisions under uncertain conditions and accept responsibility for the consequences.

E. *Conviction (beliefs)*

Commit to the vision, demonstrate passion, and show confidence in the direction being pursued.

F. *Care for People*

Know people (family, aspirations, current and future needs, favored learning modes, upward mobility, etc.), treat people with respect and dignity (listening to understand), comment only on issues and situations and not on the person, and be a team player.

2. **Personal power (independent of position power)**

Personal power is affected by the following three factors:

A. *Personal attributes*

Physical appearance and size, drive, dedication, and personal values.

B. *Knowledge*

Common sense, historical perspective, political knowledge needed by others (how to get things done through which doors, by what means, and with whom—otherwise known as tricks of the trade).

C. *Relationships*

Business connections and power by association.

3. **Variable leadership style**

Leadership style needs to be varied in accordance with the circumstances involved. In general, there are four situations that each requires a different style of leadership. (See discussion in Section 4.2.)

The influence exerted on people by an engineering manager is affected by his or her credibility, personal power and leadership style. Engineering managers need to do the right thing at the right time and place, to the right people, for the right people, or with the right people.

APPENDIX 4.B. MOTIVATION OF MISSION-CRITICAL PEOPLE

In the competitive world today, all companies struggle to attract and retain innovative knowledge workers who are critical to the mission of their operations. To be successful, companies in general, and engineering managers in particular, need to tailor specific motivation strategies to the needs of individuals. In general, most knowledge workers have the following types of needs:

A. **Need for power (40 percent)**—setting goals and offering positive recognition to allow one to stand out and be unique

B. **Need for affiliation (40 percent)**—focusing on mission, vision, and the difference the individual can make in teams

C. **Need for self-achievement (20 percent)**—offering task variety, learning, development, and growth opportunities

Nortel Networks and Cisco Systems, Inc. experienced high turnover (about 40 percent) in their information technology sector. Surveys indicate that people with mission-critical jobs left because of three specific deficiencies:

1. **The work itself**—meaningfulness, relevancy, learning opportunity, enjoyability, variability, etc.

2. **Appreciation**—thanks, recognition; they were never told that their jobs were mission critical

3. **Money**—many left for more money

The preceding list of needs and the types of deficiencies causing knowledge workers to want to leave are consistent with the Maslow need hierarchy model discussed in Section 4.6.3. It is obvious that unsatisfied needs will strip employees of motivation if they stay unsatisfied for long. Under such circumstances, knowledge workers are likely to migrate to places where they can satisfy these unmet needs.*

4.13 QUESTIONS

4.1 Preparation of the company product that was promised to a major customer is running late, and there is intense pressure on the production team to deliver the product. The director of production is eventually told by the company president to deliver, "or else." The director therefore decides to ship the product, even though it had not gone through all of its testing procedures. Members on the production team become upset due to their uncertainty about the functionality and reliability of the shipped product. The director, however, insists that "We will just have to take that chance."

As the director of production, how would you have acted differently?

4.2 As advised by the company president, the sales department received a set of specific recommendations provided by an outside management firm to reorganize for maximum effectiveness. The sales manager believes that a few of the sales staff may disagree with the recommended changes. The sales manager herself is also not fully convinced of the merits of all of the recommendations, but she wants to implement them, at least in part.

How should she proceed?

4.3 The engineering director of the company is called upon to send one engineer abroad to assist in the installation of equipment. There are three qualified candidates, each working for a different manager under the director. The director knows that all three engineers will want to go, but their superiors will oppose any of them going for fear of losing time in completing their own critical projects. How should the director make the choice?

Sources: P. McKenna and D. H. Maister. *First among Equals: How to Manage a Group of Professionals.* New York: Free Press, 2002; J. M. Ivancevich and T. N. Duening. *Managing Einstein: Leading Edge High-Tech Workers in the Digital Age.* New York: McGraw-Hill, 2002; D. M. Soat. *Managing Engineers and Technical Employers: How to Attract, Motivate and Retain Excellent People.*" Norwood, MA: ArTech House, 1996.

4.4 The marketing department needed to submit a proposal to a global customer, and it called a review meeting the next morning. By the time Bill Taylor, the design manager, was informed in the late afternoon, all of his design staff had left for the day and there was no one available. Bill Taylor decided to work on the proposal himself through the night so that he could talk with his design staff the next morning, one hour before the marketing review meeting.

All of the staff agreed with the proposed design except Henry King, a senior staff member who is recognized as the most experienced and best designer in the group. His objections were that the current design was too complex and that it would take another week to modify on the design to ensure its functional performance.

In order to pacify him, Bill Taylor invited Henry King to come along to the marketing review meeting so that Henry King would feel the pressure that marketing was exerting on design. Unexpectedly, Henry King stood up at the marketing review meeting and reiterated all of his design objections, causing tremendous embarrassment to Bill Taylor and his superior, Stanley Clark, the design director. Bill Taylor became furious.

What should Bill Taylor and Stanley Clark do?

4.5 Jerry Lucas is the division director. As branch chief, Bob Sanford reports to Jerry Lucas. Bob Sanford has four section chiefs reporting to him.

Bob Sanford is technically competent, with extensive experience in solid rocket propulsion; he is also regarded as the best expert in this field. He is highly dedicated to his work, but inexperienced in managing technical people, as he has been on the job for only two years. Bob Sanford handles his subordinates quite roughly. He reverses his section chiefs' decisions without prior consultation with them. He demands that no information or data be transmitted to persons outside the group without his knowledge and concurrence. He also bypasses his section chiefs to go to people and encourages them to come to him directly with problems. Rumors have it that he places spies or informants within the group. As expected, he delegates no decision-making authority to his section chiefs and regards all of his section chiefs as technically incompetent. He creates an atmosphere of fear and suspicion, with low group morale.

Bob Sanford does not report to Jerry Lucas candidly on project progress and on difficulties he encounters. He does not understand his own responsibility of building teamwork, enhancing group morale, and creating employee satisfaction while achieving the goals of his group. He is lacking the skills and willingness to resolve conflicts within the group.

Finally, the section chiefs as a group go in to see Jerry Lucas and complain about the lack of authority and the oppressive atmosphere in the section.

What should Jerry Lucas do?*

4.6 The board of directors receives a proposal from a business partner to jointly set up an assembly plant in a third-world country. This new plant will assemble final products with key components made by the company. Financial terms are attractive and the future marketing outlook is bright. There is just one problem. The third-world country is not a democracy, has a poor human rights record, neglects to protect its own environment, and does not safeguard workers' rights. An investment placed by the company would boost this country's economy and thus the political position of its current dictator. Should the company accept the proposal, and why?

4.7 What are some important characteristics of effective leaders? Which of these characteristics are more difficult for engineers to acquire?

*Condensed and adapted from Robert E. Shannon, *Engineering Management*, New York: John Wiley, p. 227. This material is used by permission of John Wiley & Sons, Inc.

4.8   The plant manager noticed a need to lessen the amount of waste materials, which occurred in the production process. A task force was set up, composed of the plant manager herself and two of her supervisors, to examine the problem. They met for three months and regularly published the task force objectives and findings on the plant bulletin board.

The plant manager found, to her surprise, that the workers on the shop floor exhibited limited interest in the task force and ignored the bulletin board entirely. At the end of the three-month period, the task force came up with several excellent recommendations, which require changes in work practices. Most of the workers implemented the recommended changes very reluctantly, and some even secretly worked to sabotage the new practices. Eventually, all recommendations were withdrawn.

What went wrong? How should the plant manager have handled this case?

4.9   The project was running late and the section manager thought that it was time for a pep talk with his staff. He realized that he was considered to be somewhat autocratic by his staff, but this time he thought that he would impress on them that he was really one of the members of the team and that they would work together as one in order to succeed.

The section manager thought he made quite a good speech. He pointed out that the project was running late and that, if they failed, the customer could cancel the contract. He explained further that, as manager, he was responsible for the success of the project, and so everyone would be equally to blame for the failure of the project.

Unexpectedly, a group of staff came in to see him a few days later to clarify whether they were all under threat of unemployment should it turn out in the future that they were indeed late and the contract was cancelled by the customer.

What went wrong? What would you have done differently?

4.10   A regional sales manager suspected that one of her customers was having financial troubles. However, she was reluctant to mention it to her superior because she felt that she could be wrong. She kept quiet for several months, continuing to take large orders from this customer and hoping that the customer could recover from their troubles. Eventually, the customer went bankrupt and defaulted on the payment of several large bills. What went wrong? What would you have done differently?

4.11   Company X selects someone who is weak technically, but very strong in group-process skills to lead a team in developing a new engineering product. Would such a person be successful as a team leader? What can be done to ensure that the engineering product developed by the team will be satisfactory from the technical standpoint?

4.12   Conflicts between technologists and managers may arise when the technical professionals with the skills to make a decisions have to deal with a manager, who has the right to decide. Why do such conflicts often exist in organizations wherein everyone works toward the same common goal?

4.13   Company X makes the decision to substitute aluminum for steel in a component of its product. What factors probably have contributed to this decision? At what managerial level would this decision most likely have been made?

4.14   As the department head you urgently need to find an experienced person to fill a vacancy. The work involves close cooperation and coordination with others inside and outside of the department. Candidate A has exactly the experience required, but appears to be very unsociable. Candidate B has experience in a related job and seems to have a pleasant personality, though not an extrovert. Candidate C has business experience in a different industry and is extremely sociable. All three candidates have scored sufficiently high on intelligence tests to qualify for the job in terms of general ability. Which candidate would you choose, and why?

Chapter 5

Controlling

5.1 INTRODUCTION

The function of controlling in engineering management refers to activities taken on by an engineering manager to assess and regulate work in progress, evaluate results for the purpose of securing and maintaining maximum productivity, and reduce and prevent unacceptable performance.

Although the bulk of the controlling activities appear to be primarily of administrative and operational nature, controlling has strategic importance. To efficiently implement any assignment project, program, or plan, managerial control is crucial. Any forward-looking strategic plan becomes useless if its implementation is poor. Furthermore, without adequate control, managerial delegation is ineffective, rendering the managerial leadership of questionable quality. The function of controlling also contributes to corporate renewal by pruning the dead wood, if needed.

Engineering managers exercise control by carrying out the specific tasks of (1) setting standards, (2) measuring performance, (3) evaluating performance, and (4) controlling performance. In addition, this chapter addresses the manager's control of time, personnel, business relationships, projects, and company knowledge.

5.2 SETTING PERFORMANCE STANDARDS

To set standards is to specify criteria by which work and results are measured and evaluated. Setting performance standards defines the expected results explicitly. It is important for both the company and its employees to distinguish between performance grades of "outstanding," "better than average," "average," "below average," and "unacceptable" and to understand how performance is measured. Here, "average" is defined as the generally expected performance level for the job at hand by persons with adequate training and dedication.

Setting standards provides specific guidelines for exercising authority and making decisions. Standards represent a yardstick for measuring the performance of employees and units (such as cost centers and profit centers). Proper standards facilitate employee self-evaluation, self-control, and self-advancement.

Standards are typically established in the form of how many (number of units), how good (quality, acceptance), how well (user acceptance), and how soon (timing), as imposed by the company management, the customers, or the marketplace. It is worth noting that current trends of setting standards emphasize the inclusion of customer viewpoints.

There are technical, historical, market, planning, safety, and equal employment opportunity (EEO) standards. Technical standards specify metrics related to quality, quantity, mean time between failure (MTBF), maintenance requirements, etc. Historical standards are based primarily on past records. Market standards are those related to competition, sales, return on investment (ROI), earnings expectation by securities analysts, and other factors. Planning standards mostly relate to the strategic and operational planning needs of the company. They address topics such as objectives, programs, schedules, budgets, and policies. Safety standards refer to the metrics related to the safe operation of the company's facilities. Government programs such as the U.S. Occupational Safety and Health Administration (OSHA) promulgate some of these safety standards. EEO standards are specific practices related to affirmative action for the purpose of achieving workforce diversity.

Effective standards are characterized as being company sponsored, measurable, comparable, reasonable, and indicative of the expectation of work performance on an objective basis in order to promote worker growth and as being considerate of human factors by soliciting the workers' inputs and securing their understanding and acceptance.

Example 5.1.

The engineering director asks her managers if they have any nominations for promotions from within their respective departments. The maximum number of promotions allowed for the entire division is 2. The nominations must be selective and only for people whose performance has been outstanding. One manager thought his whole team had been outstanding, so he recommended all 10 for promotion. He reasoned, "It is better for the morale of the team that they know that I support all of them fully. If the director now promotes 1 of the 10 or none at all, then they will not feel so bad knowing that at least I have thought them all worthy of being promoted."

What should the director do?

Answer 5.1.

The director should call the manager in and reprimand him for (1) not following the instructions given of being selective in nominating candidates for promotion, (2) neglecting the delegated responsibility of evaluating the performance of his staff fairly and objectively, and (3) wanting to pass on his own responsibility to his director so as to be the "good guy."

She should order the manager to repeat the nomination process and come up with a rank-ordered list of outstanding performers for possible promotion within two days. If he refuses to do so, this will be registered as one incidence of disobedience. Repeated offenses of this type will lead to immediate discharge.

There are quite a few barriers that prevent good standards from being readily developed. Standards may be too subjective, as technically strong engineers tend to set unrealistically high standards for themselves and others. If standards are set too high, workers may become demoralized and fearful of not measuring up. On the other hand, if standards are set too low—not challenging enough for the workers—management will lose the respect of employees. Standards may be confusing and may not clearly indicate the criteria for excellence. Standards may be qualitative and vague and thus subject to different interpretations by different people. Standards may also be set without proper consideration for the constraints related to resources and implementation.

A useful way for companies to set proper standards is the practice of benchmarking, discussed in the next section.

5.3 BENCHMARKING

Benchmarking is a method of defining performance standards in relation to a set of internal and external references. There are two types of benchmarking:

1. **Internal Benchmarking** uses references internal to a company to set performance standards, as illustrated by the following example:

Company X has achieved a productivity of $150,000 per employee in 2003 and is determined to continue improving performance on a yearly basis. The company president sets a new performance standard for the year 2004 at $165,000 per employee, a level 10 percent higher than the previous year.

Internal benchmarking is convenient to apply, as it offers a reasonable, short-term performance assessment. However, it may create a false sense of corporate well-being, as the resulting performance standards, in the absence of an external reference, may be deficient in the long run. That is, the 10 percent productivity increment may be inadequate, or even outright dangerous to the long-term health of the company, if the industrial average productivity gain has been 12 percent per year in the past, and some competitors of Company X are aiming at 15 percent for the coming year.

2. **External Benchmarking** uses references external to the company to set performance standards. The following examples illustrate its applications:

- **Financial ratios.** Assuming that the gross margins of many service-oriented public companies are in the range of 35 to 40 percent for the last several years, as published regularly by Wall Street securities firms, then setting the gross margin target for a specific company at the 40 percent level for 2004 is a performance standard defined by external benchmarking. Setting the gross margin at 40 percent will ensure that the company will be among the best performers in the industry, provided that the standards are met.

- **Performance metrics.** Many performance metrics are published in business literature to evaluate production, product delivery, quality control, time to market, customer service, reliability, customer problem solving, and others. Using these known metrics as references to set performance standards makes explicit the relative competitive position of a company.

- **Best practices.** Another set of important external benchmarks is called "best practices" (Anonymous 2003A; Rodier 2000). These are work processes or procedures perfected and upgraded by various companies over the years to address specific problems and issues in engineering, production, marketing, strategic management, business development, and other areas of corporate governance. These are extremely valuable "tried-and-true" methods to achieve useful results and add value to the companies. Those companies too undisciplined to consistently apply "best practices" in carrying out their corporate activities may discover that they have become less and less competitive over time.

- **Critical success factors.** Serving well as external benchmarks are critical success factors, the necessary and sufficient conditions for achieving success in specific business, engineering, production, and marketing domains. These factors are derived from the successes and failures experienced by companies in various industries. Having a clear understanding of these factors allows a company to make strategic choices to move in new directions, initiate new products to gain advantages, and capture opportunities in new markets by optimally applying corporate strengths while minimizing any exposure of weaknesses.

- **Target pricing.** In recent years, Japanese companies have successfully applied the technique of target pricing, another external benchmarking method, to achieve success in the marketplace. First, the company conducts a survey to determine the current prices of products that are in direct competition with the new products that the company is planning to market. Using these prices as references, the company sets a target selling price (such as 80 percent of the competitive product prices) for its new product. Then, the company deducts the required gross margin for the company to sustain itself while meeting the shareholders' expectations. The resulting dollar amount is the *cost of goods sold* target (the cost target) for the new product. The cost target is then imposed on the production and engineering departments as performance standards for developing the new product. Funds are available for developing the new product only if the cost target can be met. This is tantamount to an *innovation under duress* model of management control. The resulting new product is ensured a competitive edge in the marketplace.

  Traditionally, bringing a new product to the marketplace follows a sequential process. First, the product is designed by engineers on the basis of inputs from marketing. Then, it is redesigned by service engineers for serviceability. Afterwards, it is redesigned again by production engineers for manufactureability. Finally, it is mass produced, and its final product cost is accurately estimated. The company management adds the gross margin to set the product prices. At each step, engineers tend to introduce contingencies, or "cushions," to ensure that the work is done properly within their respective units. However, oftentimes this sequential process leads to products that are not cost competitive in the marketplace.

  In fact, the basic concept of target pricing can be applied to many other corporate activities to ensure that the company remains competitive. This is particularly pertinent to the current business environment, which is becoming increasingly globalized.

- **Balanced Scorecard.** In recent years, business researchers have noted the biases many companies have against monitoring corporate performance primarily by using financial metrics only on activities related to the past. (See Section 7.6.) Progressive management needs to devise a balanced set of performance metrics to properly monitor other company activities that also have a profound impact on its future success.

  Examples of such forward-looking performance metrics include the establishment of (1) the percentage of company business generated by products introduced in the last five years; (2) the percentage or amount of corporate funds spent on projects initiated in the last five years; (3) the number of patent disclosures, patent applications, and patent awards in a given year; (4) the number of new supply-chain partners engaged in the last five years; (5) the percentage of sales realized from new customers acquired during the past year; and (6) the fraction of product cost arising from new technologies adopted in the last five years.

It is self-evident that setting adequate, forward-looking standards based on both internal and external benchmarking helps guide the company to success in the future.

5.3.1 Sample Benchmarking Metrics

Many sets of performance metrics are available from business literature. Table 5.1 contains selected samples of metrics used in various business domains.

TABLE 5.1   Sample Performance Metrics

Domains	Metrics
Financial	ROI (return on investment)
	ROA (return on assets)
	ROS (return on sales)
	Debt to equity ratio
	Number of inventory turns each year
	Number of units produced per employee
	Number of units produced per hour
	Sales per employees
	Profit per unit of production
	Break-even volume
Nonfinancial	Average number of defects detected by customers in the first month of ownership
	Average number of defects detected during manufacturing and repair
	Hours lost to production due to unscheduled maintenance
	Work in progress in the plant
	Number of machines per worker
	Length of time to change a machine or introduce a new operation
	Number of job classifications in the plant
	Amount of materials made obsolete by model changes
	Average energy consumption per unit of production
	Rate of absenteeism in the workforce
	Number of months required to introduce a new product model
	Number of engineering change requests during a new product development program
	Number of units produced prior to a model change (batch production)
	Time metrics (response time, lead time, uptime, downtime, etc.)
Product Related	Parts count
	Number of material types used
	Material utilization in each component
	Assembly process used in production
	Service quality—field repair versus field replacement
	Failure-mode effect analysis
	Quality of product as experienced by customers
	Long-term durability of product
	Fraction of sales to repeat customers
	Company responsiveness to service requests

5.3.2 Limitations of Benchmarking

Benchmarking is useful, but has certain limitations. For example, some reference data may not be available, and in such cases, estimates must be made. This may cause the value of such benchmarks to be less robust. Benchmarking metrics are always based on past performance, and they do not predict the future. Neither can they be used to predict new competition. However, even with these limitations, past-oriented benchmarks are still valuable. As Confucius says, "Studying the past will lead to an understanding of the future."

Example 5.2.

Product quality has many dimensions. Which dimensions of product quality have the most impact on the product's success in the marketplace?

Answer 5.2.

According to Garvin (1984), product quality has eight dimensions:
1. Performance (operational characteristics)
2. Features
3. Reliability
4. Conformance to design specification
5. Durability
6. Serviceability
7. Aesthetics (look and feel)
8. Perceived quality (affected by brand name and company reputation)

Which of these quality dimensions affect the product's success in the marketplace will depend on the type of products involved:

A. Automobiles—(7), (8), (1), (6)
B. Consumer products—(8), (1), (6), (3)
C. Industrial gases—(1), (3), (6), (5)
D. Office supplies—(1), (8), (7)
E. Home appliances—(3), (6), (1), (5)

Reference: D. A. Garvin, "What Does 'Product Quality' Really Mean?" *Sloan Management Review*, Vol. 26, No. 1, Fall 1984, pp. 25–43.

5.4 MEASURING PERFORMANCE

After the performance standards are set, the next step of management control is to measure performance. Performance measurement refers to the recording and reporting of work done and the results attained. Engineering managers take the following steps to measure performance: (1) collect, store, analyze, and report information systematically; (2) compare the performance against established standards; and (3) issue reports such as data, results, and forecasts to document results.

Engineering managers may use techniques of time study, work sampling, and performance rating, among others, (Dhillon 1987) for measuring the performance of routine work, such as that on factory floors. The performance of professional workers needs to be measured with respect to the contributions made toward the attainment of the company's short- and long-term objectives. All measurements must be factual and accurate in order to be justified as valid basis for performance evaluation.

5.5 EVALUATING PERFORMANCE

To evaluate performance is to appraise work in progress, assess job completed, and provide feedback. Engineering managers take the following steps to evaluate employee performance: (1) establish limits of tolerance, (2) note variations (deviation within the tolerance limits) and exceptions (deviation outside of the tolerance limits), and (3) provide recognition for good performance and give timely, proper credit, if justifiable. Paying attention to deviations encourages employee self-appraisal, fosters initiative, and enhances managerial efficiency.

A rating method often used in industry is to rank an employee in one of five categories: (1) outstanding, (2) above average, (3) average, (4) poor, and (5) failure. Category ranking is based on performance metrics that are specific to the individual. More often than not, in order to indicate the importance of these performance metrics to the company, top management will assign and publicize the weight factors associated with all performance metrics. A weighted score, similar to that calculated by the Kepner–Tregoe method discussed in Section 4.4.2, is then determined for each employee. The weighted score and a written statement are then submitted to superiors at the next management level for review. Once the approval of higher management is secured, these evaluation results become the official basis for salary administration and promotional considerations in the future.

It is quite obvious that this type of rating system has basic weaknesses. Some managers may suffer from a "halo effect" and assign an employee the same rating for all performance metrics. Others may be handicapped by a "recent effect" in that they are predominately influenced by the recent events. By nature, some managers are more lenient than others, resulting in different interpretations of "outstanding performance." Furthermore, the competitive nature of getting one's own employees promoted sooner tends to cause inflated ratings. To exert some control over this potentially chaotic situation, a few industrial companies are known to have advised department managers to steer the overall evaluation results toward a Gaussian type distribution, which places the majority of employees in the "average" group, and only about 5 percent and 15 percent, respectively, in the excellent and above average groups.

At the annual appraisal time, the individual employees will be notified of the approved evaluation results. Feedback from the individual employees is solicited and documented. If the individual employee disagrees with some or all parts of the evaluation, their written comments are incorporated into the official evaluation files, which are reviewed again by superiors at the next management level. If deemed proper, the approved evaluation results may be modified. The needs for the individual's future development are also discussed and noted. Specific goals of development are then agreed upon as a gauge for monitoring the individual employees' progress at the next annual appraisal time.

The level of "average" performance is defined as the level of performance which can be generally expected of a person with adequate training and dedication at a given position. To be rated as excellent or above average, one must perform extraordinarily well and produce an unexpected positive outcome which has a recognizable impact on the company's objectives. Poor performance is usually associated with work that is not meeting the acceptable quality standards, exceeding the approved budget, not completing the work on time, and/or suffering from problems related to communications, personality conflicts, work devotion, interpersonal skills and other deficiencies.

Should the performance of an employee be rated as poor or failure, engineering managers must initiate action to correct such performance in a timely manner.

5.6 CORRECTING PERFORMANCE

To correct performance is to rectify and improve work done and results obtained. The performance evaluation may show that the quality of the employee's work is below expectation. Engineering managers must understand that there are reasons for performance deficiencies. Some employees may not know that their performance is deficient because of a lack of known performance standards or feedback. Others may not possess the required technical capabilities to perform the tasks at hand. Still others lack devotion to the profession or do not possess a good work ethic and personal initiative.

Engineering managers should correct an employee's mistakes by focusing on future progress and growth. They should take short-term action to overcome variances, such as getting assistance from outside consultants or hiring temporary workers. They should also consider long term management action to avoid repeating the noted deficiencies by improving training, modifying procedures and policies, transferring employees, or recommending dismissals.

When correcting performance, engineering managers should also offer negative feedback without attacking the employee's self-esteem. Feedback must be focused on results and outcome—not the person—and directed toward the future, not the past. Engineering managers need to avoid upsetting employees or harboring punitive motives. Managers should demonstrate a helpful and sincere attitude and pose no threat to their workers.

Generally speaking, it is not good to make mistakes with the fundamentals of engineering. It is acceptable to fail in new and risky development projects, but making the same mistakes again and again is viewed negatively by management.

5.7 MEANS OF CONTROL

Engineering managers have a number of tools available to exercise control. They may perform personal inspections, review progress, and define any variance to plans. This is the strategy of management by exception.

Managers may set priorities with respect to job assignment, resource deployment, and technology application. They may also exercise control by managing resources.

5.8 GENERAL COMMENTS

Engineering managers must constantly define which tasks should be performed, and have employees perform these tasks correctly. The principle of critical few (the Pareto principle) says that, as a rule, 20 percent of factors may affect 80 percent of results.

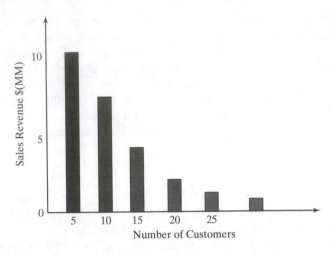

Figure 5.1   The Pareto Principle.

(See Figure 5.1.) The key is, of course, for the engineering manager to define these critical few, allocate resources to pursue them, and achieve the desirable outcome.

Control should be focused on where action takes place. In general, self-control imposed by the persons involved is the most effective type. However, by and large, people also resent control, and extensive control may lead to loss of motivation. Therefore, engineering managers must manage both the positive and negative exceptions. With information available and mechanisms in place, the preferred type of control is flexible and coordinated.

5.9 CONTROL OF MANAGEMENT TIME

Time is a valuable and limited resource for everyone, including managers. Management tasks have several common characteristics: Important tasks often arrive at unpredictable times, trivial tasks often take up a disproportionately large amount of time, and interruptions are common to a manager's schedule.

Engineering managers may waste a lot of time if the roles and responsibilities of employees are not clearly defined. Oftentimes, a lack of self-discipline—procrastination, confused priorities, lack of personal drive, or lack of planning—waste time. Other engineering managers suffer from a lack of effective delegation; for example, there may be a delegation of responsibilities without the commensurate authority, or the application of too little or too much control. Still other managers waste time because of poor communication related to policies, procedures, meetings, and other subjects (Anonymous 2003B).

The following techniques may help the engineering managers avoid wasting time:

A. Set goals for the day, the week, the month, and the year.
B. Prioritize tasks to be done, beginning with the most important tasks (A tasks), then the less important ones (B tasks), leaving the least important ones (C tasks)

TABLE 5.2 Time-Saving Tips for Engineering Managers	
1. Set goals	Write specific, measurable outcomes that you want to achieve in the next week, month, year, and five years. Consider your work, relationships, play, and well-being. Progress from goals to plan to work.
2. Use a master "to-do" list	Categorize all "to-do" ideas according to which goal each serves. Estimate all others.
3. Get the big picture	Plan your priorities so that you work foremost on whatever gives the biggest payoff and potential.
4. Cluster common tasks	Do similar tasks in the same time block (e.g., a batch of letters, several phone calls, etc.)
5. Create systems	Keep tools, forms, checklists, and information handy and organized for repetitive tasks.
6. Establish place habit	Keep everything in its predetermined place.
7. Delineate time blocks	Schedule blocks of uninterruptible time (2–4 hours) to work on projects requiring concentration. Assure colleagues of availability otherwise.
8. Design your environment	Make your setting conducive to concentration (e.g., sit with your back to traffic passing your office, and screen calls).
9. Cut meeting time	Use proven meeting time savers (e.g., go to other's offices for meetings, do stand-up meetings, set an agenda and follow through rigorously).
10. Lessen panic	Handle what worries you the most. Ask yourself, "Will this matter seem urgent 10 years from now?"
11. Take the one-minute test	Periodically take a minute to ponder, "Am I doing this in the best way to meet my goals, serve others, and take care of myself?"

Condensed and adopted from Cottringer (2003), Casavant (2003), and Anonymous (2003B).

untouched if time does not allow. Make sure that the tasks are relevant and add value. Reserve blocks of time to pursue A tasks.

C. Plan each task beforehand and group some of them together.

D. Minimize interruptions by keeping the office door closed for a specific time period and asking the secretary to hold all phone calls.

E. Make use of waiting time at the airport, on the train, in the doctor's office, etc.

F. Keep reports and memoranda short and to the point. Some managers in industry have the habit of browsing over only the first page of any report and stopping if the information is uninteresting, of secondary importance, or irrelevant to their current needs.

Enumerated in Table 5.2 are some time-saving tips that are adopted from several published sources: Cottringer (2003), Casavant (2003) and Anonymous (2003B). Engineering managers may find them useful.

Example 5.3.

The customer service manager is a busy person. He rushes from one problem to another without actually taking time to complete any job and solve any problem properly.

What control problem does the customer service manager have? What can he do to enhance his job effectiveness?

Answer 5.3.

The customer service manager has a time-control problem. He reacts to problems and does not discharge his job responsibilities effectively. He can do several things to rectify the situation:

A. Organize the customer service department into groups (e.g., repairs, parts supply, warranty, problem solving by phone, etc.) and know the capabilities of his support staff for these groups of services.

B. Set up the call center operation to automatically channel customer calls to the respective service groups.

C. Delegate the responsibilities of providing customer services to the leaders of these groups, requiring them to follow through on each and every one of the customer problems and keeping good records of all services rendered.

D. Refer specific problems (e.g., nasty customers) that the group leaders cannot resolve, to the customer service manager to handle personally.

E. Assign the analysis of service records to someone who could apply statistics and define trends suggested by these data (e.g., parts with highest failure rate, nature of frequent complaints, average time spent on problem solving, number of calls from customers unhappy with company's services, etc.).

F. Establish a system to solicit customer feedback on service and suggestions to improve.

G. Set up metrics to measure service quality (e.g., number of complaints per week, average service time spent per customer, cost per service call, etc.), making use of "best practices."

H. Monitor progress and seek ways to constantly augment the service operation.

5.10 CONTROL OF PERSONNEL

Managerial control is exercised primarily for the purposes of maximizing company productivity and minimizing potential damages arising from ethics, laws, safety, and

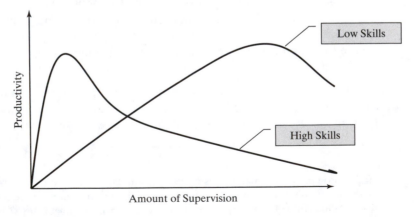

Figure 5.2   Supervision curve.

health issues. For highly skilled personnel, less control is more effective and acceptable. Excess control induces undesirable reactions and produces adverse effects. This is best illustrated by the supervision curve displayed in Figure 5.2.

To manage creative people, or those who are able to produce new and useful results, managers need to set targets, monitor the employees periodically, apply a low level of supervision, and maintain a collaborative and creativity-inducing work environment (McKenna and Maister 2002; Ivancevich and Duening 2002).

Example 5.4.

Mary Stevenson, the shop manager, works well with all of the staff members. She regards Mike Denver, who has the longest tenure and most extensive experience in the group, as the second in command for the day-to-day operation. The shop is modernizing its operation with the use of computers. Mary Stevenson and her boss, Craig Martin, decide to bring in a young computer specialist, Janet Carter, from the outside.

To make sure that the shop modernization process moves forward, Mary Stevenson spends a lot of time with Janet Carter. Mike Denver sees less and less of Mary Stevenson, although Mary still depends on Mike for the day-to-day operation. Mike resents being shut out from the work done by Mary and Janet. Mike does not complain, but after six months, he tenders his resignation and goes to work for a competitor. Mary Stevenson is shattered. She deeply regrets this major loss to the shop.

What went wrong? What was not controlled? What would you do differently?

Answer 5.4.

The key issue at hand is that Mike perceived a loss of trust and respect from Mary. Mike also incorrectly perceived an approaching obsolescence in the shop's computer technology. Both of these misperceptions caused Mike to have serious doubts about his future standing in the shop. Mary did not anticipate these perceptions and misunderstandings and did nothing to correct them in a timely manner.

Mary Stevenson should have done the following:

A. Bring Mike Carter into the loop of hiring a computer specialist to broaden the skill sets of the shop. Allow Mike to participate in the selection and interview process.

B. Announce in a staff meeting that Mary needs to spend time to bring Janet along initially and that Mike will actively assist in taking care of the day-to-day operations in the shop.

C. Get Mike involved in the work done by Janet, so that Mike, as the second in command, is kept up to date with this new type of computer work planned for the shop. Should Mike need to take over the management of the shop at some future point in time, he would have been given time to familiarize himself with the computer-related operation. Grooming a candidate in the shop for possible succession in the future should have been Mary's responsibility anyway.

D. Maintain a balance of management attention given to both computer related and other chores in the shop.

5.11 CONTROL OF BUSINESS RELATIONSHIPS

As industrial markets become increasingly global, business relationships (defined by whom the company knows and how well) represent an increasingly important competitive factor in the marketplace (Kuglin with Hook 2002; Slowinski and Sagal 2003).

It is highly advisable that engineering managers acquire the habit of proactively forming, maintaining, and controlling new business relationships for the benefit of their employers and themselves.

Contacts may be established with noncompetitors. At proper occasions, engineering managers should be accustomed to introduce themselves to others with a five-second "commercial"—a brief self-description of their key areas of expertise. Making note of others' professional specialty areas and following up with periodical exchanges will nurture the relationships. Over time, such a network of professional contacts may become a very powerful business asset to engineering managers and their employers.

5.12 CONTROL OF PROJECTS

Engineering managers exercise control over projects when serving as project leaders (Evans 2002; Oliver 2000; Katz and Thompson 2003). Tools for project control include PERT, CPM charts, or suitable computer or Internet-based project-management software. (See Section 12.3.2.) Project control focuses on several key issues, as indicated in Table 5.3.

To manage and control projects, engineering managers need skills related to organization and planning, people management, problem solving, communication, and change management. They also need to have drive and energy, broad technical knowledge, and an optimistic outlook. They must also be goal oriented and customer focused.

TABLE 5.3   Project Control Issues

1. Cost control	Monitor the actual versus projected percentage of cost expenditures and take proper actions to minimize deviations.
2. Schedule control	Monitor the actual versus projected percentage of completion time and take the proper actions to minimize deviations.
3. Critical path activities	These are activities without time slacks, which must be managed with extra care to avoid schedule delay.
4. Task deviation from plan	Delays arise from slow equipment delivery or installation; equipment damage in transportation; construction delay due to labor action, weather, utilities, and other causes; and changes of personnel.
5. Collaboration	Securing collaboration among team members is a key success factor for any project.
6. Conflict resolution	Resolving instantly all conflicts and problems among team members will assure a smooth progress toward achieving the project goals.

5.13 CONTROL OF QUALITY

To achieve success in the marketplace, companies focus on product and service quality as two of several attributes deemed important to customers. For some companies, to plan and implement quality control programs represents a major engineering management undertaking.

In the automotive industries, product quality is a well-recognized competitive factor. A recent study by J. D. Power Associates indicates that Toyota, Honda, and Hyundai are the top three carmakers in North America, with the fewest problems per 100 vehicles than the industrial average of 119 problems in 2004. In 2003, the industrial average was 133 problems per 100 vehicles.

Most noteworthy is the significant improvement in quality achieved by Hyundai, which slashed its number of problems per 100 vehicles from 272 in 1998 to 117 in 2004.

Many years ago, Deming (2000) promoted the concept of product quality in the United States and got few followers. He went to Japan and was enthusiastically welcomed there. Since then, a number of quality-control practices have been created by the Japanese, such as quality circles, Kaizen, Kanban, just in time (JIT), lean production, Taguchi, Ishikawa, and the 5S campaign. Kanban means looking up to the board in order to adjust to a constantly varying production schedule. The 5S campaign includes (1) Seiri—arrangement, (2) Seiton—tidying up, (3) Seisou—cleaning, (4) Seiketsu—cleanliness, and (5) Shukan—customizing.

Kaizen means change for the better or continuous sustainable progressiveness. It includes a number of quality practices such as customer orientation, total quality control, robotics, quality circles, suggestion system, quality betterment, JIT, zero defects, small group activities, productivity furtherance, and corporate labor-management relationships. Kaizen begins and ends with people. An involved leadership guides workers to strive for lower cost, higher quality, and faster delivery of goods and services to customers (Imai 1986; Laraia, Moody, and Hall 1999). The elimination of all nonvalue adding activities is a key emphasis in the Kaizen approach. Table 5.4 presents numerous sample Kaizen steps taken by various manufacturers. Many of these steps are based on common sense.

These quality concepts are logical, reasonable, practical, and, above all, obvious. In fact, there is really nothing new or novel in them. Many of these quality concepts were subsequently "reimported" back to the United States with Japanese labels. Thereafter, American managers started paying attention to quality. As of today, many American automotive companies are still in the mode of catching up with their Japanese competitors.

Knowing what to do is useful, but it is not enough. Pronouncing a few quality terms in Japanese will impress no one. Practicing the quality concept meticulously is the only key to success in quality enhancement and control. To achieve useful results in quality, management commitment is essential. Management commitment is reflected in company value, vision, resources assignment, customer focus, long-term strategic orientation, and rewards systems. In addition, worker dedication—drive to excellence, attention to details, and continuous betterment—must be assured.

It is interesting to note that, during the last 20 years, foreign automakers opened a total of 17 factories in the United States to produce cars and using American workers.

TABLE 5.4   Sample Kaizen Steps

Sample Kaizen Steps
Get the most useful ideas to improve operations from the workers involved.
Conduct time studies and observe the actual work activities to evaluate productivity.
Use questionnaires to collect data and small group discussions to encourage worker participation.
Workers should put everything they need next to them to minimize time (JIT and lean production).
Use a checklist that is constantly revised to inspect and study the shop floor activities.
Combining several process steps to lessen efforts required saves resources.
Incorporate prefabricated component modules to cut cost.
Use lighter and more versatile manufacturing equipment to whittle away manpower and utility costs.
Use flexible welding jigs and general purpose pallets to cut welding costs.
Use electric-driven robots and adopt a new server gut welder to increase spot welding speed.
Use general purpose carriers with common pickup points to handle all car models.
Combine primer and top coats in a single operation to gain 15 percent in speed and cut down energy consumption based on the use of a resin (a polyacetal), which would arise from the primer coat to the surface and separate the primer from the top coats.
Decrease the number of seat sets and apply more common components in design.
Use a combination of automatic, semiautomatic, and small lot stations to make products of different models and volumes.

Their superior product quality output can only be attributed to superior management practices, as there is no difference in culture, language, or value systems between American workers employed by General Motors and Ford versus those working in U.S. factories for Toyota, Nissan, and other foreign carmakers. Maynard (Maynard 2003) pointed out recently that General Motors had 60 percent of the American car market in 1960 and only about half of that today. The light-vehicle market owned by foreigners is now already 40 percent. Toyota is predicted to overtake General Motors in becoming the largest car company in the world in the near future. Maynard further pointed out that there is a difference in the background of lead managers involved. General Motors and Ford have been headed by financial professionals, whereas Toyota, Volkswagen, BMW, and Mercedes are led by engineers who are "passionately interested in everything to do with cars." This could be a factor affecting the varying extent of management commitment devoted to product quality.

Another case in point is the implementation of the quality program. Ford Motors Company is credited for having spearheaded the well-known "failure mode and effect analysis" (FMEA) method in the U.S. automotive industry around 1977. FMEA may be applied to design, process, service, and other engineering or business activities that may go wrong. Murphy's law says, "If something can go wrong, it will." FMEA is a

TABLE 5.5   FMEA Worksheet

A	Step number
B	Process description
C	Potential failure mode
D	Potential effects of failure
E	Severity
F	Potential causes of failure
G	Occurrence
H	Current process control and detection
I	Detection
J	Risk response number (RPN)
K	Recommended actions
L	Responsibility and target completion date
M	Actions taken
N	New severity
O	New occurrence
P	New detection

proactive program intended to catch all possible failures modes before they actually occur. It is solidly based on the understanding that the correction of potential failures before they actually occur will be much less costly than the required remedial actions after the fact.

The FMEA concept is rather straightforward. Table 5.5 includes the headings of an FMEA worksheet. For a given step number, the process description, potential failure mode, and potential effects of the failure in question are to be entered. Then, the severity factor (Row E) is assessed, using a number between 1 and 10 (10 being the most severe), as to impact on customers. Row F registers the potential root cause for the failure. Row G defines the probability of its occurrence, again using a number between 1 and 10, with 10 being the most likely to occur. Row H specifies the current control and detection practices. The detection factor (Row I) is assigned a value between 1 and 10, with 10 being the most difficult to detect.

The factor RPN (risk response number, Row J) is the product of three numbers, namely severity (Row E), occurrence (Row G), and detection (Row I). Remedial or preventive actions (Row K) are to be taken according to the priority order based on the RPN. Row L tabulates the person responsible for taking the recommended actions. Row M documents what actions were in fact taken.

Rows N, O, and P contain the results of reevaluation of the failure mode in question after remedial and preventive actions have been taken. The new RPN number is expected to be significantly lower than its corresponding preaction number, and it documents the corrective impact on the failure mode. Such documents may be used to satisfy the customer's requirements and serve as guidelines to drive continuous improvement.

Applying FMEA systematically should lead to a continuous reduction of the effects of failure modes in product design, product manufacturing, service, and other engineering or business applications. Many U.S. automotive companies and their tiered suppliers have applied FMEA with varying degrees of success. Although Ford initiated the use of FMEA back in 1977, it is still behind other industrial leaders today, in terms of problems per 100 vehicles, according to the 2004 study by J. D. Power Associates. This drives home the point that knowledge of quality is useful, but actual implementation is the key in achieving quality performance. Successful implementation requires management commitment and worker dedication. Four American automotive nameplates (Cadillac, Buick, Mercury, and Chevrolet) scored above the industrial average, compared with 11 foreign nameplates, as shown in the 2004 study of J. D. Power Associates. Technically, FMEA can be successfully implemented. The management challenge is how to achieve better quality performance consistently for all nameplates or products.

Like Kaizen and FMEA, total quality management (TQM) is also a very powerful program addressing the issues related to product quality and organizational productivity (Besterfield 1995; Ross 1999; Rampersad 2001; Goetsch and Davis 2001). In a typical academic course at the graduate level, TQM covers the concepts of (1) customer satisfaction, (2) empowerment of employees in problem solving, (3) continuous improvement, and (4) management excellence by creating and implementing corporate visions. To achieve TQM success, management commitment and worker dedication are also required.

Quality control is an important function in which engineering managers play a key role. They need to be actively and persistently involved with workers to apply common sense in eliminating wastes, speeding delivery, simplifying processes, paying attention to the gritty details of practice, and continuously improving the way work gets done.

Example 5.5.

What is the generic problem-solving approach applicable to solve most engineering and management problems?

Answer 5.5.

The generic problem-solving approach may look like the following:

1. Perform a situation analysis (e.g., assess strengths, weakness, opportunities, and threats).
2. Formulate a statement of the problem.
3. Define performance standards that are observable, measurable, and relevant to the goal.
4. Generate alternative solutions to the problem.
5. Evaluate these alternatives in terms of their consequences to the organization.
6. Select the best alternative solution with the most value and least adverse effect on the organization.
7. Implement a pilot test of the proposed solution and revise as indicated from practical experience.
8. Implement the solution.
9. Evaluate the outcome.
10. Revise the process as necessary.

Note that there is no one universal solution to all management problems, but that the correct solution will depend on the unique needs of the situation.

The problem-solving process practiced by Xerox Corporation is quite representative of the approaches taken by many companies in industry (Garvin 1993). The Xerox problem-solving process contains the details shown in Table 5.6.*

TABLE 5.6   Xerox's Problem-Solving Process

Step	Questions to Be Answered	Expansion/ Divergence	Contraction/ Convergence	What Is Needed to Go to the Next Step
1. Identify and select the problems.	What do we want to change?	Multiple problems for consideration.	One problem statement, one "desired state" agreed upon.	Identification of the gap and description of the "desired state" in observable terms.
2. Analyze problems.	What's preventing us from reaching the "desired state"?	Multiple potential causes identified.	Key cause(s) identified and verified.	Key cause(s) documented and ranked.
3. Generate potential solutions.	How could we make the change?	Multiple ideas on how to solve the problem.	Potential solutions clarified.	Solution list.
4. Select and plan the solution.	What's the best way to do it?	Multiple criteria for evaluating potential solutions. Multiple ideas on how to implement and evaluate the selected solutions.	Criteria to use for evaluating solutions agreed upon. Implementation and evaluation plans agreed upon.	Plan for making and monitoring the change. Measurement criteria to evaluate solution effectiveness.
5. Implement the solution.	Are we following the plan?		Implementation of agreed-upon contingency plans (if necessary).	Solution in place.
6. Evaluate the solution.	How well did it work?		Effectiveness of solution agreed upon. Continuing problems (if any) identified.	Verification that the problem is solved, or agreement to address continuing problems.

5.14 CONTROL OF KNOWLEDGE

Knowledge refers to the sum total of corporate intellectual properties that are composed of (1) patents, (2) proprietary know-how, (3) technical expertise, (4) design procedures,

*Reprinted by permission of *Harvard Business Review*. From "Building a Learning Organization" by D. A. Garvin, Vol. 71, No. 4, July–August, 1993. Copyright © 1993 by Harvard Business School Publishing Corporation, all rights reserved.

TABLE 5.7   Knowledge Management

1. Experimentation	Put systems and processes in place to facilitate the search and test of new knowledge.
2. Benchmarking	Learn from one's own experience, and best practices of others in industry, by reflection and analysis.
3. Preserve knowledge	Set policies concerning the preparation of reports, design procedures, and data books. Apply knowledge acquisition tools to preserve valuable heuristic knowledge.
4. Dissemination of knowledge	Rotate experts to different locations or jobs so that knowledge may be shared with and learned by others. Make knowledge or data available electronically companywide.

(5) empirical problem-solving heuristics, (6) process operational insights, and others (Gautschi 1999). Some of these knowledge chunks are documentable. Others typically reside in the employees' heads.

Engineering managers are responsible for developing, preserving, safeguarding, and applying corporate engineering and technology knowledge. They also need to draft policies to facilitate the control of knowledge. Some examples of knowledge management strategies are listed in Table 5.7.

One of the major problems in knowledge management is that many knowledge chunks are dispersed throughout various documents within the company. Data-mining software products represent new tools to help extract and group together related information from diverse sources for wide dissemination and effective utilization by others. Another major problem in knowledge management is that most experts do not like to share their knowledge with others, rendering the use of knowledge acquisition tools somewhat ineffective.

Knowledge control is particularly important from the viewpoint of countering industrial espionage. Special policies regulating employee contact with competitors at neutral sites (such as professional meetings, university environments, and industrial seminars) need to be defined to safeguard company knowledge.

Preserved or acquired knowledge adds little value to the company if there is no accompanying refinement in the way work gets done. Engineering managers also need to focus on effectively transferring the knowledge gained to cause a modification of the company's behavior that reflects the new knowledge and insights. Doing so will make the company a true learning organization that steadily enriches itself.

5.15 CONCLUSION

Control is another important function of engineering management, which focuses mostly on the administrative and operational aspects of the job.

Particular attention should be paid to external benchmarking when setting standards in order to avoid causing a company to lose its long-term competitive strengths. Controlling performance, especially correcting poor performance, could result in unpleasant circumstances for engineering managers.

5.16 REFERENCES

Anonymous. 2003A. *Best Practice: Ideas and Insights from World's Foremost Business Thinkers*. Cambridge, MA: Perseus Publishing.

Anonymous. 2003B. "Key Tips for Effective Time Management." *Logistics and Transport Focus*, January–February.

Besterfield, D. H. 1995. *Total Quality Management*. Englewood Cliffs, NJ: Prentice Hall.

Casavant, D. A. 2003. "Effective Time Management." *Buildings*, January.

Cottringer, W. 2003. "How to Save 12 Hours a Day." *SuperVision*, March.

Deming, W. E. 2000. *The New Economics: For Industry, Government, Education*. Cambridge, MA: MIT Press.

Dhillon, B. C. 1987. *Engineering Management: Concepts, Procedures, and Models*. Lancaster, PA: Technomic.

Evans, P. M. 2002. *Controlling People: How to Recognize, Understand, and Deal with People Who Try to Control You*. Avon, MA: Adams Media Corp.

Garvin, D. A. 1993. "Building a Learning Organization." *Harvard Business Review*, July–August.

Garvin, D. A. 1984. "What Does 'Product Quality' Really Mean?" *Sloan Management Review*, Vol. 26, No. 1, Fall.

Gautschi, T. 1999. "Knowledge as Advantage," *Design News*, Vol. 54, August 2, p.156.

Goetsch, D. L. and S. B. Davis. 2001. *Total Quality Handbook, Upper Saddle River*, NJ: Prentice Hall.

Imai, M. 1986. *Kaizen, The Key to Japanese Competitive Success*. New York: McGraw-Hill.

Ivancevich, J. M. and T. N. Duening. 2002. *Managing Einstein: Leading High-Tech Workers in the Digital Age*. New York: McGraw-Hill.

Katz, E., A. Light, and W. Thompson. 2003. *Controlling Technology: Contemporary Issues*. 2d ed. Prometheus Books.

Kuglin, F. A. with J. Hook. 2002. *Building, Leading and Managing Strategic Alliances: How to Work Effectively and Profitably with Partner Companies*. New York: AMACOM.

Laraia, A., P. E. Moody, and R. W. Hall. 1999. *The Kaizen Blitz: Accelerating Breakthroughs in Productivity and Performance*. New York: John Wiley.

Maynard, M. 2003. *The End of Detroit: How the Big Three Lost Their Grip on the American Car Market*. New York: Currency/Double.

McKenna, P. and D. H. Maister. 2002. *First among Equals: How to Manage a Group of Professionals*. New York: Free Press.

Oliver, L. 2000. *The Cost Management Toolbox: A Manager's Guide to Controlling Costs and Boosting Profits*. New York: AMACOM.

Rampersad, H. K. 2001. *Total Quality Management: An Executive Guide to Continuous Improvement*. New York: Springer.

Rodier, M. M. 2000. "A Quest for Best Practices." *IIE Solutions*, Vol. 32, February, p. 36.

Ross, J. E. 1999. *Total Quality Management: Text, Cases and Readings*. 3d ed. Boca Raton, FL: St. Lucie Press.

Shannon, R. E. 1980. *Engineering Management*. New York: John Wiley.

Slowinski, G. and M. W. Sagal. 2003. *The Strongest Link: Forging a Profitable and Enduring Corporate Alliance*. New York: AMACOM.

5.17 QUESTIONS

5.1 A number of years ago, ISO Standards 9000 series were developed to promote work quality by standardizing engineering design, testing, production, and other procedures. How many ISO standards are there and how well have these standards been accepted in the United States?

5.2   A company decides to offer an average annual raise of 8 percent, although the current inflation rate is 10 percent. Each engineering manager decides on the best way to distribute the salary increases to his or her staff. However, if everyone gets an increase of 8 percent, then there will be no differentiation between strong and weak performers for the previous year. What should you do as an engineering manager?

5.3   A key engineer in the department handed in her resignation notice; her reason for leaving was that she was offered a much higher salary from a competitor. The manager recommends to the director that the company match the competitor's offer, even though this would allow the engineer to earn above the maximum for her grade. "We can always give her smaller increases in subsequent years to bring her salary back into line," says the manager. What should the director do?

5.4   Motivation in the assembly shop is high. However, the shop manager notices that, although the daily production is above average for the shop, it drops down to a low during the first hour after the lunch break. It is further diagnosed that the operators tend to continue socializing until well after the lunch break.

The shop manager changes the lunch break and staggers it over a two-hour period so that the operators cannot go to lunch together. To his surprise, motivation begins to fall and productivity drops dramatically.

What do you think is the problem? How do you advise the shop manager to fix this problem?

5.5   Bill Carter is an excellent hardware designer, but he wants to move into management to broaden his experience. His manager is supportive and encourages Bill to go to evening classes on management. Bill works hard for two years and graduates first in his class. Soon there is a management opening in the procurement department. Bill applies for it, but an outside applicant eventually fills the position. Bill is turned down because he does not have sufficient experience in the procurement functions, a stated key requirement for the job.

Bill protests, "But I have better technical knowledge of components than all the procurement engineers put together, and I learned about procurement in my management courses. The only way I will get experience is to work on the job."

Do you think Bill should have been given the job? How would you have handled this situation differently?

5.6   The company president has noted a constant increase of reports passed on to her, many of them through the mail, some come through the company's Intranet. Most of these reports remain unread, although the company president when traveling for business does find time to browse through some of the reports while on the airplane. It has clearly become difficult for her to keep track of all projects due to information overload. However, she does not want to abandon her personal objective of being constantly informed, and she does not want to query her vice presidents for summaries of major developments.

What are the alternatives available to the company president?

5.7   The production department is undergoing an upgrade of its automation program. It has a conflict that needs to be controlled and managed. The line supervisor wants to standardize machines supplied by an American vendor, as his people will eventually use them. The automation team leader believes, on the other hand, that the Swiss-made machines would result in greater productivity. Being a specialist in automation, she was brought into the department to find ways of significantly improving productivity.

The department head does not know what to do, as these two experts frequently fight in staff meetings. He regrets that he has not kept up with the automation technology to enable him to arbitrate and decide on the best way.

What should the department head do now?

5.8 Two junior members of the production department unexpectedly come in to see you, the production director, to complain that their manager, who reports to you, commits discrimination, practices favoritism, and misuses company facilities for possible personal gain.

How would you handle this complaint?

5.9 At present, the company is running at full capacity in developing of a new product for a major customer. The sales director has unexpectedly secured a small, but highly profitable, order, which requires some low-level development work and a minor change to the current production process.

Should the company accept the small order? If so, how should the company satisfy the small order?

5.10 Some foreign countries, particularly those in the early stages of industrial development, are known to illegally copy product designs and technologies that originated from developed countries.

What are the best ways for small businesses to protect their technical know-how in foreign countries?

5.11 John Elrod founded the Elrod Manufacturing Company 50 years ago. Vernon Scott is the vice president of plant engineering, reporting to George Elrod, who took over as the company president from his father, John five years ago. Vernon Scott and John Elrod have been good friends for many years. The company's products include automotive parts (such as gears, axles, and transmissions), metal stampings, and sheet-metal subassemblies.

Also reporting to George Elrod are six plant managers. Each plant has its sales, engineering, manufacturing, warehouse, and other functions. The plant managers are responsible for the profitability of the individual plants. The total employment of the company is 12,000 people.

The company has a standing policy on capital expenditure: Expenditures below $5,000 are to be approved by plant managers, those between $5,000 and $50,000 by Vernon Scott, and those above $50,000 by the executive committee.

Vernon Scott favors expenditures for machinery and equipment directly related to manufacturing, but not for maintenance, facility expansion, and improvements that are not related to manufacturing. Over the years, plant managers have become unhappy with Scott's refusal of expenditures for nonmanufacturing equipment, such as computers. Forced to keep their plants profitable, they cannot help, but to bypass him by breaking down the nonmanufacturing projects into many small components, each below $5,000.

Eventually, Vernon Scott finds out about the piecemeal purchase of a $27,000 computer by Paul Nelson, a very capable plant manager whose plant has become the most profitable in the company. Scott demands that George Elrod fire Nelson, citing insubordination, cheating, and dishonesty.

What would you do if you were George Elrod?*

5.12 The manufacturing manager of Company X had installed a wage incentive program. She happily reported six months later that the system was a success because production was up and unit costs were down. A quality control manager said, however, that the percentage of rejects had increased markedly and that this was creating a backlog of rework requirements. An industrial engineering manager reported that the expenditures for industrial engineering studies in the department were up by 50 percent. An industrial relations manager said that arbitration fees resulting from incentive grievances had tripled.

Discuss your observations in this case. What should be done to correct the situation in this production department?

---

*Condensed and adopted from Robert E. Shannon. *Engineering Management*. New York: John Wiley, 1980, p. 296. This material is used by permission of John Wiley & Sons, Inc.

Business Fundamentals for Engineering Managers

Part II covers the fundamentals of business, including cost accounting (Chapter 6), financial accounting and analysis (Chapter 7), managerial finance (Chapter 8), and marketing management (Chapter 9). This part is to enable engineers and engineering managers to facilitate their interactions within peer groups and units and to acquire a broadened perspective of the company business and its stakeholders.

Part II also prepares engineering managers to make decisions related to cost, finance, product, service, and capital budgets—discounted cash flow and internal rate of return analyses are reviewed. These discussions are of critical importance, as decisions made during the product design phase typically determine up to 85 percent of the final costs of products. Activity based costing (ABC) is presented to define indirect costs related to products and services, and economic value added (EVA), which determines the real profitability of an enterprise above and beyond the cost of capital deployed, is addressed.

Also discussed are capital formation through equity and debt financing, resource allocation concepts based on adjusted present value (APV) for assets in place and option pricing for capital investment opportunities. By understanding the project evaluation criteria and the tools of financial analyses, engineers and engineering managers will be in a better position to secure project approvals. A critical step to developing technological projects is the acquisition and incorporation of customer feedback. For managers to lead, a major challenge is the initiation, development, and implementation of major technological projects that contribute to the long-term profitability of the company.

The important roles and responsibilities of marketing in any profit-seeking enterprise are then introduced, along with the contributions expected of engineering managers to support the marketing effort. Many progressive enterprises are increasingly concentrating on customer relationship management to grow their businesses. Such a customer orientation is expected to continue to serve as a key driving force for product design, project management, plant operation, manufacturing, customer service, and many other engineering-centered activities.

Chapter 6

Cost Accounting for Engineering Managers

6.1 INTRODUCTION

Cost control is a very important management function in both profit-seeking and non-profit organizations.

A profit-seeking organization strives to maximize its financial gains (for example, sales revenue minus costs) for its owners. These gains can be sustained over time only if all stakeholders of the firm (e.g., stockholders, customers, employees, suppliers, business partners, and the community in which the firm operates) are reasonably satisfied. A nonprofit organization (e.g., the United Way, the Ford Foundation, government agencies, educational institutions, church organizations, etc.) seeks to maximize its service value to its respective target recipients while minimizing operations costs.

This chapter covers the basics of cost accounting. The discussions focus on the costing of products, although all cost accounting concepts are equally applicable to the costing of services (Colander 2001; Bragg 2001). First, some commonly utilized accounting terms are introduced. Then, the cost analysis of a single period versus multiple periods is explained, leading to topics such as the time value of money and compound interest formulas. The costing of products follows, including the estimation of direct costs absorbed into the company's inventory. The complex problem of assigning indirect costs to products is illustrated by the conventional method of using overhead rates, as well as by the more sophisticated method of activity based costing.

Estimation of costs with uncertainties is then presented. The Monte Carlo simulation is introduced as an effective method to account for cost uncertainties. Its superiority over the conventional estimation method, which uses deterministic data, is clearly demonstrated through examples of the output distribution functions of the Monte Carlo simulation. Finally, inventory accounting is addressed to arrive at the all-important cost of goods sold (CGS).

It is important for all engineering managers to become well-versed in cost accounting. Cost control is a basic management task actively performed by numerous technology-based organizations, such as manufacturing, engineering, construction, product development, product design, technology applications, and services.

6.2 BASIC TERMS IN COST ACCOUNTING

Engineering managers need to become familiar with the standard vocabulary used by cost accountants or cost engineers, as costs are important bases for corporate performance evaluation, product cost estimation, profitability analysis, and managerial decision making. While the cost accounting systems used by various firms do not need to strictly follow the generally accepted accounting principles (GAPP) adopted by the financial accounting profession, engineering managers are still advised to understand the meaning of various accounting terms in order to ensure that their cost-based decisions are made properly. The following is a general set of accounting terms used by many firms (Bragg 2002; Rapier 1996):

A. **Cost center**—An organizational unit responsible for controlling costs related to its functional objectives (e.g., R&D, procurement, operations, engineering, design, and marketing).

B. **Inventory costs**—The total sum of product costs, which are composed of the direct costs and indirect costs related to the manufacturing of the products involved.

C. **Direct costs**—Materials and labor costs associated with the manufacturing of a product.

D. **Indirect costs**—All overhead costs (e.g., rent, procurement, depreciation, supervision, supplies, power, and others) indirectly associated with the production of products involved.

E. **Fixed costs**—Costs that do not strictly vary with the volume of products involved, such as the general manager's salary, rent for the facility, machine depreciation charges, and local taxes.

F. **Variable costs**—Costs that vary in proportion to the volume of products involved, including, for example, material, labor, and utilities.

G. **Step function costs**—Costs that would experience a step change when a specific volume range is exceeded; for example, the factory rent that may change stepwise if new floor space must be added because of the increased production volume.

H. **Contribution margin**—The product price minus unit variable cost; the economic value contributed by selling one unit of product to defray the fixed cost already committed for the current production facility.

I. **Cost pool**—An organizational unit where costs incurred by its activities performed for specific products (or other cost targets) are accumulated for subsequent assignments.

J. **Cost drivers**—Bases used to allocate indirect costs to products. Products drive the consumption of resources and the utilization of resources incurs costs. Examples include floor space, head counts, number of transactions, number of employees, labor hours, machine hours, number of setups, and material weight.

K. **Cost objects**—Targets to allocate indirect costs, such as products and services sold by the firm.

L. **Budget**—A quantitative expression in dollar value of a project or a plan of action. Examples include production budget, product design budget, engineering budget, R&D budget, sales budget, marketing budget, and advertising budget. Typically, budgets span a specific period of time (e.g., a month, a quarter, or a year).

M. **Standard costs**—Direct and indirect costs budgeted for products. The standard costs are defined by using estimations or historical costs.

N. **Variance**—The difference between standard costs and actual costs. Such variance could be the result of price variation, quantity change, technology advancement, and other factors. Conventionally, actual quantities are used when computing price variation to easily assess the procurement performance. On the other hand, the quantity-based variance is computed by using standard costs for an easy assessment of the production performance.

O. **Current costs**—Costs for the total efforts (e.g., physical efforts, raw materials, and service fees) that must be spent in order to carry out an activity or implement a plan. Current costs form a key basis for managerial decision making.

P. **Opportunity costs**—The benefit of the second-best alternative that must be forgone because of a commitment made to the first alternative. For example, an engineering manager who quits a job paying $100,000 a year to pursue a three-semester MBA degree at a university incurs an opportunity cost at graduation of $150,000 plus an out-of-pocket cost of $90,000 for tuition fees. Opportunity costs are included in managerial decision making, but are not included in a cost accounting system.

Q. **Sunk costs**—Costs that have already been spent or incurred. Such costs are typically included in all cost accounting systems, but they are not considered in any management decision-making for the future.

6.3 COST ANALYSIS

Managers perform variance analyses and study the reasons for the deviation of actual costs from standard costs. They issue periodic and systematic reports of their findings and take proper actions to improve the efficiency and effectiveness of the organizational units (Sullivan, Wicks, and Luxhoj 2003).

The two major factors affecting cost analysis are time and accuracy. For management decisions, cost analyses may be performed for a single time period or for multiple periods. Cost data may vary or may be uncertain.

6.3.1 Single-Period Analysis

Single-period analysis applies primarily to a short period of time during which the costs involved remain essentially constant. The gross profit equation for a given product line is given by the following equation:

$$\text{Gross Profit} = \text{Revenue} - \text{Costs}$$
$$GP = P \times N - (FC + VC \times N) \qquad (6.1)$$

where

$$P = \text{Product price (dollars/unit)}$$
$$N = \text{Number of products sold during the period}$$
$$\text{FC} = \text{Fixed costs (dollars)}$$
$$\text{VC} = \text{Variable costs (dollars/unit)}$$
$$\text{GP} = \text{Gross profit (dollars)}$$

For the case of breakeven (i.e., GP = 0), the break-even product quantity is given by

$$n^* = \text{FC}/(P - \text{VC}) \tag{6.2}$$

The value $(P - \text{VC})$ is defined as the contribution margin of the product. Selling each additional unit of a product generates a contribution in the amount $(P - \text{VC})$ to defray the fixed cost (FC) that has been committed to the production line.

Organizational performance can be readily assessed as the number of cost items involved is limited. One needs to make sure that the values of these cost items are valid, although from time to time the validity of such values may be tough to verify precisely, because of joint production activities and other cost-sharing systems involved.

6.3.2 Multiple-Period Analyses

The cost analyses over a longer period of time (e.g., multiple periods) are much more difficult to calculate for two reasons. First, costs may change predictably over time due to inflation, investment return, cost of capital, and other reasons. Second, future events are unpredictable (e.g., natural disasters, labor unrest, political instability, war against terrorism, spread of disease, investment climate, etc.)(Bruns 1999).

The change of costs over time needs to be addressed by using concepts such as net present value (NPV) and internal rate of return (IRR). These concepts are built on the fundamentals of the time value of money, compound interest, and the cost of capital. These topics are introduced in Section 6.4. Depreciation accounting, an important part of the indirect costs of products, is included in Section 6.5.

In dealing with the uncertainties of future costs, risks must be included in product cost analysis. Risk analysis will be elucidated in detail in Section 6.9.

6.4 TIME VALUE OF MONEY AND COMPOUND INTEREST EQUATIONS

The time value of money refers to the notion that the value of money changes with time. This is because money at hand may lose value (purchasing power) if not invested properly. Money at hand may earn income through investment. A dollar that is to be received at a future date is not worth as much as a dollar that is on hand at the present. Thus, two equal dollar amounts at different points in time do not have equal value (purchasing power).

Before introducing basic compound interest equations useful for multiperiod cost analyses, a few definitions are reviewed next.

A. **Interest**—It represents a fraction of the principal designated as a reward (interest income) to its owner for having given up the right to use the principal. It may also be a charge (interest payment) to the borrower for having received the right to use the principal during a given interest period.

B. **Compound interest**—When the interest income earned in one interest period is added to the principal, the principal becomes larger for the next period. The enlarged principal earns additional interest under such circumstances. The interest is said to have been compounding.

C. **Nominal interest rate**—The interest rate quoted by banks or other lenders on an annual basis, also called the annual percentage rate (APR).

D. **Effective interest rate**—The interest rate in effect for a given interest period (e.g., one month). For example, if the nominal interest rate for a bank loan is 12 percent, then its effective interest rate for each month is 1 percent (= 12% / 12).

E. **Nominal dollar**—The actual dollar value at a given point in time.

F. **Constant dollar**—The dollar value that has a constant purchasing power with respect to a given base year (e.g., the reference year 1995); the value is adjusted for inflation.

G. **Consumer price index**—The index tracked by the U.S. Department of Commerce to indicate the price change for a basket of consumer products. (See Figure 6.1.) Since 1993, the inflation rate in the United State has been relatively low.

To introduce the compound interest formulas for multiple period cost analyses, the following notations are used:

P = Present Value (dollar), the value of a project, loan, or financial activity at the present time

F = Future Value (dollar), the value of a project, loan, or financial activity at a future point in time

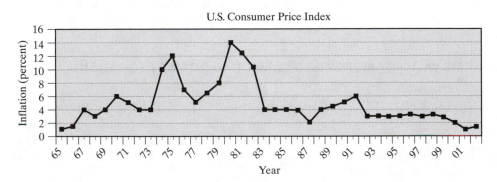

Figure 6.1 U.S. consumer price index. *Source:* U.S. Department of Commerce.

i = Effective interest rate for a given period during which time the interest is to be compounded (e.g., 1 percent per month)

A = Annuity (dollar), a series of payments made or received at the end of each interest period

n = Number of interest periods

6.4.1 Single Payment Compound Amount Factor

$$F = P \times (1 + i)^n \tag{6.3}$$

$$\frac{F}{P} = (1 + i)^n = \left(\frac{F}{P}, i, n\right)$$

Equation 6.3 defines the total value of an investment P, with periodical returns added to the principals to earn more money at the end of n periods. Its derivation is shown in Appendix 6.A.

Example 6.1.

Mr. Jones invests $5,000 at 8.6 percent interest compounded semiannually. What will be the approximate value of his investment at the end of 10 years?

Answer 6.1.

$$F = P(1 + i)^n = 5000\left(1 + \frac{0.086}{2}\right)^{20} = \$11,605.29$$

6.4.2 Present Worth Factor

$$P = \frac{F}{(1 + i)^n} \tag{6.4}$$

$$\frac{P}{F} = (1 + i)^{-n} = \left(\frac{P}{F}, i, n\right)$$

Equation 6.4 defines the present value of a sum that will be available in the future. The factor $\dfrac{1}{(1 + i)^n}$ is also called the discount factor.

6.4.3 Uniform Series Compound Amount Factor

$$F = A \times \frac{[(1 + i)^n - 1]}{i} \tag{6.5}$$

$$\frac{F}{A} = \frac{(1 + i)^n - 1}{i} = \left(\frac{F}{A}, i, n\right)$$

Equation 6.5 determines the total future value of an account (e.g., retirement, college education, etc.) at the end of n periods, if a known annuity A is deposited into the account at the end of every period. Its derivation is shown in Appendix 6.B.

6.4.4 Uniform Series Sinking Fund Factor

$$A = F \times \frac{i}{(1 + i)^n - 1} \qquad (6.6)$$

$$\frac{A}{F} = \frac{i}{(1 + i)^n - 1} = \left(\frac{A}{F}, i, n\right)$$

Equation 6.6 calculates the amount of the required annuity (e.g., a series of period-end payments) that must be periodically deposited into an account in order to reach a desired total future sum F at the end of n periods.

6.4.5 Uniform Series Capital Recovery Factor

$$A = P\frac{i \times (1 + i)^n}{(1 + i)^n - 1} \qquad (6.7)$$

$$\frac{A}{P} = \frac{i \times (1 + i)^n}{(1 + i)^n - 1} = \left(\frac{A}{P}, i, n\right)$$

Equation 6.7 defines the amount of periodical withdrawal that can be made over n periods from an account worth P at the present time, such that the account will be completely depleted at the end of n periods.

Example 6.2.

Mr. Jones wishes to establish a fund for his newborn child's college education. The fund pays $60,000 on the child's 18th, 19th, 20th, and 21st birthdays. The fund will be set up by the deposit of a fixed sum on the child's 1st through 17th birthdays. The fund earns 6 percent annual interest. What is the required annual deposit?

Answer 6.2.

The future sum of a series of annual deposits is

$$F_1 = A_1 \frac{1.060^{17} - 1}{0.06} = 28.21288 A_1$$

Annual withdrawal when the child enters college is

$$A_2 = P_2 \frac{0.06(1.06)^4}{(1.06)^4 - 1} = 0.2885915 P_2$$

$$A_2 = 60,000$$

$$P_2 = F_1$$

Answer: $A_1 = \$7369.20$ (the required annual deposit).

6.4.6 Uniform Series Present Worth Factor

$$P = A\frac{(1 + i)^n - 1}{i \times (1 + i)^n}$$

(6.8)

$$\frac{P}{A} = \frac{(1 + i)^n - 1}{i \times (1 + i)^n} = \left(\frac{P}{A}, i, n\right)$$

Equation 6.8 determines the total present value of an account to which an annuity A is deposited at the end of each period. For example, if A is the periodical maintenance costs for capital equipment, then this equation calculates the present value of all maintenance costs over its product life of n periods.

Example 6.3.

The annual maintenance on the parking lot is $5,000. What expenditure would be justified for resurfacing if no maintenance is required for the first 5 years, $2,000 per year for the next 10 years, and $5,000 a year thereafter? Assume the cost of money is 6 percent.

Answer 6.3.

Since the annual maintenance cost is the same after 15 years, the effect of resurfacing applies to the first 15 years only. The total present value of the "doing nothing" option is

$$P_1 = A\frac{(1 + i)^n - 1}{i(1 + i)^n}$$

$$= 5000\frac{1.06^{15} - 1}{0.06(1.06^{15})} = \$48,561.25$$

The total present value of the resurfacing option is

$$P_2 = P + 2000\frac{1.06^{10} - 1}{0.06 \times 1.06^{10}(1.06)^5}$$

$$= P + \$10,999.77$$

Setting $P_1 = P_2$:

$$P = \$37,561.47 \text{ (the maximum amount for resurfacing the parking lot)}$$

Example 6.4.

You need a new Pentium IV computer system and find one on sale for $3,900 cash, or $500 down and $200 monthly payments for 2 years. What nominal annual rate of return does this time-payment plan represent?

Answer 6.4.

$$\text{Loan amount} = \$3,900 - 500 = \$3,400 = P$$

$$n = 24 \text{ months}$$

$$A = \$200$$

$$P = A\frac{(1 + i)^n - 1}{i(1 + i)^n}$$

$$3400 = 200\frac{(1 + i)^{24} - 1}{i(1 + i)^{24}}$$

This is an implicit equation for the single unknown i. It may be solved by trial and error: Input a trial value of i, and compute residual $\Delta = \text{RHS} - \text{LHS}$. The i value that produces zero Δ (delta) is the answer:

Trial value of i	Δ (delta)
0.06	4.496
0.03	0.0645
0.025	−0.884985
0.029	−0.1194
0.02965	+0.0004288

Answer: Monthly rate = 2.965 percent, and the nominal annual rate is 35.58 percent.

For all multiple period problems, the timeline convention is regarded as standard. (See Figure 6.2.)

When applying these compound interest formulas, the following guidelines should be kept in mind:

(a) **P** is at present, **F** is at a future point in time, and **A** occurs at the end of each period.

(b) The periods must be consecutively and sequentially linked with the end of one as the beginning of the next.

(c) Complex problems may be broken down into time segments so that the equations may be correctly applied to each of the segments.

6.5 DEPRECIATION ACCOUNTING

In calculating indirect costs associated with production facility, equipment, and other tangible assets related to production, depreciation charges must be included. Depreciation is

Figure 6.2 Timeline convention.

a cost-allocation procedure whereby the cost of a long-lived asset is recognized in each accounting period over the asset's useful life in proportion to its benefit brought forth over the same period. This procedure is undertaken in a reasonable and orderly fashion. Specifically, the acquisition cost of an asset can be considered as the price paid for a series of future benefits. As the asset is partially used up in each accounting period, a corresponding portion of the original investment in the asset is treated as the cost incurred for the partial benefit delivered.

The U.S. Internal Revenue Service accepts three depreciation accounting methods. They are discussed next, using the following notations:

P = initial investment (dollars) at the present time

N = useful life of a long-lived asset measured in years
(e.g., $N = 25$ for buildings, $N = 15$ for equipment, $N = 5$ for automobiles, $N = 3$ for computers, etc.)

$D(m)$ = depreciation charge (dollars) in the asset's m(th) year

L = salvage value (dollars) recoverable at the
end of the equipment's useful life

$AD(m)$ = accumulated depreciation (dollars), which is the total amount
of depreciation charges accumulated at the end of the m(th) year

$BV(m)$ = book value (dollars) of an asset in its m(th) year; $BV(m) = P - AD(m)$

$P - L$ = depreciable base (dollars)

$r(m)$ = depreciation rate, a fraction of the depreciable
base to be depreciated per year

6.5.1 Straight Line

By this depreciation method, an equal portion of depreciation base $(P - L)$ is designated as the depreciation charge for each period of the assets' estimated useful life:

$$D(m) = \frac{P - L}{N} = \text{Constant}$$

$$BV(m) = P - m \times \frac{P - L}{N}; m = 1, 2, 3, \ldots$$ (6.9)

$$r(m) = \frac{1}{N} = \text{constant}$$

$$AD(m) = m \times \frac{P - L}{N}$$

More than 91 percent of publicly traded companies in the United States use this straight-line depreciation method.

6.5.2 Declining Balance

The depreciation charge is set to equal to the net book value (e.g., acquisition cost minus accumulated depreciation) at the beginning of each period (e.g., year) multiplied by a fixed percentage. If this percentage is two times the straight-line depreciation percentage, then it is called a double-declining balance method:

$$D(m) = P \times r \times (1 - r)^{(m-1)} \tag{6.10}$$

$$BV(m) = P \times (1 - r)^m$$

$$r(m) = \text{constant}; r = \frac{2}{N} \text{ (double-declining balance method)}$$

$$AD(m) = P \times [1 - (1 - r)^m]$$

Note that the salvage value is not subtracted from the acquisition cost. To make sure that the total accumulated depreciation does not exceed the depreciation base $(P - L)$, the depreciation charge of the very last period (e.g., year) must be manually adjusted.

6.5.3 Units of Production Method

The depreciation charge is assumed to be proportional to the service performed (e.g., units produced, hours consumed, etc.). Companies that are involved with natural resources (e.g., oil and gas exploration) use units of production method to depreciate their production assets. Software companies also use this method to depreciate their capitalized software development costs.

Example 6.5.

The company plans to change its depreciation accounting from the straight-line method to the double-declining method on a class of assets that have a first cost (acquisition cost) of $80,000, an expected life of six years, and no salvage value. If the company's tax rate is 50 percent, what is the present value of this change, assuming 10 percent interest compounded annually?

Answer 6.5.

$$P = 80,000; N = 6, t = 0.5, L = 0$$

$$F = (1 - t)\Delta; P = \frac{F}{(1 + i)^n}, i = 10\%$$

TABLE 6.1 Calculation of Difference Due to Depreciation Methods

Year	SL	Double Declining $\left(r = 2\frac{1}{6} = 0.333333\right)$	Δ (Delta)	$F = (1 - t)\Delta$	Present Worth
1	13333.33	26666.67	13333.34	6666.67	6060.61
2	13333.33	17778.66	4445.33	2222.66	1836.91
3	13333.33	11851.87	−1481.46	−740.73	−556.52
4	13333.33	7901.25	−5432.08	−2716.04	−1855.09
5	13333.33	5267.50	−8065.83	−4032.92	−2504.12
6	13333.33	10534.05	−2799.28	−1399.64	−790.06
			TOTAL		2191.72

Table 6.1 shows the present values of the differences between these two depreciation charges for the assets' expected life of 6 years.

Answer = $2,191.72

Example 6.6.

A new delivery truck costs $40,000 and is to be operated approximately the same amount each year. If annual maintenance costs are $1,000 the first year and increase $1,000 each succeeding year and, if the truck trade-in value is $24,000 the first year and decreases uniformly by $3,000 each year thereafter, at the end of which year will the costs per year of ownership and maintenance be a minimum?

Answer 6.6.

The average annual ownership cost is calculated as shown in Table 6.2.

TABLE 6.2   Annual Ownership Cost Computation

Year	Annual Maintenance Cost	Trade-In Value	Accumulated Depreciation	Total Accumulated Cost of Ownership	Average Annual Cost over the Ownership Period
1	1,000	24,000	16,000	17,000	17,000
2	2,000	21,000	19,000	22,000	11,000
3	3,000	18,000	22,000	28,000	9,333
4	4,000	15,000	25,000	35,000	8,750
5	5,000	12,000	28,000	43,000	8,600
6	6,000	9,000	31,000	52,000	8,667

The numbers in the 6th column are produced by dividing the numbers in the 5th column by the ownership duration in years. Answer: 5th year.

Example 6.7.

Ceramic hot-gas filters provide 2400 hours of service life. Three of the processes in a refinery are each equipped with one of these filter sets. Each filter set has an initial cost of $1,200. When production is scheduled, each process runs 24 hours per day.

TABLE 6.3   Depreciation Based on Usage

	A	B	C
Cost (dollars)	1,200	1,200	1,200
Weeks in Operation	10	10	13
Hours in Operation	1,680	1,680	2,184
Percentage of Useful Life	70	70	91
Depreciation Charge (dollars)	840	840	1,092

Total = $2,772 (zero-salvage value assumed)

In the first quarter of the year, process A did not start operating until the beginning of the 4th week. Process B terminated at the end of the 10th week. Process C was on stream for the entire period. What depreciation charge should be allocated for ceramic filters during the first quarter?

Answer 6.7.

The proper method of calculating the depreciation charge is on the basis of usage, as shown in Table 6.3.

6.6 PRODUCT OR SERVICE COSTING

Product or service costing is one of the key responsibilities of engineering managers, as the proper computation of the cost of goods sold (CGS) is of paramount importance to a product- or service-centered company.

6.6.1 Cost of Goods Sold

Product costing in a product-centered company requires the computation of costs related to direct material (DM), direct labor (DL), and factory overhead (FO) for the following three operations:

- Raw materials (Stores)
- Work in progress (WIP)
- Finished goods (FG)

Additional Definitions:

Prime cost = DM + DL

Conversion cost = DL + FO

Indirect manufacturing cost = factory overhead (FO), also called factory burden. FO includes all costs other than DM and DL.

Typically, T-accounts are set up for Stores operations, WIP operations, and FG operations.

A T-account is a tool used commonly by accountants to record transactions. As a convention, an increase in assets (e.g., cash, accounts receivables, inventories, land, machines, etc.) is debited to the left side of the T-account. A decrease in assets is credited to the right side of the T-account. The opposite is true for liabilities and owners' equities. For additional discussions on the use of T-accounts, see Section 7.3.

Here's a schematic diagram that illustrates product costing:

Stores		WIP		FG	
Debit	Credit	Debit	Credit	Debit	Credit
(1)	(2)	(3)	(4)	(5)	(6)

Explanations: (1) Purchasing raw materials (an increase in asset)

(2) Putting materials into a production process (a reduction in stores assets)

(3) Production is initiated, adding value to raw materials, while consuming labor, materials, utilities, and other resources (assets are increased)

(4) Production is complete and units shipped out to finished goods operations (a reduction of WIP assets)

(5) Receiving of finished goods in storage (an increase of FG assets)

(6) Finished goods are shipped out for sale (a reduction of FG assets)

Note that in the schematic diagram shown above (a reduction of WIP assets), materials flow from left to right, whereas cash (financial resource) flows from right to left.

Here's a specific example of product costing:

Stores		WIP		FG	
Debit	**Credit**	**Debit**	**Credit**	**Debit**	**Credit**
Beg. 75	3 (b)	Beg. 22	440(f)	Beg. 50	
198(a)	187(c)	125(d)		440(f)	430(g)
		147(e)			
End. 83(h)		187(c)		End. 60(h)	
		End. 41(h)			

(a) Purchased material for $198.00

(b) Received credit for having returned the material purchased

(c) Direct material actually shipped to WIP and used in the accounting period

(d) Direct labor used and cost assigned

(e) Factory overhead used and cost assigned

(f) Product completed and transferred out (cost of goods manufactured)

(g) Finished goods shipped out to customer, receiving the cost of goods sold (CGS) as credit

(h) The sum of ending balances in Stores, WIP, and FG represents inventory at the end of the accounting period

In computing the costs of goods sold, the inclusion of materials and labor costs is rather straightforward, as these are direct costs and quite easy to track. The difficulty in product costing is the inclusion of indirect costs, namely the factory overhead, as in the specific example just illustrated. For manufacturing operations in which the factory overhead represents a small fraction of the total product cost, the precision with which to allocate the indirect costs is irrelevant. However, in other circumstances, the allocation of indirect costs must be precise, especially when the factory overhead becomes a major portion of the product cost and the plant facility generates multiple products.

6.6.2 Traditional Method of Allocating Indirect Costs

The traditional practice of general ledger costing involves estimating all overhead costs for the upcoming year in a single cost pool (e.g., factory overhead, utilities, safety program, training, and salaries of foremen and factory managers). This total is then divided by the estimated number of labor hours to be worked. The result is an hourly overhead rate. For each product, the required labor hours are estimated. The total overhead cost for the product is then equal to the required labor hours multiplied by the hourly overhead rate.

For example, let us assume that a factory has $800,000 in overhead (e.g., salary of manager, benefits, and other general charges), 2000 direct hours per employee per year, and 20 employees. The hourly overhead rate is then $20 per hour $(800,000/(2000 \times 20))$.

The major deficiency of this method of allocating indirect costs is that it does not reflect the true relationship between the indirect costs and the cost object (such as products or services). Therefore, the allocation is often improper, as diverse types of overhead costs are lumped together, making an in-depth analysis impossible (Cooper and Kaplan 1988).

A better method of allocating indirect costs is activity-based costing, which is introduced in the next section.

6.7 ACTIVITY-BASED COSTING (ABC)

Activity-based costing (ABC) is a cost accounting technique by which indirect and administrative support costs are traced to activities and processes and then to the cost objects (e.g., products, services, and customers).

ABC is built on the notion that an organization has to perform certain activities in order to generate products and services. These activities cost money. The cost of each of these activities is only measured by and assigned to those products or services requiring identifiable activities and using appropriate assignment bases (called cost drivers). The results of ABC analyses offer an accurate picture of the real cost of each product or service, including the cost of serving customers. Nonactivity costs (such as direct materials, direct labor, or direct outside services) do not need to be included because these costs are readily attributable to the specific product or service considered.

ABC is most useful for companies with diverse products, service centers, channels, and customers, and for those companies whose overhead costs represent a large percentage of their overall product and service costs (Cokins 2001; Hicks 1999; O'Guin 1991).

All engineering managers should learn to practice ABC, because the traditional method of allocating overhead uses only high-level information about costs, and the general ledger system does not provide information related to time and resources spent on assignments and activities. In contrast, a well-practiced ABC method offers specific insights that include (a) a clearer picture for management of what generates profits and losses for the company, (b) the ability to track operating profits for specific cost objects (such as customers, orders, and products), (c) the ability to determine whether a service center is efficient or deficient, and (d) the possibility to externalize the relative profitability among products and customers.

Even a company with an overall profitability may lose money on certain products, orders, and customers in the absence of ABC. According to the published best practices of some industrial pioneers (such as Honeywell Inc. and Coca-Cola®) on the use of activity-based costing, simpler ABC models deliver better results.

ABC has become increasingly popular with industrial companies, partly because it is useful for organizations of any size and does not require a massive effort to implement, and partly because of increased processing capabilities of personal computers (PCs), reduced prices for ABC software products, and increased competition forcing companies to achieve a better understanding of their own product costs. There are several ABC products on the market. Examples include Oros®, EasyABC Plus®, EasyABC Quick®, Metify ABM, and others.

6.7.1 Sequential Steps to the Implementation of ABC

It is advisable to form a cross-functional team when implementing the ABC method of allocating indirect costs. The team should determine the cost objects. Examples include costs to serve customer; costs to purchase, carry, and process products; costs to order, receive, sell, and deliver products; and costs to perform other activities.

The team then needs to define activities that represent homogenous groups of work (such as accounting, machining, forging, and design) that lead to the cost objects.

Next to be determined are cost drivers. These are the agents that cause costs to be incurred in the activities. Cost drivers are factors that directly impact the cost of a given cost object. Examples of cost drivers are shown in Table 6.4.

The next step is to attribute activity costs to cost objects. The use of a flow chart for modeling the process is recommended. The resulting information generated by ABC is useful for management decisions.

6.7.2 Practical Tips for Performing ABC

When a company initiates the process of creating an ABC system, it is highly recommended that the company start with a small group (pilot group) of well-informed and cross-functional workers. The team should interview other workers about what they do in their jobs. The team members should be cognizant of the potential fears of job restructuring that some employees may have as a result of the ABC studies.

The team should start with the "worst" department so that immediate success may be used to get faster "buy-in" from top management. The key for ABC success is to use "close-enough" data. The team should keep the level of information manageable by avoiding being bogged down with minute details. On the other hand, an ABC system that is too broad and general will not be useful. The team may have to try out ABC cost models of different granularities on small scales. For companies attempting to employ ABC cost models for the first time, useful outputs can be expected in 6 to 12 months.

TABLE 6.4   Cost Drivers

Activity	Cost Driver
Loading	Tons
Driving	Miles
Invoice processing	Number of invoices
Machining	Machining hours
Material movement	Weight
Production	Number of products

6.7.3 Application—XYZ Manufacturing Company

XYZ is a small manufacturing company with $10 million in annual sales. It makes components for the automotive industry, and the key processes involved are forging and machining. The product-related operating activities are as follows:

A. Buying steel bars from outside vendors.
B. Testing steel bars upon delivery and moving them to storage.
C. Sending the bars to the forging area when needed for an order, at which point they are sandblasted and cut to desired lengths. Since most of the bars are large, they are then moved in bins that hold 20 to 25 pieces.
D. Sizing the bars and moving them to a forging operation where they are shaped. The bars are then moved to the in-process storage. In some cases, a steel bar may need to be forged up to three times.
E. Transferring the bars for each forging procedure from in-process storage to the forging areas and then back to the in-process storage.
F. Moving the steel bars after the final forging from the in-process storage to the machining area where they are finished. The bars are then sent to finished-goods storage.
G. Sorting, packing, and loading the bars is done in the shipping area and onto trucks for delivery to customers.

Before using ABC, the company applied the traditional costing method that included the following steps:

1. Assign manufacturing costs to products by using a plantwide costing rate on the basis of direct labor. The setup costs are included in the manufacturing overhead.
2. Assign the nonmanufacturing costs to products via a general and administrative (G&A) rate that is calculated as a percentage of the total cost.
3. Define the direct labor rate and the G&A rate on the basis of the actual results obtained for the preceding year.

The deficiencies of the traditional method are obvious. The traditional method is used because management does not know better methods.

When implementing ABC, the company did not buy any ABC-specific software. Instead, it used a standard spreadsheet program. Specifically, the company considered the following:

1. **Setup costs.** Management assigned equipment setup costs only to the steel bars in a given equipment process.
2. **Forging costs.** Depending on the weight of the steel bar involved, one or two operators may operate the forging press. Prior to forging, each steel bar must be induction heated, with the heating cost being dependent on the mass of the bar involved. Thus, the forging cost consists of three parts:

    (a) Press-operating costs on the basis of press hours
    (b) Production labor cost on the basis of labor hours
    (c) Induction-heating costs on the basis of the steel bar's weight

3. **Machining costs.** The machining centers do not require full-time operators. Once the machines are set up, workers load and unload parts for multiple centers. On average, 1 machine-worker hour is required for every $2\frac{1}{2}$ machine hours. Thus, the costs of machine-shop workers are treated as the indirect costs assigned to products on the basis of machine hours.

4. **Material movement costs.** Depending on the size of the bar, the bin size, and the required forging and machining steps, the material movement cost could vary significantly from one bar to another. Thus, the material movement cost is assigned to each bar on the cost-per-move basis.

5. **Raw material procurement** and **order processing costs.** These are readily traceable on the basis of records on hand.

The ABC cost model for the XYZ Manufacturing Company is illustrated in Table 6.5. The final results of ABC implementation are impressive. The company's sales tripled and its profit increased fivefold after the introduction of ABC. Much of this improvement came from a more profitable mix of contracts generated by a pricing and quoting process that more closely reflects the actual cost structure of the company. Particularly useful are the isolation and measurement of material movement costs that result in operational changes for increased efficiency.

TABLE 6.5   ABC Model for XYZ Manufacturing Company

	Forging-Press Hour Cost	Machine-Hour Cost	Induction-Heating Cost	Material-Movement Cost
Directly Attributable Costs	Depreciation	Depreciation	Depreciation	Depreciation
	Utilities, manufacturing supplies, outside repairs	Utilities, manufacturing supplies, outside repairs, straight-line wages, fringe benefits, payroll taxes, overtime premium, shift premium	Utilities, manufacturing supplies, outside repairs	Utilities, manufacturing supplies, outside repairs, straight-line wages, fringe benefits, payroll taxes, overtime premium, equipment leases
Distributions	Maintenance, buildings and grounds, manufacturing, engineering, commodity overhead	Maintenance, buildings and grounds, manufacturing, engineering, commodity overhead, supervision	Maintenance, buildings and grounds, manufacturing, engineering, commodity overhead	Maintenance, buildings and grounds, human resources, supervision
Total	Total costs	Total costs	Total costs	Total costs
Rate	Dollars per press hour	Dollars per machine hour	Dollars per heating hour	Dollars per move

6.8 INVENTORY ACCOUNTING

After the direct and indirect costs are estimated, the product costs can be defined. When products are transferred from work in progress operations to a finished goods warehouse, they become inventory. Inventory may be managed by one of two methods: first in and first out (FIFO) and last in and first out (LIFO). The FIFO method specifies that inventory that enters the warehouse first will leave the warehouse first. By the LIFO method, the inventory that enters the warehouse last is shipped out first.

According to the time value of money concept, these two inventory operational methods may render different costs of goods sold (CGS). The inventory accounting takes into account such possible change of product cost over time, due, for example, to inflation. In general, companies utilize one of the following three inventory accounting methods:

A. FIFO (first in and first out)

B. LIFO (last in and first out)

C. Weighted average

LIFO is most useful during periods of high inflation, as it results in less reportable earnings with lower taxes paid; LIFO is not useful, however, when prices for raw materials decrease. LIFO also provides lower inventory value, thus understating the value of the inventory in the balance sheet. Finally, LIFO is a more conservative accounting technique than FIFO. Note that LIFO is prohibited by law in some countries, such as the United Kingdom, France, and Australia.

As a product of creative accounting, FIFO defines an inventory value more closely matched with its market value. It tends to make the income statement look better than it really is. In periods when the business climate experiences stagnation or recession, innumerable companies frequently switch from LIFO to FIFO. The weighted average method represents a compromise between the two.

Table 6.6 is an illustration of the use of FIFO and LIFO accounting techniques. Assume that a manufacturing company has five units of products in inventory and each has a product cost of $100. Furthermore, the company produces five more units at

TABLE 6.6   FIFO and LIFO Inventory Accounting

	FIFO (dollars)	LIFO (dollars)	Weighted average (dollars)	
(1) Beginning Inventory 5 × 100	500	500	500	Withdrawal of 10 Units
				=================
(2) Purchase and Value Added 5 × 200	1,000	1,000	1,000	5 × 100 / 5 × 200
				5 × 300
5 × 300	1,500	1,500	1,500	=================
(3) Ending Inventory 5 ×	1,500	500	1,000	
(4) Cost of Goods Sold	1,500	2,500	2,000	

TABLE 6.7   Effect of Inventory Accounting on Net Income

	FIFO (dollars)	LIFO (dollars)	Weighted average (dollars)
Sales	10,000	10,000	10,000
CGS	1,500	2,500	2,000
Gross Margin	8,500	7,500	8,000
GS&A	2,000	2,000	2,000
Ebit	6,500	5,500	6,000
Interest	500	500	500
Taxable Amount	6,000	5,000	5,500
TAX (40%)	2,400	2,000	2,200
Net Income	**3,600**	**3,000**	**3,300**

$200 each in one period and then another five units at $300 each in a later period. During these periods, the company sells 10 units to customers. Determine the average cost of goods sold on the basis of both FIFO and LIFO and assess its impact on the company's net income.

The impact of inventory accounting on net income is quite direct, as illustrated in Table 6.7. Table 6.7 is an abbreviated income statement (see Section 7.4.1) wherein CGS is costs of goods sold, GS&A is general, sales, and administration expenses, and Ebit is earnings before interests and taxes. Upon switching from FIFO to LIFO inventory accounting, the tax liabilities are shown to have been reduced form $2.4 million to $2.0 million.

6.9 RISK ANALYSIS AND COST ESTIMATION UNDER UNCERTAINTY

When estimating product costs, some costs are well-defined and firm, while others are not. Similarly, some projects in engineering based on past experience are risk free, while others are not. Risks are defined as a measure of the potential variability of an outcome (e.g., cost or schedule) from its expected value. Risks must be properly accounted for in projects (Byrd and Cothern 2000; Bedford and Cook 2001).

6.9.1 Representation of Risks

Risks can be graphically represented by a probability distribution function. Three cases are examined next.

Figure 6.3 shows that the yield of 10-year U.S. Treasury bills has varied in the range of 8 percent (1990) to below 4 percent (2002). But once an investor purchases the Treasury bills from the U.S. Government, the yield is locked in until maturity. Such an investment is guaranteed by the assets of the U.S. Government. This return is graphically represented by a vertical line at a fixed return rate (6 percent assumed in this example) with 100 percent probability. (See Figure 6.4, Case 1.)

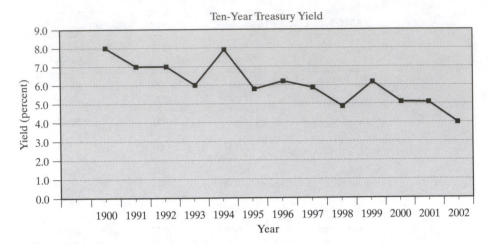

Figure 6.3 Yield of 10-year treasury bills. *Source:* U.S. Department of Treasury.

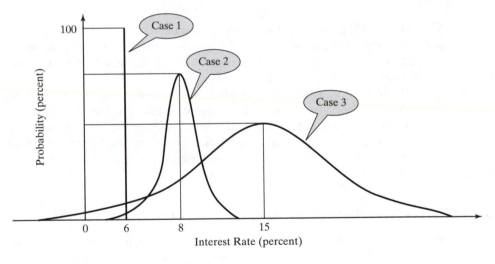

Figure 6.4 Representation of risks.

Case 2 is an investment in a blue-chip corporate stock with a most likely return of 8 percent. Due to market conditions' being usually unpredictable, the return of such an investment has some measure of risk, as represented by the bell-shaped curve centered on 8 percent in Figure 6.4. The return may vary from 4 percent (minimum) to 12 percent (maximum). Risks are measured by the standard deviation of this probability distribution curve (e.g., sigma 2). (See Figure 6.4, Case 2.)

The bell-shaped curve is mathematically represented by the normal probability density function,

$$F(x) = \frac{1}{\sigma_x \sqrt{2\pi}} e^{-\frac{1}{2}\left(\frac{x-\mu}{\sigma_x}\right)^2} \tag{6.11}$$

where

$$\sigma = \text{standard deviation}$$
$$\mu = \text{mean}$$
$$\pi = 3.14159$$

The area underneath the curve is normalized to be 1.

Case 3 is the return of an investment in real estate, centered, for example, around 15 percent. Because this investment requires tax payment, maintenance costs, and other expenditures, its minimum return may be negative. Its upside potential may be very large, however, if commercial developments and property-zoning results become favorable. This case is represented by a bell-shaped probability curve having a large standard deviation (sigma 3 being larger than sigma 2) and with its most likely return centered around 15 percent. Furthermore, the probability of achieving its most likely return of 15 percent is now only 50 percent. (See Figure 6.4.)

Risky events may be represented mathematically by the normal probability density function, which is defined by two parameters, standard deviation and mean. Besides the normal, several other probability density functions (such as triangular, Poisson, and beta) may also be used to represent risky costs. (See Section 6.9.3.)

6.9.2 Project Cost Estimation by Simulation

Recent literature outlines the advancements of PC-based techniques that estimate project costs under uncertainty (Wright 2002; Mooney 1997; Binder and Heerman 1988). The key elements of these PC-based techniques consist of the following steps:

A. Construct a *cost model* for the projects at hand with a spreadsheet program (e.g., Excel®, Lotus 123™) (Smith 2000). The spreadsheet program takes care of the required computation of the cost model, such as addition, subtraction, multiplication, and division. The numerical values entered in the spreadsheet cells are typically deterministic, each having a well-defined and fixed magnitude. The cost model encompasses all cost components and computes the total project cost.

B. Make a *three-point estimate* for each of the component costs, composed of the minimum, the most likely, and the maximum values. This is to account for the perceived-cost uncertainty. Past experience may serve as a guide in the selection of these values.

   Select a probability distribution function (e.g., triangular, normal, beta, or other distribution functions) to represent the three-point estimate of the component cost. Repeat this step for all other cost components of the project.

C. Activate *risk analysis software* to replace the deterministic values contained in the spreadsheet cells by the probability distribution functions chosen to represent the corresponding three-point estimates.

Currently, commercial PC-based software products are becoming readily available. Some examples are

- CrystalBall® for Excel
- @Risk and BestFit programs (Palisade Corporation) for Lotus 123 spreadsheet.

The BestFit program marketed by Palisade Corporation may be used to define a suitable probability distribution function on the basis of numerical three-point estimates or other user-specified input data.

The activated risk analysis software automatically converts all input probability density functions to their corresponding cumulative distribution functions. The technical fundamentals related to this conversion are illustrated in Appendix 6.E.

D. Conduct a *Monte Carlo simulation* to compute the total project cost. Upon activation of risk analysis software, a random number is first generated between 0 and 1. This random number represents a trial probability value (e.g., P1). Using this random number

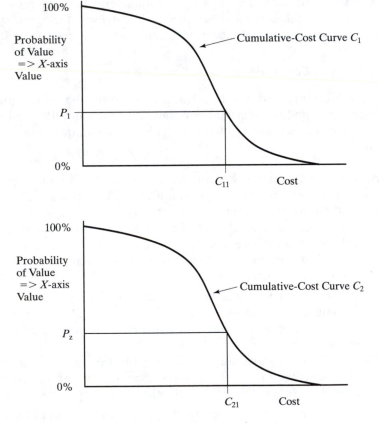

Figure 6.5 Cumulative distribution.

the specific cost value is read from the cumulative distribution of the cost component C1, which represents a random input variable) e.g., C11, see Figure 6.5). A second random number is generated (P2), which is then used to define the cost component C2 (e.g. C21). A third random number is generated to define the cost C31 of a third cost component. This process is continued until the costs of all cost components are defined.

The total project cost (e.g., TPC1) is then calculated with the spreadsheet program that contains the cost model. This is one outcome of the random output variable TPC.

The sampling process is repeated thousands of time to create a distribution of the total project costs (e.g., TPC1, TPC2, ...). These output results are then statistically grouped into bins(with zero to maximum value) to come up with a cumulative distribution. The resulting cumulative distribution for the total project cost TPC may then be converted back to its corresponding probability density function.

The total project cost so generated has a set of minimum, most likely, and maximum values.

E.   Interpret the *total project cost* represented in a cumulative distribution to arrive at the following typical results:

- There is an 80 percent probability that the total project cost will not exceed D.
- The minimum, most likely, and maximum total project costs are $A, $B,$ and C.
- The standard deviation of the total project cost is x, or the overall measure of the project risk.

Information of this type is extremely useful for decision making. It is particularly true in situations where multiple projects are being evaluated for investment purposes.

There is an additional benefit realized by using the just-described cost estimation by simulation. Because of the "risk pooling" effect due to risk sharing among all input cost components, the total project cost is expected to have a lower overall risk than the risk levels of its individual components. Various studies (for example, Canada, Sullivan, and White 1996) have confirmed that the total project cost computed by simulations requires a smaller contingency cost for a given risk level than that computed by the traditional method by using deterministic values.

Other important applications involving risk analyses include (1) project schedule and (2) portfolio optimization.

The value of risk analyses is typically to make explicit the uncertainties of input variables, to promote more reasoned estimating procedures, to allow more comprehensive analyses—or the simultaneous variation of all input variables involved—and to measure the variability of output variables. It is believed that a decision maker can make better decisions with a fuller understanding of the risk-based implications of the decision.

The use of risk analysis in the business and engineering environments is expected to become increasingly widespread in the years to come. Engineering managers are advised to become familiar with advanced tools for risk analyses.

6.9.3 Examples of Input Distribution Functions

In engineering cost estimation, several distribution functions are often used as inputs. Figure 6.6 shows the triangular probability density function. This is the easiest

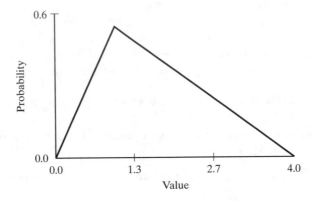

Figure 6.6   Triangular probability density function.

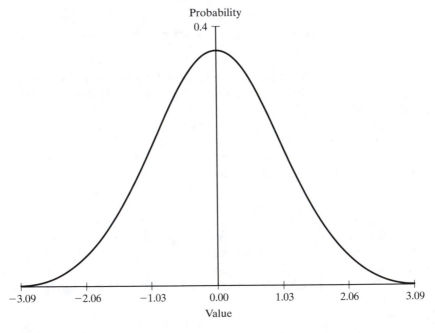

Figure 6.7   Normal probability density function.

function to apply, as the three-point estimates may be directly incorporated into this representation.

Figure 6.7 illustrates the normal probability function. Figure 6.8 displays the beta probability density function. Figure 6.9 depicts the Poisson probability function.

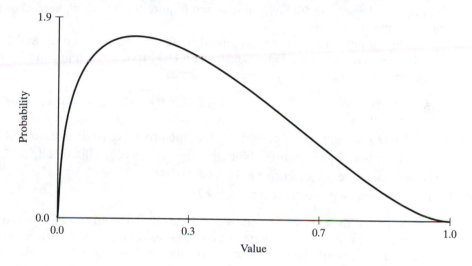

Figure 6.8   Beta probability density function.

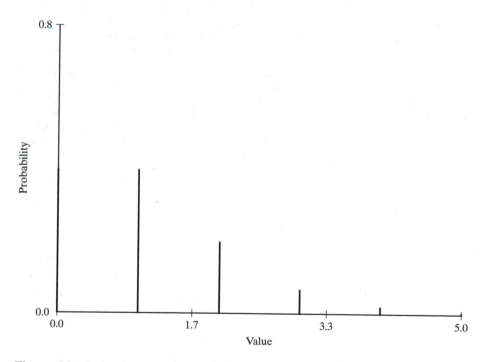

Figure 6.9   Poisson probability function.

6.9.4 Application—Cost Estimation of a Risky Capital Project

As an example, the cost estimation for a turnkey capital project is illustrated in Table 6.8.

Project managers define the base (e.g., the most likely) estimates, as well as the lower and upper bounds for each cost item in the estimate. Doing so will force them to externalize the reasons for any variance and require them to think hard about the contingency plans for each.

The output total cost is represented by the probability density and cumulative distribution functions. (See Figures 6.10 and 6.11.) From the output cumulative distribution, the following results are readily obtained:

A. The most likely total project cost is $5,136,000, which, of course, only echoes the input data.

B. There is an 80 percent probability that the project cost will exceed $5,100,000.

C. There is a 20 percent probability that the project cost will exceed $5,170,000.

D. The maximum project cost is $5,250,000.

E. The minimum project cost is $4,989.71.

Note that the information offered by items B, C, D, and E is new. In a traditional, deterministic project-cost estimate, the cost figures for items A, D, and E would be the same, and there would be no information of the kind offered by items B and C. When

---

TABLE 6.8   Cost Model of a Capital Project

#	Cost Category	Base (thousands of dollars)	Minimum (percent)	Maximum (percent)	Minimum (thousands of dollars)	Most Likely (thousands of dollars)
1XXX	Cold box	748.00	−5	5	710.60	748.00
2XXX	Rotating equipment	742.00	−3	3	719.74	742.00
3XXX	Process equipment	658.00	−2	5	644.84	658.00
4XXX	Electrical equipment	194.00	−5	3	184.30	194.00
5XXX	Instrumentation	295.00	−2	10	289.10	295.00
6XXX	Piping mat/specials	121.00	−2	10	118.58	121.00
711X	Civil construction	284.00	−2	5	278.32	284.00
712X	Mechanical construction	390.00	−1	20	386.10	390.00
713X	Electrical construction	85.00	−2	10	83.30	85.00
71?X	Other contracts	83.00	−5	10	78.85	83.00
716X	Purchased enclosures	48.00	−5	12	45.60	48.00
717X	Fabrication	179.00	−5	4	170.05	179.00
7890	Freight	80.00	−5	15	76.00	80.00
84X0	Field support	188.00	−5	7	178.60	188.00
85XX	Start-up	60.00	−10	30	54.00	60.00
81X0	Product line design	516.00	−1	15	510.84	516.00
8150	Project execution	333.00	−5	20	316.35	333.00
	Total neat	5,004.00				
	Contingency	131.90				
	Grand Total	**5,135.90**				

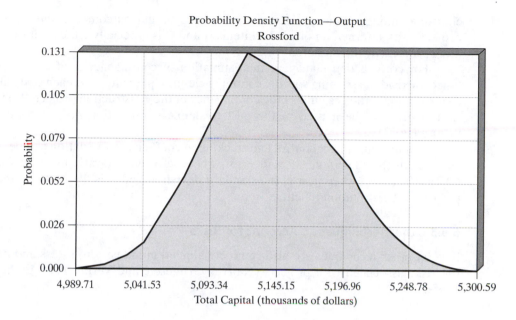

Figure 6.10 Normal probability density function for total project cost.

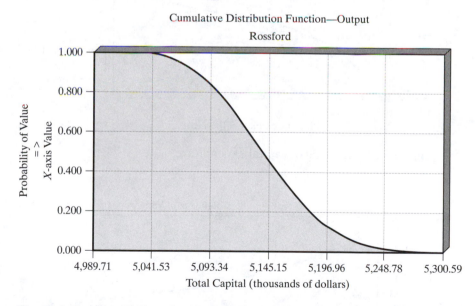

Figure 6.11 Cumulative distribution function for total project cost.

choosing among various projects that may have similar outcomes in the most likely project costs, information offered by items B and C is especially critical for differentiating projects by their inherent risks.

For construction managers, the estimation of contingency is of critical importance. Instead of assigning specific contingencies as a percentage to each cost category items, the simulation calculates the contingency of the construction project. Probabilistic models have been used in the past to define construction project contingency (Touran 2003).

In addition, project control and tracking of coefficient of variation can be readily conducted. (See Figures 6.12 and 6.13.) The coefficient of variation is defined as $CV = 100 \, (\sigma/\mu)$, wherein σ is the standard deviation and μ is the mean of the total project cost distribution function.

6.9.5 Other Techniques to Account for Risks

Several other techniques are also routinely applied in industry to assess and manage the risks associated with projects (Fabrycky, Thuesen, and Verma 1998; Eschenbach 1995).

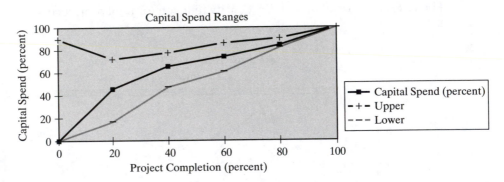

Figure 6.12   Tracking of capital expenditure.

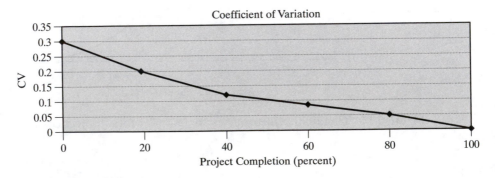

Figure 6.13   Coefficient of variation.

A. **Sensitivity Analysis.** Because of possible variation of specific input parameters, "what-if" analyses are typically performed to assess the sensitivity of the project cost and time to completion.

B. **Contingency Cost Estimation.** The cost of a risky project may be estimated by adding a contingency cost to each task (typically 5 to 7 percent of the task cost) to cover the risk involved.

C. **Decision Trees.** Decision trees are used to evaluate sequential decisions and decide on alternatives on the basis of the expected values of probabilistic outcomes.

D. **Diversification.** With this risk-management technique, several risky projects can be engaged at the same time to spread the risks by means of diversification—combining high-risk, high-return projects with low-risk, low-return projects—to achieve a reasonable overall return on investment.

E. **Fuzzy Logic Systems.** These systems of reasoning are based on fuzzy sets (Nguyen and Walker 2000; Von Altrock 1997). A fuzzy set defines the range of values for a given concept as well as the degree of membership. A membership of 1 indicates full membership, whereas 0 defines exclusion. The change of membership from 0 to 1 is gradual. For example, a fuzzy expert system employs rules such as the following:

> If the temperature T is high (A_i) and the difference in temperature is small (B_i), then close the valve V slightly (S_i).

wherein A_i, B_i, and S_i are fuzzy sets. Fuzzy logic systems have been applied to assess project risks (Cox 1995).

Example 6.8.

The company needs to decide either to develop a new product with an investment of $400,000 or to upgrade an existing product by spending $200,000, as illustrated by the decision point 1 in Figure 6.14.

A. If the new product strategy is pursued, then there is a 60-percent probability that the product will be in high demand, in which case the company will make $200,000 next year. Concurrently, there is a 40-percent chance for low demand, which will result in a loss of $100,00 for the company next year. (See Node A in the Decision Tree Diagram in Figure 6.14.)

   If the new product enjoys a high demand, then there is an 80-percent chance that the product will make $1,000,000 and a 20-percent chance it will make only $50,000 in the year after the next. (See Node a in Figure 6.14.) If the new product meets a low demand, then there is a 30-percent chance for a revenue of $500,000 and a 70-percent chance of suffering a loss of $500,000 in the year after next. (See Node b in Figure 6.14.)

B. If the company follows the strategy of improving the existing product, then there is a 60-percent chance for high demand, leading to a revenue of $100,000; and a 40-percent chance of low demand, to yield zero revenue in the first year. (See Node B in Figure 6.14.)

   If the demand is high at the end of the first year, the company needs to make a second decision (Decision Point 2) whether or not to expand the product line. The expansion

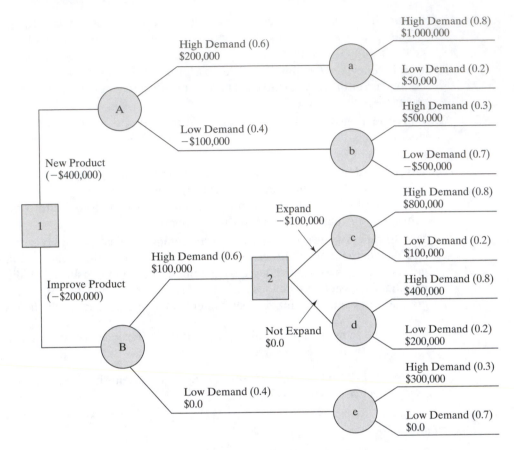

Figure 6.14   Decision tree diagram.

option will require an investment of $100,000, whereas the option of no expansion costs nothing. If the company expands the improved product line, then there is an 80-percent chance that it will reap revenue of $800,000 and a 20-percent chance of $100,000 revenue in the year after the next. (See Node c in Figure 6.14.) If the company elects not to expand at the end of the first year, then there is an 80-percent chance it will realize revenue of $400,000 and a 20-percent chance of $200,000 revenue. (See Node d in Figure 6.14.)

Should the improved product of the company see a low demand (at a probability of 40-percent) and get zero revenue in the first year, then there is a 30-percent chance that the product can generate revenue of $300,000 and a 70-percent chance of zero revenue. (See Node e in Figure 6.14.)

The interest rate is 10 percent. This problem is fully diagrammed in Figure 6.14. Determine which decisions at Decision Points 1 (product strategy) and 2 (expansion strategy) the company should make.

Answer 6.8.

This problem can be solved by using the decision tree method, which works from right to left, from future (the year after next) to present.

A. **Decision Point 2.**

At Node c, the expansion option has a total expected return of

$$ER(c) = 0.8(\$800{,}000) + 0.2(\$100{,}000) = \$660{,}000$$

At Node d, the no-expansion option has a total expected return of

$$ER(d) = 0.8(\$400{,}000) + 0.2(\$200{,}000) = \$360{,}000$$

The present value of expansion is

$$PV(\text{Expansion}) = -\$100{,}000 + \frac{ER(c)}{1.1} = \$500{,}000$$

And the present value of no expansion is

$$PV(\text{No expansion}) = 0 + \frac{ER(d)}{1.1} = \$327{,}272$$

Based on these present values, the decision should favor expansion.

B. **Decision Point 1.**

At Node a, the total expected return is

$$ER(a) = 0.8(\$1{,}000{,}000) + 0.2(\$50{,}000) = \$810{,}000$$

At Node b, the corresponding total expected return is

$$ER(b) = 0.3(\$500{,}000) + 0.7(-\$500{,}000) = -\$200{,}000$$

Thus, the present value for the high-demand case is

$$PV(\text{High Demand}) = \$200{,}000 + \frac{ER\,(a)}{1.1} = \$936{,}363$$

The present value for the low-demand case is

$$PV(\text{Low Demand}) = -\$100{,}000 + \frac{ER\,(b)}{1.1} = -\$281{,}818$$

Thus, the present value for the new product strategy is

$$PV(\text{New Product}) = -\$400{,}000 + \frac{0.6\,PV(\text{high Demand}) + 0.4\,PV(\text{Low Demand})}{1.1}$$
$$= \$8{,}264$$

On the other hand, for the product improvement strategy we have

$$PV(\text{High Demand}) = \$100{,}000 + PV(\text{Expansion}) = \$600{,}000$$

(Note that the no-expansion option is abandoned.)
The present value of low demand is

$$PV(\text{Low Demand}) = 0 + ER(e)/1.1$$
$$= 0 + \frac{0.3(\$300{,}000) + 0.7(0)}{1.1}$$
$$= \$81{,}818$$

Thus, the present value for the product improvement strategy is

$$PV(\text{Product Improvement}) = -\$200,000 + \frac{0.6(\$600,000) + 0.4(\$81,818)}{1.1}$$

$$= \$157,024$$

Since the present value for improving the product is larger than the present value for new product, the choice should be in favor of product improvement.

Even if the company decides to forgo expansion at Decision Point 2, the present value for product improvement is

$$PV(\text{Product Improvement with no expansion})$$

$$= -200,000 + \frac{0.6(427,272) + 0.4(81,818)}{1.1}$$

$$= \$62,809$$

which is still larger than that for new product development. Thus, the value of expansion is

$$\text{Value (Expansion)} = \$157,024 - 62,809 = \$94,215$$

6.10 MISCELLANOUS TOPICS

This section discusses several miscellaneous topics, including economic quantity of ordering, simple cost-based decision models, and project evaluation criteria.

6.10.1 Economic Quantity of Ordering

Ordering of parts, materials, and other supply items directly affects product costs. The ordering process must take into account the quantity needed, purchase price, order-processing fees, shipping costs, and the time value of money. Engineering managers may need to know how to arrive at the *economic quantity of ordering* in order to minimize the total cost of ordering. The next example illustrates this concept.

Example 6.9.

A manufacturing company buys 6000 steel bars a year at a fixed price of $18 each. It costs the company $85 to process and place each order. Assuming 10 percent interest compounded annually, what is the most economic quantity to order at one time?

Answer 6.9.

Let N = Number of orders placed in a year
C = Total cost of ordering at year end

$$N = 1: \quad C = 6000 \times 18(1 + 0.1) + 85$$

$$N = 2: \quad C = 3000 \times 18(1 + 0.1) + 85 + 3000(18)\left(1 + \frac{0.1}{2}\right) + 85$$

$$N = 3: \quad C = 2000 \times 18(1 + 0.1) + 85 + 2000(18)\left(1 + 0.1\frac{2}{3}\right)$$

$$+ 85 + 2000 \times 18\left(1 + 0.1\frac{1}{3}\right) + 85$$

Hence,

$$N = N: \quad C = 6000 \times 18\frac{1}{N}\left[N + 0.1\frac{1}{N}(N + (N - 1) + (N - 2) + .. + 1)\right]$$
$$+ 85N$$

$$= 6000 \times 18 \times \left(1 + 0.1\frac{1 + N}{2N}\right) + 85N$$

To find the minimum C by differentiation,

$$\frac{dC}{dN} = 0 = 5400\left(\frac{1}{N} - \frac{N + 1}{N^2}\right) + 85$$
$$N = 7.98 = 8$$
$$\frac{6000}{8} = 750$$

The economic quantity to order is 750 units, every 6.5 weeks, and 8 times per year.

6.10.2 Simple Cost-Based Decision Models

Engineering managers need to make regular choices among alternatives. In some cases, such choices may be made based on costs, as illustrated by two examples shown below.

   A. Comparison of Alternatives.   When faced with the option of purchasing one of several sets of capital equipment with similar functional characteristics, engineering managers can use the following annual cost formula to identify which has the lowest total annual cost.

   The annual cost for a long-lived asset is defined as the sum of its depreciation charge, interest charge for the capital tied down by the purchase, and its annual operational expenses; that is,

$$AC = \frac{(P - L) \times i}{(1 + i)^N - 1} + P \times i + AE \tag{6.12}$$

where

P = Initial investment (dollars)

L = Salvage value (dollars)

N = Useful life of a long-lived asset (year)

i = Interest rate (percent)

AE = Annual expenses (dollars)—taxes, supplies, insurance repairs, utilities, etc.

AC = Annual cost (dollars)

   The capital equipment with the lowest AC is preferable. (The derivation of both the approximate and exact methods is shown in Appendix 6.C.)

Example 6.10.

Your company has averaged 15-percent growth per year for the past seven years, and now you need additional warehouse space for purchased material as well as finished goods. Two types of constructions have been under consideration: conventional and air-supported fabric. (The data are available in Table 6.9.)

Which is the better economical choice?

TABLE 6.9   Warehouse Options

Type	Conventional	Air Supported
First cost (dollars)	200,000	35,000
Life (years)	40	8
Annual maintenance (dollars)	1,500	5,000
Power and fuel (dollars)	700	5,500
Annual taxes (percent)	1.5	1.5
Salvage value (dollars)	40,000	3,000
Interest rate	0.08	0.08

Answer 6.10.

Conventional:

$$AC_1 = (200{,}000 - 40{,}000) \times \frac{0.08}{[1.08^{40} - 1]} + 200{,}000 \times 0.08$$

$$+ \left(1500 + 700 + 1.5 \times \frac{200{,}000}{100}\right) = \$21{,}818$$

For Air Supported:

$$AC_2 = (35{,}000 - 3000) \times \frac{0.08}{[1.08^{8} - 1]} + 35000 \times 0.08$$

$$+ \left(5000 + 5500 + 1.5 \times \frac{35000}{100}\right) = \$16{,}833$$

Choice: Air-supported system.

Example 6.11.

A semiautomatic machine is quoted at $15,000, while an advanced machine is quoted at $25,000. The salvage value of these machines is assumed to be zero. A four-man party can produce 500 parts a day with the semiautomatic machine, by using a machinist at $200 per day, a maintenance worker at $150 per day, a parts laborer at $100 per day, and a warehouse clerk at $80 per day.

A six-man party can produce 770 parts a day with the advanced machine, using a machinist at $200 per day, an assistant machinist at $180 per day, a maintenance worker at $150 per day, two parts laborers at $100 per day each, and a warehouse clerk at $80 per day.

Material cost is $10 per part. The factory overhead is 50 percent of the direct labor cost, only when parts are being produced. Maintenance expense for the semiautomatic machine is $250 per year, and for the advanced machine is $500 per day. The estimated life of the semiautomatic machine is 20 years, and 15 for the advanced machine. The cost of money is 8 percent per year.

How many parts must be made per year to justify the deployment of the advanced machine?

Answer 6.11.

Define x = Number of parts produced per year

y = Number of working days to produce parts. For the semiautomatic machine, $y = \dfrac{x}{500}$; for the advanced machine, $y = \dfrac{x}{770}$. (See Table 6.10.)

TABLE 6.10   Comparison of Two Machines

	Semiautomatic	Advanced
First cost (dollars)	15,000	25,000
Daily production	500	770
Daily wage (dollars)	530	810
Annual maintenance (dollars)	250	500
Life (years)	20	15
Material cost (dollars per part)	10	10
Factory overhead (percent)	50	50
Interest	0.08	0.08

The annual cost of operating the semiautomatic machine is given by

$$AC_{semi} = \frac{15{,}000 \times 0.08}{(1.08)^{20} - 1} + 15{,}000 \times 0.08 + 250$$

$$+ 10 \times x + 530\frac{x}{500} \times 1.5 = 1777.78 + 11.59x.$$

The annual cost of operating the advanced machine is given by

$$AC_{adv} = \frac{25{,}000 \times 0.08}{(1.08)^{15} - 1} + 25{,}000 \times 0.08 + 500$$

$$+ 10 \times x + 810\left(\frac{x}{770}\right) \times 1.5 = 3420.74 + 11.57792x.$$

Setting $AC_{semi} = AC_{adv}$, we have

$$1775.78 + 11.59x = 3420.74 + 11.57792x$$

$$x = 136{,}172.$$

The advanced machine is justifiable if the production exceeds 136,172 parts per year.

B. Replacement Evaluation.   Engineering managers are sometimes faced with the decision of whether to replace an existing facility with a brand-new one. Again, this replacement decision may be made by identifying the option with the lowest annual cost.

In this analysis, the existing facility is treated as if it is new, in that its residual equipment life and its residual book value (initial capital investment minus accumulated depreciation) are equivalent, respectively, to the useful product life and the capital investment cost of new equipment. Thus,

$$AC_o = \frac{i \times BV(t) - L_o}{(1 + i)^{(N_o - t)} - 1} + BV(t) \times i + AE_o \qquad (6.13)$$

$$AC = \frac{i \times [P - L]}{(1 + i)^N - 1} + P \times i + AE, \qquad (6.14)$$

where

P_o = Original investment cost (dollars)

N_o = Original estimate of useful life (year)

AE_o = Annual expenses of using the original equipment (dollars)

L_o = Salvage value at the end of the equipment's original useful life (dollars)

AC_o = Annual cost of using the existing equipment (dollars)

t = Present age of equipment (years)

$N_o - t$ = Remainder life (years)

$BV(t)$ = Book value of existing equipment at the end of the t(th) year (dollars)

P = Initial investment of the replacement equipment (dollars)

AC = Annual cost of using the replacement equipment (dollars)

L = Salvage value of the replacement equipment
      at the end of its useful life (year)

N = Useful life of the replacement equipment (year)

AE = Annual expenses for using the replacement equipment (dollars)

If $AC_o > AC$, then use the replacement equipment to save costs.

Example 6.12.

A compressor air-supply station was built 18 years ago at the main shaft entrance to a coal mine at a cost of $2.6 million. The station was equipped with steam-driven air compressors that have an annual operating expense of $360,000. The salvage value at the estimated 25-year life of the station is $130,000. It can be sold now for $800,000.

A proposal has been made to replace the station with electrically driven compressors that would be installed underground near the working face of the mine for a cost of

$2.8 million. The new compressor station would have a life of 30 years and a salvage value of 10 percent. Its annual operating cost would be two-thirds of the steam-driven station. Annual taxes and insurance are 2.5 percent of the first cost of either station. The interest rate is 8 percent. Is there a financial justification to replace the steam station?

Answer 6.12.

Table 6.11 summarizes the data of these two compressors.

Assuming approximate method (straight-line depreciation):

$$\text{Steam:} \quad AC = \left(\frac{800,000 - 130,000}{7}\right) + (800,000 - 130,000) \times \frac{0.08 \times 8}{2 \times 7}$$
$$+ 130,000 \times 0.08 + 425,000 = \$561,742.86$$

$$\text{Electric:} \quad AC = \frac{2,800,000 - 280,000}{30}$$
$$+ (2,800,000 - 280,000) \times \frac{0.08 \times 31}{2 \times 30}$$
$$+ 280,000 \times 0.08 + 312,500 = \$523,060$$

Choice: Replace the old steam unit.

TABLE 6.11 Comparison of Two Compressor Drives

	Steam	Electric
Original investment (dollars)	2,600,000	2,800,000
Life (years)	25	30
Present age (years)	18	0
Remaining life (years)	7	30
Present salvage value (dollars)	800,000	
Final salvage value (dollars)	130,000	280,000
Annual expense (dollars)	360,000 + 65,000	240,000 + 72,500

Assume the exact method for calculating depreciation (sinking fund):

$$\text{Steam:} \quad AC = (800,000 - 130,000) \times \frac{0.08}{1.08^7 - 1} + 800,000 \times 0.08$$
$$+ 425,000 = 564,088.50$$

$$\text{Electric:} \quad AC = \frac{(2,800,000 - 280,000) \times 0.08}{1.08^{30} - 1} + 2,800,000$$
$$\times 0.08 + 312,500 = 558,745.13$$

Using the exact method, the same conclusion is reached—namely, to replace the old steam unit with electrically driven compressors.

6.10.3 Project Evaluation Criteria

Engineering managers are often required to make choices among capital projects that may deliver benefits and may also consume resources on an annual basis over a number of periods. Several standard methods are used in industry to evaluate such projects. These include *net present value*, *internal rate of return*, *payback*, and *profitability index*.

A. Net Present Value.

$$NPV = -P + \sum_{m=1}^{n} \frac{NCIF(m)}{(1+i)^m} + \frac{CR}{(1+i)^n}$$ (6.15)

$$m = 1 \text{ to } n$$

where

NPV = Net present value (dollars)

P = Present investment made to initiate a project activity (dollars).

$NCIF(m)$ = Net cash in-flow (dollars) in the period m, which represents revenues earned minus costs incurred ($R(m) - C(m)$) (dollars)

i = Cost of capital rate

n = Number of interest period (year)

CR = Capital recovery (dollars), which is the amount regained at the end of the project through resale or other methods of dispositions

Note that the first term on the right-hand side is the capital outlay for the project, or an outflow of value (cash). The second term on the right-hand side is the sum of discounted net cash in-flow earned over the years. The third term on the right-hand side is the discounted capital recovery of the project.

One major weakness of the NPV equation is that all benefits derived from a project must be expressed in dollar equivalents—within NCIF(m)—in order to be included. Nonmonetary benefits, such as enhanced corporate image, expanded market share, and others, cannot be represented.

For the special case of NCIF (m) = CF = constant

$$NPV = -P + CF\frac{(1+i)^n - 1}{i \times (1+i)^n} + \frac{CR}{(1+i)^n}$$ (6.16)

Projects with the largest NPV values are preferable, as NPV represents the net total value added (before tax) to the firm by the project at hand. Note that NPV may be determined only if the project's net cash in-flow NCIF (m) is known. These NPV equations may, if applicable, be readily expanded to include corporate tax rate and investment tax credits. (See Appendix 6.D.)

B. Internal Rate of Return (IRR).
Rate of return is generally defined as the earnings realized by a project in percentage of its principal capital.

The *internal rate of return* (IRR) is the average rate of return (usually annual) realized by a project in which the total net cash in-flow is exactly balanced with its total net cash out-flow, resulting in zero NPV value at the end of its project life cycle. In other words, this is the rate realizable when reinvestment of the project earnings is made at the same rate until maturity (NPV = 0).

IRR is determined by the following equations:

$$0 = -P + \sum_{m=1}^{n} \frac{\text{NCIF}\,(m)}{(1 + \text{IRR})^m} + \frac{\text{CR}}{(1 + \text{IRR})^n} \tag{6.17}$$

$$m = 1 \text{ to } n$$

For NCIF $(m) = \text{CF} = \text{constant}$

$$0 = -P + \text{CF} \frac{(1 + \text{IRR})^n - 1}{\text{IRR} \times (1 + \text{IRR})^n} + \frac{\text{CR}}{(1 + \text{IRR})^n} \tag{6.18}$$

The IRR values (before tax) of acceptable projects must be much greater than the firm's cost of capital. Projects with high IRR are preferable.

C. Payback Period.   The payback period is defined as the number of years that the original capital investment for the project will take to be paid back by its annual earnings, or

$$\text{PB} = \frac{P}{\text{CF}} \tag{6.19}$$

where

$$P = \text{Capital investment}$$
$$\text{CF} = \text{Annual cash flow realized by the project}$$

Cost reduction projects with small payback periods (e.g., less than two years) are preferable.

D. Profitability Index (PI).   Profitability index is defined by the ratio

$$\text{PI} = \frac{\text{Present value of all future benefits}}{\text{Initial investment}} \tag{6.20}$$

Projects with large PI values are preferable.

Example 6.13.

Your company is currently pursuing three cost-reduction projects at the same time.

- Project A requires an investment of $10 million. It is expected to yield a cost savings of $30 million in the first year and another $10 million in the second year.
- Project B demands an investment of $5 million. It is expected to produce a cost savings of $5 million in the first year and another $20 million in the second year.
- Project C needs an investment of $5 million. It is expected to bring about a cost savings of $5 million in the first year and another $15 million in the second year.

After the second year, there will be no receivable benefit or capital recovery from any of these projects. The cost of capital (interest rate) is 10 percent.

Determine the ranking of these projects on the basis of the evaluation criteria of NPV, IRR, PB, and PI.

Answer 6.13.

(See Table 6.12.)

$$P = \text{Present investment}$$
$$n = 2$$
$$CF = \text{Cash flow}$$
$$CR = \text{Capital recovery} = 0$$
$$i = 10\%$$

TABLE 6.12   Summary of Results

Project	Time=> 0	1	2	NPV	IRR (percent)	PB	PI
A	−10	30	10	25.5	230	0.5	3.55
B	−5	5	20	16	156	0.4	4.22
C	−5	5	15	12	130	0.5	3.39

NPV:

A) $\text{NPV} = -10 + \dfrac{30}{1.1} + \dfrac{10}{1.1^2} = 25.537$

B) $\text{NPV} = -5 + \dfrac{5}{1.1} + \dfrac{20}{1.1^2} = 16.074$

C) $\text{NPV} = -5 + \dfrac{5}{1.1} + \dfrac{15}{1.1^2} = 11.942$

IRR:

A) $0 = -10 + \dfrac{30}{1 + r} + \dfrac{10}{(1 + r)^2};\qquad r = 2.3$

B) $0 = -5 + \dfrac{5}{1 + r} + \dfrac{20}{(1 + r)^2};\qquad r = 1.56$

C) $0 = -5 + \dfrac{5}{1 + r} + \dfrac{15}{(1 + r)^2};\qquad r = 1.3$

PB:

A) $\text{PB} = \dfrac{10}{\dfrac{30 + 10}{2}} = \dfrac{10}{20} = 0.5$

B) $\text{PB} = \dfrac{5}{\dfrac{5 + 20}{2}} = \dfrac{5}{12.5} = 0.4$

C) $\text{PB} = \dfrac{5}{\dfrac{5 + 15}{2}} = \dfrac{5}{10} = 0.5$

PI:

A) $\quad PI = \left(\dfrac{3.0}{1.1} + \dfrac{10}{1.1^2}\right)\dfrac{1}{10} = \dfrac{35.537}{10} = 3.553$

B) $\quad PI = \left(\dfrac{5}{1.1} + \dfrac{20}{1.1^2}\right)\dfrac{1}{5} = \dfrac{21.074}{5} = 4.214$

C) $\quad PI = \left(\dfrac{5}{1.1} + \dfrac{15}{1.1^2}\right)\dfrac{1}{5} = \dfrac{16.942}{5} = 3.3884$

6.11 CONCLUSION

This chapter reviews basic cost accounting issues related to product and service costing. Product costs have direct and indirect cost components. While direct costs are relatively easy to assess, the indirect costs that account for overhead charges may need to be properly assessed by using tools such as activity-based costing.

Cost data may apply for a single period or for multiple periods. In the case of multiple periods, the time dependency of cost data must be considered. For time-dependent data, the concept of the time value of money and the compound interest formulas are to be applied.

Depreciation accounting affects the facility costs that are part of the indirect costs of products. Different depreciation methods will lead to more or less indirect costs for the products. Finally, the inventory costs are affected by the sequence in which the time-dependent product costs are selected.

Cost data may be uncertain because of factors related to economy, market condition, political stability, labor movement, and others. For uncertain cost data, risk analysis may be needed. The Monte Carlo simulation is an efficacious method to conduct risk analyses. Several other methods are also available to account for cost uncertainties.

Engineering managers need to become well versed in cost accounting.

6.12 REFERENCES

Bedfod, T. and R. Cook. 2001. *Probabilistic Risk Analysis: Foundations and Methods*. New York: Cambridge University Press.

Binder, K. and D. W. Heerman. 1988. *Monte Carlo Simulation in Statistic Physics: An Introduction*. New York: Springer-Verlag.

Bragg, S. M. 2001. *Cost Accounting: A Comprehensive Guide*. New York: John Wiley.

Bragg, S. M. 2002. *Accounting Reference Desktop*. New York: John Wiley.

Bruns, Jr., W. J. 1999. "Understanding Costs for Management Decisions." *Harvard Business School*, Note No. 9-197-117, January.

Byrd, III, D. M. and R. Cothern, 2000. *Introduction to Risk Analysis: A Systematic Approach to Science-Based Decision Making*. Rockville, MD: Government Institute.

Canada, J. R., W. G. Sullivan, and J. A. White. 1996. *Capital Investment Analysis for Engineering and Management*. 2d ed. Upper Saddle River, NJ: Prentice Hall.

Cokins, G. 2001. *Activity-Based Cost Management: An Executive's Guide*. New York: John Wiley.

Colander, D. C. 2001. *Microeconomics*. 4th ed. New York: Irwin/McGraw-Hill.

Cooper, R. and R. Kaplan. 1988. "How Cost Accounting Distorts Product Costs." *Management Accounting*, April.

Cox, E. 1995. *Fuzzy Logic for Business and Industry*. Rockland, MA: Charles River Media.

Eschenbach, T. G. 1995. *Engineering Economy: Applying Theory to Practice*. Chicago: Irwin.

Fabrycky, W. J., G. J. Thuesen, and D. Verma. 1998. *Economic Decision Analysis*. 3d ed. Upper Saddle River, NJ: Prentice Hall.

Hicks, D. T. 1999. *Activity-Based Costing: Making It Work for Small and Mid-Sized Companies*. New York: John Wiley.

Mooney, C. Z. 1997. *Monte Carlo Simulation*. Thousand Oaks, CA: Sage Publications.

Ngyyen, H. T. and E. A. Walker. 2000. *A First Course in Fuzzy Logic*. 2d ed. Boca Raton, FL: Chapman and Hall.

O'Guin, M. C. 1991. *The Complete Guide to Activity-Based Costing*. Englewood Cliffs, NJ: Prentice Hall.

Rapier, D. M. 1996. "Standard Costs and Variance Analysis." *Harvard Business School Note*, No. 9-196-121, January.

Smith, G. N. 2000. *Excel Spreadsheet Application Series for Cost Accounting*. Cincinnati, OH: South Western College Publishing, Thompson Learning.

Sullivan, W. G., E. M. Wicks, and J., T. Luxhoj. 2003. *Engineering Economy*. 12th ed. Upper Saddle River, NJ: Prentice Hall.

Touran, A. 2003. "Calculation of Contingency in Construction Projects." *IEEE Transactions on Engineering Management*, Vol. 50, No. 2, May.

Von Altrock, C. 1997. *Fuzzy Logic and Neuro Fuzzy Applications in Business and Finance*. Upper Saddle River, NJ: Prentice Hall.

Wright, J. F. 2002. *Monte Carlo Risk Analysis and Due Diligence of New Business Ventures*. New York: AMACOM.

6.13 APPENDICES

APPENDIX 6.A. DERIVATION OF SINGLE PAYMENT COMPOUND AMOUNT FACTOR

$F_1 = P + P \times i = P(1 + i)$ at the end of first year

$F_2 = F1 + F1 \times i = P(1 + i)^2$ (end of second year)

$F_n = P(1 + i)^n$ (End of the nth year)

Hence, $F = P(1 + i)^n$ (Equation 6.3.)

and $\quad P = F(1 + i)^{-n}$

APPENDIX 6.B. DERIVATION OF UNIFORM SERIES COMPOUND AMOUNT FACTOR

$F_1 = A$

$F_2 = F_1 + F_1 \times i + A = A + A \times i + A = A(1 + i) + A$

$$F_3 = F_2 + F_2 \times i + A = F_2(1 + i) + A = A(1 + i)^2 + A(1 + i) + A$$
$$F_n = A(1 + i)^{n-1} + A(1 + i)^{n-2} + \cdots + A(1 + i) + A$$
$$F_{n+1} = A(1 + i)^n + A(1 + i)^{n-1} + \cdots + A$$

Form a difference between these two series:

$$F_{n+1} - F_n = A(1 + i)^n$$

On the other hand, by definition,

$$F_{n+1} = F_n(1 + i) + A$$

Hence, $F_n(1 + i) + A - F_n = A(1 + i)^n$

$$F_n \times i = A[(1 + i)^n - 1]$$
$$F_n = F = A\frac{[(1 + i)^n - 1]}{i} \text{ (Equation 6.5.)}$$

The other three factors (i.e., equations 6.6, 6.7, and 6.8) are derived by substitution.

APPENDIX 6.C. DERIVATION OF ANNUAL COST COMPUTATION EQUATIONS

There are two methods to compute the annual cost: the exact method and the approximate method.

A. *Exact Method (Depreciation based on sinking fund method)*

$$AC = \frac{(P - L)i}{(1 + i)^n - 1} + Pi + AE$$

The first term on the right-hand side is the annual cost (based on sinking found depreciation method) for the $(P - L)$ amount. The second term is the annual interest charge for the investment capital P. The last term is the annual expense.

B. *Approximate Method (Depreciation based on straight-line method)*

$$AC = \frac{(P - L)}{n} + (P - L)i\frac{(n + 1)}{2n} + Li + AE$$

The first term on the right is depreciation charge based on straight-line method. The second two terms on the right are the *average annual interest charge*, which is an opportunity cost (lost interest income for having made the investment). This average annual interest charge may be derived as follows:

The interest charge for each year is

Year	Formula
1	$= Pi = (P - L)i + Li$
2	$= \left[(P - L) - \dfrac{(P - L)}{n}\right]i + Li$
3	$= \left[(P - L) - \dfrac{2(P - L)}{n}\right]i + Li$
$\vdots$	$\vdots$
n	$= \left[(P - L) - \dfrac{n - 1}{n}(P - L)\right]i + Li$

The sum of the total interest charge from year 1 to year n is

$$
\begin{aligned}
\text{Sum} &= i(P - L)\left\{1 + \left(1 - \frac{1}{n}\right) + \left(1 - \frac{2}{n}\right) + \cdots \right. \\
&\quad \left. + \left(1 - \frac{n-1}{n}\right)\right\} + nLi \\
&= i(P - L)\left\{n - \frac{1}{n}[1 + 2 + 3 + 4 + \cdots + (n - 1)]\right\} + nLi \\
&= i(P - L)\left\{n - \frac{1}{n}\frac{n(n - 1)}{2}\right\} + nLi \\
&= i(P - L)\frac{n + 1}{2} + nLi
\end{aligned}
$$

Thus, the average annual interest charge is (Eq. 6.12):

$$
\text{AAIC} = \frac{\text{Sum}}{n} = i(P - L)\frac{(n + 1)}{2n} + Li
$$

Because of this last averaging step, this method is called an "approximate method."

APPENDIX 6.D. NET PRESENT VALUE, INCLUDING CORPORATE TAX, DEPRECIATION, AND INVESTMENT TAX CREDIT

$$
\begin{aligned}
\text{NPV} &= -P(1 - \text{ITC}) + (1 - t)A\frac{[(1 + i)^n - 1]}{[i(1 + i)^n]} \\
&\quad + \text{DB}\, t\, \text{DF} + \text{SV}\frac{(1 - t)}{(1 + i)^n}
\end{aligned}
$$

where

$\quad$ NPV = Net present value

$\qquad$ A = Net uniform annual income ($=R - \text{AOC} - \text{Tax} - M - X$)

$\qquad\quad$ R = Revenue

$\qquad$ AOC = Annual operating cost

$\qquad\quad$ Tax = Tax paid for property (local, county)

$\qquad\quad$ M = Maintenance cost

$\qquad\quad$ X = Other annual expenses

$\quad$ P = Present investment

$\quad$ t = Federal corporate tax rate (e.g., = 0.46 for a large corporation)

$\quad$ i = Interest rate (or WACC)

$\quad$ n = Number of interest periods (i.e., years)

$\quad$ ITC = Investment tax credit (e.g., ITC = 0.08 or 0.1)

$\quad$ DB = Depreciable base

$\qquad$ DB = $P(1 - 0.5 \times \text{ITC})$, if ITC = 0.1

$\qquad$ DB = P, if ITC = 0.08

$\quad$ DF = Depreciation discount factor

$$DF = \sum_{m=1}^{n} \frac{DP(m)}{(1+i)^m}$$

$DP(m)$ = Depreciation percentage in period m

$$\left(\text{e.g., for straight-line } DP(m) = \frac{1}{n} = \text{constant}\right)$$

SV = Salvage value (=L), representing the capital recovery at the end of n periods

The following is a numerical example:

$$P = 10{,}000; A = 4{,}600; ITC = 0.1; n = 5; t = 0.35$$

$$SV = 0$$

$$DB = P(1 - 0.5 \times 0.1) = 0.95P$$

IRS Form 4562: 5-Year ACRS Schedule

$$DP(1) = 0.15; DP(2) = 0.22; DP(3) = 0.21$$

$$DP(4) = 0.21; DP(5) = 0.21$$

$$DF = \sum_{m=1}^{5} \frac{DP(m)}{(1+i)^m}$$

Results: NPV = 5,570 (for $i = 0.08$)

IRR = 0.2914 (for NPV = 0)

APPENDIX 6.E. CONVERSION OF A PROBABILITY DENSITY FUNCTION TO ITS CUMULATIVE DISTRIBUTION FUNCTION

The process of converting a probability density function to its cumulative distribution function is straightforward and unique. Figure 6.A1 shows a triangular probability density function for the cost of the component C1. The vertical axis represents probability, and the horizontal axis represents cost. The triangular probability density function is the easiest one to apply when a three-point estimate for a risky input variable is known.

The component C1 is assumed to have a minimum cost of $30 (point A), a maximum cost of $80 (point Z), and a most likely cost of $50 (point M). The area underneath the triangular probability density function is normalized to 1; this condition prescribes that the y coordinate for the point N is 0.04 based on the calculation of $1 = 0.04 \times 0.5 \times (80 - 30)$.

Let us insert a vertical cost line through x. With this cost line in place, we define the shaded area PNZXP as A_x, which is under the probability density function, but bound by the vertical cost line that passes through x on the left. We form a ratio of A_x to the total area APNZXA underneath the same probability density function. This ratio is designated as P_x. The value of P_x varies from 0 to 1, as x moves from Z to M and then to A. P_x is the probability for the cost of this component to be equal to or in excess of x. The pair of P_x and x represents a point in a cumulative distribution chart.

For another cost value, y, this process is repeated. A new pair of P_y and y defines another point in the descending cumulative distribution chart. After many repetitions, a descending cumulative curve is generated that resembles the one shown in Figure 6.A2. The vertical axis is the probability for the component cost to equal or exceed the value shown on the x axis. The x axis spans the minimum value of A on the left to the maximum value of Z on the right.

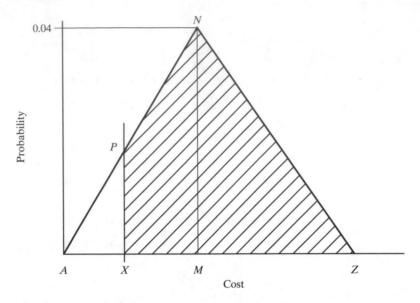

Figure 6.A1 Conversion of a triangular probability density function to its cumulative descending function.

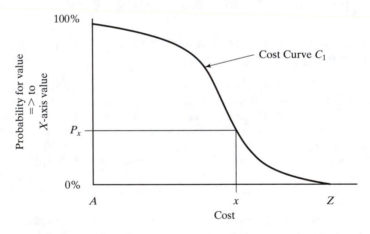

Figure 6.A2 Cumulative descending distribution function for component C_1.

6.14 QUESTIONS

6.1 The company is evaluating two specific proposals to market a new product. The current interest rate is 10 percent.

Proposal A calls for setting up an in-house manufacturing shop to make the product, requiring an investment of $500,000. The expected profits for the first to fifth years are $150,000, $200,000, $250,000, $150,000, and $100,000, respectively.

Proposal B suggests that the manufacturing operation be outsourced by contracting an outside shop, requiring a front-end payment of $300,000. The expected profits for the first to fifth years are $50,000, $150,000, $200,000, $300,000, and $200,000, respectively. The expected profits would be lower in earlier years due to third-party markup.

Which proposal should the company accept?

Answer 6.1: Proposal B.

6.2   The company's warehouse has been busy taking in and shipping out vendor-supplied automotive parts. Table 6.A1 shows the warehouse's activities in eight consecutive periods, during which time the price of the parts has steadily increased.

A.   Determine the total LIFO prices for each stock withdrawal in periods 3, 4, 6, and 8.
B.   Repeat the same price computation by using the FIFO technique.

TABLE 6.A1   FIFO and LIFO Computation

Period	Units In	Unit Price Paid (dollars)	Units Out
1	150	100	
2	250	120	
3			180
4			100
5	100	130	
6			200
7	100	140	
8			80

Answer 6.2

The stock withdrawal prices are as follows:

	LIFO	**FIFO**
Period 3	21,600	18,600
Period 4	11,400	12,000
Period 6	23,000	24,800
Period 8	11,200	11,000

6.3   A dam is being considered on a river that periodically overflows. Each time the river overflows, it causes about $600,000 in damages. The project horizon is 40 years. A 10 percent interest rate is being used.

Three different designs are available, each with different costs and storage capacities. (See Table 6.A2.)

The U.S. Weather Service has provided a statistical analysis of annual rainfall in the area draining into the river. (See Table 6.A3.)

TABLE 6.A2   Design Options

Design Alternatives	Cost (dollars)	Maximum Storage Capacity
A	500,000	1 unit
B	625,000	1.5 units
C	900,000	2.0 units

TABLE 6.A3   Annual Rainfall and Probability

Units Annual Rainfall	Probability
0	0.10
0.1 to 0.5	0.60
0.6 to 1.0	0.15
1.1 to 1.5	0.10
1.6 to 2.0	0.04
2.0 or more	0.01

Assume that the dam requires no annual maintenance, has zero salvage value at the end of its 40-year life, and is essentially empty at the start of each annual rainfall season. Which design alternative would you choose?

Answer 6.3

The alternative B should be chosen.

6.4 The NPV equation (Eq. 6.15) is described as follows:

$$\text{NPV} = -P + \sum_{m=1}^{n} \frac{C(m)}{(1+i)^m} + \frac{CR}{(1+i)^n}$$

The NPV equation is important for evaluating project-based investments. It is also a basic equation for defining the concept of "value addition," having a broad philosophical implication of what engineers do. Explain.

6.5 A manufacturing company makes three products, A, B, and C. The fixed factory overhead is $60,000, consisting of $10,000 for material handling, material waste, and procurement; $30,000 for rent and utilities; and $20,000 for safety and canteen costs. Other costs are shown in Table 6.A4.

TABLE 6.A4   Product Options

	Product A	Product B	Product C
Number of units produced per month	250	400	900
Total material costs per month (dollars)	5,000	8,000	4,000
Labor hours per unit	4	3.5	1.5
Labor rate per unit (dollars per hour)	25	20	30
Machine hour per unit (hour)	1	1	3

(A) Determine the product cost for products A, B, and C, using the activity based costing (ABC) method.

(B) If products A, B, and C are sold at $400, $350, and $150 per unit, respectively, what is the gross profit for each product?

(C) What is the company's total gross profit per month if all units produced are sold?

Answer 6.5

(A) The product costs for A, B, and C are $167.25, $134.58, and $83.16, respectively.

(B) The gross profits for A, B, and C are $223.75, $215.42, and $66.84, respectively.

(C) The company's total profit per month is $204,511.50.

6.6 A company makes and sells three technology products: A, B, and C. It has a production plant with 17,000 square feet of floor area, consisting of machine setup (2000 square feet), machining operation (9000 square feet), assembly (4000 square feet), and inspection, packaging, and shipping activities (2000 square feet).

The total annual expenditure for the plant is $200,000 for depreciation, $700,000 for utilities, $20,000 for phone and travel services, $150,000 for manufacturing supports, $200,000 for procurement, and $150,000 for supervision.

The labor hours and material costs required to manufacture the products are shown in Table 6.A5.

TABLE 6.A5 Production Hours for Three Products

	A	B	C
Machine setup (hours)	2	3	4
Machine operation (hours)	16	12	8
Assembly (hours)	4	3	2
Inspection/packing/shipment (hours)	2	2	2
Raw materials/unit of product (dollars)	950	430	640
Purchased components/unit of product (dollars)	100	80	90
Outsourced service/unit of product (dollars)	20	30	40
Number of units produced per year	700	900	550

The labor charges are $25 per hour for machine setup, $35 per hour for machining operation, $30 per hour for assembly, and $20 per hour for inspection, packing, and shipping.

The company plans to sell product A at $5,000 per unit, product B at $4,500 per unit, and product C at $4,100 per unit. All products manufactured during the year are assumed to be sold successfully. Apply the activity based costing technique to determine the product cost and individual gross margin for each product.

Answer 6.6

The product costs for A, B, and C are $2,679, $1,787, and $1,747, respectively. The individual gross margins for A, B, and C are $2,321, $2,713, and $2,553, respectively.

6.7 You are considering a good-looking Toyota hybrid car priced at $28,000 or an elegant GM luxury car at $24,000. The fuel efficiency is rated at 50 miles per gallon for the Toyota and 25 miles per gallon for the GM. The annual maintenance cost for both cars is about 0.5 percent of the car price. The gasoline in the local market is selling at $2.00 per gallon. The cars are to be driven about 10,000 miles per year. You plan to keep your car for five years only. At the end of the fifth year, the resale values of the Toyota and the GM are about 40 percent and 30 percent, respectively, of their original prices. The interest rate is 6 percent.

Which car is the better choice from the standpoint of costs?

Answer 6.7

Toyota is the better choice from the standpoint of costs.

6.8 Company X manufactures automotive door panels that may be made of either sheet metal or plastic sheet molding (glass-fiber-reinforced polymer). Sheet metal bends well to the high-volume stamping process and has a low material cost. Plastic sheet molding meets the required strength and corrosion resistance and has a lower weight. The plastic forming

TABLE 6.A6   Material Options

Description	Plastic	Sheet Metal
Material cost (dollars per panel)	5	2
Direct labor cost (dollars per hour)	40	40
Factory overhead (dollars per year)	500,000	400,000
Maintenance expenses (dollars per year)	100,000	80,000
Machinery investment (dollars)	3 million	25 million
Tooling investment (dollars)	1 million	4 million
Equipment life (years)	10	15
Cycle time (minutes per panel)	2	0.1
Interest rate (percent)	6	6

process involves a chemical reaction and has a slower cycle time. Table 6.A6 summarizes the cost components for each.

Assuming that the machinery and tooling have no salvage value at the end of their respective equipment lives, what is the annual production volume that would make the plastic panel more economical?

Answer 6.8

For production volume up to 536,156 panels per year, the plastic panels are more economical.

6.9   Company X produces two products, A and B, respectively. Table 6.A7 summarizes the cost structures of these two products per a three-month period.

TABLE 6.A7   Product Options

	Product A	Product B
Selling price (dollars per unit)	10	12
Variable cost (dollars per unit)	5	10
Fixed costs (dollars)	600	2,000
Machining time (hour per unit)	0.5	0.25

The company's manufacturing operation is limited to 30,000 machine hours available per a three-month period. Furthermore, because of a prior sales commitment, the company must produce at least 1000 units of Product B. Determine the maximum profit the company can achieve in a three-month period.

Answer 6.9

The maximum profit is $296,900.

Chapter 7

Financial Accounting and Analysis for Engineering Managers

7.1 INTRODUCTION

Financial accounting and analysis serve the important corporate functions of reporting and evaluating the financial health of a firm (Harrison and Horngren 1998).

Financial statements are prepared by certified management accountants (CMAs) and certified public accountants (CPAs), according to the generally accepted accounting principles (GAAP) in a conservative, material, and consistent manner. These financial documents provide (a) internal reporting to corporate insiders for planning and controlling routine operations and for decisions on capital investments and (b) external reporting to shareholders and potential investors in financial markets (Friedlob and Schleifer 2003).

All financial statements are designed to be relevant, reliable, comparable, and consistent. Financial accounting treats owners (shareholders) and corporations as separate entities. Owners of corporations are liable only to the extent of their committed investments. Owners enjoy a flexible tenure and participation and, as investors, may buy or sell stocks of the company at any time. On the other hand, corporations are legal entities, fully responsible for their liabilities up to the limits of their total assets. Corporations are assumed to be going concerns and in operation forever, unless they cease to exist by declaring bankruptcy or being acquired by others.

This chapter discusses the (a) language and concepts, (b) T-accounts, (c) financial statements, (c) performance ratios and analysis, and (d) balanced scorecards or tools to monitor and promote corporate productivity.

To be effective, engineering managers must know how to read financial statements; monitor the firm's activity, performance, profitability, and market position; and assess the financial health of a firm. Doing so will allow them to initiate proper projects (e.g., plant expansion, new product and technology development, new technology acquisition, strategic alliances, etc.) at the right time to add value to their employers.

7.2 ACCOUNTING PRINCIPLES

As practiced in the United States, all financial statements are formulated for a specific accounting period. A typical accounting period is three months, as the U.S. Securities and Exchange Commission prescribes that all publicly traded companies file Form 10-Q reports every quarter. All companies also need to publish their Form 10-K reports annually. The financial statements must adhere to the basic principles of accounting, discussed in the following subsections. (Anonymous 2002A; Hawkins and Cohen 2001).

7.2.1 Accrual Principle

Accounting statements include both cash and credit transactions. Revenue is recognized when it is earned. For example, a manufacturing enterprise will recognize revenues as soon as products are shipped to the customer and an invoice is sent, irrespective of any credit payment already received or yet to be collected. Sports teams are known to sell season tickets ahead of the games for cash and then recognize the applicable revenue only after each game is played. According to the accrual principle of accounting, companies recognize revenues when earned, with the assumption that the collection of this revenue from approved credit accounts and the delivery of the promised products or services are both reasonably assured.

Similarly, the accrual principle specifies that costs and expenses are established when incurred, even before actual payments are made.

7.2.2 Matching

Expenses are recognized by matching them with the revenue generated in a given accounting period. For example, the cost of goods sold (CGS) is recognized as an expense only after products are sold and revenue is recognized. Before the products are sold, CGS stays as inventory—a part of the corporate current assets—even though costs for materials, labor, and factory overhead have already been spent for these unsold products.

7.2.3 Dual Aspects

The assets of a company are always equal to the claims against it (i.e., assets equal to claims). The claims originate from both creditors and owners. Each transaction has a dual effect in that it induces two entries in order to maintain a balance between assets and claims.

7.2.4 Full Disclosure Principle

All relevant information is disclosed to the users of the company's financial reports. Extensive footnotes contained in the annual reports of numerous publicly traded companies are testimonials for such disclosure practices.

7.2.5 Conservatism

Assets are to be recorded at the lowest value consistent with objectivity (e.g., book values are often lower than market values). While profits are not recorded until recognized, losses are recorded as soon as they become known. Inventories are valued at the lower of the cost or market value.

7.2.6 Going Concern

As stated earlier, it is assumed that the company's business will go on forever. This assumption justifies the current practice of using historical data (e.g., the original acquisition costs) and a reasonable method of depreciation (e.g., straight line) by which the book value of corporate tangible assets is defined. Otherwise, liquidation accounting must be applied to define the corporate asset value by using current market prices (Hawkins and Cohen 2001).

7.3 T-ACCOUNTS

Accountants use T-accounts as tools to register transactions in preparation for creating financial statements. T-accounts are set up for any items that are assets, liabilities, equities, or other temporary holding entries. A T-account, as displayed in Figure 7.1, looks like the letter "T." On the left side of the "T," debits are recorded and on the right side credits are recorded. This type of entry is also known as double-entry book keeping.

Following the double-entry bookkeeping practice, every transaction affects at least two accounts. This is to ensure that a balance is continuously maintained between both the assets of, and the claims against, the company.

The company's assets include cash, accounts receivable, inventory, land, machines, plant facilities, marketable securities, and other resources of value. Liabilities include accounts payable, accrued expenses, long-term debts, and other claims creditors have against the company. Owners' equities include stocks, surplus, retained

Figure 7.1   Structure of a T-account.

earnings, and other claims of the owners against the company. Equation 7.1 depicts the balance between assets (A) and claims consisting of liabilities (L) and owners' equities (OE):

$$A = L + \text{OE} \tag{7.1}$$

The convention of T-accounts is as follows: to increase the amount of an asset, debit the account; conversely, to decrease an asset amount, credit the account. All liabilities and owners' equities accounts are treated in the opposite way; that is, to increase a liability or equity amount, credit the account; and to lessen a liability or equity amount, debit the account.

For accounts that do not fall directly into one of these three categories (i.e., A = assets, L = liabilities, and OE = owners' equities), we need to first define their relationships to either A, L, or OE and then treat them accordingly. Revenues, expenses, and dividends are such examples.

Revenues raise the net income of the company. The resulting net income goes into the retained earning account for the owners. Thus, an increase of revenues needs to be credited to its T-account. The company's expenses are generally deducted from its revenues in order to arrive at its net income. An increase in expenses results in a reduction of net income and consequently a reduction of owners' equities. Therefore, an increase of expenses needs to be debited to its T-Account. Similarly, an increase in dividends paid to shareholders whittles down the residual net income amount, which is then added to the retained earnings account of the owners. Thus, an increase of dividends needs to be debited to its T-Account. Table 7.1 summarizes the ways in which

TABLE 7.1   T-Account Convention for an Increase in Selected Account Items

Accounting Items	Debit	Credit
Assets Cash, accounts receivables, inventory, land machines, marketable securities, etc.	x	
Liabilities Accounts payables, accrued expenses, long term debts, etc.		x
Owners' Equities Stocks, capital surplus, retained earnings		x
Revenue		x
Expenses	x	
Dividend	x	

(*Source*: David F. Hawkins and Jacob Cohen, "The Mechanics of Financial Accounting." *Harvard Business School Note*, No. 9-101-119, June 27, 2001.)

increases in the indicated assets, liabilities, owners' equities, or other amounts should be treated in their respective T-accounts.

For engineers and engineering managers who are familiar with equations, the rule that follows may represent a convenient way of keeping them better oriented with the T-account convention. Rearranging the basic accounting equation (Equation 7.1), we get

$$\text{LHS} = A - L - \text{OE} = 0 \tag{7.2}$$

where LHS stands for "left-hand side" of the equal sign. Note that, for each financial transaction, there are two account entries. The account entry that causes the LHS to increase temporarily should be debited to its respective T-account. Examples include increases in all assets and decreases in all liabilities and owners' equities. The account entry, which leads to a temporary reduction of the LHS, should be credited to its respective T-account. Equation (7.2) remains valid after both entries of the financial transaction are entered.

Accountants use *T-accounts* to collect raw financial data that they check and recheck for validity and reliability. Then they make sure that the data are relevant to the accounting period at hand and consistent with past practices. Finally, to ensure comparability, accountants regroup them into known line items typically included in financial statements.

Example 7.1.

Study the following accounts, which contain several transactions keyed together with letters:

Cash		Office Equipment		Capital	
(a) 6,000	(b) 2,500	(d) 8,500			(a) 9,000
(e) 1,300	(c) 150				
	(f) 3,500				
	(g) 160				

Office Supplies		Law Library		Legal Fees Earned	
(c) 150		(a) 3,000			(e) 1,300
(d) 125					

Prepaid Rent		Accounts Payable		Utilities Expenses	
(b) 2,500		(f) 3,500	(d) 8,625	(g) 160	

Explain the nature of each transaction with the dollar amount involved.

Answer 7.1.

(a) Convert capital of $9,000, add $6,000 to the cash account, and purchase books worth $3,000 for the Law Library.

(b) Pay the prepaid rent of $2,500 in cash.

(c) Pay office supplies of $150 in cash.

(d) Purchase office equipment ($8,500) and office supplies ($125) by credit, creating an account payable of $8,625.

(e) Receive the $1,300 legal fees earned in cash.

(f) Pay accounts payable of $3,500 in cash.

(g) Pay utilities expenses of $160 in cash.

7.4 KEY FINANCIAL STATEMENTS

Companies in the United States use three financial statements: income statement, balance sheet, and funds flow statement (Jablonsky and Barsky 2001; White, Sondhi, and Fried 2003).

7.4.1 Income Statement

The income statement is an accounting report that matches sales revenue with pertinent expenses that have been incurred (cost of goods sold, tax, interest, depreciation charges, salaries and wages, administrative expenses, R&D, etc.). Sometimes it is also called the profit or loss statement, earnings statement, or operating and revenue statement. The income statement contains the following key entries:

A. *Sales revenue* is the total revenue realized by the firm during an accounting period. Sales revenue is recognized when earned, for example, by having goods shipped and invoices issued.

B. *Cost of goods sold (CGS)* is the cost of goods that have been actually sold during an accounting period. In a manufacturing company, CGS is calculated as the opening inventory at the beginning of an accounting period, plus labor costs, material costs, and manufacturing overhead incurred during the period, and minus the closing inventory at the end of the period.

C. *Gross margin* is the sales revenue minus the cost of goods sold.

D. *Expenses* are those expenditures chargeable against sales revenue during an accounting period. Examples include general, selling, and administrative expenses; depreciation charges; R&D; advertising; interest payments for bonds; employee retirement benefit payments; and local taxes.

E. *Depreciation* is a process by which the cost of a fixed, long-lived asset is converted into expenses over its useful life. This is in proportion to the value it has produced during an accounting period. (See Section 6.5.)

F. *EBIT* is the earnings before interests and taxes.

G. *Net income (earnings or NOPAT)* is the excess of sales revenue over all expenses (e.g., CGS, all items under D above, and corporate tax) in an accounting period. Sometimes it is also called profit, earnings, or net operating profit after tax (NOPAT).

H. *Dividend* is the amount per share paid out to stockholders in an accounting period.

I.   *Earnings per share* are the net income of a firm during an accounting period (e.g., a year), minus dividends on preferred stock, divided by the number of common shares outstanding.

J.   *Costs* can be defined as follows: While all costs are also expenditures, not all costs are expenses, which are chargeable against revenues in a given accounting period. For example, direct and indirect costs contained in the products preserved as inventory are not recognized immediately as expenses. When products in inventory are sold, then the respective cost of goods sold (CGS) is recognized as expenses in the income statement, along with other expenses.

An income statement shows the firm's activity. An example is given in Table 7.2 for the XYZ Company. In general, sales revenue is referred to as the *top line* and net income as the *bottom line* figures. These line items are examined closely by financial analysts, as are the line items of gross margin and EBIT. (For a detailed analysis and interpretation of income statement entries, see Section 7.5.2 on ratio analysis.)

Engineering managers deploy company resources to make contributions to the financial success of their employers. The impacts of their engineering activities are registered in several line items contained in the income statement:

1.   **Sales.** Engineering managers increase sales through well-designed products that satisfy the needs of customers. They introduce innovative products that address

TABLE 7.2   Example of XYZ Income Statement (Million of dollars)

	Year 2001	Year 2002
Sales (net) revenue	**8,380.30**	**8,724.70**
cost of goods sold	6,181.20	6,728.80
Gross margin	**2,199.10**	**1,995.90**
General, selling, and administrative expenses	320.7	318.8
Pensions, benefits, R&D, insurance, and others	494.6	538.7
State, local, and miscellaneous taxes	180.1	197.1
Depreciation	297.2	308.6
EBIT	**906.5**	**632.7**
Interest and other costs related to debts	82.9	114.4
Corporate tax	(32.05 percent) 264.00	(20.84 percent) 108.00
Net income (NOPAT)	**559.6**	**410.3**
Common stock dividend	151.6	172.8
Retained earnings	408	237.5

the needs of new customers in new markets. They refine products that are easy to serve and maintain, thus promoting market acceptance of the company's products. They also bring into being supply chains to increase the speed of product introduction and the extent of product customization in the marketplace.

2. **Cost of goods sold.** Engineering managers cut down product costs by innovative design, engineering, manufacturing, and quality control.

3. **R&D.** Engineering managers advance and apply new technologies to enhance product features and to foster the rapid development of new global products.

Example 7.2.

The Advanced Technologies company has had quite a successful year. At the end of the current fiscal year, its assets, liabilities, revenues, and expenses are exhibited in Table 7.3.

TABLE 7.3 Records of Financial Entries

#	Items	Thousands of dollars
1	Accounts payable	3,740
2	Accounts receivable	7,550
3	Accumulated depreciation—equipment	8,410
4	Accumulated depreciation—building	17,900
5	Advertising expense	3,340
6	Administrative expense	5,500
7	Building	36,300
8	Cash	6,320
9	Cost of goods sold	31,000
10	Depreciation expense—building	960
11	Depreciation expense—equipment	1,310
12	Equipment	14,640
13	Inventory	11,000
14	Insurance expense	840
15	Interest expense	2,100
16	Land	2,100
17	Long-term loans outstanding	42,000
18	Miscellaneous expense	1,480
19	R&D	5,200
20	Salaries payable	170
21	Sales revenue	60,300
22	Supplies expense	1,820
23	Taxes expense	2,630
24	Taxes payable	610
25	Utilities expense	2,070

Determine the net income of the company for the current year.

Answer 7.2.

To determine the company's net income, we need to create the income statement for the company. As presented in Table 7.4, only selected items of Table 7.3 are to be included in the company's income statement.

TABLE 7.4   Income Statement of Advanced Technologies

	Thousands of dollars
Sales revenue	60,300
Cost of goods sold	31,000
Gross margin	29,300
Administrative expense	5,500
Advertising expense	3,340
Supplies expense	1,820
Utilities expense	2,070
Miscellaneous expense	1,480
Insurance expense	840
Depreciation—building	960
Depreciation—equipment	1,310
R&D	5,200
Operating income	6,780
Interest expense	2,100
Taxable income	4,680
Taxes expense	2,630
Net income	2,050

7.4.2 Balance Sheet

The balance sheet is an accounting report that presents the assets owned by a company and the ways in which these assets are financed through liabilities and owners' equity. Liabilities are creditors' (such as banks, bondholders, and suppliers) claims against the company. Owners' equity stands for the claims of owners (shareholders) against the company (Hawkins and Cohen 2002). The following key entries are included in a balance sheet:

A. *Assets* are items of value having a measurable worth. They are resources of economic value possessed by the company. There are three classes of assets: current, fixed, and all others.

B. *Current Assets* are convertible to cash within 12 months. Examples include, in a descending order of liquidity, cash, marketable securities, accounts receivable, inventory, and prepaid expenses.

C. *Cash* is money on hand or in checks, and is the most liquid form of assets.

D. *Accounts receivable* is the category of revenue recognized prior to payment collection. It is money owed to the firm, usually by its customers or debtors, as a result of a credit transaction.

E. *Inventory* designates stock of goods yet to be sold that is valued at cost, including direct materials, direct labor, and manufacturing overhead. It may consist of stores, work in progress, and finished goods inventories. (See Section 6.6.) Inventory is included in the balance sheet as a current asset.

When finished goods are shipped and invoiced to customers in an accounting period, the cost of goods sold (CGS) is then recognized in the income statement as an expense.

F. *Prepaid expenses* are paid before receiving the expected benefit (e.g., rent, journal subscription fee, or season's tickets). It is a current asset.

G. *Fixed assets* are tangible assets of long, useful life (more than 12 months) such as land, buildings, machines, and equipment. (Improvement costs are added to the fixed asset value. Repair and maintenance costs incurred in a given accounting period are expensed in the income statement.)

H. *Other assets* are valuable assets that are neither current nor fixed. Examples include patents, leases, franchises, copyrights, and goodwill. Amortization accounting applies to these assets in a similar manner, as depreciation is applied to fixed assets. (See Section 6.5.)

Goodwill—a company's reputation and brand-name recognition—is recognized as an asset only if it has been purchased for a measurable monetary value, such as in conjunction with a merger or acquisition transaction.

I. *Accumulated depreciation* is the sum of all annual depreciation charges taken from the date at which the fixed asset is first deployed up to the present.

J. *Net fixed assets* are the net value of the firm's tangible assets—original acquisition cost minus accumulated depreciation. (This net fixed asset value may deviate considerably from its market value or replacement cost in a given accounting period. The conservatism principle prescribes that the net fixed asset is carried on the balance sheet even if it is lower than its current market value. Otherwise, a loss entry must be added to the balance sheet to adjust the fixed asset value downwards, should it become higher than its current market value.)

K. *Liabilities* are obligations that are to be discharged by the company in the future. They represent claims of creditors (e.g., banks, bondholders, and suppliers) against the firm's assets. Sometimes, it is also called *debt*.

L. *Current liability* describes amounts due for payment within 12 months. Examples include accounts payable, short-term bank loans, interest payments, payable tax, insurance premiums, deferred income, and accrued expenses. Accounts payable is always listed first within the category of current liabilities, with others to follow in no specific order.

M. *Accounts payable* is an expense recognized before payment. It is an obligation to pay a creditor or supplier as a result of a credit transaction, usually within a period of one to three months.

N. *Deferred income* is income received in advance of being earned and recognized (i.e., payment received before shipment and invoicing of goods). In the balance sheet, it is included as a current liability.

O. *Deferred income tax* is the amount of tax due to be paid in the future, usually within 12 months.

P. *Long-term liability* is defined as the amounts due to be paid in more than 12 months. Examples include corporate bonds, mortgage loans, long-term loans, lines of credit, long-term leases, and contracts.

Q. *Bonds* are long-term debt certificates secured by the assets of a company or a government. Bonds issued by a publicly held company are corporate bonds, and those issued by the U.S. federal government are Treasury bonds. In case of defaults, bondholders have the legal right to seize the assets of the company for recovery.

R. *Debentures* are unsecured bonds issued by the firm.

S. *Convertible bonds* are those issued by a company and allowed to be converted into common stocks according to a set of specifications.

T. *Owners' equity* is the shareholders' original investment plus accumulated retained earnings. It represents the residual value of the corporation owned by the shareholders after having deducted all liabilities from company assets. (See Equation (1) in Section 7.3.) Sometimes it is also called *net worth*.

U. *Stock* is a certificate of ownership of shares in a company. Preferred stocks have a fixed rate of dividend that must be paid before dividends are distributed to holders of common stocks.

V. *Capital surplus* is the premium price per share above the par value of the stock. It includes the increase in the owner's equity above and beyond the difference between assets and liabilities reported in the company's balance sheet.

W. *Retained earnings* are the accumulated earnings retained by the company for the purpose of reinvestment and not to be paid out as dividends.

X. *Book value* is defined as the tangible assets (such as fixed assets) minus liabilities and the equity of preferred stocks. It is the share value of common stocks carried in the books.

Z. *Stock price* is the market value of a firm's stock. It is influenced by the book value, earning per share, anticipated future earnings, perceived management quality, and environmental factors present in the marketplace.

The contributions of engineering managers affect only one line item in the balance sheet, namely, inventories. Inventories may be scaled down by applying superior production technologies, product design, and best practices of supply-chain management.

The organization of entries in a balance sheet follows the specific convention:

1. Assets are listed before liabilities, which are then followed by owners' equity.

2. Current assets and liabilities are enumerated ahead of noncurrent assets and liabilities, respectively.

3. Liquid assets are listed before all other assets with less liquidity.

4. The listing of current liabilities follows no specific order, except that accounts payable must always be listed first in this section.

Table 7.5 shows a sample balance sheet of XYZ Company.

TABLE 7.5   Example of XYZ Balance Sheet (Million of dollars)

	Year 2001	Year 2002
Assets		
Cash	231.00	245.70
Marketable securities	450.80	314.90
Accounts receivable	807.10	843.50
Inventories	1,170.70	1,387.10
Total current assets	**2,659.60**	**2,791.20**
Fixed assets	11,070.40	11,897.70
Accumulated depreciation	6,410.70	6,618.50
Net fixed assets	**4,659.70**	**5,279.20**
Long-term receivables and other investments	574.80	735.20
Prepaid expenses	260.90	362.30
Total long-term assets	5,495.40	6,376.70
Total Assets	**8,155.00**	**9,167.90**
Liabilities		
Accounts payable	571.20	622.80
Notes payable	65.30	144.50
Accrued taxes	346.30	275.00
Payroll and benefits payable	433.70	544.30
Long-term debt due within a year	30.40	50.80
Total current liabilities	**1,446.90**	**1,637.40**
Long-term debt	1,542.50	1,959.90
Deferred tax on income	288.40	405.30
Deferred credits	27.00	36.30
Total long-term liabilities	**1,857.90**	**2,401.50**
Total Liabilities	**3,304.80**	**4,038.90**
Common stock ($1.00 Par Value)	81.40	82.20
Capital surplus	1,549.10	1,589.60
Accumulated retained earnings	3,219.70	3,457.20
Total Owner's Equity	**4,850.20**	**5,129.00**
Total Liabilities and Owners' Equity	**8,155.00**	**9,167.90**

Example 7.3.

Using the data given in Table 7.3, construct the balance sheet of Advanced Technologies and determine the owners' equities at the end of the current fiscal year.

Answer 7.3.

The owners' equities are $53,990,000 at year end. (See Table 7.6.)

TABLE 7.6   Balance Sheet of Advanced Technologies

	Thousands of dollars
Cash	6,320
Accounts receivable	7,550
Inventory	11,000
Total current assets	24,870
Land	2,100
Equipment (net) ($14,640 – 1,310)	13,330
Building (net) ($36,300 – 960)	35,340
Total assets	75,640
Accounts payable	3,740
Taxes payable	610
Salaries payable	170
Long-term loans outstanding	42,000
Total liabilities	46,520
Owners' equities	53,990
Total liabilities and owners' equities	75,640

7.4.3 Funds Flow Statement

The funds flow statement compares the firm's activities in two consecutive accounting periods from the standpoint of funds. It is an accounting report that elucidates the major sources and uses of funds of the firm. It is sometimes also called *statement of changes in financial position* or the *statement of sources and uses of funds*.

The principle behind the funds flow analysis is rather simple. An increase in assets signifies a use of funds, such as buying a plant facility by paying cash or using credit. A decrease in assets indicates a source of funds, such as selling used equipment to receive cash for use in the future. An increase in liabilities produces a source of funds, such as borrowing money from a bank so that more cash is available for other purposes. A decrease of liabilities yields a use of funds, such as paying down a bank loan by using money from the company's cash reservoir. Table 7.7 presents an example of the funds flow statement of XYZ Company.

The funds flow statement shown in Table 7.7 is generated by applying the following procedure:

A. **Increase in Plants and Equipment**

   (1) Increase in fixed assets $(11897.7 - 11070.4)$ = 827.3

   (2) Increase in long-term receivable and other investments $(735.2 - 574.8)$ = 160.4

(3) As details are missing, we introduce the following reasonable assumption:

    (a) Long-term receivables   = 59.6

    (b) Increase in other investment   = 100.8

(4) Total increase in fixed asset investment $(827.3 + 100.8)$ = 928.1

B. Increase in Long-Term Debt

(1) Increase in long-term debt $(1959.9 - 1542.5)$ = 417.4

(2) Increase of long-term debt due within one year $(50.8 - 30.4)$ = 20.4

(3) Total $(417.4 + 20.4)$ = 437.8

C. Increase in Common Stock and Capital

(1) Increase in common stocks $(82.2 - 81.4)$ = 0.8

(2) Increase in capital surplus $(1589.6 - 1549.1)$ = 40.5

(3) Total   = 41.3

Most other line items in the statement are directly verifiable.

TABLE 7.7 Example of XYZ Funds Flow Statement (Million of dollars)

	2001–2002	Percentage
Sources		
Net income	410.3	24.11
Depreciation*	308.6	18.14
Decrease in marketable securities	135.9	7.99
Increase in notes payable	79.2	4.66
Increase in accounts payable	51.6	3.03
Increase in payroll and benefits payable	110.6	6.50
Increase in deferred tax on income	116.9	6.87
Increase in long-term debt	437.8	25.73
Increase in common stock and capital	41.3	2.43
Increase in deferred credit	9.3	0.55
Total Source of Funds	**1,701.50**	100
Uses		
Increase in plants and equipment	928.1	54.55
Dividend paid	172.8	10.16
Increase in cash	14.7	0.86
Increase in accounts receivable	36.4	2.14
Increase in inventories	216.4	12.72
Increase in long-term receivables and other investments	160.4	9.42
Increase in accrued taxes	71.3	4.19
Increase in prepaid expenses	101.4	5.96
Total Uses of Funds	**1,701.50**	100

*Depreciation is a noncash expenditure that must be added back here to denote a source of funds available to the firm.

7.4.4 Linkage between Statements

The three financial statements described previously are linked to one another. The net profit in the income statement is linked with the retained earning in the balance sheet. The inventory account in the balance sheet is linked with the sales revenue in the income statement. The accumulated depreciation in the balance sheet is linked with the annual depreciation charge included in the income statement. Because the depreciation charge taken in a given period affects the net profit of the company during the same period, it is thus indirectly linked to the retained earning account in the balance sheet as well.

The linkage between the funds flow statement and the other two financial statements is self-evident, as all data in the funds flow statement are derived from changes in various line items in the other two statements.

7.4.5 Recognition of Key Accounting Entries

This section offers additional notes on the recognition of several key accounting entries—assets, liabilities, revenues, and expenses—according to GAAP practiced in the United States. Other countries may have slightly different rules governing the report of these items.

A. *Assets* are the resources under company control. They have economic value and can be used to produce future benefits. Assets recognition is based on two principles: historical cost and conservatism (Healy and Choudhary 2001A). All assets are reported by using historical cost—that is, the initial capital investment value at some time in the past. The book value of a given asset is defined as its initial acquisition cost minus the accumulated depreciation. Should the asset's current market value drop below its book value, the shortfall must be reported as an expense. If its market value exceeds its book value, however, the surplus is not reported in the company's balance sheet. This is to ensure that the asset value included in the balance sheet always denotes its lower bound. Thus, the balance sheet may understate the true value of the company's assets.

Asset reporting must address the two issues of asset ownership and the certainty of its future economic benefits. If neither the ownership nor the future benefits are clearly established, an asset cannot be recognized. For example, companies routinely invest in employee training in the hope that doing so will lead to increased productivity at a future point in time. Since the completed training is really owned by the employees, and employees may leave at any time they wish, companies do not have real ownership of the training results. Thus, employee training is regarded as an expense and not an asset. When companies acquire plant facilities to make products for sales in the marketplace, its future benefits are more or less certain. Plant facilities are thus reported as assets. When companies apply resources to expand R&D and advertising, the future benefits of these investments are neither certain nor measurable. R&D and advertising are thus recognized as expenses and not assets.

GAAP accounting rules in the United States contain one exception: Generally speaking, software development costs are to be reported as expenses as they are incurred. However, once the company management becomes confident that the software development efforts can be completed and the resulting software product will be used as intended, all costs incurred from that point are to be reported as assets.

There are several ambiguities in the accounting rules practiced in the United States:

1. **Buying versus developing.** If Company A acquires Company B by paying a purchase price that exceeds Company B's net asset value, then this excess value is called goodwill. Goodwill includes the intangible assets of Company B such as its brand name, trademarks, patents, R&D portfolio, and employee skills. After the merger, the surviving company has part of its R&D (from the original company A) recognized as expenses, and part of the R&D (acquired from company B) as assets.

2. **Valuing Intangible Assets.** "If you can't kick a resource, it really isn't an asset." This saying is typically the justification used by companies to rapidly write off intangible assets from their balance sheets. Oftentimes, goodwill is significantly overvalued in a merger or acquisition transaction due to potential conflicts of interests among the parties involved. Writing off intangible assets distorts the true value of the assets reported in the company's balance sheet.

3. **Market Value.** U.S. accounting rules prescribe that marketable securities (e.g., bonds) are to be reported at their fair market values only if they are not to be held to maturity. Thus, at any given time, the real asset value of a company is distorted by not reporting the current true market values of these assets in balance sheets.

B. *Liabilities* are obligations to be satisfied by transferring assets or providing services to another entity (e.g., banks, suppliers, and customers).

A liability is recorded when an obligation has been incurred and the amount and timing of this obligation can be measured with a reasonable amount of certainty (Healy and Choudhary 2001B). For multiple-year commitments, the recordable obligation is the present value of expected future commitments wherein the discount rate is the prevailing rate when the obligation was first established. (See Equation 6.8.)

C. *Revenues* recognition must satisfy two conditions: Revenue is earned when (1) all or substantially all of the goods or services are delivered to the customers and (2) it is likely that the collection of cash or receivables will be successful. Generally speaking, the timing of product or service delivery may not be the same as that for payment collection (Healy 2001).

For magazine subscriptions, insurance policies, and service contracts, customers usually pay in advance. In these cases, payments received ahead of the service delivery dates are kept in a "deferred revenue" account. Only after the pertinent service is delivered will the applicable payments be credited to the revenue T-account during the accounting period. For products sold on credit, companies recognize the revenue as soon as the products are shipped out, and invoices are issued to customers ahead of the payment collection.

In the case of construction projects, which usually stretch out over many accounting periods, revenues are recognized in T-accounts by using the *percentage completion* method and are recognized in proportion to the expenses incurred in the project.

For products sold with money-back guarantees, companies recognize revenue at the time the product is delivered. At the end of an accounting period, management makes an estimate of the cost of returns (a liability) to adjust the revenue figure.

D. *Expenses* are economic resources that either have been consumed or have declined in value during an accounting period. Expenses are typically recorded in the form of a reduction of asset value (e.g., cash) or by a creation of liability (e.g., accounts payable).

There are three types of expenses: (1) consumed resources having a cause-and-effect relationship with revenue generated during the same accounting period (e.g., cost of goods sold); (2) other resources consumed during the same accounting period, but having no cause-and-effect relationship with revenue (such as R&D expenses, advertising expenses, depreciation charges, local taxes, pension expenses, and other general administrative expenses); (3) reduction of expected benefits of company assets generated by past investments (e.g., the write-down of production facilities and equipment that is no longer of value due to the noncompetitiveness of products or technological obsolescence) (Healy and Choudhary 2001C).

Example 7.4.

Superior Technologies sells a product at the unit price of $100. The unit cost of the product is $60. Annual sales have averaged 1 million units, and its annual selling expense has been $7 million. Market research has determined that, if the selling price of the company product is decreased to $90, there will be a 35-percent increase in the number of units sold. The engineering department estimates that, if the production volume is increased by 35 percent, it will reduce the unit product cost by 10 percent due to the scale of economies. To market the 35-percent increase in sales volume, the company's selling expense will need to increase by about 50 percent.

The company's current warehouse facilities are sufficiently large to accommodate the possible increase of 35 percent in sales volume without requiring new investment. Furthermore, regardless of the product price, the company is obliged to pay an annual loan interest of $2 million. Its corporate tax rate is 45 percent. It maintains an R&D department, whose operation is independent of the sales units, at an annual cost of $5 million. Its administrative expense is $15 million, which is also independent of the sales activities. In adition, the company incurs a pretax depriciation charge of $2 million.

Determine if the reduction of product price would increase or decrease the net income of the company and by how much.

Answer 7.4.

The reduction of product price will cause the company's net income to increase to $7.75 million from $4.95 million. (See details in Table 7.8.)

TABLE 7.8 Net Income Due to Increased Sales

	Current Operation (dollars)	Operation with Increased Unit Sales (dollars)
Units of product sold	1,000,000	1,350,000
Product price	100	90
Unit product cost	60	54
Sales revenue	100,000,000	121,500,000
Cost of goods sold	60,000,000	72,900,000
Gross margin	40,000,000	48,600,000
Selling expense	7,000,000	10,500,000
Administrative expense	15,000,000	15,000,000
R&D	5,000,000	5,000,000
Depreciation	2,000,000	2,000,000
EBIT	11,000,000	16,100,000
Interest	2,000,000	2,000,000
Taxable income	9,000,000	14,100,000
Tax (45 percent)	4,050,000	6,345,000
Net income	4,950,000	7,755,000

7.5 FUNDAMENTALS OF FINANCIAL ANALYSIS

The purpose of conducting financial analyses is to assess the effectiveness of the company's management in achieving the objectives set forth by the company's board of directors with respect to a number of critically important business factors (Friedson and Alvarez 2002). Such factors include

A. **Liquidity**—the availability of current assets to satisfy the firm's operational requirements
B. **Activity**—the efficiency of resource utilization
C. **Profitability**—the extent of the firm's financial success
D. **Capitalization**—the makeup of the company's assets and its utilization of financial leverage
E. **Stock Value**—the market price of the company's stock. The product of stock price and the number of outstanding stocks is defined as *market capitalization.*

Typically, the corporate objectives of diverse companies are growth, profitability, and return on investment (ROI).

The *growth* objective suggests that companies keep product prices low, increase marketing expenses, run plants at full capacity, take loans to keep inventory high, and strive for a larger market share and a more dominant market position. The *profitability* objective dictates that companies set prices high to maximize profits, run plants at a capacity that minimizes costs (production and maintenance), and use debt when called for. The *ROI* objective is achieved by operating the company to maximize its financial return with respect to the firm's investment (e.g., the "milk-the-cash-cow" strategy. See Section 9.5.4).

Financial analyses focus on studying period-to-period changes in key financial data and on comparing performance ratios with the applicable industrial standards.

7.5.1 Performance Ratios

In this section, we apply a specific system of calculating performance ratios by grouping together the ratios for liquidity, activities, profitability, capitalization, and stock value (Bruns 1996). In order to understand the system, we must first learn how each term used is defined.

A. **Liquidity**—The firm's capability to satisfy its current liabilities, such as buying materials, paying wages and salaries, paying interests on long-term debt, and other necessary expenditures. Without liquidity, there can be no activity.

Working capital is defined as current assets minus current liabilities. The changes in working capital over several periods provide an indication of the company's reserve strength to weather financial adversities.

Current ratio is the ratio of current assets to current liabilities. Current assets are frequently considered the major reservoir of funds for meeting current obligations. This ratio provides an indication of the company's ability to finance its operations over the next 12 months. A current ratio above 1.0 indicates a margin of safety that allows for a possible shrinkage of value in current assets such as inventories and accounts receivables. However, having a current ratio in excess of 2.0 or 3.0 may indicate a poor cash management practice.

Quick ratio is the ratio of quick asset to current liabilities. *Quick asset* is defined as cash plus marketable securities and accounts receivable. This ratio is more severe than the current ratio in that it excludes the value of inventory whose liquid value may not be certain. Thus, the quick ratio indicates the company's ability to meet its financial obligations over the next 12 months without the use of inventory that may take time to unload. It is sometimes also called the *acid test* or *liquidity ratio*.

B. **Activity**—The changes in sales and inventory. Successful activity leads to profitability.

Collection period ratio is the accounts receivable divided by average daily sales as measured in days. The average daily sales is the total annual sales divided by 360 days. This ratio measures the managerial effectiveness of the credit department in collecting receivables and the quality of accounts receivable.

Inventory turnover ratio is the cost of goods sold divided by the average inventory. It expresses the number of times during a year that the average inventory is recouped or turned over through the company's sales activities. The higher the turnover, the more efficient the company's inventory management performance will be, provided that there has been no shortage of inventory producing a loss of sales and failure to satisfy customers' needs.

Asset turnover ratio of net sales to total assets indicates the ability of the company's management to utilize total assets to generate sales.

Working capital turnover ratio is net sales to working capital. Working capital is defined as average current assets minus average current liabilities. It indicates the company's ability to efficiently utilize working capital to generate sales.

Sales to employee ratio is the company's net sales revenue divided by its average number of employees working during an accounting period. It measures the company's ability to effectually utilize human resources.

C. **Profitability**—To be profitable is the objective of most companies. Without liquidity and activity, there can be no profitability. If the company is profitable, it can readily obtain the required liquidity to keep its operations continuing.

Gross margin to sales ratio of gross margin to sales revenue measures the company's profitability on the basis of sales. Gross margin is defined as sales minus cost of goods sold. Gross margin percentage is defined as the gross margin divided by sales.

Net income to sales ratio indicates the company's overall operational efficiency (e.g., procurement, cost control, current assets deployment, and utilization of financial leverage) in creating profitability based on sales. This ratio is also known as ROS, which stands for return on sales.

Net income to owners' equity ratio measures profitability from the shareholders' viewpoint. It is also known as ROE, which stands for return on equity. This very common measure indicates the earning power of the ownership investment in the company.

Net income to total asset ratio is net income divided by total assets. It measures the management's ability to effectively apply company assets in generating profits. It is known as ROA, which stands for return on assets.

Return on invested capital is the ratio of net income divided by capital. The capital of a company is the sum of its long-term liabilities plus owners' equity.

D. **Capitalization**—The sum of the company's long-term assets and owners' equity is defined as the total capital deployed by company management to pursue business opportunities. Several ratios are in use to check this capital deployment effectiveness.

Interest coverage ratio (EBIT divided by the interest expense) calculates the number of times the company's EBIT covers the required interest payment for the long-term debt—an indication of the company's ability to remain solvent in the near future.

Long-term debt to capitalization ratio is the sum of long-term debt plus owners' equity—the total permanent investment in a company, indicating the percentage of long-term debt in the company's capital structure, excluding current liabilities. It is a measure of the company's financial leverage. Keeping this ratio small (hence, large owners' equity percentage) may not always be the smart choice, as the company will forgo the use of low-cost debt with tax deductible interest payments to enhance profitability.

Debt to equity ratio is the ratio of total liability to owners' equity. It also measures the company's financial independence and the relative stake of shareholders (insiders) and bondholders (outsiders). A low ratio indicates that the company is financially secure as far as the owners are concerned. A high ratio indicates that the firm may have difficulty borrowing money in the future.

E. **Stock Value**—This is the market price or value of the company's stock as defined by the financial markets. The company's management is obliged to pursue proper business strategies in order to steadily raise their stock value.

   Earnings per Share is the ratio of net income minus preferred stock dividends divided by the number of common stocks outstanding.

   Price to earning ratio is the ratio of the market price of common share to earning per share.

   Market to book ratio is the ratio of market price of stock to the book value per share. More precisely, the total book value of a company is defined as the total assets minus intangible assets, minus total liability, and minus the equity of preferred stocks. The book value per share is then the total book value divided by the number of outstanding common shares.

   Dividend payout ratio is the ratio of dividends per share divided by earnings per share. It indicates the percentage of annual earnings paid out as dividends to shareholders. The portion that is not paid out goes into the retained earnings account on the balance sheet.

7.5.2 Ratio Analysis

Ratios are useful tools of financial analysis (Bruns 1996; Troy 1999; Healy and Cohen 2000). Sometimes ratios of significant financial data are more meaningful than the raw data themselves. They also provide an instant picture of the financial condition, operation, and profitability of a company, provided that the trends and deviations reflected by the ratios are interpreted properly.

   Ratio analyses are subject to two constraints: past performance and various accounting methods.

   Past performance is not a sure basis for projecting the company's condition in the future.

   Various accounting methods employed by different companies may result in different financial figures (e.g., inventory accounting, depreciation, etc.), rendering a comparison that is not always meaningful between companies in the same industry. Typically, accountants try to reconcile financial statements before conducting comparative analyses. Examples of adjustments that are frequently made include

1. Adjusting LIFO inventories to a FIFO basis (Section 6.8)
2. Changing the write-off periods of intangible assets, such as goodwill, patents, trademarks, etc. (Section 6.5)
3. Adding potential contingency liabilities if lawsuits are pending (Section 6.9)
4. Reevaluating assets to reflect current market values (Section 7.4.5)
5. Changing debt obligations to reflect current market interest rates
6. Restating reserves or charges for bad debts, warranties, and product returns
7. Reclassifying operating leases as capital leases

When performing ratio analyses, it is advisable to follow this set of generally recommended guidelines:

1. Focus on a limited number of significant ratios.
2. Collect data over a number of past periods to identify the prevalent trends.
3. Present results in graphic or tabular form according to standards (e.g., industrial averages).
4. Concentrate on all major variations from the standards.
5. Investigate the causes of these variations by cross-checking with other ratios and raw financial data.

An example of finding the causes of changes noted in ratios is the performance of Dell and Compaq as studied by Healy and Cohen (2000). The return on equity ratios of Dell and Compaq in the years 1995, 1996, and 1997 were compared. Table 7.9 displays the results.

Obviously, the results raised the question by many financial analysts of why the differences were so big. Additional analyses revealed interesting details. The following DuPont equation is pertinent:

$$\text{ROE} = \text{net income/equity}$$
$$= (\text{net income/sales}) \times (\text{sales/assets}) \times (\text{assets/equity}) \quad (7.3)$$

Table 7.10 presents the ratios calculated for these two companies during the three-year period.

TABLE 7.9 Return on Equity

Company	1997 (percent)	1996 (percent)	1995 (percent)
Dell	90	58	33
Compaq	22	22	22

TABLE 7.10 Ratios for Dell and Compaq

	1997	1996	1995
Dell			
Net income/sales	7.70 %	6.70 %	5.10 %
Sales/assets	3.40	3.02	2.83
Assets/equity	3.46	2.89	2.30
Compaq			
Net income/sales	7.50 %	6.60 %	5.40 %
Sales/assets	1.82	1.99	2.38
Assets/equity	1.61	1.69	1.69

It became clear that the operational efficiency of these two companies was comparable. However, Dell had a much higher rate of efficiency in asset utilization and took a much more aggressive stand with respect to financial leverage than Compaq. Specifically, Dell was able to (1) outsource most of its manufacturing activities; (2) keep a very low inventory based on its direct sales model (which allows Dell to receive customer orders from the Internet prior to initiating the needed manufacturing operation); (3) collect accounts receivables fast, thus reducing the need for self-financed working capital; (4) practice a high-leverage financing strategy by raising debt, which in turn produced high sales for the company. In contrast, Compaq was operating much more conservatively than Dell during the same three-year period.

In financial literature, many of the ratios just defined are being systematically collected and published for U.S. companies in various industries by investment services companies. Sources of information on ratios and other financial measures are typically reported regularly and made available for use by the public from such publications, including the following:

- *Value Line® Investment Survey*
- Standard & Poor's Industrial Surveys
- Moody's Investors Services
- Specific investment letters and publications (e.g., The Zweig Forecasts)

Engineering managers should develop a habit of reading and considering such reports.

Other commercial sources are accessible through the Internet. For the 500 stocks comprising Standard & Poor's index, five specific ratios—the current ratio, long-term debt to capital, net income to sales, return on assets, and return on equity—are published in a widely available special guide for 10-year, consecutive periods (Anonymous 2002B).

Example 7.5.

For the years 2001–2002, the financial statements of XYZ Company are given in Tables 7.11 and 7.12. Define the performance ratios and compare them with industrial standards.

Answer 7.5.

The 2001 performance ratios of XYZ Corporation are displayed here:

1. **Liquidity**

   (a) Current ratio = CA/CL = 3:1

   From the creditor's standpoint, this ratio should be as high as possible. On the other hand, prudent management will want to avoid the excessive buildup of idling cash or inventories (or both).

   (b) Quick ratio = quick asset/CL = 0.93:1

   A result far below 1:1 can be a warning sign.

   (c) Interest coverage ratio = 7.833

   The EBIT of the firm could pay 7.8 times the interest and other costs associated with the long-term debts. This ratio is good.

TABLE 7.11 XYZ Balance Sheet (thousands of dollars)

	2002	2001
Assets		
Cash	18,500	17,000
Marketable securities	0	5,000
Accounts receivable	39,500	28,500
Inventories	98,000	113,000
Total current assets	156,000	163,500
Plant and equipment (net)	275,000	290,000
Other assets	3,000	8,000
Total assets	434,000	461,500
Liabilities		
Accounts payable	34,500	18,000
Notes payable	20,000	25,000
Accrued expanses	18,500	11,500
Total current liabilities	73,000	54,500
Mortgage payable	20,000	30,000
Common stock	200,000	200,000
Earned surplus	141,000	177,000
Total liabilities and equities	434,000	461,500

TABLE 7.12 XYZ Income Statement (thousands of dollars)

	2002	2001
Sales	330,000	395,000
Cost of sales*	265,000	280,000
Gross profit	65,000	115,000
Selling and administrative	95,000	88,000
Other expenses	4,000	3,500
Interest	2,000	3,000
Profit before taxes	(36,000)	20,500
Federal income tax	0	10,000
Net income (loss)**	(36,000)	10,500

*Includes depreciation of $15,500 in 2001 and $15,000 in 2002
**No dividends paid in 2002

2. **Debt versus Equity**

   (d) Long-term debt to capitalization ratio = 7.37 percent
       This debt level is prudent for firms in this industry.

   (e) Total debt to owners' equity = 22.4 percent
       Total debt = CL + long-term debt
       OE = Common stock plus capital surplus plus accumulated retained earnings

   (f) Total debt to total asset ratio = 18.3 percent

3. **Activity**

   (g) Sales to asset ratio = 0.86

   (h) Ending inventory to sales ratio = 28.6 percent

   (i) CGS/average inventory = 2.65 times
       Average inventory = The average of ending inventory of two consecutive years (e.g., 2001 and 2002)

4. **Profitability**

   (j) Net income to owner's equity ratio = 2.8 percent

   (k) Net income to sales ratio = 2.66 percent

   (l) Gross margin to sales ratio = 29.1 percent

   (m) EBIT to total asset ratio = 5.1 percent

   (n) Net income to total asset ratio = 2.3 percent

   (o) EBIT to sales ratio = 5.9 percent

Selected investment services companies track various financial measures and report them regularly for use by the public. Engineering managers should develop a habit of reading and considering such reports.

7.5.3 Economic Value Added

Developed by Stern Stewart & Company in 1989, economic value added (EVA) is an improved valuation method for asset-intensive companies or projects. EVA is defined as the after-tax-adjusted net operating income of a company or unit minus the total cost of capital spent during the same accounting period. It is also equal to the return on capital minus the cost of capital, or the economic value above and beyond the cost of capital (Ehrbar 1998; Tully 1993). Sometimes EVA is also called "economic profit." In equation form, it is defined as

$$EVA = NOPAT - WACC \times (\text{Capital deployed}) \tag{7.4}$$

where

$$NOPAT = \text{net operating profit after tax (net income)}$$
$$\text{Capital deployed} = \text{total assets} - \text{current liabilities}$$
$$WACC = \text{weighted average of cost of capital (equity and debt)}$$
$$\text{employed in producing the earnings (Section 8.3.3)}$$

If EVA is positive, the company or unit is said to have added positive shareholder value. If EVA is negative, the company or unit is said to have diminished shareholder value.

EVA may also be applied to a single project by calculating the after-tax cash flows generated by the project minus the cost of capital spent for the project. For example, the NPV equation can be modified as follows:

$$\text{NPV} = -P + \left[\sum_{t=1}^{N} \frac{C_t - P \times \text{WACC}}{(1 + \text{WACC})^t} \right] + \frac{\text{SV}}{(1 + \text{WACC})^N} \qquad (7.5)$$

Here

$$P = \text{investment capital (dollars) for the project}$$
$$C_t = \text{net after-tax cash inflow (dollars) to be produced by the project in period } (t)$$
$$P \times \text{WACC} = \text{cost of capital (dollars) spent during the period } (t)$$
$$\text{WACC} = \text{weighted average cost of capital (percent) in effect}$$
$$\text{SV} = \text{salvage value of the project at the end of } N \text{ periods}$$
$$N = \text{number of period}$$

The major advantage of EVA over return on capital (ROC) is that it may encourage managers to undertake desirable investments and activities that will increase the value of the firm. The next example shows the difference between these two methods.

ABC company has established that its WACC is 10 percent, and its ROC standard for investment purposes is 14 percent. The management is considering a new capital investment that is expected to earn a return of 12 percent. This new investment is attractive according to the EVA criterion, as 12 percent is larger than 10 percent. However, this new investment is a poor choice if evaluated with the ROC criterion, because 12 percent is less than 14 percent. Thus, by using the ROC criterion to evaluate investments, the company may lose the opportunity to create shareholder value.

A 10-year study has shown that there is a general correlation between EVA and stock returns of innumerable companies. However, the correlation between EVA and wealth creation (in the form of stock price increase) is weak.

Among users of EVA, the following firms are known leaders in industry: AT&T, Eastman Chemical, Coca-Cola, Eli Lilly, Wal-Mart, and the U.S. Postal Service.

Engineering managers should learn to apply EVA in order to strengthen their financial accounting skills.

7.6 BALANCED SCORECARD

Financial ratios are developed by accountants, who naturally emphasize the companies' financial performance of direct interest to shareholders. Nonfinancial ratios are limited in number and restricted in scope. Examples include accounts receivable, collection, inventory, utilization of fixed assets, and working capital.

All financial ratios are determined on the basis of past performance data; they are "trailing" indicators, and as such they cannot foretell the future performance of a company. Because financial ratios are oriented to the short term, usually from one quarter to another, company management is inadvertently forced to overemphasize short-term financial results, oftentimes neglecting the company's long-term growth. The narrow focus of these financial ratios makes them no longer completely relevant to today's business environment, in which customer satisfaction, employee innovation, and continuous betterment of business processes are key elements of company competitiveness in the marketplace.

These basic deficiencies in financial ratios are well-recognized in industry. Attempts have been made in the past to modify these ratios as corporate measurement metrics. Kaplan and Norton (1992) suggest that corporate measurement metrics are to be defined to cover four areas:

- Financial—shareholder value
- Customers—time, quality, performance and service, and cost
- Internal business processes—core competencies and responsiveness to customer needs
- Innovation and corporate learning—value added to the customer, new products, and continuous refinement

The significance of the balanced scorecard lies in its balanced focus on both short-term actions as well as long-term corporate growth. What you measure is what you get. Kaplan and Norton (1992) advocate the use of a total of 15 to 20 metrics to cover these four areas to guide the company as it moves forward.

As an illustrative example, the balanced scorecard metrics for a manufacturing company may contain the following:*

A. **Financial**—cash flow, quarterly sales growth and operational income, increased market shares, and return on equity

B. **Customer**—percentage of sales from new products, percentage of sales from proprietary products, on-time delivery as defined by customers, share of key account's purchase, ranking by key accounts, and number of collaborative engineering efforts with customers

C. **Internal business process**—manufacturing capabilities versus competition, manufacturing excellence (cycle time, unit cost, and yield), design engineering efficiency, and new product introduction schedule versus plan

D. **Innovation and learning**—time to develop next-generation technology, speed to learn new manufacturing processes, percentage of products that equal 80 percent of sales, and new product introduction versus competition.

In general, balanced scorecard metrics for a given company must be built up according to its corporate strategy and vision, using a top-down approach. Doing so will ensure that performance metrics at lower management levels are properly aligned with the overall corporate goals. A unique strength of balanced scorecard metrics is that they link the company's long-term strategy with its short-term actions. These metrics contain forward-looking elements at the same time that they balance the internal and external measures. The creation of such metrics provides clarification, consensus, and focus on the desired corporate outcome.

According to a recent article (Anonymous 2000), United Parcel Service has achieved an increase of 30 to 40 percent in profitability with balanced scorecard metrics. Mobil Oil's North American Marketing and Refining Division raised its standing from last to first in its industry after having implemented a balanced scorecard. Catucci (2003) recommends that managers, when implementing a balanced scorecard, take personal ownership, nurture a core group of champions, educate team members, keep the program simple, be ruthless about implementation, integrate the scorecard into their own leadership systems, orchestrate the dynamics of scorecard meetings, communicate the scorecard widely, resist the urge for perfection, and look beyond the numbers to achieve cultural transformation of the company.

A widespread use of broad-based metrics, such as those suggested by the balanced scorecard, is likely to shift the attention of corporate management from a focus primarily on financial performance to other areas of equal importance, such as customers, internal business processes, and innovation and learning. Contributions by engineering managers made in these nonfinancial areas will likely become readily and more favorably recognized in the future.

7.7 CONCLUSION

This chapter introduces the basic accrual principle of financial accounting, presents the practice of T-accounting, and discusses the workings of income statements, balance sheets, and funds flow statements, with explanations of all applicable accounting terms. Also pointed out are the ways in which contributions by engineering managers are recorded in these financial statements.

Ratio analysis uses the financial data contained in these statements to assess companies' financial health. Economic value added (EVA) is described as an upgraded method of reporting the true financial value created by a company, unit, or a specific project.

The shortcomings of ratio analysis are outlined and a broad-based system of measurement metrics (the balanced scorecard) is illustrated. An adoption of this broad-based metrics system by corporate America will likely shift corporate management's attention from being predominantly focused on short-term financial results to a balanced emphasis on corporate long-term growths. This is possible with the use of metrics that address important competitive factors such as customer satisfaction, continuous improvement of internal business processes, and innovation for growth.

As such broad-based measurement metrics become widespread, the critical contributions made by engineering managers will be increasingly recognized and rewarded.

7.8 REFERENCES

Anonymous. 2000. "The Balanced Scorecard's Lessons for Managers." *Harvard Business Management Update*, October, p.2.

Anonymous. 2002A. *Harvard Business Essentials: Finance for Managers*. Cambridge, MA: Harvard Business School Press.

Anonymous. 2002B. *Standard & Poor's 500 Guide*, McGraw-Hill.

Bruns, W. J. 1996. "Introduction to Financial Ratios and Financial Statement Analysis." *Harvard Business School Note*, No. 9-193-029, August.

Catucci, B. 2003. "Ten Lessons for Implementing the Balanced Scorecard." *Harvard Business School Balanced Scorecard Report*, January.

Ehrbar, Al. 1998. *EVA: The Real Key to Creating Wealth*. New York: John Wiley.

Friedlob, G. T. and L. L. F. Schleifer. 2003. *Essentials of Financial Analysis*. New York: John Wiley.

Friedson, M. S. and F. Alvarez. 2002. *Financial Statement Analysis: A Practitioner's Guide*. 3d ed. New York: John Wiley.

Harrison, Jr., W. T. and C. T. Horngren. 1998. *Financial Accounting*, Upper Saddle River, NJ: Prentice Hall.

Hawkins, D. F. and J. Cohen. 2001. "A Conceptual Framework for Financial Reporting." *Harvard Business School Note*, No. 9-101-118 June 27.

Hawkins, D. F. and J. Cohen. 2002. "Balance Sheet." *Harvard Business School Note*, No. 9-101-108, May.

Healy, P. M. 2001. "Revenue Recognition." *Harvard Business School Note*, No. 9-101-017, February.

Healy, P. M. and P. Choudhary. 2001A. "Asset Reporting." *Harvard Business School Note*, No. 9-101-014, January.

Healy, P. M. and P. Choudhary. 2001B. "Liabilities Reporting." *Harvard Business School Note*, No. 9-101-016, November.

Healy, P. M. and P. Choudhary. 2001C. "Expenses Recognition." *Harvard Business School Note*, No. 9-101-015, February.

Healy, P. M. and J. Cohen. 2000. "Financial Statement and Ratio Analysis." *Harvard Business School Note*, No. 9-101-029, September.

Jablonsky S. F. and N. P. Barsky. 2001. *The Manager's Guide to Financial Statement Analysis*. 2d ed., New York: John Wiley.

Kaplan R. S. and D. Norton. 1992. "The Balanced Scorecard—Measures that Drive Performance." *Harvard Business Review*, January–February.

Troy, L. 1999. *Almanac of Business and Industrial Ratios*. Upper Saddle River, NJ: Prentice Hall.

Tully, S. 1993. "The Real Key to Creating Wealth." *Fortune*, September 30 p.38.

White, G. I., A. C. Sondhi, and D. Fried. 2003. *The Analysis and Use of Financial Statements*. New York: John Wiley.

7.9 QUESTIONS

7.1   Michele Brown started up a new consulting firm to offer design and testing services to industries. During the first three months of its operation, the firm has recorded a number of transactions, shown as follows:

a.   Michele Brown initiated the business by investing $35,000 in cash; $1,200 for office equipment; and $23,500 for instrumentation.

b.   Land for an office site was purchased for $17,500; $5,000 paid in cash and a promissory note signed for the balance.

c. A prefabricated building was purchased for $7,000 cash and moved onto the land for use as an office.

d. The premium of $800 was paid on an insurance policy.

e. A consulting project for Central Construction was completed and $900 was collected.

f. Additional test equipment costing $10,000 was purchased for $2,000 in cash, and a note payable for the balance was signed.

g. A consulting project was completed on credit for Eastern Manufacturing for $1,200.

h. Office supplies were purchased on credit for $300.

i. A bill was received for $250 and recorded as an account payable for rent on a special machine used for the Western Technologies project.

j. A project for Superior Design was completed on credit for $1,000.

k. Cash was received from Eastern Manufacturing for the consulting project they received on credit.

l. The wage of engineers, $5,000, was paid.

m. Office supplies purchased earlier (for $300) were paid for.

n. Cash of $125 was paid for repairs to test equipment.

o. The company president, Larry Brown, wrote a $65 check on the bank account of the consulting firm to pay for repairs to his personal automobile, which is not used for business purposes.

p. The wage of the office workers, $3,000, was paid.

q. The maintenance expenses of the instruments, $300, was paid.

Open the following T-accounts: Cash, Accounts Receivable, Prepaid Insurance, Test Equipment, Building, Land, Notes Payable, Accounts Payable, Office Supplies, Larry Brown—Capital, Larry Brown—Withdrawals, Sales Revenue, Instrument Maintenance Expense, Test Equipment Rental Expense, Office Equipment, Instrumentation and Test Equipment Repairs.

Record the transactions by entering debits and credits directly into the accounts. Use the transactions letters to identify each debit and credit entry. What is the cash account balance at the end of this three-month period?

Answer 7.1

The balance of the cash account is $13,575.

7.2 For the years 2001–2002, the financial statements of XYZ Company are given in Tables 7.11 and 7.12. Prepare a funds flow statement that delineates the sources and applications of funds for the XYZ Company during 2001 and 2002.

Answer 7.2

The total sources is $63,500.

7.3 The XYZ Company supplies products to a number of original equipment manufacturers (OEMs). It employs 5000 mostly unionized workers and generates about $2.2 billion in revenue annually. In 2002, it won an "Excellence Award" from one of its major client companies for having supplied an OEM product at the quality level of 10 defects per million in three consecutive years, thereby exceeding the specific target of 15 defects per million set by this client.

However, the corporate parent of the XYZ Company was less than happy with its financial performance. It declared this subsidiary a "Troubled Operation" in 2002, giving notice that if its business performance did not improve within a certain timeframe, the

XYZ Company would be closed down or divested. The specific financial targets set by the corporate parent consisted of (1) net income (NOPAT) to sales ratio at 5 percent, (2) earning before interest and tax (EBIT) to sales ratio at 10 percent, and (3) return on assets at 22.5 percent.

Needless to say, management worked diligently to find ways to improve the company's business performance. One strategy was to induce 700 early retirements from the unionized workforce by November 2002. As of April 2002, only 480 workers signed up for this heavily promoted early retirement program that qualified them each to receive a $35,000 cash incentive.

Besides encouraging workers to retire earlier, what other strategies do you think the XYZ Company should pursue to achieve all three of the noted financial targets?

7.4 The company is considering the introduction of a new product that is expected to reach sales of $10 million in its first full year and $13 million of sales in the second and third years. Thereafter, annual sales are expected to decline to two-thirds of peak annual sales in the fourth year, and one-third of peak sales in the fifth year. No more sales are expected after the fifth year.

The cost of goods sold is about 60 percent of the sales revenues in each year. The sales general and administrative (SG&A) expenses are about 23.5 percent of the sales revenue. Tax on profits is to be paid at a 40 percent rate.

A capital investment of $0.5 million is needed to acquire production equipment. No salvage value is expected at the end of its 5-year useful life. This investment is to be fully depreciated on a straight-line basis over 5 years.

In addition, working capital is needed to support the expected sales in an amount equal to 27 percent of the sales revenue. This working capital fund must be made at the beginning of each year to build up the needed inventory and implement the planned sales program.

Furthermore, during the first year of sales activity, a one-time product introductory expense of $200,000 is incurred. Approximately $1.0 million had already been spent promoting and test marketing the new product.

A. Formulate a multiyear income statement to estimate the cash flows throughout its five-year life cycle.
B. Assuming a 20 percent discount rate, what is the new product's net present value?
C. Should the company introduce the new product?

Answer 7.4

The net present value is −$4.8 million, and the product should not be introduced.

7.5 The XYZ Corporation has current liabilities of $130,000 with a current ratio of 2.5:1. Indicate whether the individual transactions specified next increase or decrease the current ratio or the amount of working capital and by how much in each case. Treat each item separately.

1. Purchase is made of $10,000 worth of merchandise on account.
2. The company collects $5,000 in accounts receivable.
3. Repayment is planned of note payable which is due in current period, with $15,000 cash from bank account.
4. The acquisition of a machine priced at $40,000 is paid for with $10,000 cash, and the lump-sum balance is due in 18 months.
5. The company conducts a sale of machinery for $10,000. Accumulated depreciation is $50,000, and its original cost is $80,000.

6. The company pays dividends of $10,000 in cash and $10,000 in stock.

7. Wages are paid to the extent of $15,000. Of this amount, $3000 had been shown on the balance sheet as accrued (due).

8. The company borrows $30,000 for one year. Proceeds are used to increase the bank account by $10,000 to pay off accounts due to the supplier ($15,000) and to acquire the right to patents ($5,000).

9. The company writes down inventories by $7,000 and organization expenses by $5,000.

10. The company sells $25,000 worth (cost) of merchandise from stock to customers who pay in 30 days. Company has a gross margin of 40 percent.

Answer 7.5

The working capital is as follows: (1) unchanged, (2) unchanged, (3) unchanged, (4) $185,000, (5) $205,000, (6) $185,000, (7) $190,000, (8) $188,000, (9) $183,000, and (10) $211,667.

7.6 The income statement and balance sheet of Superior Technologies are exhibited in Tables 7.A1 and 7.A2. Conduct a ratio analysis and observe major trends and deviations from the available historical company information and industry data. (See also Table 7.A3.)

TABLE 7.A1 Income Statement of Superior Technologies (thousands of dollars)

	Year 2002	Year 2001
Net sales	193,213	91,954
Cost of goods and services	128,434	60,776
Gross profit	64,779	31,178
Selling, general, and administrative expenses	26,369	13,844
Employee profit sharing and retirement	9,831	4,167
Total overhead	36,200	18,011
Operating profit before interest and tax	28,579	13,167
Other income	956	273
EBIT	29,535	13,440
Interest paid	680	505
Taxable income	28,855	12,935
Income tax	14,712	6,934
Net income	14,143	6,001

Answer 7.6

Selected answers for 2002 are as follows:

days receivables = 54.1 days
total long-term debt to net worth = 0.77
total asset turnover = $1.8x$
gross profit (margin) = 33.5 percent

7.7 For the current fiscal year that starts on January 1, 2002, Buffalo Best Company projects its financial performance as shown in Table 7.A4.

TABLE 7.A2 Balance Sheet of Superior Technologies (thousands of dollars)

	Year 2002	Year 2001
Cash	17,856	10,841
Receivables (net)	29,053	19,350
Inventories	23,282	6,869
Prepaid expenses	664	315
Payments received on government contracts	−6,013	−405
Total Current Assets	64,842	36,970
Property, plant, and equipment	60,806	26,773
Accumulated depreciation	20,083	10,281
Net property account	40,723	17,492
Patents, etc. (net)	0	249
Other assets	429	80
Total Assets	105,994	53,791
Accounts payable	10,368	5,337
Accrued wages, pensions, taxes, etc.	21,309	10,696
Other current liabilities	5,589	2,867
Total Current Liabilities	37,266	18,900
Long-term debt	29,935	9,250
Preferred stock	3,265	0
Common stock	3,915	3,257
Paid in surplus	8,205	6,228
Retained earnings	23,408	16,156
Total Liabilities and Net Worth	105,994	53,791

TABLE 7.A3 Selected Information of Superior Technologies

	2002	2001
(Depreciation and amortization:	$8,135	$4,924)
(Preferred dividends	102	0)
(Other dividends	185	0)

	Past Ratios Fill in 2002	2001	2000	1999	1998	1997	1996
Current ratio (ratio)		2	2	2.2	2.5	2	1.9
Acid test		1.6	1.2	1.3	1.7	1.3	1
Total debt to total assets (percent)		52.3	47.9	41.2	30.4	46.0	53.3
Long-term debt as percentage of capitalization		26.5	26.3	16.8	8.4	8.3	25.4

(*Continued*)

TABLE 7.A3   (Continued)

	Fill in 2002	Past Ratios					
		2001	2000	1999	1998	1997	1996
Total debt to net worth (ratio)		1.1	0.9	0.7	0.4	0.8	1.1
Days' receivables (days)		75.7	66.3	82.2	96.7	81.9	65.5
Ending inventory turnover (sales)		13.4	7.8	7	7.6	8.1	5.7
Ending inventory turnover (cost of sales)		8.8	5.5	5.1	5.5	5.7	4.1
Net property turnover		5.6	4.4	4.8	4	5.6	6.9
Total assets turnover		1.7	1.8	1.7	1.5	1.6	1.8
Net profit to total assets (percent)	·	11.2	10.0	8.6	8.1	7.9	8.5
Net profit to net worth (percent)		23.4	19.2	14.6	11.6	14.7	18.3
Net profit to net sales (percent)		6.5	5.6	5.1	5.5	4.9	4.7
Gross profit		34.0	29.5	26.5	27.4	29.3	29.2
Industry Information							
Current ratio (ratio)		2.3	2	2.1	2.1	2.2	1.6
Acid test		1.2	1	1	1	1.1	0.6
Total debt to total assets (percent)		40.0	52.8	46.1	47.2	42.6	56.1
Long-term debt as percentage of capitalization		12.3	16.7	17.1	14.4	11.3	7.8
Total debt to net worth (ratio)		0.7	1.1	0.8	1.1	0.7	1.3
Days' receivables (days)		44	45	42	42	39	33
Ending inventory turnover (sales)		6.2	5.3	4.8	5.5	5.4	4.7
Ending inventory turnover (cost of sales)		4.8	4.1	3.8	4.2	4.4	3.8
Net property turnover		8.7	10.1	7.1	8.4	9	12.2
Total assets turnover		2	1.9	1.7	1.9	2	2.2
Net profit to total assets		6.1	5.2	4.8	5.6	4.4	5
Net profit to net worth (percent)		10.2	11.0	9.0	10.6	7.7	11.4
Net profit to net sales (percent)		3.1	2.7	2.8	2.9	2.3	2.3
Gross profit (percent)		22.8	21.7	20.6	23.2	18.6	19.9
Number of companies reported:		60	42	55	65	73	60

Note: Data derived from several literature sources, 1996–2000.

In fiscal year 2002, the products of the company are priced at $10,000 each. The company president predicts that, because of the generally anticipated recovery of the U.S. economy, there will be a 4-percent-per-year increase in each of the next four years (i.e., 2003 to 2006) in (a) price of the company's products and (b) number of products sold. However, the company's administrative and selling expenses, as well as the product cost per unit, are projected to increase by 7 percent per year. Other expenses, such as R&D, interests, depreciation, and corporate tax rate, will remain unchanged. The anticipated increase in product cost is primarily due to an increase in materials costs and a decrease in labor productivity because of high turnover rate, poor employee supervision, inadequate workforce compensation policy, cumbersome production processes, and inferior employee training. If nothing is done, the company expects its net income to decrease by as much as 50 percent around 2005 to 2006.

At a management meeting on February 12, 2002, the vice president of engineering proposed to hire a consulting firm in the second half of 2002 to develop a customized

TABLE 7.A4   Income Statement of Buffalo Best

	Year 2002 (Thousands of dollars)
Sales	20,000
Cost of goods sold	13,000
Gross margin	7,000
Administrative and selling expenses	4,000
R&D	500
Depreciation	1,000
Earning before interest and tax (EBIT)	1,500
Interest	100
Taxable income	1,400
Tax (30 percent)	420
Net operating profit after tax (NOPAT)	980

program for improving labor productivity. She noted that three options are currently available at different investment prices and projected CGS reduction levels:

Program	Investment	CGS Reduction Percentage
A	$700,000	1.0%
B	$1,100,000	2.0%
C	$1,200,000	3.0%

The investment for any one of these options will be a one-time lump sum that can be expensed as an increase in overhead in 2002. The benefits of the consultation program will be realized in a projected reduction of product cost per unit (i.e., the originally anticipated product cost before the management consultation program) by a percentage (1, 2, or 3 percent, respectively) per year for the next four years (i.e., 2003 to 2006).

The company president welcomes the idea of management consultation as a possible way to enhance workforce productivity. Assuming that the company's cost of money is 10 percent, he wants to know which management consultation program he should approve.

Answer 7.7

Program C is to be preferred.

7.8 The company's vice president of marketing proposes a new program to significantly increase the product sales by 250,000 units per year throughout the 1998–2004 period. Specifically, it is suggested that the company takes the following actions:

1. Spend $2.5 million over the period of 1998–2000 as promotional expenditures—for example, spend $1.0 million each in the years 1998 and 1999, and $0.5 million in the year 2000.

2. Make a one-time investment of $1.4 million in plants and equipment needed at the beginning of 1998 to generate these additional products. No new warehouse capability is needed. This investment is to be depreciated on a straight-line basis over the seven-year period. There will be no salvage values for these plants and equipment in 2005.

It is further assumed that the product unit cost is $8.00 in 1998, and it is estimated to increase by 3 percent per year. The product unit price is $20 in 1998, and it is estimated to change as manifested in the following table:

Items	1998	1999	2000	2001	2002	2003	2004
Unit Price	$20.00	$20.60	$21.00	$21.15	$21.25	$21.25	$21.00

The SG&A expenditure is estimated at $1.25 million in 1998, and it will increase by 3 percent per year during the six-year period. A corporate tax of 40 percent must be paid for any marginal income. There is an interest charge during this period, and the company's weighted average cost of capital (WACC) is 8 percent.

If the company's hurdle rate for this type of investment is 25 percent, would you recommend that the marketing initiative be approved?

Answer 7.8

No, the marketing initiative should not be approved.

7.9 One of the company's technology patents is about to expire, inducing a likely rush of product entries from the competition. Currently, the product line is projected to have stagnant sales of 1 million units for the next seven years from 1998 to 2005. The unit price is expected to decrease slightly by 1 percent per year during the same period.

The profitability of this product is estimated to be $1,166,000 in 1998, as indicated in Table 7.A5.

TABLE 7.A5   Simplified Income Statement (thousands of dollars)

	1998
Sales	20,000.00
CGS	10,000.00
Depreciation	1,057.00
SG&A	7,000.00
EBIT	1,943.00
Tax at 40 percent	777.00
EBIAT	1,166.00

Both the cost of goods sold (CGS) and SG&A expenses are expected to increase by 3 percent per year from 1998 to 2004. The depreciation charge is estimated to remain constant at $1 million per year, and there is no salvage value for the equipment at the end of 2004. If the product is continued as planned, the company can recover a working capital of $3.9 million at the end of 2004 from sales of residual inventory and collection of accounts receivable after having discharged all applicable short-term liabilities.

If the management decides to discontinue this product line at the end of 1997, then it can sell the fixed assets related to the product line (having a book value of $7 million) for about $3 million, and the loss of $4 million would be tax deductible. Furthermore, the company can recover the working capital (inventory plus accounts receivable, minus accounts payable and other expenses) worth about $3.9 million at the end of 1997.

Assuming that the appropriate discount rate is 12 percent, would you recommend that this product line be discontinued at the end of 1997 or be continued through 2004?

What is the next best alternative open to the company, besides either shutting it down immediately at the end of 1997 or continuing to run it until 2004?

Answer 7.9

The best option is to shut down the production line at the end of 1997. The next-best alternative is to discontinue the product line at the end of 2001.

7.10 Define the economic value added (EVA) of XYZ Company. (See income statement and balance sheet of Question 7.2 for the year 2001–2002.) Assume that the WACC is 12.35 percent for both years. Discuss the EVA results and contrast them with the EVA results for Superior Technologies in 2001 and 2002 (see Question 7.6) for which WACC is also assumed to be 12.35 percent.

Answer 7.10

The EVA of XYZ is −39,764,500 in 2001 and −80,583,500 in 2002. The EVA of Superior Technologies is $5,655,092 in 2002.

7.11 The 2000–2001 income statement and balance sheet for Buffalo Best are presented in Tables 7.A6 and 7.A7. The WACC for Buffalo Best is 10 percent for both years. Review and comment on the performance of the company based on the following:

A. Liquidity, activity, and profitability
B. Uses and sources of funds
C. Value creation based on EVA analysis

TABLE 7.A6   Income Statement of Buffalo Best (thousands of dollars)

	Year 2001	Year 2000
Sales	18,000	17,000
Cost of goods sold	11,000	10,500
Gross margin	7,000	6,500
Administrative and selling expenses	3,500	3,200
R&D	500	500
Depreciation	1,000	1,000
Earning before interest and tax (EBIT)	2,000	1,800
Interest	100	100
Taxable income	1,900	1,700
Tax (30 percent)	570	510
Net profit after tax (NOPAT)	1,330	1,190
Dividends	330	190
Retained earnings	1,000	1,000

Answer 7.11

The sources total is $10.53 million. The EVA is −$2.17 million in 2001 and −$2.46 million in 2002.

7.12 Company X manufactures technology products. It plans to expand its manufacturing operations. Based on past data, management anticipates the first project year of the as-yet-to-be-expanded operations to match the data in Table 7.A8.

A. Compute the working capital requirement during this project year.
B. Determine the taxable income during this project year.

C. Calculate the net income during this project year.

D. Define the net cash flow from this project during the first year.

Answer 7.12

(a) $290,000; (b) $680,000; (c) $408,000; (d) $608,000.

TABLE 7.A7   Balance Sheet of Buffalo Best

	Year 2001 (dollars)	Year 2000 (dollars)
Assets		
Cash and securities	5,000	6,000
Accounts receivable	15,000	10,000
Inventory	10,000	7,300
Net fixed assets		
(Investment minus accumulated depreciation)	50,000	51,000
Other	1,000	1,200
Total assets	81,000	75,500
Liabilities		
Accounts payable	20,000	15,000
Other short-term liabilities	26,000	24,000
Long-term liabilities	1,000	1,500
Total liabilities	47,000	40,500
Equities at par value	1,000	1,000
Capital surplus	12,000	14,000
Retained earnings	21,000	20,000
Total liabilities and owners' equities	81,000	75,500

TABLE 7.A8   Selected Data of Company X

	(dollars)
Sales	1,500,000
Manufacturing costs	
Direct materials	150,000
Direct labor	200,000
Overhead	100,000
Depreciation	200,000
Operating expenses	150,000
Equipment purchase	400,000
Borrowing to finance equipment	200,000
Increase in inventories	100,000
Decrease in accounts receivable	20,000
Increase in wages payable	30,000
Decrease in notes payable	40,000
Income taxes	272,000
Interest payment on financing	20,000

Chapter 8

Managerial Finance for Engineering Managers

8.1 INTRODUCTION

Managerial finance focuses on the sources and uses of funds in a company. Capital budgeting is the key responsibility of the company's top management. Capital budgeting decisions take into account the need to expand production facilities, develop new products, create new supply chains, acquire new technologies to complement the company's own core competencies, penetrate into new regional markets, and other needs. Besides defining what is worth doing, top management must also find the investment capital needed to execute the chosen short-term and long-term strategies (Davis and Pointon 1994; Weston and Brigham 1993).

Investment capital may be raised from either internal sources (retained earnings) or external sources (equity and debt financing). Equity and debt financing has advantages and disadvantages. Topics relevant to capital formation include the weighted average cost of capital (WACC), capital structure, level of leverage and risks, and impact of leverage on the company's financial performance (Shim and Siegel 2000; Brealey and Myers 2000).

Resource allocation drives the company's overall performance. Allocation decisions address three major ways of utilizing investment capital: (1) assets in place—building new facilities and developing new products, (2) marketing and R&D efforts, and (3) strategic partnerships—acquisitions and joint ventures. In deciding on resource allocation, company management needs to know what a specific endeavor is worth. Then, different ways of estimating the value of companies, projects, and opportunities need to be established.

This chapter covers the valuation of (1) assets in place by the discounted cash flow methods—both the WACC and applied present value (APV) versions—and (2) R&D and marketing opportunities by the simple option method. No discussion about the valuation of strategic partnerships is included, as engineering managers are not often involved in the financing aspects of acquisitions and joint ventures.

Engineering managers need to understand the language, principles, and practices of financial management (Droms 1998). Doing so will allow them to be ready to participate in capital budgeting decision making that involves major projects in which engineering managers provide significant inputs.

8.2 ELEMENTS OF MARKET ECONOMY

Besides the use of the net income earned by the company to finance capital projects, equity and debt financing are the typical methods of raising investment capital. *Equity financing* is defined as raising capital by issuing company stocks. *Debt financing* is defined as issuing corporate bonds to the public or taking long-term loans from financial institutions (or both).

Raising capital by either equity or debt financing requires that the company interact with the public (investors) or financial institutions (banks). To be successful in raising investment capital in a free market economy, the company needs to understand several factors that shape such a free-market economy. These factors are firms, investors, and risks.

8.2.1 Role of Firms

Firms play a central role in a free-market economy. They pursue certain activities, such as facilitating transactions by supplying goods or services and making decisions related to the use (investment) or acquisition (debt, equity) of money. They choose the most suitable sources to raise capital and decide how best to use the capital made available from these sources. By performing these managerial tasks, companies attempt to achieve their objectives of maximizing the value created for their shareholders, in both the short term and long term (van Horne 2002; Norton 2003).

In studying the management practices in Japan and the United States, innumerable business researchers have noted that a large number of American companies appear to have been managed by executives who emphasize short-term profit optimization. This short-term profit focus has been partially driven by financial incentives that are short-term oriented (bonuses, stock options, and rapid advancement tied to short-term improvements in earning per share and other such ratios, from one accounting period to the next) and are partially induced by the prevalent value system that favors quick action and fast results. Such a short-term managerial mindset tends to cause certain U.S. companies to continue losing their long-term competitiveness in the global markets (Terry 2000).

The proper roles of firms should be to perform the just-described activities in ways that achieve a balanced optimization of both the short-term and long-term values to the shareholders. (The generic roles of firms are depicted in Figure 8.1.)

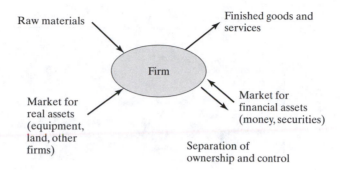

Figure 8.1   Roles of firms.

8.2.2 Investors

Investors purchase securities (i.e., stocks and bonds) to derive financial benefits and concurrently to support selected business activities. Their financial objective is primarily to maximize a return on their investment while minimizing risks. Their key concerns are the amount of return realizable by the investment and the risks incurred in preserving the investment.

Besides considering the financial objective, investors select investment targets carefully. Investors favor those companies with a good reputation, widely recognized brand names, and proven management quality. Since most investors may freely choose where to invest their money, they tend to stay away from companies that have committed socially irresponsible, environmentally unacceptable, ethically doubtful, or legally questionable practices.

Companies planning to raise capital in financial markets need to keep their investors' needs and perceptions in mind.

8.2.3 Risks

In general, the outcome (i.e., earnings) of any investment has a degree of inherent uncertainty; large in some and small in others. Investment risk is defined as the measure of potential variability of earning from its expected value. It is usually modeled mathematically by the standard deviation of the outcome when the outcome is expressed in the form of a probability density distribution function (e.g., Gaussian distribution function, see Section 6.9). For common stocks, risk is modeled by a relative volatility index, Beta, as defined in Section 8.3.1.

The rate of return on risky security can be modeled as the risk-free rate plus a risk premium:

$$r = R_f + Rp \tag{8.1}$$

The risk-free rate (e.g., R_f equals some constant rate such as 6.0 percent for 10-year U.S. Treasury bills) is the return for compensation of opportunity cost without uncertainties. The risk premium is the additional return needed to compensate for the added risks the investors undertake.

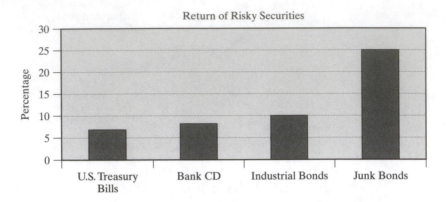

Figure 8.2   Risky securities.

Figure 8.2 illustrates several investment examples and displays the general risk-reward correlation between expected annual return and the associated risk of the investment in question. An investor may realize only 6 percent from the risk-free U.S. Treasury Bills, but a whopping 25 percent from highly precarious junk bonds.

A. *Expected Value* is the return of an outcome multiplied by its probability of occurrence. For a portfolio containing N independent investments, the total expected return is the sum of the products of individual returns and its respective probabilities of occurrence. The total expected return is the average return weighted by probability factors:

$$\text{EV} = [\Sigma P_i \times R_i; i = 1 \text{ to } N] \tag{8.2}$$

$$R_i = \text{Return of investment } I$$

$$P_i = \text{Single-valued probability of occurrence for } R_i$$

$$[\Sigma P_i; i = 1 \text{ to } N] = 1.$$

B. *Risk Aversion* delineates the generally expected unwillingness on the part of investors to assume risks without realizing the commensurate benefits. Most investors are unwilling to assume risks unless there are incremental benefits (the risk premium) that compensate them for bearing the added risks involved. Some audacious investors are more willing than others to take on additional risks to realize the added benefits.

Table 8.1 exhibits an example in which the behavior of risk-averse investors can be studied. Investment A has a nominal value of $100 when the economy is in a normal state. This investment is projected to be valued at $90 if the economy goes into a recession in the near future. On the other hand, its value may increase to $110 if the economy booms. Investment B also has a nominal value of $100 when the economy is normal, but its value decreases to zero in a recession and increases to $200 in a booming economy.

TABLE 8.1   Behavior of Investors

State of Economy	Investment A (dollars)	Investment B (dollars)	Probability
Recession	90	0	0.333
Normal	100	100	0.333
Boom	110	200	0.333
Expected payoff value	100	100	
Range	90–110	0–200	
Amount at risk	−10	−100	

Note: Conservative persons would choose Investment A, whereas risk-preferring persons would choose Investment B. Risk-neutral persons would not have a preference.

If we further assume that there is an equal probability of 33.33 percent that the state of the future economy will be either normal, in recession, or in a booming state, then the expected pay-off value of these two investments is identical (e.g., $100); however, their amounts at risk are different: −$10 for Investment A and −$100 for Investment B.

Among these two investment options, a risk-averse investor will choose Investment A for its lower downside risk; he or she will not choose Investment B because there is no gain in return for the added risks. A risk-preferring investor will choose Investment B for its reward potential of doubling the money if the economy booms in the future.

8.3 CAPITAL FORMATION

Capital formation refers to activities undertaken by a company to raise capital for short-term and long-term investment purposes.

In general, the net income earned by most companies, plus their internal resources of retained earnings, are not sufficient to finance all investments needed to achieve both their short-term and long-term corporate objectives. Even if these internal resources are sufficient, some companies still engage in external financing due to good reasons related to tax and strategic management flexibility. Many companies pursue some external resources, such as equity or debt financing (or both).

8.3.1 Equity Financing

Equity financing is the raising of capital by issuing company stocks. Stocks are certificates of company ownership, which typically carry a par value of one dollar. The price of the same stock in the the market (e.g., the New York Stock Exchange) may fluctuate in time as a consequence of economic conditions, industrial trends, political stability, and other factors unrelated to company performance. Shareholders are those who own stocks. They have the residual claims to what is left of the firm's assets after the firm has satisfied other high-priority claimants (e.g., bondholders, bankruptcy lawyers, and unpaid employees). Basically there are two kinds of stocks: *common stocks* and *preferred stocks* (whose dividends have a priority over that of common stocks).

Companies may issue new stocks to raise capital. The process requires the approval of the company's board of directors (which mirrors the interests of all shareholders) because such a move may result in the dilution of company ownership. In addition, a public offering needs to be registered with the U.S. Securities and Exchange Commission and organized by an underwriting firm. The underwriting firm helps set the offering price, prepares the proper advertisements, and assists in placing all unsold issues to ensure a successful completion of the equity financing process. Companies issuing new stocks will incur an issuing expense.

The capital received by issuing stocks has a cost with the following components: the cost of equity capital, which includes the stock issuing fees, the dividends to be paid in future years, and the capital gains through stock price appreciation expected by investors in the future. Typically, this cost of equity is set to equal the return of an equity calculated by using the capital asset pricing model (CAPM).

This well-known capital asset pricing model (CAPM) defines the return of an equity as (see Figure 8.3)

$$r = R_f + \beta(R_m - R_f) \tag{8.3}$$

where

$R_f =$ Risk-free rate (e.g., 6.0 percent of 10-year U.S. Treasury bills).

$R_m =$ Expected return of a market portfolio, a group of stocks representing the behavior of the entire market (e.g., S&P 500 Index). R_m depends on conditions not related to the the individual stocks. The typical value for R_m is in the range of 15 percent, based on long-term U.S. market statistics.

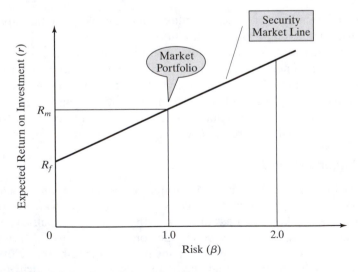

Figure 8.3   Security market line.

$\beta =$ Relative volatility of a stock in comparison to that of a market portfolio, which by definition has a Beta of 1.0. The Standard & Poor's 500 stocks serve as a proxy for the overall market. If the Beta of a stock is 1.5, then its price will change by 1.5 percent for every 1.0 percent price change of the market portfolio in the same direction. An issue with a Beta of 0.5 tends to move 50 percent less. A stock with a negative Beta value tends to move in a direction opposite to that of the overall market. Stocks with large Beta values are more volatile and hence more risky. Beta values are published in the financial literature for most stocks that are publicly held.

$R_m - R_f =$ Market risk premium.

Since the value of Beta is based on historical data, numerous financial analysts view it as a major deficiency of the CAPM model. Another deficiency is its lack of timing constraints. The same cost is assumed to be valid for capital projects of a 2-year or 10-year duration. In practice, when evaluating a specific capital investment project, some companies adjust the value of Beta manually in order to arrive at a pertinent cost of equity capital.

Recently, McNulty, Yeh, Schulze, and Lubatkin (2002)* proposed the market-derived capital pricing method (MCPM) to calculate the cost of capital. According to this method, the cost of equity capital consists of two parts: the risk-free premium and the equity return risk premium. The risk-free premium is set to equal the yield of corporate bonds currently traded in the bond market. MCPM uses a *put option* to secure the capital gain in stock price within a fixed-duration period (e.g., 5 years); this is needed to ensure that the expected return for investors is greater than the dividend yield. The annualized cost of the put option is then divided by the company's current stock price to arrive at an *equity return risk premium*. The sum of corporate bond yield and the equity return risk premium then becomes the cost of the company's equity capital for the fixed-duration period involved. The McNulty et al. article presents a specific numerical example in which the cost of equity for General Electric was calculated. Based on a comparison of many sets of company data, McNulty and his coauthors further claim that MCPM yields more realistic costs of equity than CAPM.

8.3.2 Debt Financing

Debt is the liabilities incurred by the company to make contractual payments (e.g., interest payments) under specified terms. Debt represents a fixed prior claim against the company's assets and poses a financial risk to the company. Debt financing by issuing industrial bonds or taking loans from financial institutions produces creditors. Debts are usually secured by a certain part of the company's assets. Creditors have the legal power to enforce payments and thus potentially drive companies into bankruptcy. When a company declares bankruptcy, it must satisfy the claims of creditors in a specific order: (a) secured debts (bonds or loans), (b) lawyers' fees, (c) unpaid wages, and

---

(d) stockholders. Note that bankruptcy lawyers have a payment priority ahead of the hard-working employees and risk-averse shareholders.

Companies seeking debt financing through the issuance of bonds also need to engage an underwriting firm and follow a specific set of steps. There are several types of bonds: (a) corporate bonds, (b) mortgage bonds, (c) convertible bonds (convertible to stocks at a fixed price by a given date), and (d) debentures (unsecured bonds). The length of debts may be short term (less than one year's duration), intermediate (one to seven years), or long term (more than seven years). There is also a fee associated with this type of public offering.

Companies seeking debt financing through loans will typically negotiate for terms and conditions directly with the lending financial institutions involved. The interest rate charged is usually the prevalent prime rate plus a surcharge rate. The prime rate is published by the Federal Reserve Board on a regular basis as a means to control the liquidity of the financial markets. The surcharge rate varies with the company's credit rating, which in turn depends on the company's past financial performance and future business prospects. The higher is the company's credit rating, the lower is this surcharge rate. Hence, better performing companies may borrow money at cheaper rates. This reflects the lower risks involved for the lenders.

The cost of debt capital includes the issuing fee, bond rate, or interest rate to be paid in future periods, the opportunity cost associated with the diminished company growth opportunity, and other costs.

The opportunity cost related to reduced growth opportunity results from the fact that highly leveraged companies can no longer be as aggressive in pursuing new growth business opportunities. The obligatory interest burdens tend to temper the company's otherwise bold investment strategies. These burdens also constrain the company's investment flexibility and thus cause the company to lose the potential benefits which they could have otherwise realized from such new opportunities. Examples of such new opportunities include the development of new products, engagement in leading-edge R&D, entrance into new global markets, and creation of new and innovative supply-chain partnerships.

The other costs could result from one or more of the following: (1) suboptimal operational policies that aim at the lower end of a range of sales forecasts, (2) vulnerability of the company to attacks by competitors, (3) the company's inability to access additional debt capital, if needed, and (4) the cost of bankruptcy.

Example 8.1.

> Innovative Products is a company that has enjoyed a high growth rate in recent years. Its growth has been largely financed by the retained earnings that belong to the common stockholders. (See Table 8.2.)
>
> For the last three years the company has earned an average net income of $75,000, after having paid an annual interest of $8,000 and the annual taxes of $50,000. The company's tax rate is 40 percent.
>
> Company management is considering the strategy of raising $750,000 to double its production volume. Of this amount, $500,000 would be used to (1) build an addition to the current office building, (2) purchase new IT equipment, and (3) install an advanced ERP software system. The remaining amount would be needed for working capital to add inventories and to enhance marketing and selling activities.

TABLE 8.2   Balance Sheet of Innovative Products (2002)

	Year 2002 (dollars)
Cash	40,000
Accounts receivable	160,000
Notes receivable	70,000
Inventories	260,000
Prepaid items	16,000
Total Current Assets	546,000
Land	25,000
Building (net)	165,000
Equipment (net)	350,000
Total Assets	1,086,000
Accounts payable	87,000
Notes payable	80,000
Taxes payable	60,500
Total Current Liabilities	227,500
Long-term loan (due 2007, 5 percent interest)	100,000
Common stocks	400,000
Retained earnings	358,000
Total Liabilities and Owners' Equities	1,086,000

Market research suggests that, in spite of a doubling of the company's sales volume, the product price can be kept at the current level. The EBIT is projected to be $275,000. Two specific financing options are to be evaluated closely:

1. Sell enough additional stock at $30 per share to raise $750,000.
2. Sell 20-year bonds at 5 percent interest, totaling $500,000.

Determine which financing option is to be favored from the standpoint of earning per share.

Answer 8.1.

Earning per share is defined as the company's net income divided by the number of outstanding common stocks. To compare the earning-per-share data for these two financing options, the income statement must be constructed as in Table 8.3. The following explanations may be helpful:

1. From the known values of net income ($75,000), tax rate (40 percent) and interest payment ($8,000) of the present operation, the EBIT is calculated to be $133,000.
2. For the present operation, the number of outstanding common stock is 400,000, because the common stock (usually at the par value of $1) is listed as $400,000 in the company's balance sheet. (See Table 8.2.)

TABLE 8.3   Income Statement of Innovative Products

	Present Operation No Financing (dollars)	Option 1 Equity Financing (dollars)	Option 2 Equity & Debit Financing (dollars)
EBIT (earnings before interest and tax)	133,000	275,000	275,000
Interest	8,000	8,000	25,000
Taxable income	125,000	267,000	250,000
Taxes (40 percent)	50,000	106,800	100,000
Net income	75,000	160,800	150,000
Outstanding shares of stocks	400,000	425,000	410,000
EPS (earning per share)	0.1875	0.3769	0.3658

3. For Option 1 and Option 2, the EBIT is known to be $275,000.
4. For Option 1, the interest charge remains at $8,000, but the number of outstanding stock is increased to 425,000.
5. For Option 2, the interest charge is increased by $25,000 annually due to the new loan. However, the old loan is being paid off, reducing the annual interest payment by $8,000. Thus, the net interest payment is $25,000 per year.
6. For option (2), the number of outstanding stock is increased to 410,000.

Based on this analysis, the earning per share of Option 1 is larger than those of the present operation and Option 2. Equity financing is to be preferred.

8.3.3 Weighted Averaged Cost of Capital (WACC)

The weighted average cost of capital (WACC) is a very important cost figure for any company. It is defined as

$$\text{WACC} = K_e\left(\frac{E}{V}\right) + K_d(1 - t)\left(\frac{D}{V}\right) \tag{8.4}$$

where

$D =$ debt (long-term loans, corporate bonds, etc.) (dollars)
$E =$ equity (stocks) (dollars)
$t =$ corporate tax rate (percent)
$V = E + D$ (dollars)
$K_e =$ cost of equity capital (e.g., $= 0.15$ to 0.18)
$K_d =$ cost of debt capital [e.g., $0.08 =$ yield to maturity (YTM)
rate for bonds, plus cost associated with lost growth opportunity]

In general, an increase in leverage (e.g., adding more debts) reduces the firm's WACC. This is due to the tax-deductibility of interest payments associated with the debt. For many U.S. companies, WACC is typically in the range of 8 to 16 percent.

8.3.4 Effect of Financial Leverage

Financial leverage denotes the use of debts in financing corporate projects. A measure of financial leverage is given by the leverage ratio D/V. The company is said to be highly leveraged if its leverage ratio is more than 0.5.

Leverage ratio is known to have an impact on both the variability of reportable earning per share and the return on equity values. This is illustrated by the following example.

Assume that the total assets of a company are $1000. These assets may be financed 100% by equity (Case A) or by a combination of 40 percent equity and 60 percent debt (Case B). It is further assumed that under normal circumstances the company's EBIT is $240. The EBIT value may be reduced to $60 under bad economic conditions, but it could be increased to $400 under good conditions. The corporate tax rate is assumed to be 50 percent. There is no interest payment in Case A. However, $48 must be paid in Case B as an interest charge, which is tax deductible. The net incomes in these cases are different. The earning per share and return on equity data reported out by the company varies accordingly. (See Table 8.4.)

Under normal economic conditions, the earning per share is 1.20 in Case A (no debt) and 2.40 in Case B (with debt). Thus, companies engaged in debt financing will be able to report higher earning per share data than others which carry no debt, assuming every thing else being equal.

In the absence of leverage (no debt, Case A), the earning per share varies from 0.3 to 2.0 and the return on equity from 3.0 to 20 (both from 25 to 167 percent). When the company engages in debt financing, these same ratios vary more widely from 0.15 to 4.40 and from 1.5 to 44.0 (both from 6.25 to 183 percent), respectively. Financial leverage compounds the variability of companies' financial performance.

TABLE 8.4   Impact of Leverage on EPS and ROE

		CASE A			CASE B	
	Bad	Normal	Good	Bad	Normal	Good
Assets		1000			1000	
Debt		0			600	
Equity		1000			400	
Leverage ratio (percent)		0			60	
EBIT	60	240	400	60	240	400
Interest (8 percent)	0	0	0	48	48	48
Taxable income	60	240	400	12	192	352
Tax (50 percent)	30	120	200	6	96	176
Net income	30	120	200	6	96	176
Number of shares	100	100	100	40	40	40
EPS	0.30	1.20	2.00	0.15	2.40	4.40
ROE (percent)	3.0	12.0	20.0	1.5	24.0	44.0

Notes: (1) When ROE exceeds interest (cost of debt), leverage has a favorable impact on EPS and vice versa.
  (2) Financial leverage increases the variability of EPS positively as well as negatively.

8.3.5 Optimum Leverage

The correct selection of the amount of leverage (i.e., debts) is of critical importance to the economic health of a firm. Incurring an excessive amount of debt (high leverage) will push the company to generate cash for meeting the interest payments, skimp on quality, keep inventory low, invest little in maintenance and capital expenditures, and make short-term-oriented decisions, since the company is not able to capitalize on future-growth opportunities. On the other hand, if the debt level is too low (low leverage), the firm does not feel any real pressure to be as efficient as possible (e.g., management will tend to waste resources, tolerate excessive scrap, commit capital expenditures loosely, and initiate R&D indiscriminately) and will suffer from having a higher cost of capital than otherwise because they did not take advantage of the tax deductibility of interest payments on debt.

Most firms attempt to assume an optimum amount of leverage. This is because, as the leverage exceeds the optimum level, the valuation of the firm experiences a reduction due to (a) the cost of bankruptcy, (b) agency costs (e.g., lawyers, courts, and others), and (c) preemptive costs (i.e., loss of growth opportunities). Figure 8.4 illustrates such a leverage optimum, where Vu is the value of the firm with no leverage.

8.4 CAPITAL ASSETS VALUATION

Financial management deals with three general types of capital assets valuation problems: assets in place (operations), opportunities (R&D and marketing), and acquisitions or joint ventures.

Capital budgeting problems related to assets in place are those which deliver a predictable string of cash flows in the immediate future. Examples include building a new plant facility, developing new products, and entering a new regional market. Sometimes these problems are grouped under the heading of operations, as investment in operations usually leads to immediate cash flows. Problems related to opportunities arise from decisions that do not generate an immediate flow of cash, but preserve a likelihood that future gains may be realized. Examples include R&D and marketing

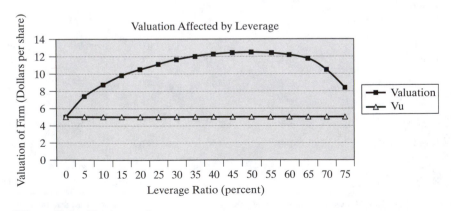

Figure 8.4 Optimum leverage.

TABLE 8.5   Recommended Capital Assets Valuation Methods*

Valuation Problems	Recommended Methods	Alternative Methods
1. Assets in place (operations)	Adjusted present value (APV)	Multiples of sales, cash flows, EBIT, or book value; DCF (based on WACC), Monte Carlo simulations
2. Opportunities (R&D, marketing)	Simple option theory	Decision tree, complex option pricing, simulations
3. Equity claims	Equity cash flow	Multiples of net income; P/E ratios; DCF (based on WACC minus debt), simulations

efforts. The third type of problems is related to acquisition, joint venture, formation of supply chains, and others, all of which may require the company to participate in equity investment and to share future equity cash flows with its business partners.

Generally speaking, each of these types of valuation problem is best handled by different valuation methods. Luehrman (1997A) suggests a list of recommended methods, as shown in Table 8.5.

8.5 OPERATIONS—ASSETS IN PLACE

There are several evaluation methods currently in use to assess capital projects in the investment category of operations.

8.5.1 Discount Cash Flow (Based on WACC)

Since 1980 or so, most companies have been using the discounted cash flow (DCF) method to determine the net present value of an operation with assets in place and WACC to specify the discount factor:

$$\text{NPV} = -P + \left[\sum_{t=1}^{N} \frac{C_t}{(1 + \text{WACC})^t} \right] + \frac{\text{SV}}{(1 + \text{WACC})^N} \qquad (8.5)$$

Here

$$
\begin{aligned}
\text{NPV} &= \text{net present value (dollars)} \\
P &= \text{initial capital investment (dollars)} \\
C_t &= \text{cash flows} - \text{future net benefits (dollars)} \\
\text{SV} &= \text{salvage value} = \text{capital gain (dollars)} \\
N &= \text{total number of periods (year)} \\
\text{WACC} &= \text{weighted average cost of capital (percent)}
\end{aligned}
$$

The net present value (NPV) is equal to the present value of all future net benefits (e.g., income minus relevant costs), plus discounted capital gain, if any, and minus the initial investment capital. It represents the net financial value added to a firm by a given capital investment. Projects with large positive NPV values are favored. This method is sometimes called the *DCF (based on WACC) analysis*, as it is based on the use of the weighted cost of capital in the all-important discount factor.

Companies accept capital projects if the NPV is greater than zero, meaning that an initiation of such projects leads to net positive value added to the companies.

8.5.2 Internal Rate of Return (IRR)

A popular variation to DCF (based on WACC) is the internal rate of return (IRR). When applying Equation 8.5 to evaluate projects, IRR is the discount rate that is realizable when the present values of all discounted cash flows balance the initial investment (NPV equals zero). IRR represents the reinvestment rate, which is held constant over the duration of project. For example, assuming no salvage value, the IRR of a $10,000 investment that yields a revenue of $5,000 per year for three years is 23.35 percent.

In general, companies specify a hurdle rate that must be met or exceeded by the IRRs of all acceptable capital projects. By adjusting the hurdle rate according to conditions related to market, economy, and environment for a specific period involved, companies exercise control over the capital investment criteria. The hurdle rate is typically three to four times WACC in value.

8.5.3 Adjusted Present Value (APV)

In recent years, objections have been increasingly raised in financial literature against the use of DCF (based on WACC) analysis for capital budgeting purposes. One of the principal objections is that WACC only insufficiently captures the tax-shield effects of debt financing in the real-world environment. As defined in Section 8.3.3, the tax-shield effects are incorporated within WACC in a single term shown as the following equation:

$$\text{WACC} = K_e\left(\frac{E}{V}\right) + K_d(1 - t)\left(\frac{D}{V}\right) \tag{8.6}$$

Other organizations are involved in much more complex debt financing situations than the one depicted by this equation, which is limited to the special case of having a constant debt-to-equity ratio.

As a refinement of the DCF (based on WACC) method, the *adjusted present value* (APV) method is suggested. This new method segregates evaluation on the basis of equity from evaluation on the basis of debt financing and then recombines the results. Specifically, the *real cash flow* is first estimated and then discounted on the basis of the cost of equity capital only; that is,

$$\text{NPV}_1 = -P + \left[\sum_{t=1}^{N} \frac{\text{RC}_t}{(1 + K_e)^t}\right] + \frac{\text{SV}}{(1 + K_e)^N} \tag{8.7}$$

where

$$K_e = \text{cost of equity capital}$$
$$\text{RC}_t = \text{real cash flow (dollars) for period } t$$

Then the "side effects" are estimated on the basis of such factors as issuing cost, tax shields, subsidized financing, hedge, and cost of financing distress:

$$NPV_2 = \text{cash flow due to interest tax shield is discounted}$$
$$\text{by using cost of debt capital } (K_d) \text{ only.}$$
$$NPV_3 = \text{sum of net present values due to all other side effects}$$

The final net present value is then the sum of these individual items:

$$NPV = NPV_1 + NPV_2 + NPV_3 \tag{8.8}$$

By itemizing all side effects, the APV method provides additional information that is valuable to management. Because the tax shield is calculated separately, the debt-to-equity ratio need no longer be kept constant for all projects within a company, as it was so presumed in the DCF (based on WACC) method.

Luehrman (1997B) works out a numerical example of using both DCF (based on WACC) and APV in estimating the value of ACME Filters. The results show that*

$$NPV \text{ (DCF)} = 417.1 \text{ million}$$
$$NPV \text{ (APV)} = 346.0 \text{ million}$$

Not only is the NPV based on APV smaller in value, but it also contains additional useful information not unbundled by the DCF (based on WACC) method:

NPV_1 (Baseline):	$157 million
NPV_2 (Interest tax shield):	$102 million
NPV_3	
Higher growth:	$34 million
Asset sales:	$16 million
Networking capital improvement:	$16 million
Margin enhancement:	$21 million
Total	$346 million

See Luehrman (1997B), for a detailed discussion on how each of these side effects are to be incorporated into the evaluation.

8.5.4 Multipliers

Another method to estimate the proper capital investment in a project is to use numerical multipliers that are based on historical data. Specifically, average multipliers are defined for use in conjunction with commonly available financial data such as sales, book value, earning before interests and taxes (EBIT), and cash flow.

In general, the financial data of many publicly held U.S. companies in various industries is widely available in literature, including Standard & Poor's Industry surveys and

---

Value Line Industrial surveys. Sales figures of numerous companies are readily obtained, and their relationship with company assets is typically recorded as the *asset turn ratios*. (See Section 7.5.1.) The reciprocal of this ratio is a multiplier that, when used together with the known sales figure, provides a rough estimate of the company's asset value.

For the XYZ Company described in Tables 7.11 and 7.12, the sales to total asset ratio for the year 2001 is 0.856 ($= 395,000/461,500$). The reciprocal of this number is 1.168, which is the sales multiplier to determine assets. If a sufficient number of other companies in the same industry are surveyed, the resulting average industry-based multiplier can be used to generate a preliminary estimate of the assets employed to produce known sales revenue.

To determine the capital investment value of a plant expansion, new product development, and other products, the company's existing total sales to total asset ratio may serve as a good yardstick to ascertain a reasonable capital investment level, but only if the project outcome in terms of future sales can be estimated.

How much debt should the company incur to finance a specific project? The *debt to asset ratio* is linked to the debt to equity ratio. (See Section 7.5.1.) For the XYZ Company described in Table 7.11, the debt to asset ratio for 2001 is 0.08. Again, if an industrial average is found for this multiplier, it can serve as a useful tool to estimate a reasonable debt level for a project on the basis of its known asset value.

Similarly, industrial average multipliers may be found for cash flow. Cash flow for a given accounting period t is defined as

$$C_t = X_t - I_t \tag{8.9}$$

where

X_t = accounting income (not including financial charges and depreciation)

I_t = investment

Specifically, cash flow is related to a number of financial accounting entries in the income statement, as follows:

Let

Revenue	R
CGS	C
Gross margin	$R - C$
SGA	SGA
Depreciation	Dep
EBIT	$R - C - \text{SGA} - \text{Dep}$
Interest	Int
Tax (tax rate T)	$T\,(R - C - \text{SGA} - \text{Dep} - \text{Int})$
NI	$(1 - T)(R - C - \text{SGA} - \text{Dep} - \text{Int})$

Then

$$\text{CASH} = (1 - T)(\text{EBIT}) + \text{Dep} \tag{8.10}$$

Note that financial charges (e.g., interest payment) are usually ignored in the cash flow computation. Depreciation, a noncash expense, is added back. The following numerical example offers additional clarification:

Sales	$200,000
Manufacturing costs (including a depreciation of $20,000)	80,000
Sales and administrative expenses	40,000
Equipment service charges	10,000
Decrease in contribution of existing product	5,000
Increase in accounts receivable	15,000
Increase in inventory	20,000
Increase in current liability	30,000
Income tax	12,000
Interest paid for bonds used to finance projects	18,000

$$X_t = 200,000 - 80,000 - 40,000 - 10,000$$
$$- 5,000 - 12,000 = \$53,000$$
$$\text{Dep} = \$20,000$$
$$C_t = X_t + \text{Dep} = 53,000 + 20,000 = \$73,000$$

For the XYZ Company in 2001, the cash flow is $10.5 million, calculated from data contained in Table 7.12.

Yet another multiplier is related to EBIT. This multiplier is estimated to be 14.49 for the XYZ Company in 2001, based on the EBIT to asset ratio of 0.051 ($= 23,500/461,500$) evident in Tables 7.11 and 7.12.

The use of some of these multipliers in combination is likely to generate figures that can serve as an acceptable ballpark estimate of the required investment for a new project.

8.5.5 Monte Carlo Simulations

Monte Carlo simulations refer to a sampling technique that processes input data presented in the form of distribution functions. All input variables are simultaneously varied within each of their respective ranges, as defined by their distribution functions. The mathematical operations (e.g., addition, subtraction, multiplication, and division) of a given cost model are readily specified in spreadsheet programs (e.g., Excel or Lotus 123). (See Smith 2000.) Upon activating a suitable risk-analysis software program, Monte Carlo simulations produce one or more outputs that are also in the form of distribution functions (Wright 2002). (See Section 6.9.)

In evaluating operations, Monte Carlo simulations may be usefully applied to the DCF (based on WACC) method. All future net cash flows and discount rates are modeled by distribution functions (e.g., triangular or Gaussion), as they are indeed expected to vary within ranges. The DCF (based on WACC) equation is readily modeled in a spreadsheet program. As the sole output, the net present value will also vary within a lower and upper bound. The following results will be helpful to decision makers of capital budgeting:

1. The maximum probability at which the net present value is projected to be negative
2. The probability at which the net present value is projected to exceed a given value (e.g., $10 million)
3. The standard deviation of the net present value output
4. The minimum net present value
5. The maximum net present value

Monte Carlo simulations are also applicable to the calculation of internal rate of return (IRR) for the evaluation of operations.

8.6 OPPORTUNITIES—REAL OPTIONS

The second category of valuation problems is related to opportunities such as R&D and marketing projects. These problems do not lend themselves to DCF or APV analyses, both of which require the estimates of projected future cash flows. If there is no cash flow, there can be no net positive present value. Financial analysts and researchers in the literature recommend that the *European simple call option* method be used to price these opportunities.

Option is a common tool frequently used for trading securities in the financial markets. There are *call* and *put* options. A *call option* provides its holder with the rights, but not the obligation, to buy 100 shares of an underlying company (e.g., IBM, Eastman Kodak, or General Electric by a certain expiration date (typically three months from the present) at a specific price (*strike* or *exercise* price). The holder pays a fee, or premium, to buy the call option, which he or she may exercise on any business day up to and including the contract expiration date.

A *put option* gives its holder the right, but again not the obligation, to sell 100 shares of stocks within a period of time at a predetermined strike price. Investors who predict that the stock price of a given company is going to decline in the future will want to sell the stocks today and buy a call option to recover the stocks at a lower price in the future. The premium for an option is dependent on five factors: (1) current stock price, (2) strike price, (3) length of option contract, (4) stock price volatility, and (5) current opportunity cost (e.g., bank interest).

For the purpose of evaluating capital project opportunities, the European simple call option is more appropriate in that the call option can be exercised only on the expiration date specified in the contract and no sooner. Table 8.6 shows the equivalence between financial calls and real calls.

Companies with new technologies, product development ideas, defensive positions in fast growing markets, and access to new markets have valuable opportunities to explore. When dealing with opportunities, three possible scenarios exist: (1) to invest immediately, (2) to reject the opportunity right away, and (3) to preserve the option of investing in the opportunity at a later time.

TABLE 8.6   Equivalence of Call Options

Financial Calls	Real Calls
Current stock price	Underlying asset value
Length of contract	Length of time to invest
Volatility	Volatility of future project value
Strike price	Capital investment for the project
Current bank interest	Cost of equity capital

This problem may be studied by using the Black-Scholes option-pricing model (BSOPM) (Black and Scholes 1973). The Black-Scholes option-pricing model is defined as

$$C = V[N(d)] - e^{-rT} \times [N(d - sigma\sqrt{T})] \qquad (8.11)$$

$$d = \frac{\ln\left(\dfrac{V}{X}\right) + T\left(r + \dfrac{sigma^2}{2}\right)}{sigma\sqrt{T}} \qquad (8.12)$$

where

$$
\begin{aligned}
C &= \text{Option price} \\
V &= \text{Current value of the underlying asset} \\
X &= \text{The exercise price of the option} \\
Sigma &= \text{Annual standard deviation of the returns on the underlying asset} \\
r &= \text{The annual risk-free rate}
\end{aligned}
$$

TABLE 8.7   Values of Cumulative Normal Distribution Function

d	$N(d)$	d	$N(d)$	d	$N(d)$
−3.00	0.0013	−1.00	0.1587	1.00	0.8413
−2.90	0.0019	−0.90	0.1841	1.10	0.8643
−2.80	0.0026	−0.80	0.2119	1.20	0.8849
−2.70	0.0035	−0.70	0.2420	1.30	0.9032
−2.60	0.0047	−0.60	0.2743	1.40	0.9192
−2.50	0.0062	−0.50	0.3085	1.50	0.9332
−2.40	0.0082	−0.40	0.3446	1.60	0.9452
−2.30	0.0107	−0.30	0.3821	1.70	0.9554
−2.20	0.0139	−0.20	0.4207	1.80	0.9641
−2.10	0.0179	−0.10	0.4602	1.90	0.9726
−2.00	0.0228	0.00	0.5000	2.00	0.9772
−1.90	0.0287	0.10	0.5398	2.10	0.9821
−1.80	0.0359	0.20	0.5793	2.20	0.9861
−1.70	0.0446	0.30	0.6179	2.30	0.9893
−1.60	0.0548	0.40	0.6554	2.40	0.9918
−1.50	0.0668	0.50	0.6915	2.50	0.9938
−1.40	0.0808	0.60	0.7257	2.60	0.9953
−1.30	0.0968	0.70	0.7580	2.70	0.9965
−1.20	0.1151	0.80	0.7881	2.80	0.9974
−1.10	0.1357	0.90	0.8159	2.90	0.9981

$N(d)$ = Cumulative standard normal distribution function evaluated at d

$\ln(x)$ = Natural log function of x

T = Time to expiration of the option (years)

Table 8.7 exhibits the representative data of the $N(d)$ function.

Example 8.2.

If $1 million is invested immediately, there will be a loss of $100,000 due to the current economic condition and marketing environment. However, if the company waits for two years, things may be different. What should the company do, assuming that the current risk-free rate is 7% and the project volatility is 0.3?

Answer 8.2.

The problem may be studied by using BSOPM. Let us define the following equivalents:

$$\text{Sigma} = 0.30$$
$$r = 0.07$$
$$X = \$1,000,000$$
$$V = \$900,000$$
$$T = 2$$

From equation (8.12),

$$d_1 = \frac{\left\{ \ln(900,000/1,000,000) + 2\left(0.07 + \frac{0.3^2}{2}\right) \right\}}{0.3\sqrt{2}} = 0.2938$$

$$d_2 = 0.2938 - 0.3 \times 1.41456 = -0.1306$$

From Table 8-7, we get by interpolation:

$$N(0.2938) = 0.6155$$
$$N(-0.1356) = 0.4462$$

From Equation 8.11, the option price becomes

$$C = 900,000 \times 0.6155 - \exp(-0.07 \times 2) \times 1,000,000 \times 0.4462$$
$$= \$166,042$$

Now the company has the alternative of investing $166,042 to preserve investment opportunities for two years, in addition to deciding for or against it right away. If this option is preserved, additional information that will cause a change in the underlying asset value may become available during the ensuing two years. The company may still decide not to invest in two years, but preserving the option to invest is still a valuable alternative.

8.7 ACQUISITIONS AND JOINT VENTURES

When considering companies as candidates for acquisition, the evaluation of the assets of the target companies becomes critically important. A number of methods are practiced to assess the value of a company (Brenan 1996; Norton 2003).

The value of a company is defined by its equity and debt, that is,

$$V = E + D \tag{8.13}$$

where

$$V = \text{firm's value in the market}$$
$$E = \text{equity (stocks)}$$
$$D = \text{debt outstanding (i.e., bonds, loans, etc.)}$$

The company management continues to maximize the candidate firm's value for its shareholders. Two factors may affect this value maximization attempt: (1) *Takeover bids* (when acquiring firms are enticed to pay a higher than normal premium to absorb the acquisition candidates) tend to raise the stock price; and (2) *stock options* (rights to buy stock at a fixed price awarded to company's management personnel and new hires) tend to dilute the shareholder value.

Example 8.3.

XYZ Company is considering the acquisition of Target Company, a smaller competitor in the same industry. The income statement of Target Company is displayed in Table 8.8. As a stand-alone company, its sales, cost of goods sold, depreciation, selling, and administrative expenses are all projected to increase by 4 percent per year.

To maintain the projected sales growth of Target Company, XYZ Company must make additional working capital investments. (See Table 8.8.)

(A) Assuming a 10 percent discount rate, what is the maximum price XYZ Company should be willing to pay for this acquisition if it is to be run as a stand-alone subsidiary of XYZ Company?

TABLE 8.8   Income Statement of Target Company (1999–Future)

	1999	2000	2001	2002	2003	Growth to Infinity (percent)
			(dollars in thousands)			
Sales	61,000	63,440	65,978	68,617	71,361	4
Cost of goods sold	29,890	31,086	32,329	33,622	34,967	4
Depreciation	4,000	4,160	4,326	4,499	4,679	4
Selling and administrative	21,010	21,850	22,724	23,633	24,579	4
IT services	0	0	0	0	0	4
EBIT	6,100	6,344	6,598	6,862	7,136	4
Tax (35 percent)	2,135	2,220	2,309	2,402	2,498	4
EBIAT	3,965	4,124	4,289	4,460	4,638	4
Cash flow	7,965	8,284	8,615	8,960	9,318	4
Additional investments Working capital	8,000	8,320	8,653	8,999	9,359	4

TABLE 8.9   Present Value of Cash Flows

	(growth rate = 4 percent, discount rate = 10 percent)				
	2003	2004	2005	2006	Infinity
Cash flow	A	$A(1.04)$	$A(1.04)^2$	$A(1.04)^3$	
Present value (2004)		$\dfrac{A(1.04)}{(1.10)}$	$\dfrac{A(1.04)^2}{(1.10)^2}$	$\dfrac{A(1.04)^3}{(1.10)^3}$	
		$A(1-r)\ldots$	$A(1-r)^2$	$A(1-r)^3$	
PV of all future cash flows at start of 1999		$\dfrac{A(1-r)}{(1.1)^5}$			
r	0.05454545				
$\dfrac{1-r}{r}$	17.3333				

TABLE 8.10   Income Statement of Target Company (stand-alone subsidiary-thousands of dollars)

	1999	2000	2001	2002	2003	Growth to Infinity (percent)
Sales	61,000	63,440	65,978	68,617	71,361	4
Cost of goods sold	29,890	31,086	32,329	33,622	34,967	4
Depreciation	4,000	4,160	4,326	4,499	4,679	4
Selling and administrative	21,010	21,850	22,724	23,633	24,579	4
IT services	0	0	0	0	0	4
EBIT	6,100	6,344	6,598	6,862	7,136	4
Tax (35 percent)	2,135	2,220	2,309	2,402	2,498	4
EBIAT	3,965	4,124	4,289	4,460	4,638	4
Cash flow	7,965	8,284	8,615	8,960	9,318	4
PV (cash flows)	132,750					4
Additional investments						
Working capital	8,000	8,320	8,653	8,999	9,359	4
PV (WC)	$133,333					
(A) NPV of target company	($583) (Not to acquire)					

(B) XYZ Company could also integrate Target Company into its existing corporate IT operations. Web-based customer services, inventory management, order processing, and other activities can be readily added to cut down the required working capital by 50 percent from its stand-alone values. There is, however, an increased IT service charge

TABLE 8.11   Income Statement of Target Company (integrated operations-thousands of dollars)

	1999	2000	2001	2002	2003	Growth to Infinity (percent)
Sales	61,000	63,440	65,978	68,617	71,361	4
Cost of goods sold	29,890	31,086	32,329	33,622	34,967	4
Depreciation	4,000	4,160	4,326	4,499	4,679	4
Selling and administrative	21,010	21,850	22,724	23,633	24,579	4
IT services	1,000	1,040	1,082	1,125	1,170	4
EBIT	5,100	5,304	5,516	5,737	5,966	4
Tax (35 percent)	2,135	1,856	1,931	2,008	2,088	4
EBIAT	2,965	3,448	3,586	3,729	3,878	4
Cash flow	6,965	7,608	7,912	8,228	8,558	4
PV (cash flows)	121,598					4
Additional investments Working capital (50%)	4,000	4,160	4,326	4,499	4,679	4
PV (WC)	$66,667					
(B) NPV of integrated company	$54,932 (To bid for no more than $55 million)					

of $1 million for the first year, and this charge increases by 4 percent per year. Again, assuming a 10 percent discount rate, what is the maximum price XYZ should pay for Target Company under the integration scenario?

Answer 8.3.

For estimating the present values of cash flows for the period 2004 to infinity, the computations illustrated in Table 8.9 are needed. In Table 8.9, A denotes a constant but unknown cash flow defined at the beginning of year 2003.

The same factor $(1 - r)/r$ is applicable for the estimation of the present values of working capital for the period of 2004 to infinity. The spreadsheet represented by Table 8.10 enumerates the results:

The NPV of Target Company, as a stand-alone operation, is negative, not justifying its potential acquisition by XYZ Company.

On the other hand, if XYZ integrates Target Company into its IT operations, then the value of this target company is significantly improved, as displayed in Table 8.11.

Under the integration scenario, Target Company is worth about $55 million. Any price below this figure will denote a net gain for XYZ Company.

8.7.1 Common Stock Valuation Model (Dividend Valuation Model)

The stock price of the acquisition candidate depends on the dividend stream it is able to generate. The *dividend valuation model* offers an estimate of the acquisition candidate's stock price as the sum of the present values of its future dividends:

$$P_o = \sum_{t=1}^{\infty} \frac{D_t}{(1 + r)^t}$$

(8.14)

Here

P_o = equity price
D_t = dividend payout (DPS) per period.
$r = K_e$ = Cost of equity capital incurred by the firm
Upper limit = infinite (going concern)

Two special cases may be considered:

1. **Constant Dividend**
   If $D_t = D_o$ = constant, then

$$P_o = \frac{D_o}{r} = \text{capitalized value of dividend}$$

The stock price P_o is equivalent to a single present value that is capable of producing a stream of constant dividends of value D_o indefinitely. (The derivation of this closed-form solution is included in Appendix 8.A.)

2. **Finite Holding Period**
   For investors who hold a given stock for only N periods and plan to sell the stock at the price P_n thereafter, the stock price may be calculated as

$$P_o = \sum_{t=1}^{N} \frac{D_t}{(1 + r)^t} + \frac{P_n}{(1 + r)^N}$$

(8.15)

where

r = effective rate of return required by the market of the firm's stock equals the cost of equity capital incurred by the firm (K_e)
P_n = investment recovery at the end of N periods.

Once the stock price is known, the total value of the acquisition candidates is then equal to the stock price multiplied by the outstanding number of its stocks.

8.7.2 Dividend Growth Model

If the target company is capable of paying out dividends that grow at an annual rate of g percent, then its stock price is calculated as the capitalized dividend value at a rate equal to the cost of capital minus the dividend growth rate:

$$P_o = \frac{D_1}{(K_e - g)}$$

(8.16)

In this equation,

$$g = \text{growth rate of dividend per share (percent)}$$
$$K_e = \text{cost of equity capital } (\textit{note: } K_e > g)$$
$$D_1 = \text{dividend for the next period (dollars/share)}$$

(The derivation of this equation is given in Appendix 8.B.)
The growth rate of annual dividend is calculated by the equation

$$g = (1 - b)\,\text{ROE} \tag{8.17}$$

where

$$\text{ROE} = \text{return on equity} - \text{net income/average equity capital}$$
$$b = \text{pay-out ratio of dividend} = \text{dividend paid out/net income}$$
$$1 - b = \text{retained dividend ratio} = \text{dividend retained/net income}$$

In Equation 8.17, the assumption is made that the company management is capable of applying the retained earning to foster dividend growth in the current year at the same rate as that of ROE that was accomplished by the company in the previous fiscal year.

8.7.3 Modified Earning Model

The stock price of a company that reinvests retained earning to generate dividend growth is delineated by the equation

$$P_o = \frac{\text{EPS}_1}{K_e} + \text{PVGO} \tag{8.18}$$

where

$$\text{EPS}_1 = \text{earning per share in the next period}$$
$$K_e = \text{cost of equity capital}$$
$$\text{PVGO} = \text{present value of growth opportunities}$$

Now, EPS_1/K_e is the capitalized value of EPS, and PVGO is the net present value of all returns (on the per-share basis) generated by having reinvested the retained earning at the rate of ROE. Specifically,

$$\text{PVGO} = \frac{\text{EPS}_1(1 - b)(\text{ROE} - K_e)}{K_e(K_e - g)} \tag{8.19}$$

Typically, growth stocks have large PVGO values that arise from the reinvestment of continuously increased earnings at the rate of

$$g = (1 - b)\,\text{ROE} \tag{8.20}$$

(For a derivation of Equation 8.20, see Appendix 8.C.)

Example 8.4.

The company's stock is selling now at $50 per share. Its expected dividend next year is $2.00 per share. A 20-percent annual growth of dividend is anticipated for the next three years. From the fourth year on, its dividend growth rate will be reduced to only 6-percent annually. What is the expected long-term rate of return (R) from buying this stock at $50?

Answer 8.4.

On the basis of the dividend model, we would formulate the following equation for evaluating the overall rate of return:

$$P_o = \frac{\text{Div}_1}{(1 + R)} + \frac{\text{Div}_2}{(1 + R)^2} + \frac{\text{Div}_3}{(1 + R)^3} + \frac{\text{Div}_4}{(1 + R)^4} + \frac{\text{Div}_5}{(1 + R)^5(R - g)}$$

$$50 = \frac{2}{(1 + R)} + \frac{2(1.2)}{(1 + R)^2} + \frac{2(1.2)^2}{(1 + R)^3} + \frac{2(1.2)^3}{(1 + R)^4} + \frac{2(1.2)^3(1.06)}{(1 + R)^5(R - 0.06)}$$

By trial and error, $R = 0.11153$. The long-term rate of return of this stock is 11.153 percent.

Example 8.5.

The company expects total sales revenue of $20 million and a total cost, including tax, of $15 million in the forthcoming year. During the subsequent five years (e.g., year 2 to year 6), the revenues and costs will increase 25 percent per year, and all profits will be reinvested back into the business. Thereafter, the company's growth will be decreased to only 5 percent per year and the company will need to reinvest only 40 percent of its profits.

If the company is offered $75 million in cash by a major competitor, is this a fair acquisition price, assuming the opportunity cost of capital is 12 percent?

Answer 8.5.

Stock price is the present value of expected future dividend. On the basis of this model, we have

$$P_o = \frac{\text{Div}_1}{(1 + R)} + \frac{\text{Div}_2}{(1 + R)^2} + \frac{\text{Div}_3}{(1 + R)^3} + \frac{\text{Div}_4}{(1 + R)^4}$$

$$+ \frac{\text{Div}_5}{(1 + R)^5} + \frac{\text{Div}_6}{(1 + R)^6} + \frac{\text{Div}_7}{(1 + R)^7(R - g)}$$

$$\text{Div}_1 = \text{Div}_2 = \text{Div}_3 = \text{Div}_4 = \text{Div}_5 = \text{Div}_6 = 0.0$$

$$\text{Div}_7 = (20 - 15)(1.25)^5(1.05)(1 - 0.4) = \$9.613 \text{ Million}$$

For the sixth year, $g = (1 - b)$ ROE; $0.25 = (1 - 0)$ ROE

$$\text{ROE} = 0.25$$

For the seventh year,

$$g = (1 - b) \text{ ROE} = (1 - 0.6)\, 0.25 = 0.1$$

$$P_o = \frac{9.613}{(1 + 0.12)^7(0.12 - 0.1)} = \$217.4 \text{ Million}$$

The offered acquisition price is much too low, and the offer should be rejected.

8.7.4 Equity Cash Flows

For the valuation problems related to joint ventures and partnerships, ownership claims and equity cash flows must be evaluated. Financial specialists are typically involved in evaluating such problems by using sophisticated computer programs and models. For detailed discussions of this type of evaluations, see Luehrman (1997A). Other advanced corporate finance issues are discussed by Odgen, Jen, and O'Conner (2003).

Capital budgeting is a managerial responsibility related to the company's capital investments. Capital budgeting decisions are made on the basis of certain rational decision criteria. Engineering managers are advised to acquire the necessary background knowledge in order to become effective contributors to such decision-making processes.

8.8 CONCLUSION

This chapter reviews the basic elements of financial management: raising and applying investment capital. Equity and debt financing are the two most common ways of obtaining financing. There are costs involved in each: the cost of equity capital and the cost of debt capital.

There are different types of capital projects in which a company might invest: operations (assets in place), opportunities (R&D and marketing), acquisitions, and joint ventures.

Different evaluation methods are used for these capital projects. For example, DCF (based on WACC), IRR, APV, multipliers, and Monte Carlo simulations are useful for evaluating operations. Option pricing is suitable for evaluating opportunities for which there are no predictable cash flows. Acquisitions and joint ventures are advanced financial topics, the evaluation of which should be deferred to knowledgeable financial specialists on the subject.

Engineering managers are advised to become well prepared to actively participate in the company's capital budgeting process by becoming familiar with the sources and costs of capital as well as the evaluation criteria adopted for capital budgeting. Doing so will enable them to constantly bring forth and screen useful projects and valuable opportunities (including risks assessment) and to be in a position to initiate winning capital project proposals on a timely basis.

8.9 REFERENCES

Anonymous. 1998. "Valuing Capital Investment Projects." *Harvard Business School Note*, No. 9-298-092, December 4.

Anonymous. 2000. "Capital Budgeting: Discounted Cash Flow Analysis." *Harvard Business School Note*, No. 9-298-068, June 28.

Black F. and M. Scholes. 1973. "The Pricing of Options and Corporate Liabilities." *Journal of Political Economy*, Vol. 81, May–June, pp. 637–654.

Brealey, R. A. and S. C. Myers. 2000. *Principles of Corporate Finance*. 6th ed. New York: Irwin/McGraw-Hill.

Brenan, M. J. (Editor). 1996. *The Theory of Corporate Finance*. Cheltenham, UK: Edward Elgar Publishing.

Davis, E. W. and J. Pointon. 1994. *Finance and the Firm: An Introduction to Corporate Finance.* 2d ed. Oxford, UK: Oxford University Press.

Droms, W. G. 1998. *Finance and Accounting for Nonfinancial Managers: All the Basics You Need to Know.* 4th ed. Reading, MA: Addison-Wesley.

Edleson, M. E. 1999. "Real Options: Valuing Managerial Flexibility." *Harvard Business School Note,* No. 9-294-109, June.

Luehrman, T. A. 1997A. "What's It Worth? A General Manager's Guide to Valuation." *Harvard Business Review,* May–June.

Luerhman, T. A. 1997B. "Using APV: A Better Tool for Valuing Operations." *Harvard Business Review,* May–June.

Norton, III, G. M. 2003. *Valuation: Maximizing Corporate Value.* New York: John Wiley.

McNulty, J. J., T. D. Yeh, W. S. Schulze, and M. H. Lubatkin. 2002. "What Is Your Real Cost of Capital?" *Harvard Business Review,* October.

Ogden, J. P., F. Jen, and P. F. O'Conner. 2003. *Advanced Corporate Finance: Policies and Strategies.* Upper Saddle River, NJ: Prentice Hall.

Shim, J. K. and J. G. Siegel. 2000. *Financial Management.* 2d ed. Hauppauge, NY: Barron's.

Smith, G. N. 2000. *Excel Spreadsheet Application Series for Cost Accounting.* Cincinnati, OH: South-Western College Publishing.

Terry, B. J. (Editor). 2000. The *International Handbook of Corporate Finance.* Chicago: Glenlake Publishing, New York: AMACOM.

Van Horne, J. C. 2002. *Financial Management and Policy.* 12th ed. Upper Saddle River, NJ: Prentice Hall.

Weston, J. F. and E. F. Brigham. 1993. *Essentials of Managerial Finance.* 10th ed. Fort Worth, TX: Dryden Press.

Wright, J. F. 2002. *Monte Carlo Risk Analysis and Due Diligence of New Business Ventures.* New York: AMACOM.

8.10 APPENDICES

APPENDIX 8.A. DERIVATION OF AN INFINITE SERIES

Let

$$A = \sum \frac{1}{(1+r)^m} \qquad m = 1, 2, \ldots \infty$$

$$B = \sum \frac{1}{(1+r)^{(m+1)}} \qquad m = 1, 2, 3, \ldots \infty$$

Then

$$B - A = \frac{1}{(1+r)^{(m+1)}} \tag{8.A1}$$

Furthermore,

$$B(1+r)^{(m+1)} - A(1+r)^m = (1+r)^m \tag{8.A2}$$

Substituting Equation (8.A2) into Equation (8.A1), we have

$$A(1+r)^{(m+1)} + 1 - A(1+r)^m = (1+r)^m \tag{8.A3}$$

or

$$A(1 + r) + \frac{1}{(1 + r)^m} - A = 1$$

$$A = \frac{1 - \dfrac{1}{(1 + r)^m}}{r} \qquad (8.\text{A}4)$$

As m approaches infinite, we obtain

$$A = \frac{1}{r} \qquad (8.\text{A}5)$$

This relation is used in Section 8.7.1 of this chapter.

APPENDIX 8.B. DERIVATION OF THE DIVIDEND GROWTH MODEL

$$P_o = \sum_{t=1}^{\infty} \frac{D_t}{(1 + r)^t} \qquad (8.\text{B}1)$$

where

$$D_1 = \text{Next year's dividend}$$
$$D_2 = D_1(1 + g)$$
$$D_3 = D_2(1 + g) = D_1(1 + g)^2$$
$$D_4 = D_1(1 + g)_3$$
$$\vdots$$

$$\begin{aligned}
P_o &= \frac{D_1}{(1 + r)} + \frac{D_2}{(1 + r)^2} + \frac{D_3}{(1 + r)^3} + \cdots \\
&= \frac{D_1}{1 + r}\left[1 + \frac{1 + g}{1 + r} + \left(\frac{1 + g}{1 + r}\right)^2 + \left(\frac{1 + g}{1 + r}\right)^2 + \cdots\right] \\
&= \frac{D_1}{1 + r}\left(\frac{1 + r}{r - g}\right) \qquad (8.\text{B}2)
\end{aligned}$$

$$P_o = \frac{D_1}{r - g} \qquad (8.\text{B}3)$$

(See Section 8.7.2.)

APPENDIX 8.C. DERIVATION OF PRESENT VALUE OF GROWTH OPPORTUNITY (PVGO)

The net present value of reinvested retained earning is as follows:

$$\text{NPV} = -(1 - b)\,\text{EPS} + (1 - b)\,\text{EPS} \times \frac{\text{ROE}}{k_e} \qquad (8.\text{C}1)$$

where the first term on the right-hand side (RHS) of the equal sign is the reinvested retained earning, with b being the payout ratio. The second term on the RHS describes the capitalized return, assuming at the same effective rate as ROE. By substitution, we have

$$PVGO = \frac{NPV}{[K_e - g]} = EPS(1 - b)\frac{ROE - K_e}{K_e(K_e - g)} \tag{8.C2}$$

(See Section 8.7.3.)

8.11 QUESTIONS

8.1 The company has 10,000 shares of common stock outstanding, and the current price of the stock is $100 per share. The company has no debt. The vice president of engineering discovers an opportunity to invest in a new technology project that yields positive cash flows with a present value of $210,000. The total initial capital that is required for investing and developing this project is only $110,000. It is proposed that capital be raised by issuing new equity. All potential purchasers of common stock will be fully aware of the new project's value and cost, and are willing to pay "fair value" for the new shares of the company.

A. What is the net present value of this project?
B. How many shares of common stock must be issued, and at what price, to raise the required capital, assuming the costs of underwriting these new shares are negligible?
C. What is the effect, if any, of this new project on the value of the stock of existing shareholders?

Answer 8.1
   A. $100,000; B. 1111.11 shares at $99 per share; C. slight dilution.

8.2 XYZ Company is financed by debt (50 percent), preferred stocks (20 percent) and common equity (30 percent). Its common stock price is $43 per share. It pays a dividend of $3.00 and has a growth rate of −2 percent. Its annual preferred stock dividend is $82 per $1,000 share with a flotation cost of 7.5 percent per share. The interest for long-term debt is 11 percent. Its corporate tax rate is 30 percent.
   What is the company's weighted average cost of capital (WACC)?

Answer 8.2
   WACC = 12.33 percent.

8.3 XYX Company receives a contract from one of its major customers. The contract calls for 20,000 hours of work billed at $75 per hour to be completed within one year. The normal billable work for each engineer is 2000 hours per year. Thus, the contract requires the involvement of 10 full-time engineers. At the present time, the company has only 6 full time engineers who may be able to work on this contract. The engineers' wages average $40 per hour. If the engineers are to work overtime, their overtime pay is time and a half (i.e., $60 per hour).

(a) What would be the contribution margin if 10 engineers were available with no need to work overtime?
(b) What would be the contribution margin if all 6 engineers are used full time and the deficiency is made up through 8000 hours of overtime?

(c) What would be the contribution margin if the company hires two more full-time engineers and makes up the rest with overtime?

(d) If the cost of recruiting and training each new engineer is $15,000, what would be the contribution margin on hiring two new engineers, after factoring in recruiting and training costs?

Answer 8.3
(a) 0.4667; (b) 0.36; (c) 0.4133; (d) 0.3933.

8.4 XYZ company plans to install a new production line consisting of several precision machines costing a total of $800,000. The installation of these machines requires another $150,000. The products made by the machines are projected to deliver a net income after tax of $400,000 per year for the next 10 years. The useful life of each machine is estimated to be 10 years. At the end of 10 years, these machines have a salvage value of $20,000 each.

Compute the cash flows generated by this new production line.

Answer 8.4
The annual cash flow is $493,000.

8.5 Using the data from Question 7.6, conduct an evaluation of the Superior Technologies' common stock in 2002, using the three different methods specified, as follows:

A. The market value of the company's net property has risen and it is now about two times the value reported in the balance sheet. Calculate the stock price by using its *net asset value* as a basis.

B. Assuming that the company's cost of equity (K_e) is 16 percent, determine the stock price by applying the *dividend growth model*.

C. Price-to-earning (P/E) ratio reflects the general sentiment of the securities market toward a specific company or the industry in general. Assuming that the average P/E ratio is about 10 for the industry, of which the company is a member, define the stock price, using the *earning model*.

Answer 8.5
A. $19.47/share
B. $25.61/share
C. $24.29/share

8.6 Company A has been performing reasonably quite well over the last several years. Table 8.A1 exhibits an abbreviated set of its financial data for the years 1998 to 2003.

1. Analyze the company's dividend-payout ratio to common stockholders and comment on the suitability of this dividend policy.

2. Discuss the debt-financing policy of this company over the years. In your opinion, does the long-term debt of the company represent a percentage too high (aggressive) or too low (conservative) in its capital structure? Why?

3. Do you regard the percentage of common stock equity in the company's capital structure as adequate, and why?

4. For 2004, the company needs an influx of $30 million to finance business expansion. Which financing option should the company pursue? Why?

Answer 8.6
The option of debt financing is recommended.

TABLE 8.A1   Financial Data of Company A (Millions of dollars)

	2003	2002	2001	2000	1999	1998
Net income	41	33	34	35	27	22
Preferred dividends	3	3	3	1	0	0
Common dividends	18	18	18	18	13	12
Total assets	492	455	417	403	280	258
Current liabilities	68	57	75	68	51	43
Long-term liabilities	113	114	75	83	57	57
Preferred stock ($100 par)	82	82	82	82	0	0
Common stocks ($6.25 par)	45	45	45	45	38	38
Capital surplus	6	0	0	0	34	34
Retained earnings	177	157	140	126	100	86

TABLE 8.A2   Financing Options of Company B

	Base Case (dollars)	Option A (dollars)	Option B (dollars)
Capitalization	30,000,000	30,000,000	30,000,000
Common stock equity	30,000,000	6,000,000	6,000,000
Preferred stock equity	0	24,000,000	0
Long-term debt	0	0	24,000,000
EBIT (operating income)	3,000,000	3,000,000	3,000,000

8.7   Company B is currently financed by common stock equity. It is considering two alternative ways of financing in order to increase the return on common equity. Table 8.A2 lists the two options under consideration, along with the base case.

   The company's capitalization and EBIT (earning before interest and tax) remain constant at $30 million and $3 million, respectively. The composition of the capitalization changes from 100 percent common stock equity (base case) to a mix of common and preferred stocks (Option A) and to a mix of common stock and long-term debt (Option B).

   The corporate tax rate is 40 percent. Dividends of preferred stocks are paid at a 5 percent rate. The interest charge for the long term debt is 4 percent.

1. Compute the rate of return on common stocks equity for the three cases. Explain why these numbers change from one case to another.
2. Compute the rate of return on capitalization for the three cases.
3. Among the three cases indicated, which financing option is to be preferred by the company, and why?

Answer 8.7

   None of the stated options is to be recommended. Instead, the company should pursue Option C, which limits the company's debt-to-capitalization ratio to no more than 50 percent, resulting in a return to common stock equity of 9.6 percent.

Chapter 9

Marketing Management for Engineering Managers

9.1 INTRODUCTION

Companies have essentially two major activities: marketing and innovation. Marketing is the whole business of the enterprise seen from the viewpoint of customers. The purpose of marketing is to provide products and services that meet the needs and wants of customers. Marketing impacts the top line (i.e., sales revenue) of any enterprise. Innovation strengthens the enterprise's competitive marketing position to sustain profitability by way of timely application of unique technologies and other core competencies (Kotler, Jain, and Maescincee 2002; Kotler 1999; Kotler 1997).

Engineers are known to be technologically innovative. If they can also master marketing, the resulting combined capabilities will likely enable them to contribute significantly toward creating business success for their employers and thus propel them to higher leadership positions.

The objective of this chapter is to prepare engineering managers to interact effectively with marketing and sales personnel in companies offering technology-based products and services. Emphasis is given to the introduction and application of marketing concepts for engineering and technology products and services (Paley 2001). Such concepts and applications include marketing management processes in profit-seeking organizations, identification of opportunities and threats facing an organization, and marketing tasks.

Engineering managers are encouraged to continue studying additional current references on specific issues in marketing management.

9.2 THE FUNCTION OF MARKETING

Marketing and sales are critically important to profit seeking companies because they strive to ensure satisfaction in the exchange of values between the producers and consumers of products and services, as exhibited in Figure 9.1.

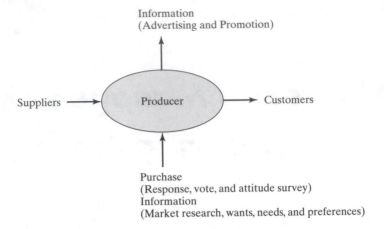

Figure 9.1   Marketing interactions.

9.2.1 Sales versus Marketing

Sales is a process by which producers attempt to motivate target customers to buy the available products. The mentality behind sales, as illustrated by Figure 9.2, is that "someone out there will need the products." At the time that Ford Motor Company was the dominant carmaker in Detroit, the well-known saying attributed to Henry Ford was, "You can have any color you want for your car, as long as it is black." Sales does not take the customers' concerns into account.

In contrast, companies with a marketing orientation offer something customers want by seeking feedback from the marketplace, adjusting the product offerings, and increasing the value to consumers. (See Figure 9.3.)

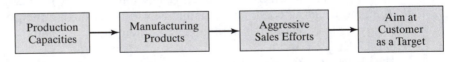

Figure 9.2   Sales orientation.

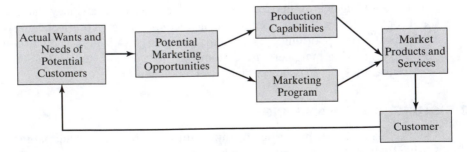

Figure 9.3   Marketing orientation.

In the pursuit of marketing strategies, companies solicit intelligence, financial data, and customers' responses to constantly reassess the market. They evaluate such factors as changing needs, competition, cost effectiveness, product substitution, and maturity of products. A long-term orientation ensures that benefits for both producers and customers will be sustained. Sales strategies are only a part of marketing.

9.2.2 The Marketing Process

The marketing efforts of companies are typically focused on four specific aspects (Sutton and Klein 2003):*

A. **Customer focus.** The purpose of a profit-seeking business is to understand customers' needs, deliver value to customers, and offer services to ensure that customers are satisfied. In other words, the customer comes first.

B. **Competitor focus.** Companies seek advantages relative to competitors, monitor competitive behaviors, and respond to the strategic moves of competitors.

C. **Interfunctional coordination.** Companies integrate all functions, share information, and organize themselves to provide added value for the customers.

D. **Profit orientation.** Companies focus on making profits in both the short term and the long term.

To achieve business success, companies must search constantly for future markets, in addition to actively serving the markets of today. The primary responsibility of marketing is to scan the relevant business environment for future opportunities (such as what bundle of products and services to offer to whom, at what price, at what time, and in which market segments) and to provide insight into the needs of current customers and the intentions of competitors.

Presented in Figure 9.4 is the marketing process, which is iterative in nature. This process defines opportunities (unsatisfied needs) in the marketplace; the products or

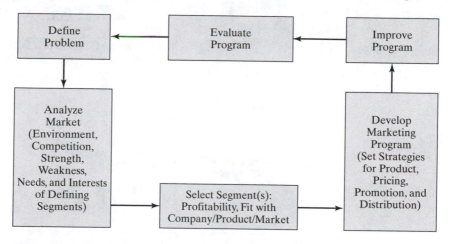

Figure 9.4   Marketing process.

*Reprinted with permission of John Wiley & Sons, Inc. From *Enterprise Marketing Management: The New Science of Marketing*, by D. Sutton and T. Klein; Copyright © 2003 John Wiley.

services with features to satisfy these needs; and the product pricing, distribution, and communications strategies to reach the target market segments. Market segments refer to specific groups of customers who share similar purchasing preferences.

For companies to succeed in the marketplace, marketing must be a core value, central to the company's strategy formulation and execution. Through marketing, companies identify and satisfy the needs of customers and achieve long-term profitability by attracting and retaining customers. Specific tasks undertaken to attain these objectives include (1) interacting with and understanding the market and its customers, (2) planning long-term marketing strategies, and (3) implementing short-term tactical marketing programs.

The effectiveness of a marketing program is often measured by two metrics: (1) how attractive the company's products and services are to the target customers and (2) how successfully the company can satisfy and retain these target customers. Figure 9.5 illustrates the marketing effectiveness diagram.

The marketing program of a company is regarded as a *total success* if both the customer retention and product attractiveness to customers are high. This is when high profitability can be achieved at a maximum growth rate. The marketing program scores a *partial success* if the product attractiveness to the customer is high, but the customer retention is low. While lost customers are typically replaced by new customers, the total customer base will show little growth. *Partial failure* of a marketing program is when product attractiveness to the customer is low, but the customer retention is high. Under this scenario, business remains stagnant because it relies on

Figure 9.5 Marketing effectiveness diagram.

loyal customers to repeatedly buy mature, noncompetitive products. The company's sales may slow down or fall as few new customers are added. The marketing program is a *total failure* if both product attractiveness to the customer and the customer retention level are low. Customers are leaving, and the company's sales are falling.

Marketing strategies are implemented at the corporate, business, and operational levels. Top management provides inputs to identify future opportunities, and addresses such questions as what business the company is in and what business the company should be in. The business managers then specify their ideas, bring out products or services, and strive to create and maintain a sustainable competitive advantage in the marketplace. At the operational level, managers and support personnel conduct the planning for specific marketing programs, and implement and control marketing efforts related to segmentation, products, pricing, distribution channels, and communications.

9.2.3 Key Elements in Marketing

Those who plan and implement marketing programs are called *marketers*. Marketers consider various influence factors and make diverse decisions to penetrate a specific marketplace (Ferrell 1999). Marketers pay attention to several key elements of marketing, such as the market itself, the environment, the customer, the product, pricing, promotion, and distribution.

The *market* is made of buyers who are expected to purchase certain products and services, and also buy substitutes that offer similar values. Of great importance to the marketers is the size of the market, measured in millions of dollars per year, its growth rate, its location characteristics, and its requirements. The market must be large, stable enough with a reasonable growth rate, and relatively easy to reach and serve in order for it to be a worthwhile target for the marketers.

The *environment* refers to competitors, barriers to entry, rules and regulations, resources, and other such factors affecting the marketers' success in a given market segment. Marketers must also understand the opportunities and threats present in the environment.

The *customers* consist of all potential buyers of a given product or service. Companies need to understand why they buy, how they buy, who makes the decision to buy, who will use the product, in what specific way the use of the product will contribute value to the user, what might be the buyer's preference related to service and warranty, what other product features the customer may want, and other factors. The more a company knows about its customers, the better the company serves the customers in order to build and maintain competitive advantages in the marketplace. (Appendix 9.B contains additional sample customer-survey questions.)

To become customer-oriented, companies need to (1) define the generic needs of customers through research (such as the buyer's perception of an automobile's status, safety, or cost), (2) identify the target customers by segmentation (including which selected groups of customers have shared needs), (3) differentiate products and communications (for example, offering special reasons for customers to buy through unique product attributes or unique communications), and (4) bring about differentiated values for customers.

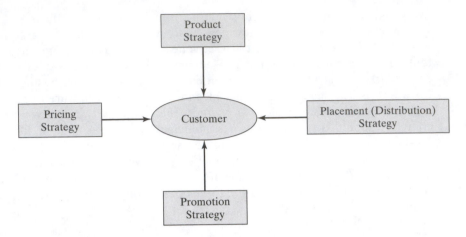

Figure 9.6   Marketing mix.

The *product* or *service* symbolizes the actual "bundle of benefits" that is offered to customers by the marketers. Factors considered include functional attributes, appropriateness to customer needs, distinguished features over competition, product-line strategy, and product-to-market fit.

Promotion and *communication* consider strategies of product and brand promotion, options to use a push–pull strategy, selection of advertising media, and the choice of promotional intensity. These ensure that the selected means for communication are compatible with the target market segments.

The *pricing strategy* concerns itself with (1) the choice of either a skimming or a penetrating strategy to set the price, (2) the use of value-added pricing, and (3) the fit of a chosen pricing strategy to the target segment.

The *placement (distribution) strategy* defines such areas as (1) the product delivery options of either an intensive, exclusive, or selective distribution system; (2) the company's relationship with dealers; and (3) the changes in distribution systems.

These elements characterize the multidimensional nature of marketing and are centered on customers, who are the focus of any successful marketing program.

9.2.4 Marketing Mix (or the Four P's)

Four of the seven key elements of marketing centered around customers form the *marketing mix* (Francese 1996). (See Figure 9.6.) Specifically, the marketing mix consists of price, promotion, product, and placement (distribution)—the *four P's* of marketing.

To help the reader understand the market and the customer, market forecast and market segmentation are introduced in Sections 9.3 and 9.4. The four marketing mix elements shown in Figure 9.6 are addressed in detail in Sections 9.5 to 9.8.

Example 9.1.

A company has divided the market for its existing product into three segments: (1) massmarket applications, (2) applications requiring a quality product, even though consumers continue buying on price, and (3) critical applications to which both quality and reliability are important. Advise the company on the marketing mix that should be applied to these three segments.

Answer 9.1.

For the company to be successful, a different set of marketing strategies needs to be applied to each of these segments, as suggested in Table 9.1.

TABLE 9.1   Marketing Strategies for Specific Segments

	Segment 1	Segment 2	Segment 3
Price	Low	Low	High
Product	Standard	Quality	Quality and reliability
Promotion	Broad	Limited	Focused
Place	Multiple distributions	Multiple distributions	Direct

Multiple distributions are recommended, including mass-merchandise department stores for wide distribution. Direct distribution should include catalogs, specialty stores, and upscale department stores.

9.3 MARKET FORECAST—FOUR-STEP PROCESS

The purpose of conducting a market forecast is to define the characteristics of the target market as to, for example, size, stability, growth rate, and serviceability. Market size and growth rate must be large enough to warrant further consideration by marketers.

Any future market is always uncertain due to potential changes in end-user behavior, global economics, new technologies, competition, and economic and political conditions.

Barnett (1988) emphasizes that the key to successfully forecasting market size is to understand the underlying forces behind the demand. Barnett proposes a four-step process, as follows: (1) define the market, (2) segment the market, (3) determine the segment drivers and model its changes, and (4) conduct a sensitivity analysis. These steps are explained below.*

9.3.1 Define the Market

On the basis of customer interviews, the market should be defined broadly to include the principal product to be marketed, its "bundle of benefits" to customers, and product substitution.

---

*Reprinted with permission of *Harvard Business Review*. From "Four Steps to Forecast Total Market Demand" by F. W. Barnett, *Harvard Business Review*, July–August, 1988. Copyright © 1988 by Harvard Business School Publishing Corporation, all rights reserved.

9.3.2 Segment the Market

In segmentation, the potential customers for the principal product are divided into homogeneous subgroups (segments) whose members have similar product preferences and buying behavior. (Market segmentation is elucidated in detail in Section 9.4.)

9.3.3 Determine the Segment Drivers and Model Its Changes

Segment drivers may be composed of macroeconomic factors, such as the increase in white-collar workers and in population, as well as industry-specific factors, including the industrial growth rate and business climate. Possible sources of information related to segment drivers are industrial associations, governments, industrial experts, marketing data and service providers, and specialized market studies.

9.3.4 Conduct a Sensitivity Analysis

Sensitivity analyses are conducted to test assumptions. Monte Carlo simulations may be performed to generate the maximum—most likely—and minimum total market demand values, as well as an assessment of the risks involved. (See Section 6.9.)

The following are illustrative examples in which the total market demand for a product is estimated by (a) defining the industrial segments that purchased the product in the past, (b) determining the future growth rates of these industrial segments, and (c) calculating the total market demand for the product with these industrial segment growth rates as the segment drivers. The assumption here is that the product demand of a given industrial segment is in direct proportion to its segment growth rate.

For example, to predict the demand of electricity in future years, a utility company has subdivided its consumers into three segments: industrial, commercial, and residential. The need for electricity by the industrial segment depends on its future production level and business climate. The electricity demand by the commercial segment is related to retail sales that in turn are negatively affected by retail stores consolidating and by growing Internet sales. (Web-based sales increased by 28.5 percent to $14.8 billion in 2002.) The residential electricity demand is affected positively by new home sales and home sizes and negatively by the increased energy efficiency of home appliances.

A second example is a paper-producing company that has determined the total market demand for uncoated white paper by deconstructing the market into end-use segments such as business forms, commercial printing, reprographics, envelopes, stationery, tablets, and books. The drivers in each segment are then modeled in terms of macroeconomic and industrial factors, using regression analysis and statistics. Examples of applicable drivers include growth in the use of electronic technology, white-collar workers, the present level of economic activity, the growing use of personal printers, population growth, demand growth induced by price reduction, and the practice of paying bills online and not by checks stuffed into envelopes.

Market forecast is a difficult, but critical first step to take when developing a marketing program. Companies routinely engage both internal and external resources to assess the principal characteristics of the target market for their products.

9.4 MARKET SEGMENTATION

Once it is determined that the target market is worth pursuing, (i.e., the market is large and stable enough with a high growth rate), then it is useful to understand the potential customers so that they can be served well. Market segmentation is a process whereby companies recognize the differences between various customer groups and define the representative group behaviors. Members in each of these customer groups respond to product or service offerings in similar manners and have comparable preferences with respect to the price–quality ratio, reliability, and service requirements.

9.4.1 Purpose of Market Segmentation

By dividing consumers into groups that have similar preferences and behaviors with respect to the products being marketed, companies achieve four specific purposes: (1) matching products and services to appropriate customer groups, (2) creating suitable channels of distribution to reach the customer groups, (3) uncovering new customer groups that may not have been served sufficiently, and (4) focusing on niches that have been neglected by competitors.

Overall, segmentation allows companies to realize the following benefits: developing applicable marketing strategies and objectives, formulating and implementing marketing programs that address the needs of the different customer groups, tracking changes in buying behavior over time, understanding the enterprise's competitive position in the marketplace, recognizing opportunities and threats, and utilizing marketing resources effectively.

9.4.2 Steps in Marketing Segmentation

Displayed in Figure 9.7 is a segmentation flow diagram that illustrates the key steps in segmenting a market.

Companies need to classify consumers into segments by understanding their individual, institutional, and product-related characteristics. *Individual* characteristics include culture, demographics, location, socioeconomic factors, lifestyles, family life cycle and personalities. *Institutional* characteristics include the type of business, its size, and extent of global reach. *Product-related* characteristics include type of use [original equipment manufacturer (OEM) versus end user], usage level, product knowledge, brand preference, and brand loyalty. Also to be studied are benefits sought by consumers such as psychological and emotional benefits, functional performance, and price.

Millions of consumers purchase cars every year. To some, cars are a status symbol; to others, cars are simply a means of transportation. A large number of car buyers emphasize safety and reliability, while others focus on fuel economy. Socioeconomic factors, demographics, personalities, and family life are all known to influence the behavior of car buyers. These consumers are extensively segmented by all major carmakers.

9.4.3 Criteria for Market Segmentation

To be effective, the segmentation of a market needs to satisfy several requirements.

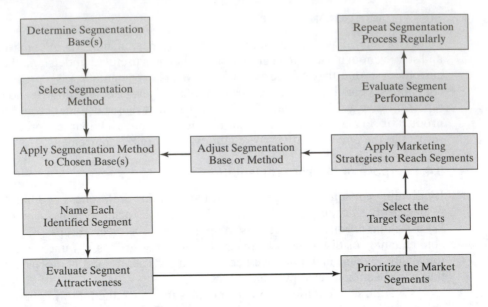

Figure 9.7   Segmentation flow diagram.

The segmentation should be measurable. It should result in readily identifiable customer groups. The identified customer groups should also be homogeneous. Each group's members should possess more or less unified value perceptions and display compatible behavioral patterns. These customer groups are reachable by promotion and distribution means. Above all, the segments should be large enough in size to justify marketing efforts, and they should have a high-growth rate to allow the company to achieve long-term profitability.

9.4.4 Pitfalls of Market Segmentation

There are pitfalls to market segmentation. Certain "old economy" companies adopt the asset-rich business strategy of pursuing the scale of economy advantages. For these companies, a potential pitfall is oversegmentation, because the selected segments may be too small or fragmented to serve effectually. Such an oversegmentation is not a pitfall for other "knowledge economy" companies that form partnerships to establish supply chains for "build-to-order" products. To foster differentiation, knowledge economy companies pursue product customization as the basis for their business strategies (Gilmore and Pine 2000). Examples of these products and services are minibrewers, computers, and custom cosmetics.

For other companies, overconcentration (lack of balance between segments) could have a negative impact on their overall marketing effectiveness (Hartley 1998).

Market segmentation is a prerequisite to developing a workable marketing program. Knowledge derived from customer groups serves as valuable inputs to product design, pricing, advertising, and customer services, all of which are yet to be finalized.

9.5 PRODUCT STRATEGY

The product and service strategy takes the center stage in any marketing program (Brethauer 2002). If marketed properly, products that offer unique and valuable functional features to consumers are expected to enjoy a strong marketplace acceptance.

Products may be generally classified as either industrial or consumer oriented. Their characteristics are different, as shown in Table 9.2. Marketing programs for consumer products are quite different from those for industrial products, even though the same basic marketing approach applies to both (Miller and Palmer 2000; Kirk 2003).

A good marketing program must take into account the consumer's perception of products. Indeed, consumers perceive products in different ways than the producers and marketers do. When buying products, consumers look for "bundles of benefits" that satisfy their immediate wants. Products that producers regard to be different because of physical embodiment (e.g., input materials), production process, or functional characteristics may in fact be equivalent in how consumers perceive them, provided that the same or similar benefits are advantagious (substitute products) to the customers. Table 9.3 contains illustrative examples of these different perceptions.

Companies must define competition based on the way customers perceive their products. Note that products that appear to be physically different to marketers may appear to be the same to users.

Product strategy must also be established with respect to competition. A company may decide to market premium products, characterized by having features that are outstanding or superior to those offered by the competition. Such outstanding product

TABLE 9.2   Industrial Versus Consumer Products

	Industrial Products	Consumer Products
1. Number of buyers	Few	Many
2. Target end users	Employers	Individual
3. Nature of products	Tailor-made, technical	Commodity, nontechnical
4. Buyer sophistication	High	Low
5. Buying factors	Technical, quality, price, delivery, service	Price, convenience, packaging, brand
6. Consumption	OEM parts for reselling, own consumption	Direct consumption
7. Producer end-user contact	Low	High
8. Time lag between demand and supply	Large	Small
9. Segmentation techniques	SIC (standard industrial classification), size, geography, end user, decision level	Demographics, lifestyle, geography, ethnic, religious, neighborhood, behavior
10. Classification of goods	Raw materials, fabricated parts, capital goods, accessory equipment, MRO supplies	Convenience (household supplies, foods), shopping (cameras, refrigerators), specialty (foods, brand-name clothing)

Note: MRO is the abbreviation for maintenance, repair, and operations.

TABLE 9.3   Product Perception

Products	Producers and Marketers Perception	Consumers and Users Perception
Cosmetics	Chemicals, colors, creams, etc.	Hope (to look younger and more attractive)
Washing machine	Steel, assembly, capital, distribution	Convenience (in doing household chores)

features may be possible because of the company's innovative capabilities, technological superiority, and other core competencies. Companies with such "high-road" brands (see Section 9.5.5) tend to enjoy and sustain high profitability. Other companies may elect to make commodity-type (value) products with commonly available features so that they compete head-on against their competitors on the basis of price, service, distribution, and customer relations management. They pursue the option of "low-road" products. Product positioning is the step that addresses such competitive issues related to products.

9.5.1 Product Positioning

An important question that companies should answer is, Which product attributes should be included? A *perceptual map* is a useful tool to position the company's products in relation to existing competitive products in the marketplace. It enables companies to select the correct set of product attributes to maximize its marketing advantages. It also articulates customer preferences and identifies gaps; these are useful steps in positioning new entries or repositioning existing products.

Figure 9.8 is an example of a perceptual map for automobiles regarding the product attributes of price and styling. Only the relative magnitudes of the attributes are

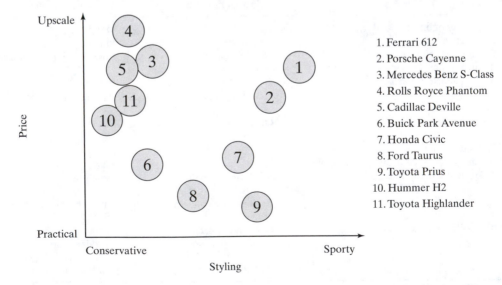

1. Ferrari 612
2. Porsche Cayenne
3. Mercedes Benz S-Class
4. Rolls Royce Phantom
5. Cadillac Deville
6. Buick Park Avenue
7. Honda Civic
8. Ford Taurus
9. Toyota Prius
10. Hummer H2
11. Toyota Highlander

Figure 9.8   Perceptual map.

TABLE 9.4   Inputs to Six-Dimensional Perceptual Map

Product	F_1	F_2	F_3	F_4	F_5	F_6
P_1						
P_2						
P_3						
P_4						
Your Product						
P_5						
P_6						
P_7						

emphasized in such a map. However, the map helps to identify which car models are in direct competition and which ones are not. It is also possible to link customer segments to these pairs of product attributes, thus enabling companies to refine their advertising strategies for these customer segments.

Products with more than two important attributes are readily mapped into an n-dimensional perceptual map. A product (e.g., P_1) is designated by a single point having the coordinates F_1, F_2, F_3, through F_n, with each representing an independent product attribute. This representation is complete if the elements of the attributes set $(F_1, F_2, \ldots, F_n)$ are mutually exclusive and collectively exhaustive. For example, for automobiles, these attributes include price, styling, fuel economy, driving comfort, safety, brand prestige, power, and longer-term dependability (number of problems per 100 three-year old vehicles). The spacing between two neighboring points (each identifying a product) as depicted in this n-dimensional map is equal to the square root of the sum of the individual attribute differences, squared. Presented in Table 9.4 is an example of the description of products with six distinguishable attributes.

9.5.2 Product Life Cycles

Every product goes through a number of important stages throughout its useful life (Hayes and Wheelwright 1979; Capon 1978). These stages include:

1. The *initiation stage*—product testing, market development, and advertising
2. The *growth stage*—product promotion, market acceptance, and profit growth
3. The *stagnation stage*—price competition, substitution, and new technologies
4. The *decline stage*—cash-cow strategy with no more investment.

Companies need to understand which phase a given product is in. (See Figure 9.9.)

From the standpoint of the product life cycle, an important product strategy is to sequence the introduction of new products so that a high average level of profitability can be maintained for the company over time.

Engineering managers are particularly qualified to constantly come up with innovative products so that their employers may introduce these products at the right intervals.

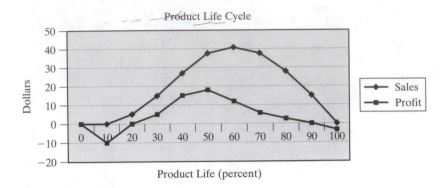

Figure 9.9   Product life cycle.

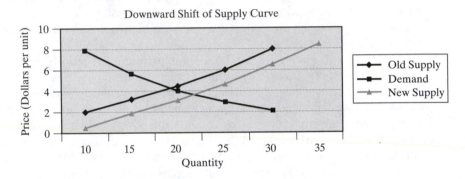

Figure 9.10   Impact of product innovation on supply curve.

9.5.3 Product Supply Curve

The product supply curve portrays the market behavior of companies tending to generate a larger quantity of products if the product price rises. In a price-versus-quantity plot, this curve rises from the lower left corner to the upper right corner. (See the "Old Supply" curve in Figure 9.10.)

Product innovations (e.g., better production methods, lighter materials, and simpler designs) may lower the product cost. Companies usually lower the price to achieve a larger market share because of their strengthened competitive positions. Thus, product innovations are likely to promote an increased market demand for the products and a downward shift in the supply curve, as exhibited in Figure 9.10.

Product innovations afford a major opportunity for engineering managers to shape the company's product strategy.

9.5.4 Product Portfolio

Another product strategy issue is related to the types of products concurrently being marketed. With the exception of a few, most companies market a group of products at the same time, referred to as a *product portfolio* (Yip 1981).

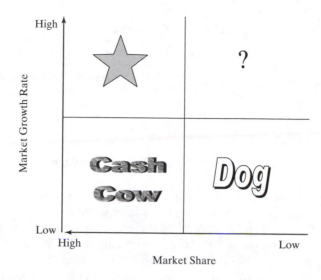

Figure 9.11   Product portfolio.

Products in a portfolio are usually not "created equal." From the company's standpoints of profitability and market-share position, some are more valuable than others. Boston Consulting Group (BCG) of Boston, Massachusetts, developed a portfolio matrix based on the two measures of growth rate and market share. (See Figure 9.11.) According to this classification scheme, products are regarded as *stars* if they enjoy high growth rate and high market share and *question marks* if they have high growth rate, but low market share. *Cash Cows* are those products with low growth rate and high market share. Products are designated as *dogs* if both growth rate and market share are low.

Figure 9.11 indicates that companies need to differentiate the products they market by strategically emphasizing some and deemphasizing others, according to the responses from the marketplace. For example, a useful strategy to manage a product portfolio is to milk the *cows* to provide capital for building *question marks* into *stars* that will eventually become *cash cows*. Divest the *dogs*.

9.5.5 Products and Brands

Numerous high-tech companies operate in a "product-centric" business model, in that they market products on price and performance. Recent market studies show that the success of technology-based products in the marketplace is not purely dependent on the price–performance ratio, but also on the trust, reliability, and promised values the customers perceive in a given brand.

According to Ward, Light, and Goldstine (1999) brand is "a distinct identity that differentiates a relevant, enduring and credible promise of value associated with a product, service, or organization that indicates the source of that promise."

The brand of a company is more than a name. It stands for all of the images and experience (e.g., products, service, interactions, and relations) that customers associate with the organization. It is a link forged between the company and the customers. It is

---

TABLE 9.5   Promises of Value

Corporate Brands	Promises of Value
IBM	Superior service and support
Apple	Simple and easy to use
Lucent	Newest technologies
Gateway	Friendly service

---

a bridge for the company to strengthen relationships with customers, according to Keller (2000).

A promise of value is an expectation of the customer that the company is committed to deliver. Examples of such promises of value from several companies are listed in Table 9.5. They must be relevant to the enduring needs of the targeted customers and made credible by the persistent commitments of the company. To be competitive, the promise of value must be distinguishable from those offered by other brands.

Research by Ward, Light, and Goldstine (1999)* indicates that customers consider questions at five levels when purchasing both high-tech and consumer types of products. These questions may be grouped into a brand pyramid, as illustrated in Figure 9.12.

Technology-oriented buyers are typically focused on questions at Levels 1 and 2. However, higher level business managers who make purchase decisions are also known to address questions at Levels 3–5. These decision makers are interested in what the product or service will do for them, not just how it works. Consequently, to project a trustworthy and reliable image, to build strong relationships with customers, and to enhance brand loyalty and customer retention, companies are well advised to pay attention to questions at all five levels. This is the emphasis of brand management.

Brand is a major asset that must be properly managed and constantly strengthened. Useful inputs for brand management are typically secured from customer feedback. Once the market is properly segmented, the promise of values is specifically designed to address the needs of the target segments involved. Actions are then taken to deliver the stated promise of values, and results are constantly measured to monitor progress.

Brand is evaluated on how well it is doing with respect to a number of metrics: delivering according to the customer's desire, relevance to the customers, value to the customers, positioning, product portfolio management, and integration of marketing efforts, management, support, and monitoring.

In the past, brand management has been focused primarily on points of difference, such as how a given brand differs from the other competing brands within the same category. Maytag is known to emphasize "dependability." Tide® focuses on "whitening power." BMW stresses "superior handling." Recently, Keller, Sternthal, and

---

*From S. L. Ward, L. Light, and J. Goldstine, "What High-Tech Managers Need to Know About Brands," *Harvard Business Review*. July–August, 1999. Copyright © 1999; reprinted by permission of Harvard Business School Publishing Corporation, all rights reserved.

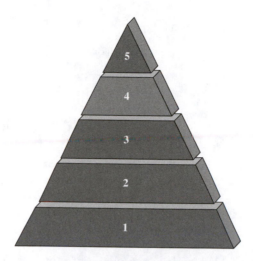

Level 5	What is the personality of the brand (aggressiveness, conservatism, etc.)?
Level 4	Does the value offered by the product reflect that favored by the customers (conservatism, family values, achievement)?
Level 3	What are the psychological or emotional benefits of using the product in question? How will the customers feel when they experience its technical benefits?
Level 2	What are the technical benefits to customers (solutions to problems, cost saving benefits, and speed of production)?
Level 1	How does the product work (product characteristics, technical features, and functional performance)?

Figure 9.12   Brand pyramid.

Tybout (2002) suggested that attention should be paid to points of parity and the applicable frame of reference, in addition to the points of difference, when marketing a given brand. Emphasizing the frame of reference is intended to help customers recognize the brand category comprising all of the competing brands. Focusing on the points of parity will ensure that customers recognize a given brand as a member of the identified brand category.

Brand may be classified with respect to the two dimensions of category and relative market share. The brand category is defined as *premium* if the category is dominated by premium brands—those with high values to customers. Examples of premium brands include BMW, Mercedes Benz, Jaguar, and other luxury and specialty cars that each have special, high-value attributes. The brand category is defined as *value* if it is dominated by value brands—those with basic, minimum, low-end attributes. Examples of value brands include Chevy, Saturn, and other compact and four-door family cars. Gillette markets its Mach-3 Turbo shaving system as a premium brand, whereas the cheap disposable razors from its own company as well as its competition are the value brands. The relative market share refers to the percentage of market share a given brand is able to attain.

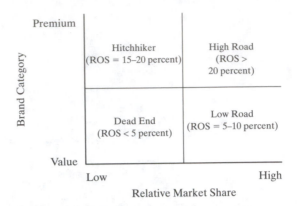

Figure 9.13 Brand Classes.*

In Figure 9.13, brands are grouped into four classes: high-road, low-road, hitch-hiker, and dead-end brands (Vishwanath and Mark 1997). ROS is return on sales, which is defined as net income divided by the sales revenue. (See Section 7.5.1.)

High-road brands are those with products that offer premium features, options, qualities, and functions to command high selling prices while attaining a leadership position in the market share. Examples of such high-road brands are Coca-Cola, Frito-Lay, and Nabisco. These brands enjoy excellent profitability that may be sustained for long periods. The key success factors for high-road brands are technological innovation (constantly adding new product features and values), time to market, flexible manufacturing, and advertising.

Low-road brands are those that offer value brands while enjoying a high market-share position. Because of marketplace competition and a lack of distinguishable product features, these brands can be successfully managed by emphasizing cost reduction, production efficiency, product simplification, and distribution effectiveness.

Hitchhiker brands are those with premium product values and low market share. For these brands to become high-road brands, management must emphasize cost reduction, flexible manufacturing, and product innovation.

Dead-end brands are value brands with low market share. These brands attain only marginal profitability. There are several strategies to grow the profitability of these brands: (1) Reduce the price to penetrate the market and thus move these brands to the low-road category. (2) Increase the scale of economies by applying the "string-of-pearls" strategy: producing and marketing several products together to cut costs. (3) Finally, introduce a supererior, premium product to "trump" this brand into the hitch-hiker category. Failing all of these attempts, dead-end brands should be discontinued. Table 9.6 summarizes the strategies that deal with these four classes of brands.

The preceding discussion on product brands should assist engineering managers in understanding the significant value added by brands to the success of the company's

TABLE 9.6   Strategic Options for Brands*

Brands	Strategic Options
High road	Apply R&D to constantly innovate to make products premium—adding new product features and changing forms and functions
	Expand product lines (product proliferation)
	Initiate media campaign
	Capital investment
	Decrease time to market
	Flexible manufacturing
	Direct store delivery to preoccupy shelf space
Low road	Pursue cost reduction aggressively
	Lessen product proliferation (SKUs) by simplifying product types and designs
	Consolidate production facilities to improve efficiency and cut wastes
	Use realized cost savings to slash price
	Consider ways to add premium products (advancing to high road)
Hitchhiker	Apply R&D to constantly innovate to make products premium—adding new product features and changing forms and functions
	Cost reduction
	Reduce time to market
	Flexible manufacturing
	Initiate media campaign
	Consider capital investment
Dead end	Cut price (advancing to low road)
	Outsource in areas with economies of scale
	Apply the "string-of-pearls" strategy to enhance scale
	"Trump" the category by introducing a supererior, premium product that resets consumer's expectation (advancing to hitchhiker)
	Do not spend on marketing
	Make no capital investment

*Source: Vijay Vishwanath and Jonathan Mark. "Your Brand's Best Strategy." *Harvard Business Review*, May–June 1997. Copyright © 1997; reprinted by permission of Harvard Business School Publishing Corporation, all rights reserved.

marketing program (Bedbury with Fenichell 2002; Perry with Wisnom 2003). Such an understanding should make it easier for them to channel their support efforts to actively enhance the company's brand strategy.

9.5.6 Engineering Contributions to Product and Brand Strategy

The product is a key element in the marketing mix. Engineers and engineering managers have major opportunities to add value by (1) understanding the customers' perceptions of products, (2) designing products with features that are wanted by customers, (3) helping to position the company's products strategically to derive marketing advantage, (4) practicing innovations in the design, development, production, reliability, serviceability, and maintenance of products to differentiate them from others, (5) sustaining the company's long-term profitability by creating a constant flow of new products for introduction on a timely basis, (6) assisting in managing companies' product portfolios by adding premium features to some and reducing costs to others, and (7) ensuring commercial success of the high-road and hitchhiker brands in the marketplace.

In the "knowledge economy" of the 21st century, time to market is an increasingly important competitive factor. Once the desirable set of product attributes is known from market research, those companies that bring the suitable products to the market first will enjoy preemptive selling advantages and will recover the product development costs faster than others.

Engineering managers should also be well prepared to contribute in shortening the products' time to market by utilizing advanced technologies to create modular design, eliminate prototyping, whittle away design changes, foster parts interchangeability, ensure quality control, and other innovations.

9.6 PRICING STRATEGY

Price is a very important product attribute (Corey 1980). Companies pay a great deal of attention to the setting of product prices. Setting the price too high will discourage consumers from buying, whereas setting it too low will not assure profitability for the company. Generally speaking, the two major strategies for setting the product prices are the skimming strategy and the penetrating strategy.

9.6.1 Skimming and Penetration Strategies

Companies applying the *skimming strategy* set the product price at the premium levels initially and then cut the product price in time to reach additional customers. In other words, they "skim the cream" first. An example is the marketing of a new book with hardcover copies selling at a high price (e.g., $29.95) followed by the paperback version at a low price ($4.95). New technology products are also typically sold at high prices initially in the absence of competition. As competitors enter the market with products of similar features, product prices are lowered accordingly.

In contrast, companies pursuing the *penetration strategy* set product prices low to penetrate a new market and rapidly acquire a large market share. A high market-share position sets forth a barrier of entry for other potential competitors. In general, companies use a penetration pricing strategy to enter an existing, but highly competitive market. An example is the marketing of Japanese motorcycles in the United States.

9.6.2 Factors Affecting Price

In setting product prices, besides using the skimming and penetration strategies, companies broadly consider a number of other factors: financial aspects, product characteristics, marketplace characteristics, distribution and production capabilities, price–quality relationship, and the relative position of power. These factors will be discussed next.

Financial Aspects—The more solid the company's financial position, the more capable it is at initially setting the product price low. Companies strong in finance stay afloat for a long period of time even with low profitability. Companies that desire high, short-term profitability tend to set the product prices high.

Product Characteristics—The product price may be set in direct proportion to the value and importance of the product to users, as well as the income levels of its target customers. Usually, a new product in its early life cycle sells at a high price; the company benefits from the product's novelty.

Marketplace Characteristics—Companies set product prices in reverse proportion to the level of competition in the marketplace. The level of competition refers to the number of direct competitors, the number of indirect competitors marketing substitution products that offer similar value to customers, and the competitive counter-strategies (speed and intensity) that these competitors may exercise. Companies tend to set the product price high if the barriers to market entry are high. The barriers to market entry depend on lead time and resources—technical and financial, patents, cost structure, and production experience. In addition, products with inelastic price-demand characteristics tend to carry a high price. A product has inelastic price-demand characteristics if a large price increase induces a small change in the quantity of the product demanded in the marketplace.

Distribution and Production Capabilities—Product availability to consumers depends on the company's product distribution capabilities. With strong distribution channels in place, companies may set the product price high, as quickly making products available to consumers represents a competitive strength.

Sales volume impacts the company's production experience. Companies with extensive production experience are known to produce products at a low unit cost. A lower product unit cost enables these companies to set a lower product price to gain market share.

The Boston Consultant Group studied manufacturing operations and discovered that there is a correlation between production volume and product unit cost. For every doubling of the production volume, the unit cost is whittled down by about 15 percent—or the 85-percent experience curve. (See Figure 9.14. Note that the horizontal axis in the figure is nonlinear.)

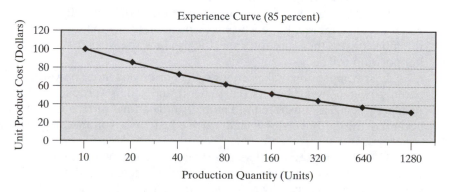

Figure 9.14   Experience curve.

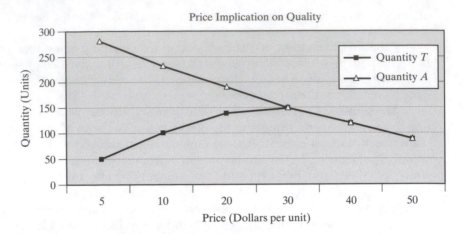

Figure 9.15   Price-quality relationship.

Companies with a faster time-to-market strategy are able to accumulate production experience more quickly, attain a lower product unit cost sooner, and sustain company profitability for longer periods of time.

Price-Quality Relationship—One important consideration in setting the product price is the perceived cost–quality relationship by customers. There is substantial evidence in business literature to indicate that customers tend to believe that "low-priced items cannot be good." Price is perceived to be an indicator of quality.

Product prices should not be set too low. There is a price threshold below which customers may raise questions regarding the quality, as indicated in Figure 9.15. The demand curve "Quantity T" illustrates a normal price-demand relationship in the absence of a price threshold, whereas the demand curve "Quantity A" contains a price threshold at about $30 per unit, below which the demand for the products starts to drop off as the perception of poor quality related to low price sets in.

Relative Position of Power—Consumer products are typically marketed by a few major companies to a very large number of customers. For innumerable industrial products with high technological contents, the number of both producers and customers may be limited. The greater the number of sellers there are available for a given product, the weaker each seller's position in the marketplace will be. Similarly, the more buyers there are for a specific product, the weaker the buyers' relative market position will be.

Less competition makes either sellers or buyers more powerful. The relative position power between buyers (customers) and sellers (producers) has an impact on product pricing, as illustrated in Figure 9.16. The final price offered by the sellers and accepted by the buyers is usually arrived at by a suitable negotiation or auction process.

If both buyers and sellers are strong—for example, when the U.S. government (customer) procures fighter airplanes from the defense industry (producer)—a final price is typically reached by a *negotiation* made up of a series of offers and counteroffers. A typical pricing arrangement may be cost plus a fixed percentage of gross margins.

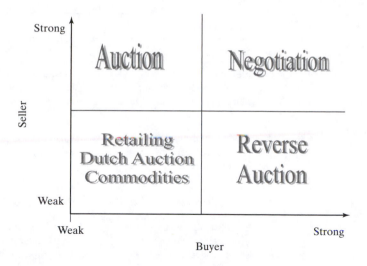

Figure 9.16   Processes of setting product prices.

When the sellers are strong (e.g., selling an original master painting, a porcelain vase from the Ming Dynasty, or some other type of unique physical asset) and the buyers are weak, sellers tend to take advantage of their dominant supplier position by employing an auction. An *auction* is a bidding process by which buyers are forced to compete against each other by committing themselves to consecutively higher prices, with the final price being set by the highest winning bid.

If buyers are in a strong position (e.g., due to large transaction volumes), they force weak sellers to compete against each other in a reverse auction. A *reverse auction* requires the prequalified sellers to submit increasingly lower bid prices within a fixed period of time. The lowest bid defines the final price and the ultimate winner of the sales contract (Smock 2003). Some large companies employ such pricing tactics to purchase high-volume supply items such as computers, paper and pencils, tires, batteries, and maintenance, repair, and operations (MRO) goods.

Finally, when both sellers and buyers are weak, products are usually not differentiable, and the product prices are highly depressed and fixed. Examples include various commodity products sold in retail stores. Some sellers (e.g., Land's End) may activate a Dutch auction to compete. In a *Dutch auction*, sellers slash the product prices by certain percentage at a regular time interval (e.g., every week) until the products are sold or taken off the market. In this case, buyers compete against other "sight-unseen" buyers to seize the lowest possible selling prices.

The Internet has made many of these pricing processes much more practical and efficient to implement (Zimmerman 2000). Because of its ability to allow sellers and buyers to rapidly reach other buyers and sellers, respectively, the Internet tends to weaken the relative power positions of both the sellers and the buyers, causing products to become increasingly commoditized, thus depressing product prices and intensifying competition.

Table 9.7 enumerates a number of other factors that have an impact on setting the product price.

TABLE 9.7   Factors Affecting Product Price

Factor	Skimming	Penetration
Demand	Inelastic Users know little about product Market segments with different price elasticity	Elastic Familiar product Absence of high-price segment
Competition	Few competitors Attracts competition Market entry difficult	Keep out competition Market entry easy
Objective	Risk aversion Go for profits	Risk taking Go for market share
Product	Establish high-volume image Product needs to be tested Short product life cycle	Image less important Few technical product problems Long product life cycle
Price	Easy to go down later More room to maneuver	Tough to increase later Little room to maneuver

Distribution and
    Promotion
	No previous experience
	Need gross margin to
	    finance its development
		Existing distribution system
		    and promotion program

Financing	Low investment Faster profits	High investment Slower profits
Production	Little economy of scale Little knowledge of costs	High economy of scale Good knowledge of costs

9.6.3 Pricing Methods

In setting product prices, companies broadly consider a number of factors (some of which were discussed in Section 9.6.) and methods. Several of these methods are briefly discussed next.

Cost Oriented—Some companies set prices by adding a well-defined markup percentage to the product cost. This is to ensure that all products sold generate an equal amount of contribution margin to the company's profitability:

$$\text{Price} = \text{Cost} + \text{Markup (e.g., 35 percent of cost)} \qquad (9.1)$$

Cost-plus contracts are often used for industrial products related to R&D, military procurements, unique machine tools, and other uses. Small sellers use cost-plus pricing to ensure a fair return while minimizing cost factor risks. Larger buyers favor this type of pricing so that they can push for vendor cost reduction via experience. Larger buyers may optionally offer to help absorb the cost risks related to inflation.

Often, sellers and buyers enter a target-incentive contract, which prescribes that, if actual costs are lower than the target costs, sellers and buyers split the savings at a specific ratio. On the other hand, if the target costs are exceeded, then both parties pay a fixed percentage of the excess; the buyers pay no more than a predetermined ceiling price.

Profit Oriented—Other companies prefer to require that all products contribute a fixed amount of profit. This pricing method ensures that sellers realize a predetermined return on investment goal:

$$\text{Price} = \text{Cost} + \text{Profits (e.g., in term of ROI)} \tag{9.2}$$

Market-oriented—Some companies set prices of certain industrial products, such as those requiring customization, to what the buyers are willing to pay. For example, the companies strive to negotiate for the highest price possible and take advantage of the fact that product and pricing information may not be easily accessible. The continued advancement of Web-based communication tools tends to make information just one click away, rendering this type of pricing method no longer practical in today's marketplace.

Companies may also price the products slightly below *the next-best alternative* products available to the customer. The companies that have exhaustively studied the next-best alternative products available to their customers garner advantages in price negotiations.

Competitive bidding is used often by governments and large buyers. Usually three bids are needed for procurements exceeding a specific dollar value. Sealed bids are opened at a predetermined date, and the lowest bidder is typically the winner. Some industrial companies may engage in *negotiated bidding*, wherein they continue negotiating with the lowest one or two bidders for additional price concessions after the bidding process (e.g., a reverse auction) has been completed.

Value-Added Pricing—Companies with extensive application know-how related to their industrial products may set product prices in proportion to the products' expected value to the customer. The product's value to the customer depends on the realizable improvement in quality, productivity enhancement, cost reduction, profitability increase, and other such benefits attributable to the use of the product. Producers set the product prices high if there is a large value added by these products to the customer.

Competition Oriented—A common pricing method is to set prices at the same level as those of the competition. Doing so induces a head-on competition in the marketplace. In oligopolistic markets (typically dominated by one or two major producers or sellers and participated in by several other smaller followers), the market leader sets the price.

One well-known example of a competition-oriented pricing practice is target pricing. *Target pricing* was initiated and applied by many Japanese companies. Some American companies have now started to successfully apply this method. Target pricing is briefly addressed as one of several external benchmarking strategies in Section 5.3. The process of target pricing (see Figure 9.17) is as follows:

(a) Determine the market prices of products that are similar or equivalent to the new product planned by the company. Find all product attributes customers may desire. This is usually accomplished by a multifunctional team composed of representatives of such disciplines as design, engineering, production, service, reliability, and marketing. Select a product price (e.g., at 80 percent of the market price) that makes the company's new product competitive in the marketplace. This is then the target product price.

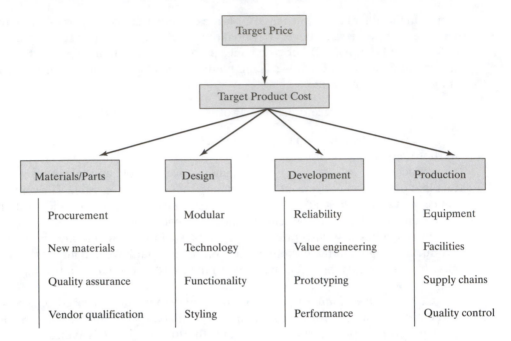

Figure 9.17   Target pricing.

(b) Define a gross margin that the company must have in order to remain in business.

(c) Calculate the maximum cost of goods sold (CGS) by subtracting the gross margin from the target product price. This is the target product cost, which must not be exceeded.

(d) Conduct a detailed cost analysis to determine the costs of all materials, parts, sub-assemblies, engineering, and other activities required to make the new product. Usually, the sum of these individual costs will exceed the target product cost previously defined.

　　Apply innovations in product design, manufacturing, procurement, outsourcing, and other cost reduction techniques to bring the CGS down to or below the target product cost level.

(e) Initiate the development process for the new product only if the target product cost goal can be met.

The target pricing method ensures that the company's new product can be sold in the marketplace at the predetermined competitive price, with features desired by consumers, to generate a well-defined profitability for the company. This method systematically evaluates low-risk, high-return investment opportunities because it forces the company to invest only when the commercial success of the product is more or less assured. Furthermore, it also focuses the company's product innovations on finding ways to meet specific and well-defined target product cost goals. It avoids the potential of wasting its precious intellectual talents in chasing ideas with no practical value.

Numerous companies use the pricing methods just discussed. Product prices are usually set by the marketing department in consultation with business managers. Engineering managers are advised to become aware of these methods, but to defer pricing decisions and related discussions to the marketing department.

9.6.4 Pricing and Psychology of Consumption

Recent studies indicate that buyers are more likely to consume a product when they are aware of its cost. The more they consume, the more they will buy again and thus become repeat customers. One useful way to induce them to repeatedly consume the products is to remind them of the costs committed through the choice of payment methods. This is based on the assumption that the more often the customers are reminded of the payments, the more they feel guilty if they do not fully utilize the products they have paid for.

Gourville and Soman (2002) point out that time payment better induces regular consumption of a product than lump-sum prepayment (e.g., prepaid season tickets) at the same total value. This is because the time payments remind the buyers of the costs periodically and thus invoke the *sunk-cost effect* on a regular basis. The psychology of the sunk-cost effect is that consumers feel compelled to use products they have paid for to avoid the embarrassing feeling that they have wasted their money.

Credit card payments are less effective in inducing consumption than cash payments because cash payments require the buyers to take out currency notes and count them one by one; thus, they experience the "pain" of making payments.

In price-bundling situations, the more clearly the individual prices of products are itemized, the better the perceived sunk-cost effect will be. Breaking down large payments into a number of smaller ones, thus clearly highlighting the costs of individual products sold in the bundle, can enhance this effect.

Studies of membership rates at commercial wellness and fitness centers support this logic. It has been well documented that those members who pay the membership fees once a year use the facilities only occasionally. These members are the least likely to renew, in comparison with those who pay on a monthly basis. Similar observations are made in sports events in which holders of season tickets show up less frequently than those who buy tickets for specific sets of events.

These examples point out that companies can induce customers to become repeat customers by focusing on ways to encourage consumption. Only consumption lets customers experience the benefits of the products they have purchased. Without such favorable experience, they may not feel that they have good reasons to buy the products again in the future. Hence, besides providing a good bundle of value made up of price, product features, convenient and efficient order processing and delivery, and quick-response after-sales services, companies should devise ways to stimulate consumption as a strategy to cultivate and retain repeat customers.

Example 9.2.

The company has been selling a number of products to customers. It is about to launch a new product with features far superior to any products currently on the market. One option is to price this new product at a large premium above the current price range so that the heavy development expenses can be readily recouped. The other option is to set the price comparable with the existing range in order to retain customer loyalty. What is your pricing advice to the company?

Answer 9.2.

Hold a focus group to find out the potential response of customers to the new product's features. Are these features of real value to them? How much more are they willing to pay for these features? Exciting new features from the manufacturer's viewpoints may not be as exciting to customers. Should customers appreciate the new features, then it is advisable to apply the skimming strategy and set a high price for the new product. This is also the principle of value pricing. Furthermore, doing so will avoid "cannibalizing" the existing products of the company.

An efficacious promotional campaign is essential to heighten product awareness. Keep monitoring the response of the market. If the market response is poor, cut down the product price slowly to induce more demand.

9.7 MARKETING COMMUNICATION (PROMOTION)

Marketing communication is intended to make the target customers aware of the features and benefits of the company's products. Product promotion follows a well-planned process (see Figure 9.18) for who says what to whom, in what way, through what channel, and with what resulting effect.

9.7.1 Communication Process

Companies select communicators who are publicly recognized and who have trustworthy images, as these speakers tend to induce public acceptance of their messages. Examples include Bob Dole for Viagra®, Tiger Woods for golf products, and Yao Ming for basketball-related items.

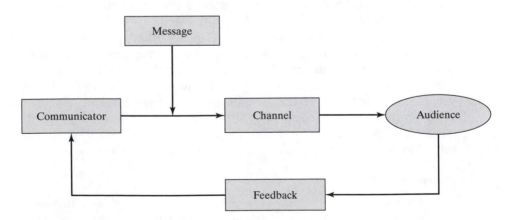

Figure 9.18   Market communications.

Messages may be in various forms including slogans. A slogan is a brief phrase used to get the consumer's attention and acts as a mnemonic aid. Successful slogans typically represent a symbolization of product features in terms of the customser's wants and needs (such as information, persuasion, and education). Examples include "Ring around the Collar"; "Where's the Beef?" "You are what you know"; and "One investor at a time."

Channels of communication are specific means to foster market communications. In general, there are two channels of communications: the marketer controlled and the consumer controlled. The *marketer-controlled* channels include advertisements placed in trade journals, national television programs, distribution of specific product brochures, promotion by technical salespeople, industrial exhibitions, and direct-mail marketing. The *consumer-controlled* channels include interpersonal communications by word of mouth, news reports, and others sources of information perceived to contain no conflicts of interest.

The audience is the target for marketing communications. When selecting communications channels to reach specific consumers, the consumers' characteristics, media habits, and product knowledge must be taken into account. A consumers' characteristics include socioeconomic status, demographics, lifestyle, and psychology. For industrial customers, characteristics include big versus small firms, large versus small market shares, and stable versus unstable financial position. Media habits point to sources of information preferred by the customers (e.g., types of magazines and TV programs). Product knowledge is the consumers' understanding and appreciation of the values offered by the products.

Some companies invest a considerable amount of effort into educating their consumers. A case in point is the known practice of some drug companies of sponsoring large-scale clinical studies conducted by universities and other independent organizations. The purpose of such funded studies is to produce results for publication in technical journals from which consumers gain product knowledge in ways preferred by the sponsoring drug companies.

The impact of marketing communications on products may be short term or long term. The short-term impact is related to recall, recognition, awareness, and purchase intention with respect to the products in question. The long-term impact is reflected in the purchase by customers and brand loyalty with repeat purchase. Several factors influence the effectiveness of marketing communication, such as timing, price, product availability, responses by competition, product warranty conditions, and service.

Marketing communication brings about heightened product awareness. An improved familiarity with the product induces more people to buy the product at the current price, thus causing an up-shift of the product demand curve (as illustrated in Figure 9.19).

9.7.2 Promotion Strategy

Product promotion may be pursued by either causing the consumers to want to pull the products from the supply chain or pushing the products to the consumers through the supply chains. Many companies practice both strategies.

In a *pull* strategy, the consumers go to retail stores to query about the products because they have been informed of the product values by advertisements and other

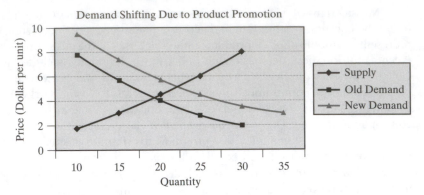

Figure 9.19 Up-shift of demand curve.

promotional efforts of the sellers. In this case, the product or service is presold to the consumers, who practically pull the products through its distribution channels. (See Figure 9.20.)

In exercising a *push* strategy, sellers introduce incentive programs (e.g., factory rebate, sales bonus, telemarketing, rebate selling, door-to-door sales, or discount coupons) to push products onto the consumers. Figure 9.21 illustrates the push strategy. Table 9.8 compares these two promotional strategies.

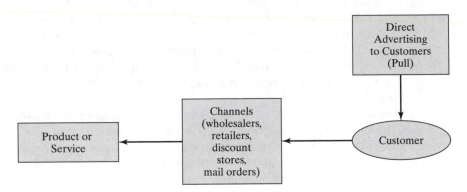

Figure 9.20 The pull strategy.

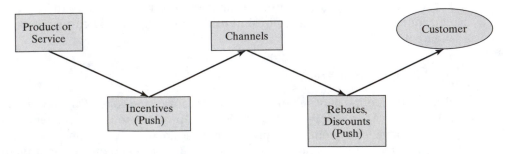

Figure 9.21 The push strategy.

TABLE 9.8   Comparison of Pull and Push Promotional Strategies

	Push	Pull
Communication	Personal selling	Mass advertising
Price	High	Low
Product's need of special service	High	Low
Distribution	Selective	Broad

TABLE 9.9   Promotion of Products

	High-Tech Products	Consumer Products
Marketing costs	Low	High
Consumer segments	More	Less
Focus	More	Less
Advertising	Less important	More important
Marketing channels	Trade shows Users groups Trade journals Internet	TV Print media Internet Radio
Brand	Important	Critically important

Source: Scott Ward, Larry Light, and Jonathan Goldstine, "What High-Tech Managers Need to Know about Brands." *Harvard Business Review*, July–August 1999.

9.7.3 Promotion of High-Tech and Consumer Products

High-tech and consumer products are promoted differently. To bring the most convincing marketing messages to the intended users, marketers for high-tech and consumer products use different channels. Table 9.9 summarizes the major differences in tactics applied in marketing these products.

9.7.4 Internet-Enabled Communications Options

Communications among sellers, intermediaries (e.g., distribution partners), and buyers have been significantly enhanced by the Internet (Zimmerman 2000). Figure 9.22 presents four specific modes of communication.

Manufacturers and suppliers usually set up the intranet to communicate with intermediaries (business to business, or B-to-B). Intermediaries may create their own websites and other tools to communicate with customers in a business-to-customers (B-to-C) mode. A direct communication between manufacturers suppliers and customers can be readily fostered by the company's websites, call centers, and other means for order processing, inquiry coordination, problem solving, and additional mission-oriented activities such as customer surveys, focus groups, and product testing.

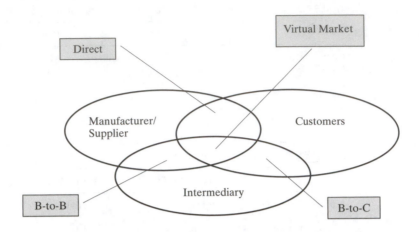

Figure 9.22 Modes of communications enhanced by the Internet. *Source*: Griffith, David A. and Jonathan W. Palmer, "Leverage the Web for Corporate Success," *Business Horizons*, January–February 1999.

The virtual market is a segment of the Internet domain wherein third-party portals (e.g., Google™, Yahoo®, and other search engines), auction sites (e.g., eBay®), and eMarketplace (e.g., ChemConnect®) actively provide channels to access information useful to all parties involved (manufacturers, suppliers, intermediaries, and customers).

In the B-to-B markets, businesses buy essentially two kinds of goods from other businesses: manufacturing inputs (raw materials, equipment, and components) and operational inputs (MRO goods, office supplies, spare parts, travel services, computer systems, cleaning, and other services). They buy these goods by systematic sourcing or spot sourcing. Electronic hubs provide the useful functions of aggregation and matching.

9.7.5 Contextual Marketing

Companies invest effort into creating websites that offer product information and facilitate sales transactions. However, studies show that these efforts have not yet returned the high profitability generally anticipated from such a marketing approach. The basic reason is that it remains unpredictable how frequently new and repeat customers visit these websites and then actually place orders.

A new way of thinking is offered by Kenny and Marshall (2000),* who suggest that the focus of e-commerce should be shifted from contents to context. They believe that contextual marketing (bringing the marketing message directly to the customer at the point of need) is the key. A number of contextual marketing examples are described next.

---

*From D. Kenny and J. E. Marshall, "Contextual Marketing: The Real Business of the Internet," *Harvard Business Review*, November–December 2000. Copyright ©; reprinted by permission of Harvard Business School Publishing Corporation, all rights reserved.

Johnson and Johnson's banner ads for Tylenol® (an over-the-counter pain-killing drug) show up on e-brokers' websites whenever the Dow Jones Industrial Average fails by more than 100 points on a given business day. They anticipate that investors will have headaches and thus will need Tylenol when they see their stocks lose money. The marketing message is brought out in the correct context as a way to reinforce its relevance and to offer transactional convenience at the right time.

CNET and ZDNET web sites attract diverse visitors interested in computers. Instead of placing banner ads in these CNET and ZDNET sites to redirect visitors to its own website, Dell offers product information directly within the CNET and ZDNET websites. Dell piggybacks on CNET's and ZDNET's relationship with its computer-savvy customers in order to promote Dell's own customer-acquisition economics. Doing so holds the customer's attention on computer design and offers competitive design choices and speedy order processing at the optimal moment. Here, the tactic used by Dell is to insert itself into a preexisting customer relationship at the right time and place.

Several search engines (e.g., Google, Yahoo, AOL®, MSN®, Lycos®, AltaVista™, and HotBot) practice contextual marketing. When a user conducts a keyword search, the output is typically placed in a left-aligned column under the heading "Matching Sites" and rank ordered according to hit frequency. Often several items under the heading "Sponsored Links" are placed on top of the "Matching Sites" column. These are paid advertisements related to the keywords entered by the user. They are there to offer contextual marketing messages relevant to the expressed interest of the user.

As Web-based technologies continue to advance, the Internet will become more accessible by many more users from almost anywhere and, at any time, causing them to become overwhelmed by information and choices. Bringing the right marketing information to the customer at the point of need is likely to become a critical success factor for various companies in marketing communications.

9.8 DISTRIBUTION (PLACEMENT) STRATEGY

Numerous organizations are involved in moving products and services from the points of production to the points of consumption. As indicated in Figure 9.23, some companies may engage intermediaries (e.g., wholesalers and retailers) to distribute their products, while others may elect to interact directly with their customers.

Distribution channels serve very useful functions, including

Transportation—Overcoming the spatial gap between the producer and the user
Inventory—Bridging the time gap between production and usage
Allocation—Assigning quantity and lot size
Assortment—Grouping compatible products for the convenience of users, since technical representatives sell several product lines
Financing—Facilitating timely possession of products
Communication—Providing product information and feedback to and from consumers

In recent years, distribution channels for some products have experienced significant changes due to upgraded logistics, transportation technologies, and advancements

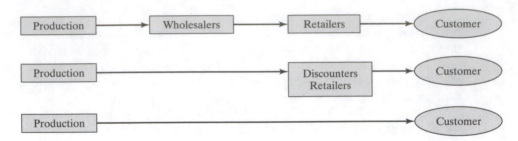

Figure 9.23   Distribution channels.

in communications technologies. For example, the Internet has enabled many producers to deliver digitized products—books, newspapers, magazines, music products, video products, and travel services—directly to consumers, thus bypassing the traditional intermediaries. Because of the use of sophisticated websites from which extensive product catalogs may be accessed, retail stores have also lost some of their traditional importance for selling physical goods—clothing, cars, appliances, etc.

Furthermore, logistics companies such as the United Parcel Service (UPS) are constantly improving their transportation capabilities and satellite-based communications system technologies in order to deliver physical goods anywhere in the world and to constantly track the status of consumer orders.

Warehouse design is expected to increasingly involve gantry robots and complex process optimization for constantly improving the efficiency of automated high volume operations.

9.8.1 Types of Distribution

Traditionally, distribution is classified as either intensive, exclusive, or selective.

In *intensive distribution*, products are stocked in diverse outlets, such as hardware stores, department stores, and catalog rooms, for wide distribution. This type of distribution is particularly suitable for consumer products of low technology and differentiation features. Examples include films, calculators, electric fans, books, and CDs.

With *exclusive distribution*, certain products are distributed only through exclusively designated outlets. This allows producers to retain more control over price policy, promotion, credit, and service, as well as to enhance the image of the products. Examples include dealerships for specific cars and qualified product centers for brand-name PCs.

The *selective distribution* is suitable for certain other products, the sales and service of which require special technical know-how and training. Examples include electronic instruments, high-tech equipment, and custom software.

9.8.2 Organizational Structures

In order to enhance distribution effectiveness, some companies elect to exercise more control over the supply chain by integrating forward. Others have elected to integrate backward.

A *forwardly integrated* organization strives to control the distribution channels leading to the customers. For the purpose of securing a larger market share and exercising more direct control, a producer may attempt to build its own retail outlets. Doing so allows the producer to gain a direct access to customers and thus to benefit from their feedback.

On the other hand, a *backwardly integrated* organization seeks to control the value chain leading backwards to production. For example, some retailers or wholesalers may attempt to own specific production facilities or to outsource production for creating private-label products in order to market products with their own brand names, reduce costs, ensure supply, and control quality.

9.8.3 Impact of E-Commerce on Distribution

The Internet has significantly modified the traditional classifications of distribution. Many consumer products, as well as certain high-technology products, are now marketed directly through the company's websites, including order processing and after-sales services. As a consequence, many intermediate companies currently involved in distributions—wholesalers, discounters, and some retail stores—are gradually being forced out by the Internet-enabled e-commerce companies and by the increased involvement of efficient and fast-responding logistics providers.

One immediate impact of e-commerce is the reduction of the final product price and product delivery schedule, both of which are beneficial to endusers.

Example 9.3.

Customers' wants and needs are regionally different for products intended for global markets. How can a centralized, concurrent engineering team develop a product that will serve as the common "platform" for global markets?

Answer 9.3.

One option is to segregate the mechanical aspects (functionality) of the products from their aesthetic aspects (look and feel). General Motors is accomplishing this challenging objective by

A. Building identical assembly plants for Buick® cars at four global locations
B. Outsourcing major subassemblies to local industries to lessen import duties and to satisfy local content laws
C. Standardizing the technical specifications so that parts supplied by one region can be readily rerouted for use by another, in order to balance loads due to market demand, labor disputes, governmental regulations, and other unpredictable events
D. Modifying design to account for local market conditions relative to cultural preference (e.g., car names in local languages, styling preferences, purchase habits, colors, etc.)
E. Retaining centralized concurrent engineering approach to facilitate global business strategy and scale of economy, while being flexible enough to adjust to local needs

---

Example 9.4.

The company plans to enter a new global market. It has three products currently selling well in its home country. The company's brand name is strong and internationally well recognized.

TABLE 9.10 Product Characteristics

	Product A	Product B	Product C
Segment size (dollars)	Small	Medium	High
Segment growth rate	Medium	Medium	Low
Profitability	High	High	Medium
Product value to customers	High	Medium	Medium
Brand strength	High	High	High
Delivery/distribution efficiency	Low	High	Medium
After-sales service	Medium	Medium	High

Current market research indicates that the segments for these three products in the targeted global market are of different size, growth rate, and profitability for the foreseeable future. Other product characteristics are included in Table 9.10.

Which one product should be selected to penetrate the targeted global market? Why? If the company has the required resources to market all three products in the targeted global market, in what priority order should the company proceed?

Answer 9.4.

To enter a global market, the company must examine two key questions: (1) How attractive is the target market segment to the company, and (2) how acceptable is the product offered to the customers in the target segment?

The attractiveness of a market segment to a company is generally defined by three factors: segment size, segment growth rate, and profitability. By using the information presented in the table, it becomes clear that the ranking based on "attractiveness" should be Product B first, with Products A and C sharing the second spot.

How acceptable the company's product is to the customers depends on the product value as perceived by the customers, the brand strength of the product, the delivery or distribution efficiency that affects the product's availability to the customers, and the ease with which customers obtain the needed after-sales service. Based on the "acceptability" criterion, the ranking of these products is Products B, C, and A.

Since both the "attractiveness" and "acceptability" criteria are equally important, we need to come up with a combination ranking, which says that the company should select Product B as its first choice to enter the global market, followed by Product C and then Product A.

9.9 OTHER FACTORS AFFECTING MARKETING SUCCESS

There are several other factors that may affect the marketing success of any company.

9.9.1 Alliances and Partnerships

Nowadays, companies realize increasingly that they do not always have or cannot cultivate internally, with the resource and time constraints under which they have to work, all competencies needed to compete in the world markets. Either because market access may be unattainable, the technology unaffordable, resources unavailable, time to market too long, or because other barriers may exist, an individual company may find it increasingly difficult to compete alone. Partnerships and alliances have become more

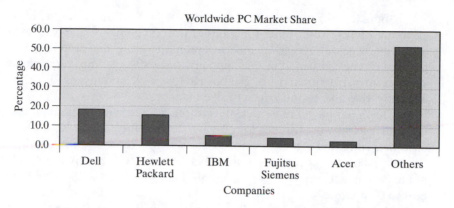

Figure 9.24   Worldwide PC market share (First quarter of 2004).
Source: IDC Press Release, Framingham, MA, April 15, 2004.

and more important for numerous companies in order to compete effectually and to constantly deliver value to customers.

For companies to succeed in the marketplace under these circumstances, the marketing concepts must penetrate to all members of the partnerships and alliances. (Kuglin with Hook 2002). The following examples illustrate the working of such partnerships and alliances:

A. Calyx and Corolla formed a network of partnerships to provide the seamless delivery of fresh flowers from grower to final customer in one-fourth of the time required by the traditional channels.

B. Dell teamed up with parts manufacturers and assembled computer systems in their own plants. Then Dell linked with logistics partners to deliver custom-designed personal computers (PC's) within three days to customers anywhere in the world. The success of the Dell model is clearly reflected in its 18.6 percent worldwide market share of PC's in the first quarter of 2004, ahead of Hewlett-Packard, IBM, Fujitsu Siemens, Acer, and others. (See Figure 9.24.)

Partners must appreciate that mutual gain results only when all members of the alliance embrace the marketing concept and come to recognize the importance of creating superior customer value by joining hands.

9.9.2 Customer Interactions and Loyalty

The interactions between customers and companies play a very important role in securing marketing success. Creating a pleasant experience for customers (in order processing, product information dissimilation, inquiry coordination, problem solving, after-sales service, and market surveys) is crucial for customer retention. Winning customer cooperation in offering much-needed feedback is vital to the company's product development. Management of customer relations is thus an important corporate responsibility.

Some lessons from the past are noteworthy. Customer interactions are not limited to marketing. Many other functions of the company are involved, including service, accounts receivable, legal, engineering, manufacturing, and shipping. Empowered employees can act on behalf of the company to satisfy customer requirements. Adequate support infrastructure must be established to enable employees to perform the tasks in an innovative and customer-responsive manner.

The major payoff of a successful customer interaction program is customer loyalty. Customer loyalty contributes to company profitability. Studies indicate that increasing customer retention rate by 5 percent raises profits by 25 to 95 percent (Reichhold and Shefter 2000). Loyal customers are valuable because they buy more, refer their pleasant experience to new customers, and offer consultations to these new customers at no cost to companies.

Dell created a *Customer Experience Council* to monitor the effectiveness of the programs that were geared to build customer loyalty. The council determined that customer loyalty is attained by (1) how the company fulfills the customer's orders, (2) how well the products perform, and the effectivenes and promptness of past sales service or support (or both). Other studies have shown that there are five primary determinants of loyalty: (1) quality customer support, (2) on-time delivery, (3) compelling product presentations, (4) convenient and reasonably priced shipping and handling, and (5) clear and trustworthy privacy policies.

To build customer loyalty, the customer interaction strategy must be focused on creating trust. Amazon.com is viewed by many as one of the most reliable and trustworthy websites on the Internet. It registers user preference, becomes smarter over time at offering products tailored to each user, provides one-click convenience for purchasing items, and delivers the ordered products free of errors. It is reported that 59 percent of Amazon.com sales are derived from repeat customers, roughly twice the rate of typical "bricks-and-mortar" bookstores.

Vanguard Group, a company that markets index-based mutual funds, offers timely and high-quality financial advice on its website and does not attempt to hard sell any specific product. Its customer interaction strategy is focused on building trust. "You cannot buy trust with advertising or salesmanship; you have to earn it by always acting in the best interest of customers," says Jack Brennan, Vanguard CEO.

eBay is known to have over 50 percent of its new customers referred by loyal customers who also serve as helpers to them. One major concern in the business of auctioning used merchandise is reliability and fraud prevention. eBay asks each buyer and seller to rate each other after every transaction. The ratings are posted on the website. Every member's reputation becomes public record. Furthermore, eBay insures the first $200 for each transaction and holds the money in escrow until the buyer is satisfied with the received product.

How is trust related to profitability? Studies show that, in some businesses, customers must typically stay on board for at least two to three years just for the companies to recoup their initial customer-acquisition investment. In other words, for companies to achieve profitability, customers must be loyal enough to stay beyond this break-even period. A large percentage of customers defects before many new companies reach this

TABLE 9.11   Customer-Loyalty Related Statistics

Products	Acquisition Cost per Customer (dollar)	Years to Break Even	Percentage of Customers Defecting before Breakeven Point (percent)
Consumer electronics and appliances	56	4+	60+
Groceries	84	1.7	40
Apparel	53	1.1	15

Source: Frederik F. Reichhold and Phil Shelfter, "E-Loyalty: Your Secret Weapon on the Web," *Harvard Business Review*, July–August 2000. Copyright © 2000; reprinted by permission of Harvard Business School Publishing Corporation, all rights reserved.

break-even point. Table 9.11 lists statistics related to customer acquisition cost, years to break even, and percentage of customers who defect before the break-even point.

A large number of companies are successful in planning and implementing strategies to interact effectively with customers. According to Peppers and Rogers (1999), these companies identify and prioritize customers, define their needs, and customize products to fit these needs. They reap the benefits of increased cross selling, reduced customer attrition, enhanced customer satisfaction, minimized transaction costs, and sped-up cycle times. Companies known for their success in relationship marketing include Pitney Bowes, Wells Fargo, 3M, Owens Corning, British Airways, Hewlett-Packard, and American Express.

Customer loyalty is won, not by technology, but through the delivery of a consistently superior customer experience. It requires a well-designed customer interaction strategy that is supported by companies with a firm corporate commitment.

9.9.3 Organizational Effectiveness

Marketing success is influenced by how effectively the company operates. In general, organizations with less rigid structure have a higher likelihood of becoming more customers focused, technologically innovative, and market responsive. Certainly, any conflicts between internal functions—manufacturing, design, engineering and marketing—must be minimized. Technology for mass customization requires an integration of R&D, procurement, customer relations management, and supply-chain management to achieve a high degree of customer satisfaction.

Above all, company management must apply discretionary resources (e.g., R&D, production capacity, human resources, organizational expertise, and information services) to the right combination of strategies (e.g., marketing, product, distribution, promotion, and price) so that maximum strategic marketing leverage can be achieved to capture opportunities offered in the marketplace. Figure 9.25 illustrates this core concept of organizational effectiveness.

Figure 9.25   Organizational effectiveness.

Example 9.5.

Engineering refined the design specifications of a product as originally recommended by marketing. Manufacturing made further changes to the product design in order to fabricate the product automatically. Unfortunately, the product did not sell in sufficient quantities to make it a success. Explain the possible reasons.

Answer 9.5.

The product features defined initially by the marketing department may not be exactly what the majority of customers wants. Product testing is a critical step to fine-tune the product design. The selected method of production does not readily accommodate an adjustment of product features, even if they are identified by feedback from the marketplace. The manufacturing of the product should be based on demand to assure market acceptance, not based on production technology, which is aimed at cost reduction.

Skipping the product testing step and applying the automatic production method too soon are two likely reasons for the noted failure.

9.10 CONCLUSION

This chapter covers many important issues related to the marketing of the company's products and services. Engineering managers should understand the overall objectives of the firm's marketing efforts and become sensitive to various marketing issues affecting engineering. They should become well versed with marketing terminology and elements of the marketing mix, namely, product, price, promotion, and distribution. It is important for engineering managers to accept the fact that marketing involves a lot of uncertainties associated with consumers' perceptions, competitive analyses, and market forecasting. They need to wholeheartedly adopt customer orientation in planning and implementing all engineering programs. They must strive to work closely with marketing personnel and remain supportive of the overall marketing efforts by providing high-quality engineering inputs to the firm's marketing program.

Obviously, the engineering inputs most useful to marketing are related to products and associated production activities. These include specifying and designing product features to be of value to customers; generating innovative ideas to offer premium product features that demand high margins; utilizing technologies to confer competitive advantages in time to market, quality, reliability, and serviceability; and delivering after-sales technical services needed to ensure customer satisfaction.

Engineers are also expected to control costs by improving and managing the production process, resources (labor and materials), and quality control and by estimating product cost accurately with the activity-based costing (ABC) method. Engineers may also get involved in training salespeople, making presentations before customers, conducting industrial exhibits, and evaluating customer feedback related to new product features.

Having learned the marketing concepts and been exposed to the complex marketing issues reviewed in this chapter, engineering managers will be able to appreciate the difficult but critically important functions of marketing and can become more effectual in interacting with marketing management.

Marketing and innovations are two principal activities of any product-based and profit-seeking organization. Engineers already know how to innovate. If they also learn how to market, this combination of capabilities will surely enable them to become major contributors in any engineering organization.

9.11 REFERENCES

Barnett, F. W. 1988. "Four Steps to Forecast Total Market Demand." *Harvard Business Review*, July–August.

Bedbury, S. with S. Fenichell. 2002. *A New Brand World: 8 Principles for Achieving Brand Leadership in the 21st Century*. New York: Viking.

Brethauer, D. M. 2002. *New Product Development and Delivery: Ensuring Successful Products through Integrated Process Management*. New York: AMACOM.

Capon, N. 1978. "Product Life Cycle." *Harvard Business School Case*, No. 579–072.

Corey, E. R. 1980. "Note of Pricing." *Harvard Business School Case*, No. 580–091.

Ferrell, O. C. 1999. *Marketing Strategy*. Fort Worth, TX: Dryden Press.

Francese, P. K. 1996. *Marketing Know-How: Your Guide to the Best Marketing Tools and Sources*. Ithaca, NY: American Demographic Books.

Gilmore, J. H. and B. J. Pine, II (Editors). 2000. *Markets of One: Creating Customer-Unique Value through Mass Customization*. Cambridge, MA: Harvard Business School Press.

Gourville, J. and D. Soman. 2002. "Pricing and the Psychology of Consumption." *Harvard Business Review*, September.

Griffith, David A. and Jonathan W. Palmer. 1999. "Leveraging the Web for Corporate Success," *Business Horizons*, January–February.

Hartley, R. 1998. *Marketing Mistakes and Successes*. New York: John Wiley.

Hayes R. H. and S. G. Wheelwright. 1979. "The Dynamics of Process-Product Life Cycles." *Harvard Business Review*, March–April, p. 127.

Kaplan, S. and Sawhney M. 2000. "E-Hubs: The New B-to-B Marketplaces." *Harvard Business Review*, May–June.

Keller, K. L. 2000. "The Brand Report Card." *Harvard Business Review*, January–February.

Keller, K. L., B. Sternthal, and A. Tybout. 2002. "Three Questions You Need to Ask about Your Brand," *Harvard Business Review*, September.

Kenny, D. and J. E. Marshall. 2000. "Contextual Marketing: The Real Business of the Internet." *Harvard Business Review*, November–December.

Kirk, B. C. 2003. *Lessons from a Chief Marketing Officer: What It Takes to Win in Consumer Marketing*. New York: McGraw-Hill.

Kotler, P. 1997. *Marketing Management: Analysis, Planning, Implementation and Control*. 9th ed. Upper Saddle River, NJ: Prentice Hall.

Kotler, P. 1999. *Kotler on Marketing: How to Create, Win, and Dominate Markets*. New York: Free Press.

Kotler, P., D. C. Jain, and S. Maesincee. 2002. *Marketing Moves: A New Approach to Profits, Growth and Renewal*. Cambridge, MA: Harvard Business School Press.

Kuglin F. A. with J. Hook. 2002. *Building, Leading, and Managing Strategic Alliances: How to Work Effectively and Profitably with Partner Companies*. New York: AMACOM.

Mello, S. 2002. *Customer-Centered Product Definition: The Key to Great Product Development*. New York: AMACOM.

Miller P. and R. Palmer. 2000. *Nuts, Bolts and Magnetrons: A Practical Guide for Industrial Marketers*. New York: John Wiley.

Monplaisir, L and Singh, N. (Editors). 2002. *Collaborative Engineering for Product Design and Development*. Stevenson Ranch, CA: American Scientific Publishers.

Paley, N. 2001. *Marketing for the Nonmarketing Executive: An Integrated Resource Management Guide for the 21st Century*. Boca Raton, FL: St. Lucie Press.

Peppers D. and M. Rogers. 1999. "Is Your Company Ready for One-to-One Marketing." *Harvard Business Review*, January–February.

Perry A. with D. Wisnom, III. 2003. *Before the Brand: Creating the Unique DNA of an Enduring Brand Identity*. New York: McGraw-Hill.

Pope, C. A., R. R. Burnett, M. J. Thun, E. E. Calle, D. Krewski, K. Ito, and G. D. Thurston. 2002. "Lung Cancer, Cardiopulmonary Mortality, and Long-term Exposure to Fine Particulate Air Pollution, "*Journal of American Medical Association,* Vol. 287(9), March 6.

Reichhold F. F. and P. Shefter. 2000. "E-Loyalty: Your Secret Weapon on the Web." *Harvard Business Review* July–August.

Smock, D. A. 2003. "Ten Commandments of Reverse Auctions." *Purchasing*, Vol. 132, No. 2, February.

Sutton, D. and T. Klein. 2003. *Enterprise Marketing Management: The New Science of Marketing*. New York: John Wiley.

Ulrich, K. T. and Eppinger, S. D. 2000. *Product Design and Development*. 2d ed. Boston: Irwin/McGraw-Hill.

Vishwanath V. and J. Mark. 1997. "Your Brand's Best Strategy." *Harvard Business Review*, May–June.

Ward, S., L. Light, and J. Goldstine. 1999. "What High-Tech Managers Need to Know about Brands," *Harvard Business Review*, July–August.

Yip, G. S. 1981. "Market Selection and Direction: Role of Product Portfolio Planning." *Harvard Business School Case*, No. 581–107.

Zimmerman, J. 2000. *Marketing on the Internet*. Gulf Breeze, FL: Maximum Press.

9.12 APPENDICES

APPENDIX 9.A. PRODUCT CONCEPT TESTING PROGRAM (AIR CLEANERS)

In 1976, Dr. C. M. Chang, currently of the University at Buffalo, developed a nonelectronic particle-filtration method while he was employed at a Fortune 500 industrial company. This technology was effective in eliminating airborne submicron particles in air streams without producing the ozone gas which is harmful to people's health. Product concepts were subsequently refined for air cleaning products for residential markets. The impetus for pursuing these product concepts is the notion that harmful respirable dust particles are expected to increase in concentration over time in the ambient air due to combustion-related emissions from cars, trucks, power plants, and factories. Respirable dust particles trapped in human lungs are known to lead to asthma, cancer, and other diseases and discomfort. Elimination of such particulate pollution in residential environments without the presence of ozone as a harmful gaseous agent should be attractive to consumers.

Before embarking on the major investment required for creating this proposed product line, the company decided to conduct a product testing program to assess the consumers' acceptance of the conceived product concepts and define additional product features that would satisfy consumers' needs to the maximum extent possible.

A. Focus Group Meetings.   A commercial product concept testing company was engaged to conduct a series of focus group meetings for the purpose of gauging customers' responses to the proposed product concepts.

In preparation for the focus group meetings, 18 versions of the products were proposed. These product concepts covered the particles' and gaseous pollutants' removal at high, medium and low performance levels. Tentative product names were assigned. Fabrication costs and features of competitive products were defined. Product brochures were prepared.

Three teams of customers were selected to form the focus groups. A product presentation was made to each focus group, including product price, pollutant removal performance, and availability. Questions asked by participating customers were answered.

Each focus group was asked to freely deliberate among its members their likes and dislikes and the advantages and disadvantages of the products. The groups were told that the products were currently in beta-testing stages in another part of the country and would soon be available in this region. The key question then was whether the customers would buy such products for their own use if they became available within a few months. If not at the proposed price, at what price would they be willing to buy the product?

All deliberations were video recorded for subsequent analyses.

B. Findings.   The key findings are as follows: Among high-efficiency filter products, consumers buy air cleaners primarily for medical reasons. Therefore, products that are equally efficient, but at a lower price are not necessarily attractive for two reasons: (a) A high price ensures the perception of superior product features. (b) There may be an insensitivity to price because of medical insurance coverage. An air-cleaner device at equal price, but with much higher particle-removal performance (for example, one with 95 percent filtration versus another with 99.95 percent filtration) does not appear to be attractive because consumers cannot notice the performance difference without employing measurement devices.

High-priced products with intermediate particle-removal efficiency are not readily accepted by consumers with medical problems. These products are priced too high for other consumers interested only in reducing the residential environmental dust.

Products with low filtration efficiency and low price may appeal to the general public interested in having an adequate furnace filter at a price "worth the risk" for them. A throwaway filter of this type appears to be worth further consideration, provided that it would

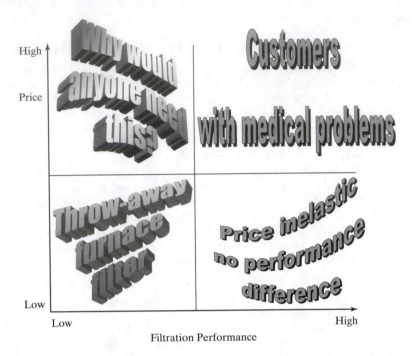

Figure 9.A1 Air cleaner categories.

cost only $5.00 or less and could fit into most existing furnaces. Figure 9.A1 summarizes the four categories of air cleaners defined by the two product features of price and particle-removal efficiency.

The capability to remove gaseous pollutants in residential air does not seem to appeal to consumers. Window add-on and self-standing units do not seem to be attractive to them either. In other words, consumers do not understand the beneficial value of the product features of ozone-free generation and efficient control of submicron particles. It would take a major advertising effort to communicate the message and educate the masses.

However, countless consumers do purchase air-cleaning equipment without really knowing the difference in brands, because most electronic air cleaners have been purchased as part of a bigger package. Retailers' reputations for price and quality (e.g. Wal-Mart Stores, Sears) have exerted a noticeable influence on consumers.

C. **Learning.** The key lesson learned from this product-testing program was that, if customers do not appreciate the value offered by a product, they will not buy it. In this case, the product developer did not fully appreciate the fact that, although it was scientifically evident at the time that inhaling submicron particles is detrimental to personal health, the majority of consumers does not appreciate the importance of such scientific facts and thus does not assign the proper importance to the product features offered. As indicated by Mello (Mello 2002), customer inputs are indeed useful.

Another lesson learned was that the timing of the products was also off. The air-cleaning products described were ahead of their time. In March 2002, studies published in *the Journal of American Medical Association* indicated that, for every 10-microgram/m^3 increase in particle concentration in ambient air, there is an 8 percent increase in risk for lung cancer death and a 6 percent

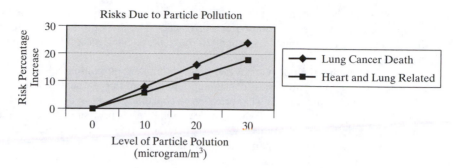

Figure 9.A2   Risk factors (*Source: Journal of American Medical Association*, March 6, 2002.)

increase in death due to other lung- and heart-related diseases. (See Figure 9.A2.) Several major U.S. cities are known to have particle-pollution levels in excess of the U.S. Environment Protection Agency (EPA) standard of 15 microgram/m³. In the absence of similar studies and authentic sources of information, consumers will not understand the long-term benefits of such products as air cleaners.

In March 2003, Sharperimage.com advertised an Ionic Breeze air purifier at $349.95. This product was built on the basis of the principle of electrostatic precipitation. Flat-plate collectors, which are easily cleaned, are used to capture most submicron particles found in residential buildings.

In summary, the product concept-testing program that Dr. Chang's company used supported the notion that product concepts developed from technology may not be as commercially marketable as those derived from the perceived needs and wants of customers. The results of the program add evidence to the known reality that marketing is as important as technology innovation for a company to achieve commercial success.

APPENDIX 9.B. CONSUMER SURVEY AND MARKET RESEARCH

To market consumer products, companies need to have a very detailed understanding of their customers, just as companies marketing industrial products also need to understand their customers, although to a much lesser extent.

When dealing with customers, the key questions typically concern what, how, where, when, why, and who.

For example, what functions does the product serve? What are the criteria to buy the product (price, color, size)? What is the value to the customer? What do they really want from the product (psychological, functional, and other benefits)?

How do customers compare products? How do customers decide to buy? How is the product used? How much are customers willing to pay for it? How much do they buy? How would the distribution mode and service center location affect the customers' buying decision?

Where is the purchase decision made (e.g., what is the customer's position in the company or household)? Where do they receive information? Where do they buy their products (retail store, mail order, department store, Internet, etc.)?

When do they buy it (weekly, monthly, special occasions)?

Why do they prefer one brand over the other (performance, price, convenience, packaging, colors, service, etc.)?

Who are the customers (age, background, sex, geographic location, members of social groups, etc.)? Who buys the competitor's products? Who does the buying (wife, husband, children, purchasing agent, engineers, others)? Who makes what decision for whom (decision making units)?

To understand the behavior of consumers in making purchase decisions, companies focus on customers' habit in purchasing, consuming, and information gathering. Who buys, how often, where, how much, when, and at what price? Who consumes, on what occasions, how do they consume, in what quantities, where, when, and with what other products? What media do they use (industrial exhibits, trade journals, TV, newspaper, radio, Internet, etc.), and when?

It is also useful for companies to understand the process by which consumers make their purchase decisions. This process typically encompasses the steps of need arousal (problems to solve; discovery from neutral sources; Jones the bragging next-door neighbor; etc.), information search (on-line resources are now one click away), and evaluation and decision making (comparative shopping, making trade-offs, brand versus product, and price versus quality and features).

APPENDIX 9.C. NEW PRODUCT DEVELOPMENT

New product development has been extensively studied in the literature.[1] Typically, companies follow a number of steps to develop new products: idea generation, idea screening, profitability analysis, product development, and commercialization.

To bring about new product ideas, companies use multiple sources. The internal sources include R&D and salespeople. The external sources include users initiated, competitors, industrial organizations, and product consulting firms. Innovation is the principal technique to identify valuable ideas by applying known principles to different processes or in unique configuration, thus producing unexpected benefits such as lower cost, smaller size, lighter weight, enhanced user convenience, better functionality, etc.

Two examples of innovative products are noteworthy. The new Segway Human Transporter is a clever combination of known technology (battery, gyroscope, wheels, stick-enabled weight-based control) and constructed in a unique way to create unexpected benefits. Another example is Bandag, a major tire retreading company that provides specialty composite materials in its retreads in response to different road conditions and travel patterns. It developed an innovation by embedding a computer chip in the tire. This chip monitors tire performance, calculates wear and tear, and determines balance. It becomes a value-added service to truck-fleet managers who are then able to track the total cost of tire purchases better than before. Its value in use is derived through information and a tracking system that provides important savings due to added life to tires and minimized fuel costs.

Now Bandag's Tire Management Solution business unit offers quality service in the acquisition, maintenance, and recycling of tires to customers, allowing customers to focus on their own core business interests.

Product ideas must be screened with respect to criteria such as marketability, market size and growth rate, production capabilities, and profitability. Marketability refers to the extent to which the proposed product ideas are acceptable to the customers in the target segments. Factors affecting marketability include price, product quality, substitution products from competition, available distribution channels, and customer preferences. Market size and market growth rate

---

[1]*Sources*: D. M. Brethauer. *New Product development and Delivery: Ensuring Successful Products through Integrated Process Management.* New York: AMACOM, 2002; L. Monplaisir and N. Singh (Editors). *Collaborative Engineering for Product Design and Develop*ment. Stevenson Ranch, CA: American Scientific Publishers, 2002; K. T. Ulrich and S. D. Eppinger. *Product Design and Development.* 2d ed. Boston: Irwin/McGraw-Hill, 2000.

must be at reasonable levels for the product ideas to be worth pursuing. Suitable production capabilities, such as raw materials, know-how, skilled labor, and equipment must be readily available. Finally, product ideas must have the potential for generating sufficient margins to justify the required investment expenditures.

In conducting profitability analysis for product ideas, several factors should be considered. The *demand factors* include users' needs, substitution products, price, advertising, segment size and growth rate, distribution, competition, and interaction with existing products. The *cost factors* are related to product cost, plant size, technology level, utilization of the plant, possible technological changes, and cost interdependency between existing and new products. The *profit factors* consist of product pricing, timing of profits (e.g., in three to five years), constraints on profits (due to plant capacity, financing limits, distribution requirements, management limits), governmental constraints (OSHA, trade, and environmental, laws and regulations), and labor restrictions (availability of trained labor and seniority). The *decision factors* assess the profit levels, risks, uncertainty (as in estimating the risks in the market), and capital budgeting (investment, cost of capital, timing, and cash-flow pattern).

Once product ideas are deemed worth pursuing, some companies invoke the target pricing approach (see "Pricing Methods" in Section 9.6.3) to determine the maximum cost of goods sold (target CGS) that would ensure commercialization and profitability of these product ideas.

A very detailed cost analysis follows for the requisite functions of *product design* (product attributes by a perceptual map, modular concepts, interchangeable components, subsystem compatibility, advanced materials, emerging technologies) and *fabrication and prototyping* (flexible production, fabrication technology, automated processes, quality control, supply chains, maintenance requirements). The full-scale product development work is to be authorized only if the target CGS will not be exceeded.

It is generally advisable for product development to be entrusted to a multifunctional team in which all major disciplines (such as marketing, sales, design, value engineering, production, distribution, procurement, customer service, and finance) are represented. Doing so will cut down the needs of subsequent design changes, improve the all-important time-to-market performance, and ensure a high probability of product acceptance in the marketplace.

One important area in which engineers should make significant contributions is cost reduction in the design and manufacturing of products. For example, developing innovative techniques to combine production steps, simplifying design, using lower cost materials, and creating interchangeable parts for groups of products can all lower costs. Another area in which engineers can add value is the improvement of product reliability, serviceability, and maintainability.

For certain products, companies may elect to conduct a test marketing program. The purpose is to seek feedback from users regarding such variables as service quality, likes, dislikes, desired new product attributes, relative ranking of existing product features, and price levels that users are willing to pay. Also useful to collect are data related to sales results from the targeted segment, adequacy of selected distribution channels, effectiveness of the promotional message, and service quality. Such information will allow the company to augment its product attributes before launching a major product.

Product commercialization involves final preparations before launching the product. These preparations include specifying a *product name* that is easy to pronounce and distinguish; a *value package* with price, features, value, and product functionality; *product styling* (e.g., industrial design is important to car buyers); *production* (financing, labor and materials resources, technologies, and supply chains); *sales and marketing* (advertising, distribution, sales force training, and product brochures), and *service* (customer call centers, websites, maintenance and repair, and problem solving).

According to published survey results, the introduction of new products in different industries has resulted in quite different product failure rates. (See Table 9.A1.)

TABLE 9.A1   Product Failure Rate

Product Types	Failure Rate (percent)
New industrial goods	20–40
New technology products	90
New packaged consumer goods	40–80
New food and drug items	50–80

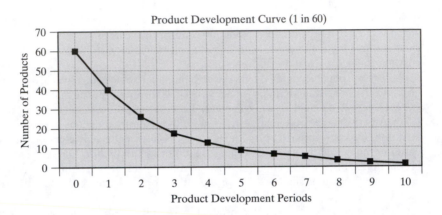

Figure 9.A3   Product development curve.

Figure 9.A3 presents the "1 in 60" product-development curve, in which 60 product ideas pursued initially whittle down to only one product that will become commercially successful, if the company follows the traditional, sequential product-development process. However, if a concurrent product development process is employed, the success rate will increase to 1 in 7.

There are a large number reasons products fail in the marketplace. The following list summarizes the principal reasons:

1. Lack of meaningful uniqueness in product features ("me-too" products)
2. Poor planning (no distribution, service, spare parts, inadequate advertising)
3. Wrong timing (economy, season)
4. Excess optimism
5. Poor product performance (functional deficiency)
6. Product champion failure
7. Company politics
8. Unexpectedly high product cost

9.13 QUESTIONS

9.1   What are the bases for tradeoffs between conflicting wants and needs of different customers with respect to the same product? How important is it to emphasize product quality when a new and unique product is launched?

9.2 Is it better to market a new product quickly and then upgrade the design later or to incorporate all design modifications or improvements before launching the product?

9.3 How can product-development costs be minimized by entering the market late?

9.4 Customers' wants and needs are regionally different for products intended for global markets. How can a centralized, concurrent engineering team develop a product that will serve as the common "platform" for global markets?

9.5 ABC company wishes to enter a new market arena on the basis of its strength in core technologies and financial staying power. However, the market arena in question is currently dominated by a major competitor with 80 percent of the market share, and a number of smaller competitors are each focused on small niche segments. How should ABC Company enter this market?

9.6 A company makes a range of products and sells to several large, loyal customers to achieve a healthy market share. A new competitor has emerged to offer equivalent products at much lower prices. What should the company do?

9.7 The company wants to develop a new product for a high-end consumer market. It is known that customers in this market are difficult to identify and are geographically dispersed. How should the company plan for product distribution and promotion?

9.8 The company wishes to sell its current product in a new market segment. At the same time, it wants to launch a new product in the existing market segment. How should the company handle the product promotion?

Part III

Engineering Leadership in the New Millennium

Part III of this book addresses five major topics: (1) engineers as managers and leaders (Chapter 10), (2) ethics in engineering and business management (Chapter 11), (3) Web-based enablers for engineering and management (Chapter 12), (4) globalization (Chapter 13), and (5) engineering management in the new millennium (Chapter 14). These discussions provide additional building blocks to prepare engineers and engineering managers to assume technology leadership positions and to meet the challenges of the new millennium.

Engineers are known to possess a strong set of skills that enable them to do extraordinarily well in certain types of managerial work. However, some may also exhibit weaknesses that prevent them from becoming effective leaders in engineering organizations, or even from being able to survive as engineers in the industry. The expected norms of effective leaders are described. Steps are discussed that enable engineering managers to enhance their leadership qualities and attune themselves to the value-centered business acumen. Certain outlined steps should be of great value to those engineering managers who want to become better prepared to create new products and services based on technology, integrate technology into their organizations, and lead technology-based organizations.

Many tried and true rules are included that serve as suitable guidelines for engineering managers to becoming excellent leaders. Above all, engineering managers are expected to lead with a vision of how to apply company core competencies to create value, insights into how to capture opportunities offered by emerging technologies, and innovations in making products and services better, faster, and cheaper, so that they constantly improve customer satisfaction. The concepts of value addition, customer focus, time to market, mass customization, supply chains, and enterprise resources integration are also discussed.

Although engineers are known to be ranked high in trustworthiness and integrity (ahead of businessmen, bankers, certified public accountants, and lawyers), it is important for all engineers and engineering managers to remain vigilant in observing a code of ethics along with taking seriously other topics related to ethics. (See Table 11.2.)

The changes wrought by the Internet are transforming most aspects of company business, including information dissemination, product distribution, and customer service. As processor design, software development, and transmission hardware technologies continue to advance, their roles in business will surely steadily grow and affect many functions of engineering management in the future. Progressive engineering managers need to know what Web-based enablers of engineering and management are currently available and which ones can be applied effectively to promote product customization, expedite new products to market, align supply chains, optimize inventory, foster team creativity and innovation, and improve customer service. Presented in considerable length is a comprehensive set of Web-based tools related to product design, manufacturing, project management, procurement, plant operations, knowledge management, and supply-chain management.

Globalization is further expanding the perspectives of engineers and engineering managers with respect to divergence in culture, business practices, and value. Globalization is a major business trend that will affect many enterprises in the next decades. Engineers and engineering managers must become sensitized to the issues involved and prepare themselves to contribute to enterprises wishing to capture new business opportunities offered in the global emerging markets. They need to be aware of the potential effects of job migration due to globalization and prepare to meet such challenges. The major hurdle for engineers and engineering managers to overcome is the creation of global technical alliances that will enable them to take advantage of new technological and business opportunities.

Engineering management will face external challenges in the new millennium. What these specific challenges are, how engineering managers need to prepare to meet these challenges, and how to optimally make use of location-specific opportunities to create competitive advantages will be examined. Progressive companies will also change organizational structures, set up supply chains, expedite e-transformation, and apply advanced tools to serve customers better, cheaper, and faster. Globalization is also expected to constantly evolve. The United Nations has predicted that, by the year 2020, three of the five biggest national economies will be located in Asia. There will certainly be winners and losers as businesses become more and more global. It is important for future engineering managers to explore prudent corporate strategies for engineering enterprises in the pursuit of globalization, while minimizing any detrimental impact on the environment and maintaining acceptable human rights and labor conditions.

How should engineering managers prepare themselves to add value in the new millennium? What are the success factors for engineering managers in the new century? What might be the social responsibilities of engineering managers in the decades ahead? These questions are addressed in the final chapter of the book. Globalization will create ample opportunities for those who know how to properly prepare and equip themselves with the required global mindset, knowledge, and savvy.

To foster the leadership roles of engineering managers, a six-dimensional model is proposed to emphasize the inside, outside, present, future, local, and global dimensions. The management challenges for engineers in these dimensions are discussed.

Chapter 10

Engineers as Managers and Leaders

10.1 INTRODUCTION

Engineering managers lead teams, groups, units, or enterprises to generate products and services that are highly technical in nature. The importance of developing engineering managers is well recognized. Thamhain (1990) reports that 85 percent of engineering managers surveyed believe that the development of new engineering management talent is crucial to the survival and growth of their companies' businesses.

Technical talents are important prerequisites for becoming leaders, as most engineering professionals do not readily accept superiors whose engineering credentials they do not respect. For this reason, countless organizations choose only top-flight engineers as engineering managers.

Typically, engineers are well trained for certain managerial functions. They are known to have the following skills and attributes:

1. Thinking logically, methodically, and objectively while making unemotional decisions based on facts
2. Analyzing problems and defining technologically feasible solutions
3. Understanding what motivates other engineers
4. Evaluating work with highly technological contents
5. Planning for the future, taking into account technology, productivity, and cost effectiveness
6. Discussing technical information with customers
7. Possessing technical expertise that enhances high quality leadership

Statistics indicate that, whereas engineers are highly trained professionals, as a group, they play only a limited leadership role in U.S. industry and economy. A survey conducted by *Business Week* in 1990 indicated that only 26 percent of CEOs in the top 1000 companies had their first degrees in engineering; about 46.7 percent were business majors.

Does the engineering mindset represent a disadvantage for top-level managerial jobs? Some people say there may be certain personal attributes, traits, and habits that tend to cause engineers to be poor managers. This chapter addresses some of these potential drawbacks.

The first part of this chapter covers (1) the differences in work done by engineers and managers, (2) the career path of a typical engineer, (3) the factors affecting the promotion of engineers to managers, and (4) the factors causing engineers to fail as managers. The second part of the chapter discusses (5) leaders and managers; (6) leadership styles, qualities, and attributes; (7) leadership skills for the 21st century; (8) unique contributions expected of engineering managers; and (9) a career strategy for the new millennium. The chapter concludes with a "take-charge formula" to inspire engineers to be proactive in managing their own professional lives.

10.2 DIFFERENCES IN WORK DONE BY ENGINEERS AND MANAGERS

As a technical contributor, the engineer focuses primarily on the operational aspects of the work—what it takes to get a technical assignment accomplished. Once the engineer moves to a managerial position, he or she will focus on the strategic aspects of the work, such as what work should be done, why it should be done, who should be assigned to do the work, what resources should be used to do the work, and in what sequence, etc.

As deliberated in earlier chapters, the functions of engineering management include (1) *planning*—forecasting, action planning, administering policy, and establishing procedures; (2) *organizing*—designing a team structure, delegating, and establishing working relationships; (3) *leading*—decision making, communicating, and selecting, motivating, and developing people; and (4) *controlling*—setting standards and measuring, evaluating, and controlling performance. Some of these functions are proactive and others are reactive in nature. Engineering managers should get involved in the following tasks:

1. Setting goals for the group, department, or enterprise
2. Establishing priorities
3. Defining policies and procedures
4. Planning and implementing projects and programs to add value
5. Assigning responsibilities and delegating the commensurate authorities to others while maintaining control
6. Attaining useful results by working through people
7. Processing new information and handling multidisciplinary issues
8. Making tough decisions under uncertain conditions
9. Finding the proper solution quickly among several feasible alternatives
10. Doing things right the first time, with a sense of urgency
11. Coaching, teaching, and mentoring others
12. Dealing with people—handling conflicts, motivation, and performance correction

TABLE 10.1   Work Done by Engineers and Managers

Characteristics	Engineers	Managers
Focus	Technical/scientific tasks	People (talents, innovation, relationships); resources (capital, knowledge, process know-how); projects (tasks, procedure, policy)
Decision-Making Basis	Adequate technical information with great certainty	Fuzzy information under uncertainty (people's behavior, customer needs, market forecasts)
Involvement	Perform individual assignments	Direct work of others (planning, leading, organizing, controlling)
Work Output	Quantitative, measurable	Qualitative, less measurable, except financial results when applicable
Effectiveness	Rely on technical expertise and personal dedication	Rely on interpersonal skills to get work done through people (motivation, delegation)
Dependency	Autonomous	Interdependent with others
Responsibility	Pursue one job at a time	Pursue multiple objectives concurrently
Creativity	Technology centered	People centered (conflict resolution, problem solving, political alliance, networks building)
Bottom Line	"How" (operational)	"What" and "why" (strategic)
Concern	Will it work technically?	Will it add value (market share, financial, core technology, customer satisfaction)?

Adapted and revised from P. Morrison, "Making Managers of Engineers." *Journal of Management in Engineering*, Vol. 2, No. 4, 1986.

Table 10.1, which is adopted and revised from Morrison (1986), illustrates the fundamental differences between the work done by engineers and that performed by managers.

It is important for engineers who aspire to become managers to fully understand these differences and the requirements associated with the management work. An individual should assess the compatibility of these implied requirements with his or her own personality, aptitude, value system, personal goals, preparations, and other factors. The individual should be convinced that taking on managerial responsibilities will indeed lead to long-term happiness.

According to a survey reported by Badawy (1995),* engineers move into management because of one or more of the following reasons:

1. Gaining financial rewards
2. Exercising authority, responsibility, and leadership

*From M. K. Badawy, *Developing Managerial Skills in Engineers and Scientists: Succeeding as a Technical Manager*, 2d ed, New York: John Wiley & Sons, 1995. Copyright © 1995 John Wiley; this material is used by permission of John Wiley & Sons, Inc.

3. Acquiring power, influence, status, and prestige

4. Receiving career advancement, achievement, and recognition

5. Combating fear of technological obsolescence

6. Responding to a random circumstance—an opportunity that is suddenly available

10.3 CAREER PATH OF A TYPICAL ENGINEER

In any engineering organization, people accept different positions; some are classified as line positions, whereas others are staff positions. Holders of line positions perform various activities in order to produce and market products or services. Holders of staff positions provide advice, services, and support to the holders of line positions. A career path represents a planned progression of jobs within an organization, each of which develops the necessary business or technical skills for the next-higher level position. The interrelationship between the jobs is the main difference between a business position that is a part of the career path (e.g., line positions) and the business position that does not qualify the individual for major advancements (staff positions).

10.3.1 The Engineer as a Technical Contributor

At the beginning of an engineer's career, his or her job is mostly technical. Tasks performed include the application of engineering principles and know-how to specific technical problems in order to verify or predict a specific technical outcome. The engineer is expected to be able to "do things right" on the basis of his or her college education, which may be supplemented by on-the-job training.

Over time, the engineer may build a reputation for being particularly knowledgeable in one or more technical domains (e.g., component design, process development, quality control, reliability analysis, and others) and becomes recognized as an expert for others to consult with, should such a need arise. Because of his or her expertise, the engineer may be asked to serve as a technical contributor on teams involved in product development, technology-based projects, feasibility studies, and other such multidisciplinary group endeavors. Still, the work done by the engineer remains mostly technical in nature.

On a technical career path, the engineer should accomplish technical work of high quality based on solid understanding of engineering fundamentals, efficacious application of basic principles, and sound technical judgment. As seen by coworkers and managers alike, the engineer needs to be easy to work with. The engineer must exhibit strong motivation to learn and improve, in addition to a high degree of maturity and professionalism. In well-managed enterprises, the promotion of such engineers to the next higher levels along the technical career ladder is usually assured.

Appendix 10.A contains a number of success factors affecting the career of an engineer in industry. Appendix 10.B itemizes some common reasons for career failures.

10.3.2 Midlevel Positions

After a couple of initial promotions along the technical career ladder, the engineer is now ready to assume larger responsibilities and will be asked to make an important choice between staying on the technical ladder and moving forward on the managerial career path.

The midlevel positions along the technical career path include such titles as senior engineer, consultant, engineering/development associate, and corporate fellow. In the managerial career path, these positions are nominally equivalent to group leader, section engineer, manager, and director, respectively. Some companies may use slightly different titles, but the basic organizational design concept is that there are equivalent positions along these two career ladders leading up to the director or fellow level.

On the technical career path, the engineer may assume larger responsibility for projects of highly technical contents for which specific funds and people are allocated. The managerial responsibilities of an engineering or development associate or corporate fellow are limited. The expected key contribution is to add value through technological leadership related to new technology products and processes by invention and unique application of emerging technologies. A corporate fellow is usually recognized as a technical leader both inside and outside the company. Note that for numerous well-established companies, the technical ladder has a glass ceiling that tops out at the corporate-fellow level. It is rare for companies to have another technical position at the corporate vice-presidential level.

More conventional is the managerial ladder, which requires the engineer to take on responsibilities that deal with nonengineering issues affecting the profitability of the organization. As the individual progresses up this career ladder, the engineer spends less and less time on engineering and more and more on managerial assignments. This career path continues beyond director to reach vice president of engineering/technology and chief technology officer. The expected key contribution is to add value by deciding on the correct technology-focused projects for the company to pursue and by securing the needed resources to complete them.

In a 2002 report, the National Science Foundation described the employment situation of U.S. engineers and scientists. Out of a total of 2,343,600 engineers and scientists, 46.1 percent held management and administrative positions. Figure 10.1 indicates

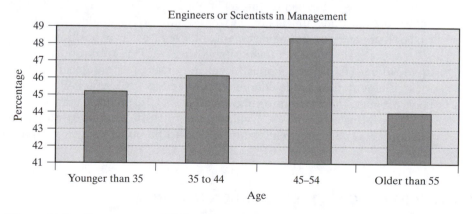

Figure 10.1   Percentage of U.S. engineers or scientists in management.
Source: National Science Foundation, 2000.

TABLE 10.2   Changing of Work Content

	First-Line Supervisor (percent)	Midmanager (percent)	Executive (percent)
Technical	70	25	5
Managerial	25	50	25
Visionary	5	25	70

that this percentage varies only slightly with age. About every one out of two engineers or scientists has taken on managerial or administrative responsibilities.

Whether a technical or managerial career ladder is suitable for an engineer is very much a personal choice, as there are advantages and disadvantages to both.

Appendix 10.C includes some tips for first-time supervisors and managers to cope with their new responsibilities. At any of the previously described position levels, engineers will need to effectually collaborate with others, especially with their own superiors. Appendix 10.D offers some hints on how to properly manage one's own superiors.

10.3.3 Promotion to the Next Higher Levels

Generally speaking, promotion along the technical career path is primarily based on company needs. A firm may have five corporate fellows covering five different technology domains, but have only one chief technology officer. If the individual's expertise serves as a core competency of the company, and the engineer's qualifications and the importance of responsibilities so justify, companies are known to have brought into being new technical positions to fit such individuals.

Promotion along the managerial ladder is subject to tough competition as the available opportunities become more and more limited and carry increasing executive-level responsibilities.

Once in executive positions, engineering managers are expected to exert leadership in creating and implementing technological strategies to realize business successes. Table 10.2 illustrates the change in work content of an engineering manager who advances from entry to midmanagement and executive level positions.

Specifically, the chief technology officer (CTO), who reports to the chief executive officer (CEO) and hopes to get important technological strategies accepted for implementation by the company, will need to work with and solicit support from peer officers: the chief information officer (CIO), chief financial officer (CFO), chief operating officer (COO), chief marketing officer (CMO), and chief knowledge officer (CKO). It takes leadership talents to be efficacious in such a collaborative setting.

10.4 FACTORS AFFECTING THE PROMOTION TO MANAGER

Generally speaking, there are three basic prerequisites for engineers to receive promotions to the managerial ranks. These prerequisites will be discussed next.

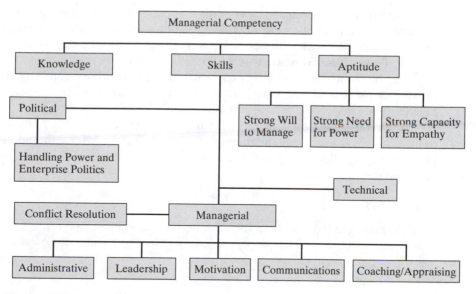

Figure 10.2   Managerial competence.

10.4.1 Competence in Current Technical Assignments

The engineer must be able to master the duties and responsibilities of his or her current position, have the respect of his or her coworkers, and receive favorable recommendations from his or her superiors.

10.4.2 Readiness and Desire to Be Manager

The candidate must have demonstrated the readiness to handle larger and more challenging responsibilities, as well as have gained the required skills and knowledge via courses, seminars, on-the-job training, professional activities, teamwork, volunteer work, tasks related to proposal development, feasibility studies, technology assessment, and other avenues. The candidate must possess the skills to manage time and have the desire to seek leadership positions and opportunities to exercise power and manage people, spearhead change, and resolve conflicts.

10.4.3 Good Match with Organizational Needs

Being competent and ready for promotion to managerial positions is a necessary, but not sufficient, condition for being promoted. The candidate's ambition, desires, and capabilities must also be a good match with the current and long-term needs of the company. Well-organized enterprises constantly need new managers and leaders because of the dynamics in the business environment, advancement of technology, competition in the marketplace, and the mobility of people.

Managerial competency may be classified according to the categories illustrated in Figure 10.2. Specifically, it is helpful for engineers who aspire to become managers to focus on the following general capabilities:

A. **Engineering Management Skills**—These skills include the engineering management functions of planning, organizing, leading, and controlling (see previous chapters), the ability to work with people, and excellent communication skills.

B. **Power Base Formation**—Candidates must nurture the ability to build personal power by technical know-how, experience, and networking. Promoting and marketing one's own achievements are important, following the well-known saying, "Early to bed, early to rise, work like hell, and advertise!"

C. **Assertiveness**—Candidates should demonstrate the ability to become and remain assertive in exercising judgment and making decisions. They should be proficient in resolving conflicts and problems of a technical, political, conceptual, and people-centered nature.

Example 10.1.

According to H. C. Hoffman (1989), engineering managers must possess a set of skills not taught in a typical college engineering curriculum. These skills include the ability to

1. Deal with ambiguity and uncertainty
2. Lead (people related and technological)
3. Take risks
4. Delegate
5. Be a team builder
6. Communicate
7. Initiate

How do you propose that an engineer acquire these skills?

Answer 10.1.

To acquire the previously described skills, engineers could follow these steps:

1. Understand why each skill set is important, and verify its importance by talking with trusted partners (e.g., parents, close friends, relatives, professional acquaintance, and mentors). Other useful steps to enhance understanding and building leadership skills are to

    A. Browse technical, business, and managerial publications (e.g., technical journals, *BusinessWeek*, *Fortune*, *Wall Street Journal*, and *Harvard Business Review*).

    B. Keep informed of new advancements in the field, such as business strategies, market expansion, technologies, innovations, customer relations management, enterprise integration systems, supply-chain management, business models, lean manufacturing, and e-business.

    C. Absorb new concepts and practices, and become proficient in identifying *best practices*, *success factors*, and other benchmarks.

    D. Recognize new opportunities in technologies, business, and products potentially valuable to the organization.

2. Understand the metrics (standards) for measuring and monitoring progress made in acquiring these skill sets.
3. Conceive a plan for action, including specific action steps and milestones.
4. Make a commitment by setting aside time and effort to implement the plan.

5. Take courses and training seminars, observe the experienced managers and leaders in action, and ask questions of qualified people to acquire the specific techniques needed to facilitate technical and managerial growth and build and maintain skills.

   Training programs are available at professional societies, companies offering training services, American Management Association courses, and university-based training programs.

6. Seek proactively opportunities to practice the learned techniques. Volunteer for team assignments, become an officer in a student organization. Do volunteer work in church, scouting organizations, benefits, the United Way, the Rotary Club, or political groups. Spend time in professional societies or industrial committees. Join Toastermasters International to practice public speaking skills.

10.5 FACTORS CAUSING ENGINEERS TO FAIL AS MANAGERS

Certain articles address the factors that cause engineers to fail as managers. Engineers may be handicapped by a number of perceived shortcomings that do not allow them to become good managers. These perceived shortcomings are reviewed in the following subsections.

10.5.1 Lack of Political Savvy

Engineers tend to be straightforward, honest, and open, and have strong views based on verifiable facts and data. According to Broder (1992), some engineers

1. Hate company politics. They tend to be technically intelligent, but sometimes politically amateurish.
2. Do not build a personal network.
3. Are uneasy trying to fit into an organizational culture because of strong beliefs, unique value systems, rigid principles, and inflexible attitudes.
4. Have an engineering mindset that is rational, efficient, and introspective. They may see things as either right or wrong, and may not be willing or able to accept different shades of gray. For holders of midlevel and top-level managerial positions, this mindset may confer disadvantages when attempting to resolve conflicts, handle disagreements, and foster alliances.

10.5.2 Uncomfortable with Ambiguous Situations

The technical training of engineers is based on equations, logic, experimental data, and mathematical analyses; this tends to make engineers see the world as orderly, certain, and black and white. The business environment in the real world causes some engineers to

1. Be uncomfortable with approximate or incomplete answers, since they have been trained well to recognize indeterminate problems and declare them unsolvable. They are not used to the idea of introducing additional assumptions and making such problems solvable. Some engineers

   (a) Hate problems with inaccurate or unknown factors.
   (b) Dislike planning with uncertainty (e.g., strategic planning).

2. Want to avoid using intuitive knowledge. They prefer cognitive knowledge, which is based on facts and data, and thus they lack the ability and willingness to make tough decisions by using intuition and gut feelings.

10.5.3 Tense Personality

Some engineers are too serious in their approach to professional life. They may be unable to say no and unable to ask for help, allowing their personal ego and pride to get in the way. Afraid to be wrong, they may have the tendency to take mistakes personally.

10.5.4 Lack of Willingness to Take Risks

Some engineers are conservative in nature. They may have a low tolerance for risks. Because of their tense personalities, they may not be comfortable taking risky opportunities to reach for higher levels of rewards. Some graduates with a masters degree in business administration (MBA) are said to be "often wrong, never in doubt"; they continue, in spite of often being wrong, to try new risky approaches until they reach their goals. Engineers are quite different in this respect. While having strong self-confidence in their own technical capabilities, many engineers do not like to take risks.

10.5.5 Tendency to Clinch on Technology

Some engineers do not feel comfortable leaving their fields of technological strengths when assuming managerial responsibilities. They tend to lean on technology as a safety net. From their viewpoints, technology is more readily controllable, and they are fearful of losing their sense of control. Some of them even have the uninformed notion that technology is the only thing worthy of respect, valuable, intellectually pure, and deserving of their efforts. Their perspectives are limited, causing them not to recognize that other functions that may be technically less intensive, such as customer service, marketing, procurement, production, and supply-chain management can also contribute equal, or in some cases, more value to the organization than engineering and technology.

10.5.6 Lack of Human Relations Skills

Because of their conservative nature, some engineers may be reactive in social settings and remain inflexible in dealing with a diversity of issues and people. Some of them may be readily argumentative and self-righteous when confronted with viewpoints radically different from their own. Over time, some of them may be perceived as suffering from a lack of human relations skills and the inability to become good team players.

10.5.7 Deficiency in Management Skills and Perception

One of the noted shortcomings of highly talented engineers is their lack of willingness to delegate. Some of them are not able to work through people and help others to succeed. They would prefer to ensure high quality by doing the projects or tasks themselves. Many are unwilling to develop subordinates for fear that one day the trainees may become technically more talented than themselves.

Other engineers may have the tendency to apply self-imposed, ultrahigh standards in appraising employees. They have difficulty tolerating below-par performance by teammates or coworkers.

Still other engineers have the tendency to overmanage and overcontrol their subordinates.

10.5.8 Not Cognitive of a Manager's Roles and Responsibilities

Some engineers are unexpectedly promoted to managerial ranks because they performed well as technologists. Due to a lack of preparation on their part, these newly promoted engineers have either limited or no understanding of what managers are supposed to do—that is, to add value by efficiently applying resources to the right projects. They lack the preparation to do a manager's job. They are not aware of the fact that people problems require more time and attention than technical problems. Because of a lack of exposure to nontechnical, but equally important issues, they have not acquired the background required to develop a well-rounded business sense.

10.5.9 Narrow Interests and Preparation

Some engineers are specialists in narrow technical fields. As a consequence, they have narrow technical viewpoints and limited vision and perspectives beyond technologies. They are not prepared to deal with accounting, marketing, production, finance, and other broad-based corporate issues outside of technology.

Numerous engineers may suffer, to varying extents, from some of the shortcomings just cited. Engineers who aspire to become managers should carefully examine their strengths and weaknesses and commit their efforts to making sure that all of these factors for failure are minimized over time.

10.6 LEADERS AND MANAGERS

Companies need both leaders and managers. According to Kotter (1990), management focuses on the steps of (1) choosing goals and targets typically set by existing corporate strategies; (2) planning action steps to achieve the goals; (3) allocating required resources; (4) creating organizational structures; and (5) exercising control, measuring results, and correcting performance to ensure the attainment of goals. The primary duty of management is to keep the organization functioning properly. Management planning is deductive in nature, since it aims at creating orderly results.

Leadership, on the other hand, focuses on the steps of (1) setting a vision and direction for the future in response to changes imposed on the organization externally, (2) creating strategies to produce the changes needed to achieve the vision, (3) designing action steps to accomplish the strategic goals, (4) aligning people and forming coalitions, and (5) motivating and inspiring people to move in the right direction. The function of leadership is to produce change. Setting direction by leadership is inductive in nature, as it attempts to produce a new vision and new strategies. Vision is the picture of what the company will be in the long run with respect to its business portfolio, market position, technological prowess, and company culture.

The emphasis placed and approach taken by managers are different from those of leaders. Table 10.3 summarizes such differences, based on published work by Zaleznik (1992).

Kotter (1990) further believes that strong leadership with weak management is no better—and is sometimes actually worse—than the reverse. A competitive organization needs both strong leadership and strong management and should use each to balance the other. Management is needed to ensure that complex organizations

TABLE 10.3   Differences between Managers and Leaders*

Characteristics	Managers	Leaders
Focus	Do things the right way Administration, problem solving Reconcile differences Seek compromises Maintain balance of power	Do the right thing Direction setting Creativity and innovation
Emphasis	Rationality and control Accept and maintain status quo Put out fires	Innovative approach Challenge status quo Blaze new trails
Targets	Goals, resources, structures, people	Ideas
Orientation	Tasks, affairs Persistence Short-term view	Risk taking Imagination Long-term perspective
Success Factors	Tough-mindedness Hard work Tolerance Goodwill Analytical capability	Perceptual capability
Points of Inquiry	How and when	What and why
Preference	Order, harmony	Chaos, lack of structure
Aspiration	Classic good soldier	Own person
Favor	Routine Follow established procedure	Unstructured
Approach with People	Using established rules	Intuitive and empathetic
Personality	Team player	Individualist
Relevance	Necessary	Essential
Thrust	Blend in Bring about compromise Achieve win–win	Stand out Lead changes
Mentality	"If it isn't broken, don't fix it"	"When it isn't broken, this may be the only time you can fix it."

Adapted from Abraham Zaleznik, "Managers and Leaders: Are They Different?" *Harvard Business Review*, March–April 1992. Copyright © 1992; reprinted by permission of Harvard Business School Publishing Corporation, all rights reserved.

operate properly, including attaining incremental improvements. Leadership is needed to cope with changes that will be thrust upon the organization due to advancements in technology, changes in market environments, and competition at the international level.

The transition from managers to leaders has been the subject of a few studies (Conger 1992; Zenger and Folkman 2002). Engineers interested in becoming managers and leaders should also review the 78 questions raised by Clark-Epstein (2002), study the 12 principles delineated by Billick and Peterson (2001), and master the 6 competencies described by Pelus and Horth (2002). They may benefit from the lessons learned by other leaders (Krames 2003). Those who regard General Electric's Jack Welch as a professional to emulate have several resources to consult (Welch 2001; Krames 2002; Slater 2001).

Example 10.2.

Every engineering manager has strengths and weaknesses. The key to continuous improvement is, of course, to identify one's own weaknesses and do something about them. How should an engineering manager study oneself and systematically discover opportunities for improvement?

Answer 10.2.

Henry Mintzberg (1990), in his article "The Manager's Job: Folklore and Fact," proposed a large number of insightful questions for managerial self-study. These questions are worth studying.

10.7 LEADERSHIP STYLES, QUALITIES, AND ATTRIBUTES

10.7.1 Leadership Styles

The consulting firm Hay Group conducted a survey on leadership styles with a sample of 3871 executives, from a worldwide database of 20,000 executives. According to Goleman (2000),* the following styles were identified by the survey as particularly useful for generating results: authoritative ("come with me"), affiliate ("people come first"), democratic ("what do you think"), and coaching ("try this"). Goleman also advises that leaders should minimize the use of two additional styles, namely, coercive ("do what I tell you") and pacesetting ("do as I do now").

Effective leaders are said to switch constantly between the authoritative, affiliative, democratic, and coaching styles, depending on the situation at hand. The key concept here is flexibility. Those engineers who are weak in any of these four styles are advised to seek improvement.

---

*From J. J. Babarro and J. P. Kotter, "Managing Your Boss," May–June, 1993, Copyright © 1993; reprinted by permission of Harvard Business School Publishing Corporation, all rights reserved.

Example 10.3.

The NASA space shuttle *Columbia* broke up during its descent on February 1, 2003, killing all crew members on board. Immediately after its launch, it was made known to NASA Mission Control Center that a piece of solid foam broke off from the shuttle's external fuel tank during its ascent and smashed into its left wing at high speed. During Columbia's 16-day space journey, NASA management dismissed the significance of this mishap and did nothing. Instead, NASA insisted that the incident was a routine affair, as foam pieces had been broken off in a number of previous space flights. Furthermore, NASA overturned the requests of its lower level engineers to obtain U.S. Department of Defense pictures of the wing that were photographed by military satellites. Low-level engineers e-mailed a query to their superiors about the possibility of getting the astronauts to take a space walk, which was safe and easy to do, to inspect the wing. That e-mail was never answered.

NASA management lost two precious opportunities to avert the disaster. Furthermore Columbia could have had its mission extended by another three to four weeks, allowing the Shuttle Atlantis, whose launch schedule could have been accelerated to February 10 (from its original launch date of March 1) to rendezvous and rescue the crew. The other option would have been for two Columbia spacewalkers to repair the damaged wing with heavy tools and metal scraps scavenged from the crew compartment.

The Columbia Accident Investigation Board (CAIB) conducted interviews, shot foam pieces experimentally at test wing targets, and analyzed a large volume of data. Its final report pointed clearly to a NASA management failure, attributable to several specifically named managers.

What were the principal modes of management failure deemed responsible for the shuttle Columbia disaster?

Answer 10.3.

The Columbia shuttle disaster was the result of several factors (Langewiesche 2003):

1. **Culture.** NASA claims that it is a "badgeless society," meaning that it does not matter what title of position and responsibility is on one's name badge; everyone is equal when it comes to concern about shuttle safety. NASA has an open-door and free communications policy.

   The opposite was true, according to Langewiesche. He found that many NASA employees were afraid to speak up. In fact, Langewiesche claims in his article that many employees feared that their position in the organization predetermined how they themselves were viewed, as well as how their opinion was welcomed or accepted by the higher-ups. Many employees felt NASA's leadership style to be coercive, not democratic.

2. **Overconfidence and Personal Arrogance.** The NASA managers had shown arrogance and insularity, while exhibiting a tough and domineering management style. Langewiesche says that the CAIB report singled out a specific NASA decision maker as "intellectually arrogant and an abysmally failed manager." Moreover, Langewiesche claims that the NASA managers demonstrated imperious and self-convinced attitudes, suffered from a lack of curiosity, and believed in themselves blindly. It is not difficult to see and understand that, as NASA had completed over 100 successful space flights, managers could blithely accept the notion that there would be no fatal risk upon the foam being struck at lift-off. Previously, when lift-off had created similar mishaps, there were no serious repercussions.

   Ultimately, it was conclusively substantiated by CAIB tests that the falling foam punched a hole about 10 inches wide into the wing's leading edge. This hole allowed the hot gases of reentry to enter the shuttle compartment and burn it from the inside. Unfortunately, there was no other past experience upon which the higher-ups could base their decisions, although they perhaps should have erred on the cautious side, rather than the less costly side.

3. **Pressure to Meet Deadline.** The NASA administrator set stringent performance goals related to the International Space Station Project. The strict deadline for completing the "core," of which Columbia was a part, was February 19, 2004. As a consequence, organizational and bureaucratic concerns weighed heavily on the managers' minds.

A combination of the foregoing three factors led the key managers to stubbornly believe that the foam strike was insignificant and also led them to forego the opportunity to collect more data and hence fail to initiate emergency steps to save the crew. It was a colossal engineering management failure.

10.7.2 Emotional Intelligence

According to Goleman (1998),* all effective leaders have a high degree of emotional intelligence. There are five components of emotional intelligence: self-awareness, self-regulation, motivation, empathy, and social skills. Each of these components can be learned and enhanced by coaching, observation, training, and practice.

Engineering managers are advised to nurture these components of emotional intelligence so that they can lead effectually.

10.7.3 Inspirational Leadership Qualities

Leaders typically have vision, energy, authority, and strategic direction. However, leaders will not succeed if there are no followers.

Followers are hard to find in these "empowered" times. Goffee and Jones (2000) point out that there are four specific qualities needed by leaders who want to be truly inspirational:

1. **Approachability and Humanity**—It is a human quality to show personal humility and vulnerability. Doing so will demonstrate authenticity, promote trust building, enhance solidarity, and foster collaboration. Acknowledging one's own shortcomings opens up opportunities for improvement. (See the lesson contained in "The Wisdom of the Mountain," Appendix 10.E.) However, the authors advise that the weaknesses shown should be tangential flaws (such as hardworking habits) only and not fatal ones or character-related issues.

2. **Tact**—It is the ability to know when and how to act based on intuition. It is a situation sensor that is capable of reading underlying currents, detecting subtle cues, and gauging unexpressed feelings. (See the lesson contained in "The Sound of the Forest," Appendix 10.F.) This quality is widespread among excellent business leaders. However, one needs to make sure that reality testing is done frequently with trustworthy friends or confidants, to avoid disasters due to one's own inability to evaluate faulty situations.

*Reprinted by permission of *Harvard Business Review*. From "What Makes a Leader" by D. Goleman, issue November–December/1998. Copyright © 1998 by Harvard Business School Publishing Corporation, all rights reserved.

3. **Tough Empathy**—It is the ability to establish a balance between respect for individuals and the task at hand, in order to impel leaders to take risks and to care about the people and the work they do. Leaders do well when they close the distance between themselves and their employees. Leaders must give employees what they need by helping, coaching, and participating in what they do. (See the lesson contained in "The Wheel and the Light," Appendix 10.G.)

4. **Uniqueness**—It is the ability to maximize the benefits derivable from the leader's own uniqueness (e.g., dress style, physical appearance, imagination, loyalty, expertise, handshake, humor, etc.) and to use this separateness to motivate others to perform better.

Leaders need to be themselves, acquire more skills, and apply these four qualities to fit their own personality styles to successfully inspire others.

10.7.4 Leadership Attributes

Leaders are those who have special knowledge, are accessible, exhibit charisma (the natural ability to attract followers), and possess the authority to delegate. Table 10.4 contains a set of basic guidelines for engineering managers to apply to leading.

Various research articles on leadership indicate that efficacious managers possess a set of common attributes. They have unquestionable character. Their creditability is high because of their technical skills, ethics, fairness, and moral standards. They master the functions of management, such as planning, organizing, leading, and controlling. They constantly perfect their skills in dealing with people, managing time, and controlling their own stress. They communicate effectually both in oral and written forms (Mai and Akerson 2003). They have learned to listen well. (See the lesson contained in "The Sound of the Forest," Appendix 10.F.) They are full of energy and in good health. They are enthusiastic and positive about things and people. They are self-motivated, flexible, and independent. They take the initiative to originate actions to influence events. They take prudent risks (Nelson 1999). They have superior conceptual skills in reviewing data, solving problems, taking action, and planning strategically. They are both persistent and persuasive in achieving the goals they set out to achieve. Because of these attributes, they create good impressions and build confidence in others' minds. Effective leaders attract those who are willing to follow them.

In fact, these attributes are not dissimilar to the profile of top executives commonly noted in business literature. (See Table 10.5.)

TABLE 10.4 Guidelines for Leading as Engineering Managers

1. Prepare oneself (e.g., study the rules, policies, and objectives of your organization)
2. Understand all jobs under one's direction
3. Be observant of things going on around oneself (e.g., managing by walking around)
4. Pay attention to details
5. Pose questions
6. Keep things in perspective (e.g., avoid being too close to the trees and unable to see the forest)
7. Be anticipatory of the future conditions

TABLE 10.5   Profile of Top Executives

---

1. Are able to work with people
2. Possess social poise (self-assured and confident)
3. Are considerate of others
4. Are tactful and diplomatic
5. Practice self-control
6. Are able to analyze facts, to understand and solve problems
7. Are able to make decisions
8. Are able to maintain high standards
9. Are tolerant and patient
10. Are honest and objective
11. Are able to organize time and priorities
12. Are able to delegate
13. Are able to generate enthusiasm
14. Are able to be persuasive
15. Possess a great concern for communication

Managers and leaders have different mental orientations. These mental orientations make leaders as important as managers in adding value to their organizations. Leadership talents are defined as having a natural predisposition for recurring patterns of thoughts, feelings, and behaviors that can be applied productively. The Gallup Organization (www.gallup.com) has identified 20 key leadership talents by interviewing more than 40,000 leaders and top-tier managers over a period of 30 years (Fulmer and Wagner 1999). These 20 leadership talents are classified in the following four groups:

A. **Ability to Provide Direction**

1. Vision: is able to build and project beneficial images
2. Concept: is able to give the best explanation for most events
3. Focus: is goal oriented

B. **Drive to Execute—Related to Motivation**

4. Ego driven: defines oneself as significant
5. Competitive: has the desire to win
6. Achiever: is energetic
7. Courageous: relishes challenges
8. Activator: is proactive

C. **Capacity to Develop Relationship with Others**

9. Relater: is able to build trust and be caring
10. Developer: desires to help people grow
11. Multirelater: has a wide circle of relationships
12. Individuality perceiver: recognizes people's individuality
13. Stimulator: is able to create good feelings in others
14. Team leader: is able to get people to help each other

D. **Management System—Related to Management Abilities**

15. Performance orientated: is results oriented
16. Disciplined: is able to structure time and work environment
17. Responsible and ethical: is able to take psychological ownership of one's own behavior
18. Arranger: is able to coordinate people and their activities
19. Operational: is able to administer systems that help people to be more effective
20. Strategic thinker: is able to do "what-if?" thinking and create paths to future goals

It is obvious that a good leader must be a good manager, but a good manager may not be a good leader if he or she lacks some of the leadership talents indicated.

Stogdill (1981) states that a successful leader is characterized by the attributes outlined in Table 10.6.

Cheves (1992) suggests 11 attributes that an effective leader must possess. These are delineated in Table 10.7.

TABLE 10.6   Leadership Attributes (Stogdill)

1. A strong drive for responsibility and task completion
2. Vigor and persistence in pursuit of goals
3. Adventuresome and original in problem solving
4. Drive to exercise initiative in social situations
5. Self-confidence and sense of personal identity
6. Willingness to accept consequences of decisions and actions
7. Readiness to absorb interpersonal stress
8. Willingness to tolerate frustration and delay
9. Ability to influence another person's behavior
10. Capacity to structure social interaction systems to suit the purpose at hand

---

TABLE 10.7   Leadership Attributes (Cheves)

1. Communicate well
2. Are good listeners
3. Are approachable
4. Delegate
5. Lead by example
6. Read situations and people well
7. Are good teachers
8. Care about people and display it
9. Are fair, honest, and consistent
10. Know how to criticize
11. Know how to accept criticism

The 20 leadership talents described by Gallup, the 10 characteristics of successful leaders by Stogdill (Table 10.6), and the 11 attributes of effective leaders by Cheves (Table 10.7) are consistent with one another. In fact, all of them asymptotically converge to a finite set of common leadership attributes that all future engineering leaders should feel comfortable assimilating. Engineers who aspire to become leaders are encouraged to understand and display these attributes so that, over time, the leadership talents and attributes will become second nature to them.

Example 10.4.

Highly talented technical professionals may have academic training (advanced degrees), experience (company tenure), professional credentials (societal committee activities, awards, business connections), and accomplishments (patents, publications, completion of major projects) superior to the engineering manager. They could be difficult to manage. What are some of their characteristics and working habits? What strategies are effective in managing them?

Answer 10.4.

Highly talented technical professionals tend to have the following work-related preferences:

A. They favor individual assignments with clearly recognizable responsibility. They do not prefer teamwork, wherein individual contributions may be crowded out.
B. They tend to strive for perfection, as they view the technological output as a reflection of themselves.
C. Technical professionals typically become easily frustrated by unexpected changes in program priority or resource allocation strategies for approved action steps deemed essential to achieve the program objectives.
D. They hate management jargon.
E. They find happiness in technical work, without being constrained by other nontechnical concerns or the involvement of low-skill people.
F. They are readily turned off by administrative details, restrictive policies and guidelines, poor quality decisions based on questionable data and assumptions, excessive reporting requirements, and overly tight management control.
G. These workers assign high value to independence, self-motivation, self-direction, and fairness.
H. They demonstrate a reserved attitude in social interactions.

Highly talented technical professionals may be managed by adopting the following strategies:

1. Decide on objectives of technical programs or tasks, define the funding priority, understand the reasons for decisions, and secure the company's commitment to the chosen programs.
2. Assign technology programs and tasks to specific individuals by clearly communicating the objectives, budget constraints, expected results, and other details. Suggest specific ways to measure outcome.
3. Solicit comments from the individual technical professionals involved regarding project value, interest, readiness to perform tasks, and other issues that may have been of concern to them.

4. Invite the individual technical professionals to:

   (a) Outline specific technical methods to accomplish the program or task at hand.

   (b) Produce an action plan and define budget requirements (accounting for man-hours, equipment, supplies, computation resources, outside resources, etc.).

   (c) Specify preliminary milestones of when interim results are to be reported out (monthly, biweekly etc.).

   (d) Define deliverables.

5. Review and accept the plans. Authorize the individuals to commence programs and tasks.

6. Be available for any unexpected problems encountered by the individuals pursuing these assignments (e.g., practicing the concept of management by exception). Must be helpful and leave sufficient room for independent work by the individuals.

7. Acknowledge receipt of the final report after it is submitted. Read the report, review results with individuals, and invite comments on any work extension needed or desired and on how to enhance the management aspects of the program.

8. Evaluate the work performed in terms of its expected value to the organization and offer feedback, including any responses from top management and other parties affected by the accomplished programs or tasks.

9. Praise the individuals appropriately whenever good work is done, by, for example, practicing the concept of motivation by positive reinforcement and recognition. Offer improvement suggestions if performance is to be improved.

10. Seek and arrange opportunities for the individuals to make prepared presentations on technical programs and tasks during review meetings with upper management groups.

11. Document tasks and evaluations (including specific contributions made and their significance to the company) to be in a position to provide an instant report to upper management, if required, and to form a basis for the annual appraisals of the individuals involved.

10.8 LEADERSHIP SKILLS FOR THE 21ST CENTURY

The 21st-century arena is characterized by a high pace and level of education, low patience and compliance with authority, close relationships with customers, and a brisk speed to market. Continued changes due to global competition, environment, technological advances, and population diversity are expected to be very rapid. Jack Welch, a former CEO of General Electric, said, "When the rate of change on the outside exceeds the rate of change on the inside, the end is in sight." (Krames 2002). Business success in the 21st century requires global connectivity, obsession with customer satisfaction, enhanced performance of people and technologies, alternative organizational frameworks, real-time responses, and enduring self-examination.

In the new century, leaders need a new set of skills to help exercise their leaderships centered on result. These skills include

1. Leading with a strategic focus and vision. Advancing and articulating a value proposition which represents a proposed model to create value to companies' stakeholders.

2. Managing multiple points of view simultaneously, such as those from customers, suppliers, shareholders, and employees. Remaining flexible and adaptable in

dealing with technology, working with people, and forming business networks. Being capable of negotiating for solutions that are acceptable to parties involved.

3. Keeping high-level goals in sight, while managing and tracking day-to-day success. Keeping the spirit of the enterprise alive.

4. Fostering productive changes. The "boiled frog syndrome" is explained as follows: If you put a frog in a pan full of cold water and slowly turn up the heat, the frog will boil to death rather than jump out. If you drop a frog in a pan full of boiling water, however, it will jump out immediately. The moral of the story is that, if people do not sense a significant need to change, they may not get out of their comfort zones and change what they are doing. Usually, effective leaders are needed to convince people of the need to change.

5. Being inspirational, technologically savvy, entrepreneurial, and devoted to service.

6. Investing in a business model that guides employees' decision making at all levels.

7. Devising and maintaining transformational knowledge systems.

8. Accessing relevant information rapidly in light of the explosion of available information.

9. Understanding how global business practices have evolved.

10. Learning quickly while not relying on what is already known or understood.

In his May 1999 speech at the University of Richmond, Virginia, Gen. H. Norman Schwarzkopf, U.S. Army (Ret.) (1999) pointed out that leadership is a whole combination of different ingredients; but, by far, the single most important ingredient of leadership is character. During the last 100 years, about 99 percent of leadership failures have been due to faulty character, not incompetence. His remarks are particularly pertinent today in view of the alleged unethical financial practices committed by leaders in companies such as Enron, Adelphia Communications, Global Crossing, WorldCom, and others in 2002.

Engineering managers need to have business savvy in order to lead in the 21st century. The combination of technological know-how and business savvy is powerful. The following guidelines may help engineering managers to acquire the needed business acumen:

1. Become well versed in the business issues faced by the company. This includes a thorough understanding of the corporate vision; the company's priorities, strengths, and weaknesses; the current market position; business processes; and engineering and technology factors driving shareholder value. In other words, constantly sharpen one's own business sense.

2. Know how to define proper metrics to measure company financial and cost performance. These include income statements, balance sheets, funds flow statements, and various ratio analyses.

3. Recognize that technology is to be viewed as a tool to achieve business success. That is, technology can make a business become more profitable and productive. Delivering value to customers remains the key to achieving business success.

4. Be able to recommend suitable emerging technologies to enhance shareholder value and to mobilize resources (including the engagement of networked partners) to turn these visionary goals into reality.

5. Be persistent in pursuing vision, which is based on legacy and not on activity. Winston Churchill said, "The further backward you look, the further forward you can see" (Jackson 2004).

Example 10.5.

Negotiating for agreements between employees, departments, suppliers, production partners, and networked distributors is part of a manager's job. Explain the guidelines for efficacius negotiations.

Answer 10.5.

There are numerous excellent books on negotiations. Walker (2003) offers a summary of key guidelines for effective negotiation:

1. Focus on the merits by attacking the underlying issues involved, not the opponent or his or her position.
2. Look for creative solutions with which you both can win.
3. Prepare yourself well beforehand. The prenegotiation preparation centers on standards that suggest the best deal and available alternatives.

    Sample questions related to preparation include the following: (1) How much are the other vendors selling that brass dish for? (2) What do your competitors charge for the service you are offering? (3) How much does a person with your experience get paid? (4) What is your best alternative to a negotiated agreement (BATNA)?
4. Raise questions to find out what your opponent really wants, and prepare clever arguments to support what you need.

    The *parable of the orange* says that two parties each want an orange and agree finally to split it in half. But it turns out that one side simply wanted the juice and the other side wanted the rind. If only they had worked together to solve the problem, each side could have gotten what it wanted. According to Walker, situations similar to that in the parable of the orange pop up a lot.

    Be prepared to ask questions pertaining to who, where, what, why, and how as they tend to drive your opponent to disclose more information than the yes–no questions (e.g., How did you arrive at that figure?). Posing questions to your opponent is also useful for fending off your opponent's questions to you that you may not be prepared to answer. Table 10.8 presents some examples of such exchanges.
5. Listen intently, as the power rests with the listeners. Silence is one of the best weapons available to negotiators. Keeping silent will force the opponent to talk more and, as a consequence, to revise his or her position and reveal useful information in the process.

TABLE 10.8   Sample Questions and Answers

Question	Answer
What is the most you would pay if you had to?	If you think that no agreement between us is possible, perhaps we should get someone trustworthy to arbitrate.
If your company agrees to be merged into ours, how many of your employees can be laid off to achieve economics of scale?	Which of your branch offices would you be keeping and which would you close?

6. Make use of the principle of consistency. The *principle of consistency* is that people have the need to appear reasonable. This can be used skillfully to make your opponent feel that, to appear reasonable, he or she needs to use your standards that have been determined during your prenegotiation preparations. The more authoritative your standards seem, the better. An example of such a standard is the price charged for similar goods by the competitors of your opponent.

7. Let your opponent make the opening offer. Studies indicate that people often underestimate their own strengths and exaggerate those of the rivals.

8. Take a psychological test (e.g., the Thomas-Kilmann Conflict Mode Instrument) to understand your own style, be it "competitor" or "collaborator." Taking such a test will help define any aspects of your negotiation style that should be fine-tuned.

9. Be aware of some tactics employed by your opponents. Some may flinch at your proposals on purpose. Others, with the intent to mislead, may exaggerate things they do not really care about. At the close of negotiation, some opponents may say something like, "Wow, you did a fantastic job negotiating that. You were brilliant." Yet others may take advantage of *the Columbo effect* by lulling you into underestimating them and becoming overconfident.

10. Practice makes perfect. Effectual negotiation is 10 percent technique and 90 percent attitude. Attitude is affected by realism, intelligence, and self-respect.

10.9 UNIQUE CONTRIBUTIONS EXPECTED OF ENGINEERING MANAGERS

In what way is the technical background of a manager important to today's executives? If an engineering manager does only what a typical nontechnical manager does, then the engineering manager does not earn his or her keep. Specifically, what can a technically trained manager do that a nontechnical manager cannot? Technological intuition and innovation are the areas engineering managers can and should excel at (Tucker 2002; Hesselbein and Johnston 2002, Betz 2003).

How are innovation and creativity measured? A commonly accepted performance yardstick is counting the number of patents a company receives. More recently, a new measure was proposed to assess the relative value of patents. If a new patent application cites certain prior art patents as background on which the new patent is based, then the prior art patents are regarded as *forward citations*. The value of a patent is said to be directly proportional to its number of forward citations, as more forward citations indicate a broader significance.

For the moment, let us stay with the patent number as a measure of innovation and creativity (Gibbs and Dematteis 2002). According to statistics published by the U.S. Patent and Trademark Office, only two U.S. companies are among the top 10 global companies receiving the most U.S. patents in 2001. (Figure 10.3 shows these comparisons.) One of several likely reasons for this outcome is that, compared with their American counterparts, Japanese companies may have appreciated to a greater extent the strategic importance of patents to their long-term competitiveness and hence devoted more management attention and critical resources to innovation and creativity. U.S. managers do have their work cut out for them in the years ahead.

Besides the aspect of relative competitiveness, the new era is full of challenges due to rapid advancement in technologies, Internet-based communications techniques, and globalization. The need for technological leadership is becoming increasingly pronounced

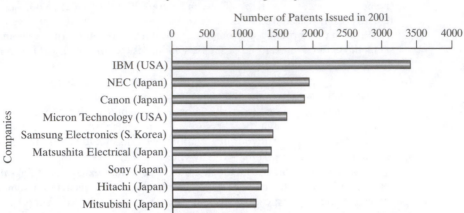

Figure 10.3   Top 10 companies receiving the most U.S. patents in 2001.
Source: U.S. Patent and Trademark Office.

(Hamel 2002). The areas in which engineering managers are expected to make significant contributions are discussed in the next sections.

10.9.1 Technologists as Gatekeepers

A gatekeeper's primary job is to inspect and authorize the entry of people or materials into a gated organization. Technically capable people are usually entrusted with this important corporate activity to systematically scout, evaluate, and introduce new technologies for use by the enterprise.

Engineering managers are in the best position to mobilize capable technologists who understand the new or emerging technologies available in the marketplace and their relevant value to company's products, operations and services. Capable technologists can also bring in and selectively apply new technologies.

10.9.2 Technological Intuition

According to Thurow (1987), nontechnical managers do not have enough background to develop intuitions about which of the possible technologies now on the horizon are apt to advance further and which are apt to be discarded.

Nontechnical managers cannot judge the merits of revolutionary changes in technology; this is known as a factor of ignorance. They have no choice but to procrastinate and wait for it to become clear which technology is the best; this is known as a factor of risk aversion. By the time the answer is clear, foreign firms may already have a two- to three-year lead in understanding and employing the new technologies. Thus, the nontechnical managers cannot exert technological leadership.

This is also evident in diverse technical start-up companies, which, by the way, are typically headed by technically talented people. These talented entrepreneurs are able to invent new technology to serve as the basis for a new business. Eventually, some of these start-up companies are bought up by big companies that cannot develop the technologies on their own.

Technological intuition is most needed in the strategic planning process conducted by a company. Hence, engineering managers have an important role to play here.

10.9.3 Technological Innovations

Another area of technological leadership expected of engineering managers is in the management of technological innovation (White and Wright 2002; Dundon 2002; Gaynor 2002). Technological innovation is the process by which technological ideas are generated, strengthened, and transformed into new products, processes, and services that are used to make a profit and establish a marketplace advantage.

The following statistics were published by Pearce and Robinson (1994):

1. Out of 58 initial product ideas, only 12 survive the business-analysis screening for compatibility with the company's mission and long-term objectives. This step uses 8 percent of the total development time.
2. Only 7 of the 12 ideas remain after an evaluation of their potential. This step uses 9 percent of the total development time.
3. Three of the seven remaining ideas survive after development work is completed. This step uses 41 percent of the total development time.
4. Only two of the original ideas remain past the pilot and field testing involved in the commercialization step, which uses 19 percent of the total development time.
5. Eventually, only one idea results in a commercially successful product. This step uses 23 percent of the total development time.

Within the product development process (see Figure 10.4), the most time-consuming and resource-intensive steps involve development and testing. Engineering managers can make significant contributions to shorten development time and reduce costs while assuring technical quality.

In heading up concurrent engineering teams in product development, engineering managers can excel by

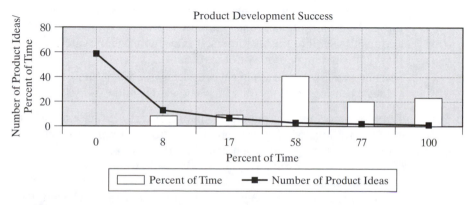

Figure 10.4   Product development.

- Asking pertinent technological questions
- Applying their interdisciplinary background to set technological priorities
- Incorporating new technologies to achieve competitive advantages and to satisfy customers' needs

Exerting strong technological leadership is where technically trained managers must shine. This is the uniquely attractive niche for engineering managers. Engineers do not have serious competition from nontechnical majors here, as it is relatively easy for engineers to learn how to manage, but not so easy for nontechnical managers to learn engineering. However, those engineering managers who cannot exercise technological leadership will be no better than nontechnical managers as far as the value added to their companies is concerned.

Innovation requires knowledge, ingenuity, and predisposition. Innovation cannot succeed without hard work. Purposeful work demands diligence, persistence, and commitment. "Nothing great was ever achieved without enthusiasm," said Ralph Waldo Emerson (Anonymous 1996). Innovations need to be built on the company's strengths and core technologies. They should focus on opportunities that are "temperamentally fit"—that is, exciting and attractive to the innovators—inspiring them to do the required hard work. In addition, innovation must be market driven and focused on customers.

Engineering managers can benefit from taking a systematic approach to enhance individual innovation. Such an approach could include

1. Analyzing innovative opportunities systematically, focusing on (a) the unexpected successes or failures of the company and its competitors, (b) incongruities in processes (production, distribution, and customer behavior), (c) process needs, (d) changes in industry and market structures, (e) changes in demographics, (f) changes in meaning and perception, and (g) new knowledge.
2. Being observant (asking, looking, and listening). The types of questions to pose may include these: (a) What engineering processes or technologies from the past should be kept because they have future value? (b) What past engineering and technological practices should be modified to be more relevant? (c) What activities should be eliminated because they have no future value? (d) What needs to be performed to ensure future success?
3. Recognizing that innovations must be simple and focused, application specific to the present marketplace, and pinpointed on satisfying a need and producing an end result useful to the customers. It is not wise to innovate for the distant future markets, which may or may not materialize.
4. Starting small scale and aiming at producing a series of small, but useful incremental values. Focusing on areas for which knowledge and expertise are available.

Equally important for engineering managers is to exercise leadership in fostering corporate innovations. Managing group innovation is closely linked to managing group creativity. Implementing some well-established techniques, such as those enumerated in the following list, may enhance the creativity of groups:

A. **Brainstorming (for Groups of 6–12 People).** Many important corporate business or engineering issues may be cast in the form of problems. Examples of such issues include product design simplification, product or component cost reduction, and improvement of operations.

By using brainstorming techniques, the leader defines a specific problem and requests each participant in the group to take a turn proposing possible solutions. No criticisms are allowed during these exchanges in order not to impede the free flow of ideas. After all of the ideas are generated and recorded, the group then carefully evaluates each solution and jointly defines the best solution to the problem at hand.

B. **Nominal Group Technique (Small Groups).** The leader defines a specific problem. Each member is encouraged to generate as many written solutions as possible during the group meeting. Each member is then invited to present his or her solution and to elucidate the relevant rationale behind the proposed solution. No criticism is allowed.

After all of the proposed solutions are presented and recorded, each solution is thoroughly discussed, evaluated, and criticized. The participants are then requested to anonymously rank all of the solutions in writing. The final results are presented to management for action.

C. **Delphi Technique (for Identifying Future Trends)**

The leader defines a specific problem and a set of questions and sends them to a panel of geographically dispersed domain experts who do not have contact with one another. Each expert then answers the questions individually and anonymously. A summary of all of the answers is documented by the leader and sent back to the experts. By reviewing the comments and the possible criticisms, the experts, again anonymously, modify their original answers. No one knows who proposed or criticized what specific solutions. The focus is on the merits of the ideas, not on the personality of those who advanced the ideas. The leader again prepares a summary and returns it to the experts, offering additional explanations and justifications. In the end, every solution is justified. Each time the experts respond to a summary, it is called a "wave." After the third "wave," a summary is prepared and the leader makes a forecast. This method is time consuming, but it is particularly useful for predicting the future course that a company may take in technology and business.

Engineering managers ought to be well versed in many of these techniques to manage and promote creativity and innovation both from individuals and groups.

10.9.4 Specific Contributions by Engineering Managers

Engineering managers are capable of adding value to their employers in diverse ways. (Some of these value-adding activities related to product development are delineated in Section 9.5.6.) The following are additional broad-based contributions expected of engineering managers:

1. Use of specific new technologies in product design—novel use of materials, parts, subassemblies, production technologies, and other components.
2. Application of web-based technologies to e-transform the enterprise in order to achieve refinements in process efficiency, quality, speed, or customer satisfaction.
3. Selection of enterprise integration tools for expediting business information collection, transmission, and processing in order to realize speed, cost, and quality advantages.
4. Alignment of networking partners to secure competitive advantages in supply chains, production systems, and customer service.
5. Looking out for new technology-based tools that could facilitate serving customers better, cheaper, and faster, with products that have a larger degree of customization.

6. Employment of new technologies and innovations to add value to stakeholders other than customers (e.g., investors, employees, suppliers, and the communities in which the companies operate).

7. Scanning literature to constantly learn the best practices of technology management in the industry.

10.10 CAREER STRATEGIES FOR THE 21ST CENTURY

Today's leaders adopt specific guidelines in managing their own careers. Among many such guidelines published in the literature on career management, the eight specific guidelines offered by Kacena (2002) are particularly comprehensive and insightful. See Table 10.9. Fernandez (1999) offers a slightly different set of career strategies. (See Table 10.10.)

TABLE 10.9 Guidelines for Managing Your Own Career

---

1. Think, speak, act, and walk like an entrepreneur. Adopt an entrepreneurial mindset, as if your own investment is involved. Accept the notion that jobs exist so that problems can be solved.

2. Make chaos a friend. Embrace change as an opportunity for growth. The mantra for today's career advancement is "eager to stay, yet ready to leave."

3. Don't be afraid to break the rules. Attempt to be a visionary, as innovation is possible only in cultures that tolerate mistakes. Be detail oriented.

4. Know your own strengths and weaknesses. Set high standards for yourself and affiliates. Be very competitive. Market yourself. Express commitment, passion, and excitement about your own work.

5. Be nonlinear. Radical career shifts will become commonplace. Companies are ignoring specialization in favor of adaptability, cross-functionality, people skills, and a rock-solid customer focus. Follow the new paradigm that anyone who does not know how to do something must learn or partner with someone who does know.

6. Maintain balance. Set your own priorities with respect to health, family, and business—in that order—and have fun. "Earn a living, make a life."

7. Stay connected. Building alliances by networking is essential. Establish reciprocal relationships with colleagues, clients, customers, and competitors. Seek to be helpful and supportive so that they become resources for ideas, skills, and knowledge.

8. Always keep your options open. Keep abreast of the market, nurture skills that are marketable, stay professionally active, and avoid becoming complacent.

TABLE 10.10 Career Strategies

---

1. Balance your priorities between your job, personal interests, family, and the community. This is the key to having a full and meaningful life.

2. Cultivate a broad business background through education, diversified work experience, and success skills. Ranked most important among success skills are integrity and persistence. Persistence means unwillingness to accept defeat.

3. Learn leadership from proven leaders. Observe successful people and learn from their behavior.

4. Learn what the company and industry are really about. Understand the company values, aspirations, brand character, market position, organizational structure and culture, and qualities of people.

5. Make an impact. Strive to contribute from the basis of knowledge and attitude. Ask how the world would be different and just a bit better because of your efforts.

Engineering managers are advised to refine their own career strategies constantly by using the aforementioned inputs as general guidelines.

10.11 TAKE-CHARGE FORMULA

According to Roach (1998), everyone must strive to become exceptional and to "take charge" of underutilized potential:

T   Time should taken to reflect on your strengths and weaknesses, as well as to do something about the weaknesses.

A   Attitude must be fostered and modified as needed.

K   Knowledge must be updated to keep yourself marketable.

E   Empathy and consideration in caring for others' feelings should be strived for.

C   Communication must be constantly improved and perfected.

H   Health and humor must be diligently nurtured.

A   Appearance must be properly maintained.

R   Respect yourself and others and live one day at a time—enjoy life; why simply wait.

G   Goals should be set for yourself and family by, for example, creating a five-year plan.

E   Empower the possibilities by finding ways to delegate, assist, entrust, and praise others and by being generous and giving to others.

10.12 CONCLUSION

Many "rules of thumb" are derived from experience. They are common sense heuristics that are all reasonable and intuitively correct.

Knowing what is needed to become (a) an effective engineer, (b) a good engineering manager, or (c) an excellent engineering leader is a very good start. The next step is to learn the skills and capabilities to shape one's own attitudes, and to acquire the attributes needed to become an effectual engineer, good engineering manager, or excellent engineering leader. To be successful, one must practice, practice, and practice until the preferred behavior becomes second nature.

10.13 REFERENCES

Anonymous. 1996. *The Columbia World of Quotations*, New York: Columbia University Press.

Badawy, M. K. 1995. *Developing Managerial Skills in Engineers and Scientists: Succeeding as a Technical Manager*. 2d ed. New York: John Wiley.

Betz, F. 2003. *Managing Technology Innovation: Competitive Advantage from Change*. 2nd ed. New York: John Wiley.

Billick, B. and J. A. Peterson. 2001. *Competitive Leadership: Twelve Principles for Success*. Chicago: Triumph Book.

Broder, D. 1992. "Clinton's Performance Won't Mimic Carter's." *St. Louis Post— Dispatch, Editorial Page*, November 17.

Cheves, G. K. 1992. "Characteristics of Effective Leaders in Systems." *Industrial Management (Supplement)*, Vol. 4, July–August.

Clark-Epstein, C. 2002. *78 Important Questions Every Leader Should Ask and Answer*. New York: AMACOM.

Clugston, M. 1998. "Manager." *Canadian Business*, 1988, pp. 268–270.

Conger, J. A. 1992. *Learning to Lead: The Art of Transforming Managers to Leaders*. San Francisco: Jossey-Bass.

Dundon, E. 2002. *The Secrets of Innovation: Cultivating the Synergy that Foster New Ideas*. New York: AMACOM.

Felden, J. 1964. "What Do You Mean I Can't Write?" *Harvard Business Review*, May–June.

Fernandez, V. 1999. "Career Strategies for the 21st Century." *Executive Speeches*, June–July.

Gaynor, G. H. 2002. *Innovation by Design: What It Takes to Keep Your Company on the Cutting Edge*. New York AMACOM.

Gibbs, A. and B. Dematteis. 2002. *Essentials of Patents*. New York: John Wiley.

Goffee, R. and G. Jones. 2000. "Why Should Anyone Be Led by You?" *Harvard Business Review*, September–October.

Goleman, D. 1998. "What Makes a Leader." *Harvard Business Review*, November–December.

Goleman, D. 2000. "Leadership That Gets Results." *Harvard Business Review*, March–April.

Goleman, D., R. Boyatzis, and A. McKee. 2002. *Primal Leadership: Realizing the Power of Emotional Intelligence*. Cambridge, MA: Harvard Business School Press.

Hacker, D. 2000. *A Pocket Style Manual*. 3d ed. Boston: Bedford/St. Martin's.

Hamel, G. 2002. *Leading the Revolution: How to Thrive in Turbulent Times by Making Innovation a Way of Life*. Cambridge, MA: Harvard Business School Press.

Hesselbein, F. and R. Johnston (Editors) 2002. *On Creativity, Innovation and Renewal: A Leader to Leader Guide*. San Francisco: Jossey-Bass.

Hoffman, H. C. 1989. "Prescription for Transitioning Engineers into Managers." *Engineering Management Journal*, September.

Humprey, B and J. Stokes. 2000. *The 21st Century Supervisor: Nine Essential Skills for Frontline Leaders*. San Francisco: Jossey-Bass.

Jackson, Michael A. 2004. *Look Back to Get Ahead: Life Lessons from History's Heroes*. New York: Seaver Books.

Kacena, J. F. 2002. "New Leadership Directions," *The Journal of Business Strategy*, March–April.

Kim; W. C. and R. A. Mauborgne. 1992. "Parables of Leadership." *Harvard Business Review*, July–August.

Kotter, J. 1990. "What Leaders Really Do." *Harvard Business Review*, May–June.

Krames, J. A. 2002. *The Jack Welch Lexicon of Leadership*. New York: McGraw Hill.

Krames, J. A. 2003. *What the Best CEOs Know: Seven Exceptional Leaders and Their Lessons for Transforming Any Business*, New York: McGraw-Hill.

Langewiesche, W. 2003. "Columbia's Last Flight." *The Atlantic Monthly*, November.

Mai R. and A. Akerson. 2003. *The Leader as Communicator: Strategies and Tactics to Build Loyalty, Focus Effort and Spark Creativity*. New York: AMACOM.

Mintzberg, H. 1990. "The Manager's Job: Folklore and Fact." *Harvard Business Review*, Vol. 68, Issue No. 2, March–April 1990, pp. 163–177.

Morrison, P. 1986. "Making Managers of Engineers," *Journal of Management in Engineering*, Vol. 2, No. 4.

National Science Foundation. 2000. "Women, Minorities and Persons with Disabilities." *NSF* 00–327.

Nelson, B. 1999. *1001 Ways to Take Initiative at Work*. New York: Workman Publishing.

Pearce, A. J. and R. B. Robinson, Jr. 1994. *Strategic Management: Formulation, Implementation and Control*. Homewood, IL: Richard D. Irwin.

Pelus, C. J. and D. M. Horth. 2002. *The Leader's Edge: Six Creative Competencies for Navigating Complex Challenges*. San Francisco: Jossey-Bass.

Roach, M. B. 1998. "Take Charge." Speech before the 17th Turbomachinery Symposium at Dallas, Texas, November 9.

Schwarzkopf, N. 1999. "Leaders for the 21st Century." *Vital Speeches of the Day*, June 15.

Slater, R. 2001. *Get Better or Get Beaten! 29 Leadership Secrets from GE's Jack Welch*. 2d ed. New York: McGraw Hill.

Stogdill, R. 1981. *Handbook of Leadership: A Survey of Theory and Research*. New York: Free Press.

Strunk, Jr., W. and E. B. White. 2000. *The Elements of Style*. 4th ed. Needham Heights, MA: Allyn and Bacon.

Thamhain, H. 1990. "Managing Technology: The People Factor." *Technical & Skills Training*, August–September.

Tucker, R. B. 2002. *Driving Growth through Innovation: How Leading Firms Are Transforming Their Futures*. San Francisco: Berrett-Koehler Publishers.

Thurow, L. 1987. The Tasks at Hand, *Wall Street Journal*, June 12, p. 46D.

Wagner, S. 1999. "Leadership: Lessons from the Best." *Training and Development*, March.

Walker, B. 2003. "Take It or Leave It: The Only Guide to Negotiating You Will Ever Need." *Inc. Magazine*, August.

Welch, J. 2001. *Jack: Straight from the Gut*. New York: Warner Books.

White, S. P. with G. P. Wright. 2002. *New Ideas about New Ideas: Insight on Creativity from the World's Leading Innovators*. Cambridge, MA: Persus.

Zaleznik, A. 1992. "Managers and Leaders: Are They Different?" *Harvard Business Review*, March–April.

Zenger J. H. and J. Folkman. 2002. *The Extraordinary Leader: Turning Good Managers into Great Leaders*. New York: McGraw Hill.

10.14 APPENDICES

APPENDIX 10.A. TEN FACTORS FOR SURVIVAL AND SUCCESS IN CORPORATE AMERICA

To be successful in corporate America, one needs to pay attention to the following common sense success factors:

1. **Excellent Performance.** Make sure that all assignments are performed well, as "You are only as good as your last performance." Pay attention to ensure that both the performance and its impact are properly recorded and made known to people in the organization who affect your career growth. Self-promote as needed.

2. **Personality.** Project a mature, easy-to-work-with, positive, reasonable, and flexible personality. How one acts and behaves is important.

3. **Communication Skills.** Of all aspects of communication, the written form is the most difficult to master. Check the writing advice offered by Felden (1964), with respect to readability, correctness, appropriateness, and thought. Consult the books by Strunk, Jr and White (2000) and Hacker (2000), with respect to style.

4. **Technical Skills and Ability.** Keep your individual capabilities (e.g., analysis, design, integration, product development, tools application, etc.) current and marketable.

5. **Human Relations Skills.** Constantly review ways of interacting with people and make sure that you are creating and maintaining acceptable working relationships. Avoid being labeled as "not able to work well with other people."

6. **Significant Work Experience and Assignments.** Seek diversified business and engineering exposure and high-impact assignments to build up your experience portfolio. Doing so will increase your ability to add value to the organization.

7. **Self Control.** Improve your ability to stay cool and withstand pressure and stress by, for example, taking courses in leadership training. According to a *CNN News* report in 2001, a British military training camp was offering training services to business executives, subjecting the executives to a high-pressure artificial military environment to toughen them up for handling the real-world business environment.

8. **Personal Appearance.** Follow the example of superiors to fit yourself into the corporate image. "Dress for success."

9. **Ability to Make Tough Business Decisions.** Take careful risks when needed. Anyone can make the easy plays, but only great people make the tough plays.

10. **Health and Energy Level.** Take care of your health and maintain a high level of physical vitality.

APPENDIX 10.B. SEVEN MOST COMMON REASONS FOR CAREER FAILURES

Some engineers fail in their careers for one reason or another. Listed next are seven common reasons for career failure; these are relevant to technologists as well as managers.

1. **Poor Interpersonal Skills.** A lack of interpersonal skills is the single biggest reason for career failure. Few people are fired or asked to resign due to deficiencies in their technical capabilities. As a measure of social intelligence, interpersonal skills are important to achieve success in any organization. One needs to be sensitive to the feelings of others, able to listen and understand the subtext in communication, give and take criticism well, strive to build team support, and be emotionally stable.

2. **Wrong Fit.** From time to time, a person may find it hard to adapt one's abilities, styles, personalities, and values to the culture and business practices of the workplace. The workplace may assume a cultural norm that is unfamiliar to some individuals. It is well known that rigidly layered corporations operate differently from dynamic partnerships or entrepreneurial start-ups. The individual's core value system, with priorities, profit motives, and social or environmental preferences may not be fully compatible with those of coworkers on the job. In addition, the chemistry among coworkers within a unit, department, or company could also be a source of conflict. Often, the management style of the superior is difficult for the individual to adapt to. In cases of such a wrong fit, the best strategy for the individual is to move on.

3. **Unable to Take Risks.** Lack of risk-taking abilities is a major stumbling block to the advancement of one's career. For fear of failure, some engineers stay in their current positions for too long and are not willing to accept promotions that require relocation within the company or to venture out for new positions outside of the company.

   Others feel comfortable with the technical work they do because they are able to control all of the key components of their work (e.g., data, facts, analysis, procedure, and equipment) and the quality of its outcome. Naturally, some of them may feel uneasy when requested to take on managerial responsibilities that involve (1) people who may react differently; (2) data that are often incomplete and inaccurate; (3) objectives that are usually multifaceted; and (4) decision-making tasks primarily based on personal intuition and judgment. The inability to take calculated risks could lead to failure in one's career progression.

4. **Bad Luck.** Sometimes, engineers get hurt by business circumstances that are beyond their control or expectation (e.g., mergers and acquisitions, corporate downsizing, change of market conditions, change of business strategies, advancement of technologies, etc.). Career disruptions due to bad luck can happen to anyone. However, one should be able to recover quickly if one's record demonstrates that past achievements consistently created value to employers, and such value-creation capabilities are widely marketable.

5. **Self-Destructive Behavior.** Certain engineers exhibit work habits or behavior patterns that are self-destructive. Examples of self-destructive behavior include working in secret, resistance to change, being excessively aggressive, having an uncooperative attitude, picking fights with people, becoming overly argumentative, being readily excitable about trivialities, and displaying a lack of perspective. Such behavior is clearly unwanted in any group environment.

6. **Lack of Focus.** Some engineers pride themselves on being a jack-of-all-trades, getting busily involved in almost everything, but being good at nothing. Failing to focus on creating value to the employer is detrimental to one's own career.

7. **Workplace Biases.** Under ideal conditions, all workplaces should be free of biases with respect to race, gender, age, national origin, religious beliefs, and other individual qualities. In reality, some workplaces are managed more effectively and progressively than others. Individual workers need to monitor the real situation at hand and take proactive steps to avoid being hurt by such biases. Engineers serve themselves well by constantly checking against these bias-based failures and proactively managing those over which they can exercise some control.

APPENDIX 10.C. TIPS ON COPING FOR FIRST-TIME SUPERVISORS AND MANAGERS

It typically takes two to three years for a first-time supervisor or manager to become fully effectual. Here are few tips that can help the novice to cope during this initially challenging period:

1. Organize the office so that important files and project folders can be located readily.

2. Get a good perspective from one's superior in term of priorities, strategic plans, previous problems, and vision to operate the unit or department. Do the homework to learn the new languages: finance, marketing, manufacturing, and customer service. Acquire the business perspectives of markets, cost-price position, product distribution, supply-chain management, enterprise integration application, and customer relation management.

3. Obtain training in evaluating staff performance, managing time, and developing multidisciplinary teams.

4. Ready yourself mentally to assign responsibilities while maintaining control in order to achieve results through people.

5. Communicate your expectations to staff, both individually and in groups, and solicit feedback.

6. Foster relationships with peer managers in other departments.

7. Build the relationship with your own superior.

8. Start practicing and polishing your own management styles in order to become increasingly effective.

Humphrey and Stokes (2000) offer another set of nine identifiable people, technical, and administrative skills for frontline supervisors in the new century:

- **People skills**
    1. Communication—ability to adjust own style to correspond with individuals needs and an ability to listen to other people
    2. Teamwork—ability to take into account the diversity of other people's backgrounds
    3. Coaching skills—ability to assist other people

- **Technical skills**
    4. Business skills—ability to assess business performance of others
    5. Continuous improvement—ability to constantly update and refine one's own technical skills
    6. Technologically savvy—ability to use modern office equipment.

- **Administrative skills**
    7. Project management skills—ability to plan and implement new ideas
    8. Writing and documentation skills—ability to write reports and keep management informed
    9. Resource management skills—ability to network with those who control resources.

Over time, first-time supervisors and managers need to demonstrate the ability to build team spirit, create a work environment that fosters self-motivation, solve people and technical problems, and make decisions by integrating technical issues with business issues affecting the company.

APPENDIX 10.D. HOW TO MANAGE ONE'S OWN SUPERIORS

Both engineers and engineering managers need to properly manage their respective superiors. The superior needs the active support of all employees to succeed, as most of the work is done by the subordinates. On the other hand, all of the subordinates need their superiors' support to move forward along their individual career paths.

The power of a superior should be taken seriously. One of the primary reasons for job turnover is personality conflict with the individual's own superior, not because of technical performance.

It is also of critical importance that one understands the corporate mindset. Whenever the organization appoints a group leader or manager, the following unwritten rules apply:

1. The organization knows that no one is perfect and that the appointee is no exception.
2. The appointee's strengths are valued more than the trouble caused by his or her weaknesses. Even if the appointee appears to be difficult for some subordinates to deal with, the organization counts on him or her to lead the group and add value. Unless the appointee clearly violates the stated rules, the organization will back the appointee most of the time.
3. To achieve the goals of the organization, the organization trusts the views and desires of the appointee over those of his or her subordinates.

The organization also expects employees to behave in certain ways. These include being attuned to the superior and not insisting that the superior adjust to the employees. Work closely to support the superior and help him or her to succeed. Avoid questioning the superior's judgment and decisions, as the superior typically has access to more and better information and data than the employees and may not be in a position to share such information or data freely.

In readying oneself to manage superiors effectively, it is useful for employees to form the following habits:

1. Understand the business and personal pressure the superior is under, his or her values and motivators (achievement, success, recognition, money, value systems, priorities, principles, and other factors), work style (peacekeeper, conflict lover, riser or setter, channel oriented), and personal style (optimistic, fighter).

2. Expect modest help, and request it only when you really need it. It is better to get help from your own networked coworkers and friends.

3. Be sensitive to the superior's work habits. Watch how he or she receives data and information and works on it. Learn his or her preferred mode of communications—face to face, phone, e-mails, or staff meetings, for instance.

4. Stay in touch with the superior, unless he or she does not want to be bothered regularly.

5. Present materials clearly and without complex details and jargon.
   - Emphasize the significance (the benefits and realizable impact) of your technical work to the group or company, not its technological complexity, sophistication, or elegance.
   - Use concise language to elucidate ideas and recommendations clearly.

6. Do not defend a cause unless it deserves it. Keep it in perspective. Do not complain when you do not get all that you asked for.

7. Exercise self-control. Manage your own overreactions or counter productive behavior.

The following set of guidelines (Babarro and Kotter 1993)* for managing the superior–subordinate relationship is recommended:

1. Accept that your superior's support is important to you. Understand how important your support is to your superior.

2. Understand your own response to the superior's style and personality, and manage it. Respect the style and orientation of your superior to his or her work. Understand your response to your position in the hierarchy and how you feel about working within a structure.

3. Learn to take feedback objectively, not personally, and maintain your sense of self and your own uniqueness.

4. Push back when necessary, but only for business reasons and to maintain personal integrity; do not push for political gain or to embarrass the superior.

5. Learn the superior's goals, aspirations, frustrations, and weaknesses. Study and understand what the superior thinks is important. Study and be able to emulate the superior's communications style for the sake of being heard.

6. Be dependable; follow through on serious requests for information and work output.

7. Display respect to others and expect respect from others in all matters of business and on-the-job interpersonal interactions regarding time, resources, and alternative work styles.

---

*From D. Goleman, "What Makes a Leader?" *Harvard Business Review*, November–December, 1998. Copyright © 1998; reprinted by permission of Harvard Business School Publishing Corporation, all rights reserved.

8. Be honest and share all relevant data about the situations and concerns at hand.

9. Keep private any criticism and conflict that may arise between the two of you, and always work for a jointly satisfactory solution.

10. Be manageable by and available to those beneath you.

APPENDIX 10.E. THE WISDOM OF THE MOUNTAIN

For more than 20 years, Lao-li studied under the great master Hwan at a mountaintop temple. He struggled and struggled, but could not reach enlightenment. One day, the sight of a falling cherry blossom spoke to his heart. "I can no longer fight my destiny," he reflected. "Like the cherry blossom, I must gracefully resign myself to my lot." Lao-li determined to retreat down the mountain, giving up his hope of enlightenment.

Lao-li looked for Hwan to tell him of his decision. Hwan surmised his intention to resign and said to Lao-li, "Tomorrow, I will join you on your journey down the mountain." The next morning, the two started walking downhill. Looking out into the vastness surrounding the mountain peak, Hwan asked, "Tell me, Lao-li, what do you see?" "Master, I see the sun beginning to wake just below the horizon, meandering hills and mountains that go on for miles, and couched in the valley below, a lake and an old town." Hwan smiled, and they continued their long descent.

Hour after hour, as the sun crossed the sky, they pursued their journey, stopping only as they reached the foot of the mountain. Again, Hwan asked Lao-li to tell him what he saw. "Master, in the distance I see roosters as they run around barns, cows asleep in sprouting meadows, old ones basking in the late afternoon sun, and children romping by a brook."

The master, remaining silent, continued to walk until they reached the gate to the town. There the master gestured to Lao-li, and together they sat under an old tree. "What did you learn today, Lao-li?" asked the master. "Perhaps this is the last wisdom I will impart to you." Silence was Lao-li's response.

At last, after a long silence, the master continued. "The road to enlightenment is like the journey down the mountain. It comes only to those who realize that what one sees at the top of the mountain is not what one sees at the bottom. Without this wisdom, we close our minds to all that we cannot view from our position and so limit our capability to grow and improve. But with this wisdom, Lao-li, there comes an awakening. We recognize that alone one sees only so much—which is not much at all. This is the wisdom that opens our minds to betterment, knocks down prejudices, and teaches us to respect what at first we cannot view. Never forget this last lesson, Lao-li: What you cannot see can be seen from a different part of the mountain."

When the master stopped speaking, Lao-li looked out to the horizon. As the sun set before him, it seemed to rise in his heart. Lao-li turned to his master, but Hwan was gone.

So the old Chinese tale ends. But it has been said that Lao-li returned to the top of the mountain to live out his life. He became a great enlightened one himself.*

APPENDIX 10.F. THE SOUND OF THE FOREST

Back in third-century China, the King of Tsao sent his son, Prince Tai, to the temple to learn under the great master Pan Ku how to become a good ruler. When the prince arrived at the mountaintop temple, the master sent him to the Ming-Li Forest. After one year, the prince was to return to the temple to describe the sound of the forest.

When the prince returned after one year, he told the master, "I could hear the cuckoos sing, the leaves rustle, the hummingbirds hum, the crickets chirp, the grass blow, the bees buzz, and the wind whisper and holler." When the prince had finished, the master told him to go back to the forest to listen more. The prince was puzzled by the master's request. Had he not discerned every sound already?

For days and nights on end, the young prince sat alone in the forest listening. But he heard no sounds other than those he had already described. Then, one morning as the prince sat silently beneath the trees, he started to discern faint sounds unlike those he had ever heard before. The more acutely he listened, the clearer the sounds became. The feeling of enlightenment enveloped the boy. He reflected, "These must be the sounds the master wished me to discern."

When the prince returned to the temple, the master asked him what more he had heard. "Master," responded the prince reverently, "when I listened most closely, I could hear the unheard—the sound of the flowers opening, the sound of sun warming the earth, and the sound of the grass drinking the morning dew."

The master nodded approvingly. "To hear the unheard," remarked Pan Ku, "is a necessary discipline to be a good ruler. For only when a ruler has learned to listen closely to the people's hearts, hearing their feelings not communicated, pains unexpressed, and complaints unspoken, can he hope to inspire confidence in his people, understand when something is wrong, and meet the true needs of his citizens. The demise of states comes when leaders listen only to superficial words and do not penetrate deeply into the souls of the people to hear their true opinions, feelings, and desires."*

APPENDIX 10.G. THE WHEEL AND THE LIGHT

At the beginning of the Han dynasty, Liu Bang, the Chinese Emperor, had just unified China into a strong country. To commemorate this event, Liu Bang invited all of his high-ranking political advisors, military officials, and scholars to a grand celebration.

At the center table sat Liu Bang with his three staff chiefs, Xiao He, the logistics master; Han Xin, the military chief; and Chang Yang, the political and diplomatic strategist.

An another table sat Cheng Cen, the master scholar, whom Liu Bang consulted often during the war years. With him sat three of Cheng's disciples.

Everyone enjoyed the event. Food and entertainment were plentiful. Midway through the festivals, one of Cheng's disciples discreetly raised a question. "Master, look at the central table. Xiao He had the best supply logistics. Han Xin's military tactics were beyond reproach. Chang Yang had the winning diplomatic tactics. Their contributions were clearly visible and readily understood. How is it, then, that Liu Bang is the emperor?"

The master smiled and asked his disciples to imagine the wheel of a chariot. "What determines the strength of a wheel in carrying a chariot forward?" he inquired. "Is it the sturdiness of the spokes, Master?" one disciple responded. "But then, why is it that two wheels made of identical spokes differ in strength?" After a moment, the master continued, "See beyond what is seen. Never forget that a wheel is made not only of spokes, but also of the space between the spokes. Sturdy spokes poorly placed make a weak wheel. Whether their full potential is realized depends on the harmony between them. The essence of wheel making lies in the craftsman's ability to conceive and build the space that holds and balances the spokes within the wheel. Think now, who is the craftsman here?"

A glimmer of moonlight was visible behind the door. Silence reigned until one disciple said, "But Master, how does a craftsman secure the harmony between the spokes?" "Think of sunlight," replied the master. "The sun nurtures and vitalizes the trees and flowers. It does so by giving away its light. But in the end, in which direction do they all grow? So it is with a master

*From W. Chan Kim and R. A. Mauborgne, "Parables of Leadership," July–August, 1992. Copyright © 1992; reprinted by permission of Harvard Business School Publishing Corporation, all rights reserved.

craftsman like Liu Bang. After placing individuals in positions that fully realize their potential, he secures harmony among them by giving them all credit for their distinctive achievements. And in the end, as the trees and flowers grow toward the giver, the sun, individuals grow toward Liu Bang with devotion."*

10.15 QUESTIONS

10.1 How can engineering managers make the best use of tools such as the Myers-Briggs Type Indicator (MBTI) to assist in selecting project leaders or assigning employees to teams to ensure the likelihood of avoiding personality conflicts that could otherwise hinder team success?

10.2 Engineering managers may be called upon to resolve conflicts between employees, departments, vendors, and business partners, as well as to handle customer complaints. What are the recommended guidelines for handling complaints? Please elaborate.

10.3 Hoffman (1989) believes that a management education program should have three elements:

(1) *Behavioral*—people skills, motivation, team building, communications, and delegation

(2) *Cognitive*—production, marketing, finance, and control

(3) *Environmental*—markets, competition, customers, political, social, and economical environment in which the organization operates

The importance of the first two elements should be self-evident. Explain why the third element, the environmental, is important.

10.4 How is engineering management different from management in general?

10.5 How does a manager or leader become a good superior? What should the superior do and not do?

10.6 Does the job of managing a high-technology function (e.g., an engineering design department) differ from that of managing a low-technology function (e.g., a hotel)? Explain the specific details of the jobs.

10.7 What rules and principles can guide managers to have successful people management skills?

10.8 There are so-called "unwritten laws of engineering" that recommend acceptable conduct and behavior for engineers in industry. How important are these unwritten laws to individual engineers, and where can these laws be located?

10.9 Some engineers and managers are known to have more difficulties in interpersonal relations than others. These difficulties may arise due to personality, chemistry, value system, priority, tolerance, competition and other such factors. How can they improve their interpersonal skills?

10.10 In your opinion, what are the characteristics common to many future engineering leaders? Please explain.

*From W. Chan Kim and R. A. Mauborgne, "Parables of Leadership," July–August, 1992. Copyright © 1992; reprinted by permission of Harvard Business School Publishing Corporation, all rights reserved.

Chapter 11

Ethics in Engineering Management and Business Management

11.1 INTRODUCTION

Recent U.S. corporate scandals have raised serious questions about ethics in the workplace. The cases of Enron, Global Crossing, Adelphia, and others—all alleged to have falsified financial data or misused corporate funds—are still pending at the time of this writing. Steps have been taken by FASB (Financial Accounting Standards Board) to tighten the financial audit guidelines for the future. Not too long ago, two top editors of *The New York Times* resigned because they failed to reign in one of their staffers who fabricated or plagiarized three dozen stories over a six-month period. Exposures of these cases in the news media bring about public anxieties and apprehensions. But the focus of the public's attention on ethics in the workplace should help motivate managers to avoid willful wrongdoing (Elliot and Schroth 2002).

As the markets in the new millennium become more dynamic and business relationships increasingly intertwine, opportunities for conflicts of interest and ethical dilemmas are likely to become more prevalent. This chapter addresses some of the issues and solution strategies related to ethics in business and engineering management (Seebauer and Barry 2001; Johnson 1991; Boylan 2000; Boatright 2000).

Before delving into a discussion of ethics, two pieces of background information may be useful. Patterson and Kim (1992) report that, on the basis of interviews with 5000 people across the United States in 1992, 90 percent of the people surveyed decide alone—without the influence of church, governments, or family—what is moral and what is not in their lives. Seventy-four percent of those surveyed said they would steal. Sixty-four percent said that they would lie when it suited them. If any of these numbers are believable, then the moral standards of the general public in the United States are on shaky ground. Let us hope that the sample of 5000 people picked in the aforementioned survey might have included a few too many bad apples.

Recently, the Pew Research Center for the People and the Press in Washington, DC, conducted a survey by asking the following question: "Would you say most business executives try to obey the laws governing their professions, or do they try to find a way around the laws?" Thirty-five percent of respondents said they obey the laws, 58 percent said they find a way around the laws, and 7 percent offered no answer. In terms of whose interests the companies put first, Table 11.1 summarizes the drastic difference between the public's perceptions of what current practices are and what the perceived right emphasis should be.

In a recent *Gallup Poll*, the public was asked to assess the honesty and ethical standards among professions. The sum of the percentages of responses for "very high" and "high" are added together to provide a composite percentage for purposes of ranking. Table 11.2 presents the top 10 ranked professions, plus business executives and a few others for purposes of comparison. In this survey, engineers are ranked number 9 and business executives number 18 among 32 professions. Both engineers and business executives have their work cut out for them to enhance the public's perception of their honesty and ethics.

There are numerous situations in which engineers and engineering managers may encounter problems with ethics. These situations include public safety and welfare, risks, health and environment, conflicts of interest, truthfulness, integrity, choice of a job, loyalty, gift giving and taking, confidentiality, industrial espionage, trade secrets, discrimination, and professional responsibility.

Generally speaking, there are microethical and macroethical issues. In microethics, the focus is on the relationships among engineers and their coworkers, clients, and employers. In macroethics, engineers are concerned with the collective social responsibility of the profession in relation to, for example, product liability,

TABLE 11.1   Whose Interests Do Companies Put First?

	Current Practice (percent)	Right Emphasis (percent)
Top Executives	43	3
Stockholders	37	14
Employees	3	31
Customers	5	27
Communities	5	19
None	2	N/A
Don't know	5	N/A

Source: PEW Research Center for the People and the Press (2002),
<http://people-press.org/reports/display.php3?ReportID=149>,
see questions No. 19 and 20 in the questionnaire section.

TABLE 11.2   Gallup Poll on Honesty and Ethics, 2000

No.	Profession	Very High and High (percent of responses)
1	Nurses	79
2	Druggists/Pharmacists	67
3	Veterinarians	66
4	Medical Doctors	63
5	Elementary/High School Teachers	62
6	Clergy	60
7	College Teachers	59
8	Dentists	58
9	Engineers	56
10	Policemen	55
13	Bankers	37
18	Businessmen	23
25	Stockbrokers	19
28	Lawyers	17
32	Car Salesmen	7

Source: Gallup Poll <*http://poll.gallup.com/content/default.aspx?ci=1654*> Subscribers may access its detailed breakdown scores.

sustainable development, globalization, and the impact of technology. Sustainable development refers to industrial practices that minimize harmful impacts on the environment, while maximizing the efficiency of energy and material utilization. This chapter will address both types of issues.

In the research literature, two basic approaches are taken to handle ethical issues: addressing the general philosophy underlying a particular outlook, and examining specific cases in order to draw out lessons related to ethics. There are weaknesses in both approaches. The philosophical approach lacks a connectedness to the real-world environment, thus producing no guidelines to deal with actual situations. On the other hand, the case-based approach produces only "school solutions" after the fact. These may or may not generate any useful lessons to be learned (Paine 1997; Wilcox and Theodore 1998). Since there are a number of well-known cases related to ethics, this second approach is selected for use in this chapter.

Not all problems in ethics have practical solutions, just as not all product design problems have feasible solutions. Furthermore, as more and more companies pursue globalization, an additional concern is that what is "normal" in one culture may not translate acceptably to another (Machan 1999; Pava and Primeaux 2002).

In this chapter, the deliberations about ethics proceed with basic definitions, followed by discussions about engineering ethics that are then extended to business ethics. The chapter concludes with general guidelines to deal with ethical issues.

TABLE 11.3   Percentage of Companies with Codes of Ethics

Countries	Percentage of Large Corporations with Codes of Ethics
United States	90
Canada	85
United Kingdom	57
Germany	51
France	30

Source: Mark S. Schwartz, "A Code of Ethics for Corporate Code of Ethics."
Journal of Business Ethics, November–December 2002, pp. 27–43.

11.2 ETHICS IN THE WORKPLACE

Ethics is important to corporations because companies with strong, positive overall reputations attract and keep the best customers, employees, suppliers, and investors. These companies also avoid the trouble of litigation, fines, recalls, bankruptcy proceedings, and antitrust suits. Corporations are made of people; what people say and do creates the corporate reputation.

Corporations take steps to address ethics-related issues. According to Schwartz (2002), over 90 percent of large U.S. corporations have formulated company-specific codes of ethics. This trend is expanding to corporations in other countries as well. (See Table 11.3.)

A code of ethics is a formal, written document that contains normative guidelines for behavior. Publishing and implementing such a code allows a corporation to provide a set of consistent standards for employees to follow, both to avoid any adverse legal consequences due to possible wrongdoings and to promote a wholesome public image.

11.2.1 Universal Moral Standards

Whether a code of ethics contains right or wrong guidelines depends on the benchmark standards used to make the assessment. Schwartz (2002) has assembled a set of universal moral standards that appear to be intuitively correct. (See Table 11.4.)

TABLE 11.4   Universal Moral Standards

No.	Standards	Contents
1	Trustworthiness	Honesty, integrity, reliability, loyalty
2	Respect	Respect for human rights
3	Responsibility	Accountability
4	Fairness	Process, impartiality, and equity
5	Caring	Avoiding unnecessary harm
6	Citizenship	Obeying laws and protecting the environment

Source: Mark S. Schwartz, "A Code of Ethics for Corporate Codes of Ethics." *Journal of Business Ethics*, November–December 2002, pp. 27–43.

By using such a set of universal moral standards as a yardstick, different codes of ethics can be compared and evaluated.

Ethics in the workplace may be discussed within three different scopes—ethics in engineering, management, and business—depending on the expected complexity of the situation and the potential impact on stakeholders.

11.2.2 Engineering Ethics

Engineers play key roles in the advancement, production, and operation of technology. They should therefore assume a degree of responsibility for the consequences of applied technologies. Martin and Schinzinger (1996), Harris, Pritchard, and Rabins (2000), and Davis (1998) address different aspects related to engineering ethics.

Engineering companies publish codes of ethics to guide engineers in performing their work in ethically and socially responsible manners. A large number of engineering societies, such as the National Society of Professional Engineers (NSPE), American Society of Civil Engineers (ASCE), American Society of Mechanical Engineers (ASME), and the Institute of Electrical and Electronics Engineers (IEEE), have also issued codes of ethics to provide discipline-specific guidelines. In addition, the files of NSPE's Board of Ethical Review contain various types of issues in engineering ethics and their respective resolutions.

Pinkus, Shuman, Humon, and Wolfe (1997) have noted that there are three principles of engineering ethics: competency, responsibility, and public stewardship. Competency means that engineers are obliged to know as much as is reasonably possible about the technology with which they work. They should be honest and candid enough to acknowledge their own deficiencies and seek assistance from others to fill in any gaps. Responsibility requires that engineers voice their concerns when an ethical dilemma is identified. Responsible organizations must then evaluate these concerns promptly. With respect to public stewardship, engineers must understand the risks associated with the technology that they apply and deploy.

A key concept in engineering ethics is the notion of "professional responsibility," a type of moral obligation arising from the special responsibility possessed by an individual engineer. According to Whitebeck (1998), this moral responsibility requires that engineers exercise judgment and care to achieve and maintain a desirable state of affairs, as well as to protect the public health and safety.

A large number of U.S. engineering schools are now actively involved in teaching ethics in undergraduate curricula. ABET, the Accreditation Board of Engineering and Technology, reviews and accredits countless engineering school programs in the United States on a regular basis. ABET has defined a specific program outcome related to ethics in its *Engineering Criteria 2000*. This program prescribes that graduates are to have an understanding of professional and ethical responsibility and the broad education necessary to understand the impact of engineering solutions in a global and societal context. The ABET *Engineering Criteria 2000* have been implemented since the fall of 2001.

Teaching ethics to engineering students is aimed at achieving four outcomes: (1) raising ethical sensitivity, (2) increasing knowledge about relevant standards of conduct, (3) improving ethical judgment (e.g., by way of discussions of real-world cases), and (4) strengthening the ethical willpower to enable a greater capability to act

TABLE 11.5 Sample Engineering Ethics Cases

No.	Case	References
A	Hooker Chemical—Love Canal, 1978	Gary Whitney. "Case Study: Hooker Chemical and Plastics," in T. Donaldson (editor), *Case Studies in Business Ethics*. Englewood Cliffs, NJ: Prentice-Hall.
B	The Collapse of Walkways at Hyatt Regency Hotel in Kansas City, 1981	*http://ethics.tamu.edu/ethics/hyatt/hyatt1.htm* E. Phrang and R. Marshall. "Collapse of the Kansas City Hyatt Regency Walkways." *Civil Engineering ASCE*, July 1982.
C	DC-10 Cargo-Door Accident Case near Paris, 1974	P. French. "What Is Hamlet to McDonnell-Douglas or McDonnell-Douglas to Hamlet: DC-10." *Business and Professional Ethics Journal*, Vol.1, No. 2, 1982.
		J. Fielder and D. Birsch. *The DC-10 Case: A Study in Applied Ethics, Technology and Society*, Albany, NY: State University of New York Press, 1992.
D	Spiro Agnew and Construction Kickback in Maryland, 1973	Richard M. Cohen and Jules Witcover. "A Heartbeat Away: the Investigation and Resignation of Vice President Spiro Agnew," New York: Viking, 1974.
E	Space Shuttle Challenger Explosion, 1986	R. L. B. Pinkus, L. Shuman, N. P. Hummon, and H. Wolfe, *Engineering Ethics: Balancing Cost, Schedule and Risk, Lessons Learned from the Space Shuttle*. New York: Cambridge University Press, 1997.

ethically. Teaching ethics illustrates real-world complex relationships between technology development and the welfare of individuals, society, and the environment. It also promotes an understanding of the professional nature of engineering and of the responsibilities associated with a professional career. It enhances the engineers' abilities to analyze situations that raise questions about ethics and to articulate reasonable ways to respond to ethical dilemmas.

In recent years, there have been a number of major engineering ethics cases reported in the engineering and business literature. Table 11.5 lists some sample cases. The case method of teaching ethics encourages students to express ethical opinions, prompts them to identify ethical issues, and helps them formulate and justify decisions in an effectual manner. The case method also seeks to strengthen students' sense of the practical context of ethics.

A. **The Hooker Chemical Case.** Located in Niagara Falls, New York, Love Canal was 10 feet deep, 60 feet wide, and 3000 feet long and surrounded by a virtually impervious clay soil. Hooker Chemical obtained permission to use the canal for dumping waste chemicals in 1942 and subsequently acquired a strip of land 200 feet wide with the canal in the center. Hooker sealed chemical wastes into steel drums, dropped them into the canal, and covered them with a layer of clay. Approximately 22,000 tons were deposited from 1942 to 1953. At the time, there were no federal or state regulations governing the dumping of chemical wastes.

In 1953, Hooker closed the dump and sold the land to the city's board of education for one dollar after the board threatened to condemn the land. Hooker included a clause in the deed to the board describing the past use of the land and required that the board assume the risk of liability for any future claims that might result from the buried chemicals.

Subsequently, the board constructed a school on part of the land and sold the remainder to a developer who built homes for families. None of the homes were built directly over the canal. In 1978, traces of chemicals were noted on the surface of the land. Residents complained of increased rates of miscarriage, risks of birth defects, increased urinary tract infections, and other health-related problems. In August 1978, President Carter declared Love Canal a limited disaster area. Federal investigators took 5000 soil samples, but failed to establish a direct link with the buried chemicals.

B. The Hyatt Regency Walkways Case.   During a dance party in July 1981, two walkways suspended over the atrium of the hotel lobby collapsed, killing 114 people and injuring 185 others. Detailed investigations indicated that the mechanical supports for the walkways were insufficiently designed for the anticipated loads. The license of the engineering firm responsible for the design was revoked.

C. The DC-10 Case.   In March 1974, a Turkish Airlines DC-10 crashed near Paris due to a failed cargo door and the collapse of the passenger compartment floor. Investigators noted that all of the hydraulic and electric control systems were severed. A total of 346 passengers died.

The problems of the DC-10's cargo door and the buckling of the passenger compartment floor were noticed when McDonnell-Douglas pressure-tested the airplane in 1970. In 1972, an American Airlines DC-10 experienced similar problems while flying over Windsor, Ontario. (That plane was brought down safely.)

McDonnell-Douglas used an electric latching system and installed three parallel redundant control systems, instead of the industrywide standard of four. The design engineers placed all three systems in the leading edge of the wing, rather than channeling through different sections of the airplane to provide redundancy.

Records also verified that the Turkish Airlines DC-10 went in for cargo-door modifications in July 1972. Although three McDonnell-Douglas inspectors stamped the maintenance papers, the actual work to modify the cargo door was never done.

D. Spiro Agnew and the Construction Kickback in Maryland.   In 1961, an engineering firm in Baltimore noticed that county government contracts were being awarded to those firms which maintained close connections with governmental officials. The firm started to donate to the campaign funds of Spiro Agnew, who was running for county executive at the time. Agnew won, and his contractual relationship with the firm continued. From that point on, this engineering firm would provide 5 percent of its contract value as cash kickback to Agnew. This practice continued over a long period, even when Agnew was later the governor of Maryland and then vice president of the United State. The total payment made to Agnew was alleged to be over $100,000. This engineering firm accumulated the needed cash by giving key employees "bonuses" and requiring them to pay the money back partially in cash. The firm also contrived "loans" to individual employees, who then channeled cash back to the company treasury.

In October 1973, Spiro Agnew resigned as vice president after a no-contest plea. The court fined him $10,000 and sentenced him to unsupervised probation for three years.

E. **The Space Shuttle** *Challenger* **Case.**   In the well-known Space Shuttle *Challenger* explosion case, the night before the launch Morton Thiokol engineers had identified the potential danger of launching the shuttle in temperatures less than 53 degrees Fahrenheit. NASA management challenged the recommendation. During an off-line discussion among Morton Thiokol participants in Utah, a vice president of engineering was the only one among four to hold out for a launch delay. A senior vice president told him bluntly, "It's time to take off your engineering hat and put on your management hat." The vice president capitulated and the launch went forward, resulting in the disaster on record. This episode and the anecdote about engineering and management "hats" are now widespread in the literature on engineering ethics.

11.2.3 Management Ethics

Should there be any difference between engineering ethics and management ethics? Not a lot, except that managers must consider broader issues and deal with ethical situations more complicated than those typically encountered by engineers. Green (1994) and Schminke (1998) include some discussion on management ethics.

In the space shuttle Challenger case, the Morton Thiokol management standpoint was as follows: The company was a contractor to NASA and had clearly expressed its technical concerns related to the launch. Even though the recommendation was not fully supported by available data, the company did fully discharge its moral and ethical responsibility in the O-ring issue. The proper role for Morton Thiokol was indeed to respect the view of its client. If NASA, as the paying client, decided to launch the shuttle anyway, then the responsibility for any negative consequences rested entirely with NASA.

From the NASA management standpoint, the O-ring issue was technically not supportable by data. On the other hand, NASA had its mission goals to fulfill. Under such circumstances, someone made a decision under uncertainty based on gut feeling and personal experience in risk assessment. Unfortunately, the decision turned out to be wrong. However, management is paid to make such hard decisions under uncertainty.

Generally speaking, engineering managers consider factors related to the well-being of the organization, such as cost, schedule, employee morale, customers, supply chains, investors, public image, local communities, health and safety, social and environmental impacts, market development, profitability, and globalization. Engineers, on the other hand, focus on technical matters that fall within their professional engineering practices, such as product design, production, technology, public health and safety, and environmental impact.

Example 11.1.

> The ocean liner SS *United States* was a luxurious ship in the 1950s. It had approximately one-half-million square feet of harmful asbestos insulation. Initial estimates indicated that it would cost $100 million to have it refurbished in the United States. In 1992, it was towed to Turkey, where the cost of removing the asbestos was cited at only $2 million. Turkish officials, however, refused to allow the removal because of the danger of exposure to the cancer-causing asbestos. In October 1993, the ship was towed to the Black Sea port of Sebastopol, where laws were lax and the removal of asbestos would cost less than $2 million.
>
> Do you approve of the program that removed asbestos from the SS *United States* at Sebastopol? Why or why not?

Answer 11.1.

> The ethically correct answer is to disapprove. The removal of asbestos at Sebastopol was economically attractive for the shipowner, because the local laws did not require specific safeguards and cautious processes needed for effectively protecting the health of workers. Thus, the cost saving was derived by taking advantage of the ignorance of the local people and the inadequacy of the local laws. The program has the potential for damaging the health of workers at Sebastopol, a blatant violation of core human rights. It must be rejected without reservation.

11.2.4 Ethics in Business

As engineering managers move up the corporate ladder, they get more and more involved in influential decision making that extends beyond the traditional domains of engineering and technology. Guidelines related to ethics in business kick in at that time to help shape the decision making to embrace business implications.

Such situations arise from the fact that business management must take care of the broad interests of five stakeholder groups: shareholders, customers, employees, suppliers, and people in local communities in which the company operates. Business managers must also remain consistent in their professional responsibilities and ethical standards. Note that all of these stakeholders are members of the society at large, but they represent only a small part of it. (See Figure 11.1.) Thus, inherently, there will be situations in which choices made to pursue the interests of these five stakeholder groups may clash with the interests of the remaining members of society. Although an ethical company must attempt to eliminate or minimize such clashes, conflicts of these types are likely to occur.

In fact, actual situations in real environments are far worse than this. As indicated previously in Table 11.1, questions have been raised concerning the conflicts of interest

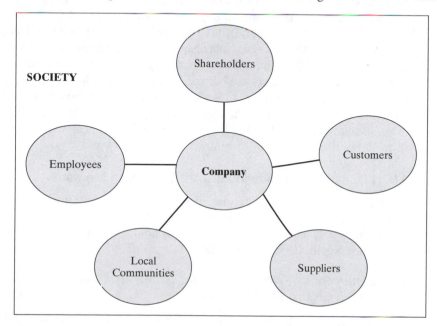

Figure 11.1   Company stakeholders within society at large.

reflected in the management of numerous U.S. corporations. The U.S. public perceives that many companies unethically put the top managers' own interests ahead of the five major stakeholder groups. Congress passed the Sarbanes–Oxley Act of 2002, the Public Company Accounting Reform and Investor Protection Act, to more tightly regulate some of the questionable accounting practices.

Damian and Stephen (1996), McHugh (1998), and Cavanagh, Moberg, and Valasquez (1995) offer additional viewpoints related to business ethics.

11.3 GUIDELINES FOR MAKING TOUGH ETHICAL DECISIONS

Everyone could use guidelines when making tough ethical decisions. Peter Drucker, a well-known business management professor, recommends a "mirror test." Ask yourself, "What kind of person do I want to see when I shave ... or put on my lipstick in the morning?" (Seglin 2000). This method may not deter those violators who are self-serving and who apply double standards to justify what they do.

Norman Augustine, former CEO of Lockheed Martin, has proposed four questions to gauge how ethical a course of action is:

1. Is it legal?
2. If someone else did it to you, would you think it was fair?
3. Would you be content if it appeared on the front page of your hometown newspaper?
4. Would you like your mother to see you do it?

If you answer "yes" to all four questions, then, according to Augustine, whatever you are about to do is probably ethical. Following this method of screening, one is then advised to always have the cell phone numbers of a lawyer, editor of the hometown newspaper, and one's own mother on hand, plus a well-calibrated, unbiased "barometer" of fairness.

Badaracco (1998) believes that character is forged in situations where responsibilities come into conflict with values. These situations are called "defining moments." At defining moments, managers must choose between right and right. Badaracco suggests a set of questions for individuals, managers of working groups, and executives of companies to answer when evaluating such defining moments. These questions are described in the next sections.

(Badaracco (1998) includes various examples in his article to demonstrate the application of these questions to real-world situations.)

11.3.1 Questions for Individuals Facing Defining Moments

What feelings and intuitions come into conflict in this situation? Which of the values in conflict are most deeply rooted in my life? What combination of expediency and shrewdness coupled with imagination and boldness will help me implement my personal understanding of what is right?

11.3.2 Questions between Right and Wrong That Managers of Working Groups Must Answer

What are the other persuasive interpretations of the ethics of this situation? What point of view is the most likely to win a contest of interpretations inside of my organization and influence the thinking of other people? Have I orchestrated a process that can manifest the values I care about in my organization?

11.3.3 Questions that Confront Company Executives

Have I done all that I can to secure my position and the strength of my organization? Have I thought creatively and boldly about my organization's role in society and its relationship with stockholders? What combination of shrewdness, creativity, and tenacity will help me transform my vision into reality?

Some managers are constantly required to resolve the conflict between discharging their responsibility to maximize shareholder value, on the one hand, and behave in an ethical manner, on the other. Bagley (2003) suggests a simple decision tree to guide managers in making ethical decisions in any corporate projects. (See Figure 11.2.)

Bagley advises managers to raise three questions:* (1) Is it legal? (2) Does it maximize shareholder value? and (3) Is it ethical? Managers should refuse to pursue

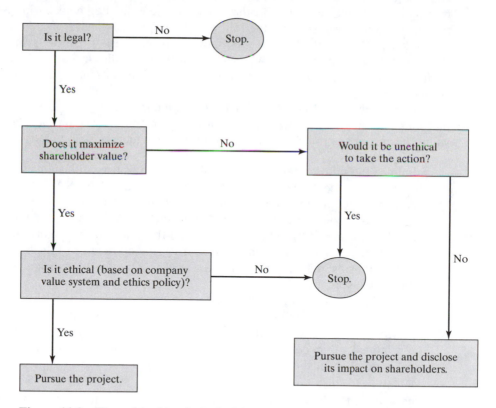

Figure 11.2   The ethical leader's decision tree. (*Source*: Bagley 2003.)

any project or take any action if the answer to any of these questions is "no." However, managers should decide to carry out projects that are legal and ethical, even if these projects do not maximize shareholder value, as long as pursuing them could benefit the other stakeholders of the company. (See Figure 11.1.) The argument proceeds as follows: Managers work for the best interests of the company, and these best interests may not always require them to maximize shareholder value.

The logic of this decision-tree model is rather compelling. But a major weakness of this model is that, for some courses of action, the answers to the questions related to shareholder values and ethics will likely be "maybe" instead of a clear cut "yes" or "no."

In the Challenger case, Morton Thiokol engineers and managers argued for a delay of the launch due to a suspected O-ring failure under low-temperature conditions. In contrast, NASA managers were under pressure to proceed with the launch as scheduled. Numerous questions were raised. Did Morton Thiokol have to prove that the flight was unsafe in order for the launch to be delayed? There were no data that could conclusively substantiate the recommendation of a launch delay. Should NASA managers have allowed factors other than engineering judgment to influence the flight-schedule decision? From the following two options, which stand should NASA have taken? (1) Don't fly if it cannot be shown to be safe. (2) Fly, unless it can be shown to be unsafe. Certainly, the NASA program objective to build the shuttle for about half of the originally proposed cost might also have contributed to the decisions that resulted in disaster.

According to the decision-tree model presented in Figure 11.2, it is clear that NASA had the legal authority to decide. When it comes to the second question, shareholder value, the answer was not clear cut. Two options were under consideration at the time:

A. Launch the shuttle as scheduled. The benefits of a successful launch were expected to be the continued enhancement of the value of NASA programs to the American people and to the scientific communities, and the preservation of public trust and confidence in the capabilities of NASA management. On the other hand, if the launch were unsuccessful, then there would be a significant sacrifice of human life, the destruction of physical assets, and the loss of the public's goodwill. Perhaps some personnel changes at NASA would have become necessary under that scenario.

The key problem was that neither Morton Thiokol nor NASA had data to confirm that the launch would definitely be unsafe. It was only the best judgment of Morton Thiokol engineers that the launch might be unsafe. The expected cost of a launch failure could not be quantitatively estimated because there was no reliable number for the probability of the O-ring to fail at low temperatures.

B. Delay the launch until temperatures were right. The cost of such a delay would have been the loss of public confidence in NASA managerial competency and certain operational and equipment maintenance costs associated with a delay.

Depending on the assumed value of the probability of occurrence for the O-ring failure, the expected cost for option A could change. Thus, the question related to shareholder value in Figure 11.2 should be answered "maybe." To move forward from this point on, it would require the introduction of an assumption (usually unproven and likely relying on gut feeling or an extrapolation from past experience) related to the probability for the O-rings to fail. Determining an accurate probability becomes a judgment call. Apparently, NASA management believed that the probability for the O-rings to fail was extremely low.

NASA's assumption of an extremely low probability for the O-rings to fail also led to a "yes" answer to the third question related to ethics in Figure 11.2. If this probability of O-ring failure were high, then it would have been unethical for NASA to decide on a course of action that would have been likely to lead to the demise of the shuttle crew. An academic ethicist (Werhane 1991) evaluated the *Challenger* disaster and concluded that it was the result of four kinds of difficulties: (1) different perceptions and priorities of engineers and management at Thiokol and at NASA; (2) a preoccupation with roles and responsibilities on the part of engineers and managers; (3) contrasting corporate cultures at Thiokol and its parent, Morton; and (4) a failure both by engineers and managers to exercise individual moral responsibility. (For a detailed discussion of the space shuttle Challenger case, see Vaughan (1996).)

While the decision-tree model is generally useful, its application can become complicated when conflicts of interest arise among the stakeholders. What happens if value is added to one group of stakeholders at the expense of others? The following examples illustrate this type of situation:

1. Some U.S. company boards routinely approve bonus and stock options to top managers to offer them extra incentives at the expense of shareholders. This practice contributes to the distorted corporate emphasis indicated in Table 11.1. Such practices are not common in other industrialized countries.

2. When involved in mergers and acquisitions, some top managers negotiate for special separation contracts to benefit themselves at the expense of the surviving company.

3. Companies outsource manufacturing operations to developing countries where environmental and safety regulations are typically less stringent than in the United States; this achieves cost advantages for the companies at the expense of the communities in which they operate and of the employees in the developing countries.

4. Companies sell unsafe automobiles and drug products to consumers, causing deaths due to accidents or side effects, while realizing sales revenue and profitability for themselves.

5. The U.S. Securities and Exchange Commission (SEC) announced its settlement with 10 Wall Street security firms to restore public trust. These firms agreed, without admitting any wrongdoing, to pay a fine of $1.4 billion for giving biased stock investment advice that caused investors to lose money during the recent past and delivered huge financial benefits to the firms involved.

The troubling point here is that all decision makers involved in the preceding cases were fully aware of the ethical implications related to their actions. Thus, knowing what is ethical and unethical is clearly only the first step. The decision tree in Figure 11.2 offers a logical road map to force a needed critical review of all management decisions. However, additional steps beyond making knowledge available are required to ensure that all decisions carried out are ethical at all times and under all circumstances.

11.4 CORPORATE ETHICS PROGRAMS

Building an efficacious corporate ethics program requires leadership, commitment, planning, and execution. According to Nirvana (1997), companies need to have clear statements of their vision and values. They should have an organizational code of ethics. Creating an ethics officer position and forming an ethics committee helps communicate the company's ethics strategy, coordinate employee training, maintain a help line to offer confidential advice as needed under specific circumstances, monitor and track activities with ethics implications, and take action to reward good and punish unethical practices.

More than 90 percent of all Fortune 500 companies have codes of ethics, and 70 percent have statements of vision. Companies such as Proctor & Gamble, IBM, Johnson & Johnson, Texas Instruments, John Deere, Cummins Engine, Eaton, and Dow Corning publish codes of ethics on-line. The Center for the Study of Ethics in the Professions at the Illinois Institute of Technology received a grant from the National Science Foundation to design and maintain a website, *http://www.iit.edu/ departments/csep/PublicWWW/codes/index.html/*. The site stores 850 codes of ethics on-line, including those issued by engineering associations (Section 11.2.1) and corporations.

A corporation's statement of ethics serves as a behavioral compass for the employees. Kinni (2003) recommends a set of guidelines to formulate a code of ethics. Companies must first establish their values and direction. To be relevant, a statement of ethics must be connected to the core direction in which the organization is going. The code of ethics should be centered on such values as integrity and truthworthiness, instead of being merely a compliance document. It should include specifics unique to the business under consideration or details about the company philosophy. The company should then go public with the code of ethics so that all employees, customers, suppliers, and any other stakeholders are fully informed of it. The code of ethics must be updated regularly to help guide the day-to-day behavior and decision making of the employees. The overall effectiveness of a statement of ethics depends on the company leaders' commitment to its disciplined enforcement. (Examples of ethics statements can be found in Murphy, 1998.)

It is useful to issue guidelines to handle potentially unethical situations and to set clear standards of conduct applicable to daily responsibilities. Employees are advised to (1) analyze carefully the situation at hand; (2) list all possible failings and downsides of the potentially unethical practice in question; (3) compile all possible benefits that could accrue if the practice in question ceases and is admitted now, rather than discovered by someone else later; (4) issue a memo; and (5) attempt to make a full disclosure to coworkers, the superior, the superior's superior, the company president, the customers, the public, and the press.

Besides defining what the employees are expected to do, it is equally important to spell out what they should not do. Of specific value is the description of actions that are deemed unethical. Setting such lower bounds ensures that there is no ambiguity in interpreting what is not allowed. As sample unethical cases from both internal and external sources are continuously added to the code of ethics, the resulting casebook will become an increasingly important benchmark reference.

Honeywell has put teeth into its ethical principles by making it mandatory for all employees to adhere to the company's code of conduct. Starbucks Coffee introduced its Framework for a Code of Conduct for coffee-producing countries, in order to standardize ethics practices among coffee retailers, exporters, and growers.

Some people believe that it is ineffective for companies to self-police their own codes. Instead, there should be an independent monitoring of ethical practices. To deter wrongdoing, the penalties must be high and the enforcement disciplined. Among examples of such penalties are dismissal with charges and forfeiture of all pension rights, a jail term with no opportunities for parole, severe financial penalties, and denial of reentry to industries involved.

There should probably also be more recognition from society for good actions taken by companies or individuals. Making ethically sound decisions may have been taken for granted by innumerable people for a long time. In 1998, *Deloitte/Management Magazine* created a Business Ethics Award program in New Zealand and had recognized the following companies with best practices in ethics since that time: Norske Skog Tasman in 2002, Methanex NZ in 2001, New Zealand Post in 2000, and 3M New Zealand in 1999. More awards of this type should be established to honor such people and to reflect society's appreciation for these exemplary ethical behaviors!

In managing corporate ethics programs, the key is to make sure that everyone is honest and will disclose fully all of the details of every situation in question. A well-known publisher suggests a single question as the basis for assessing an ethical situation: "If what I just said or neglected to say, did or neglected to do, saw and failed to report, or heard and failed to mention, were disclosed openly by someone else in reputable communications channels, would it embarrass me, my organization, or my family?" If the answer is "yes," then the action or inaction in question is unethical. This question is likely to be useful for honest people with self-respect. However, embarrassment is a personal perception based on value. It may not be as useful to white-collar crooks who are driven by greed and who are willing and able to circumvent the laws to act unethically. Generally speaking, individuals who offer a full disclosure to all concerned are usually ethical (Johnson and Phillips 2003).

The commitment of top management to the corporate ethics program is of critical importance to its success. A case in point is Johnson & Johnson. The company is known for its Credo Challenges sessions in which employees and managers talk about ethics related to current business problems and offer criticisms of existing policies and ideas for improvement. The company achieved excellent business results by using this industrial best practice in the field of ethics.

An opposite case in point is Enron, which had a 65-page code of ethics. Yet some top Enron managers allegedly entered into special business deals with off-balance-sheet financing that resulted in a falsely inflated corporate profitability that permitted selected management personnel to cash out stock options while siphoning out special bonus payments to individuals, all eventually at the expense of the company shareholders. Reports of internal whistleblowers were simply ignored by Enron top managers who elected to take no corrective actions. It was clear to everyone involved that these were unethical and illegal management actions. Allegedly acting as its partner in crime, the auditing firm Arthur Andersen was also accused of committing the criminal offense of willfully destroying Enron papers relevant to the case.

No ethics program will be efficacious unless company top management supports its implementation.

Example 11.2.

A global job migration trend is starting to develop. Significant numbers of jobs are being exported from the United States and other developed countries to such Third World countries as Mexico, China, India, the Philippines, Ireland, and Bangladesh. Typical jobs are related to software programming, call-center service, financial accounting, tax preparation, selected R&D, and claims processing. Other types of jobs may be involved in the future. The principal driving force behind the job migration trend is cost: A comparable quality of workmanship may be accomplished in Third World countries at about one-third the cost of paying workers in developed countries.

Does Corporate America display an apparent indifference to its workforce at home? Is it unethical for American companies to outsource work in search of cost-effectiveness at the expense of American workers?

Answer 11.2.

Corporate America is legally empowered to seek cost-effectiveness in creating and marketing products and services, as long as it is doing so in compliance with laws and commonly accepted ethics standards. Because of the free market system the United States practices, Corporate America does not have the obligation to guarantee jobs for any sector of American workers, be they engineers, software programmers, accountants, claims-processing clerks, call-center service personnel, or any others. Individual workers, on the other hand, need to keep their own skills marketable at all times. If certain sectors of employment appear to be declining because of global competition, the workers in these sectors must be able to learn new skills quickly in order to keep themselves competitive in the job market. Past cases with textile, steel, and agriculture workers are typical examples of decline due to globalization.

The job protection concept may be appropriate in a socialist system, wherein the government exercises control over the economy, but it is not appropriate in a free-market economy. By outsourcing, Corporate America does not display an apparent indifference to its workforce at home. It is not unethical for Amercian companies to outsource work.

11.5 AFFIRMATIVE ACTION AND WORKFORCE DIVERSITY

Related to corporate ethics is the issue of affirmative action. The law is very clear on this issue. In 1964, the Civil Rights Act was enacted to ban discrimination on the basis of a person's color, race, national origin, religion, or sex. The Equal Employment Opportunity Act was passed in 1972, empowering the Equal Employment Opportunity Commission (EEOC) to enforce Title VII of the Civil Rights Act. The EEOC directives do not allow for discrimination in hiring, placement, training, advancement, compensation, or other activities against any person on the basis of race, color, religion, national origin, gender, age, disability, marital status, or any other such characteristic protected by law. These acts have been enforced, and progress in affirmative action has been made over the last several decades.

For the past 30 years or more, a number of private colleges and universities have introduced race-sensitive admissions policies to increase the number of African-American, Hispanic, and Native American students they enroll. Bowen, Bok, and Barkhart (1999) defend such policies on the ground that these institutions have the right to define the preferred composition of their student bodies. In recent years, there have been more state or institutional actions taken against such policies. In 1996, California abolished affirmative-action programs and decided to stop using race and gender as college admission factors. Washington State followed the Californian example in 1998. In 2000, Florida joined this anti–affirmative action movement.

Massachusetts Institute of Technology, Cambridge, Massachusetts, excluded white students from its annual summer mathematics and science programs for incoming freshmen and high school students. In 2002, a complaint was filed with the Department of Education's Office of Civil Rights. Subsequently, MIT opened the programs to all students.

The University of Michigan has a well-known undergraduate admissions policy that allows 12 points for a perfect SAT score and adds 20 points if the applicant happens to be black, Hispanic, or Native American, but assigns no points for being white or Asian. Figure 11.3 illustrates the changing race and ethnicity composition of undergraduate students admitted to the University of Michigan. The United States Supreme Court announced its famous "split decision" in 2003: The vote was against the point system, but in favor of a race-based admissions policy for the University of Michigan Law School to use in the absence of a point system. Proponents on both sides of this issue were able to claim victory. Is it fair to those white students whose admissions were denied just because of their race and ethnicity? Questions like this remain unresolved.

Even after companies allow more minorities and women to gain entry, there is still an atmosphere of tension, instability, and distrust between white and nonwhite managers, according to Carver and Livers (2002). Minority managers have a high turnover rate, and they encounter deep-rooted, complex, and highly personal attitudes and assumptions in their coworkers.

Thomas (1990) argues that companies need to move beyond affirmative action and to strive for equal opportunities for everyone in the workforce, as a way of creating

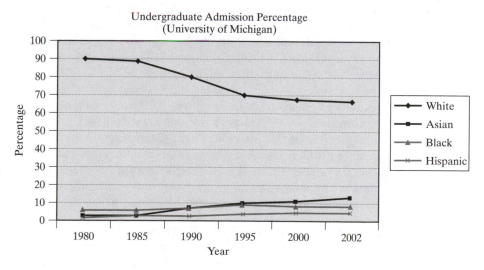

Figure 11.3   Undergraduate admission at the University of Michigan.
Source: University of Michigan. (Race and ethnicity of 6.5 percent is unknown.)

competitive advantage in a global economy. Globalization may indeed precipitate workforce diversification and force everyone to work with everyone else, regardless of values, culture, race, or gender. Thomas and Ely (1996) indicate that companies should use the integration paradigm to promote equal opportunities for all people, promote open discussion of diverse cultural issues, eliminate all forms of domination (by hierarchies, race, gender, etc.) that inhibit full contribution, and secure organizational trust.

For sure, this debate will rage on. In the meantime, engineering managers are well advised to constantly treat everyone and every situation honestly and fairly and to value each person's contributions properly.

11.6 GLOBAL ISSUES OF ETHICS

The problems of ethics are broader in complexity and scope for companies with a global reach than those for companies which operate domestically. This is due to the fact that the values, business practices, laws, environmental and safety standards, and other related references for making ethical decisions differ depending on the countries involved.

Engineering managers encounter problems with global ethics implications in a number of ways. Some problems related to product safety, plant operations, environmental discharges, work rules, and child labor are to be expected. Managers may need to interact with local governments with respect to permits, customs, transportation services, and procurement of parts and materials. They may also need to set up local supply chains and require all local participants to comply with specific codes of conduct.

Take environmental standards as an example. Countries that are economically active consume a disproportionate amount of energy and natural resources per capita compared with emerging countries. According to the World Bank's *Little Green Data Book* for 2003, America uses 16 times as much energy per person than India. About 15 percent of the world's population living in rich countries are responsible for 50 percent of the overall carbon dioxide emissions, the principal gas behind global warming. Figure 11.4 exhibits a detailed account of CO_2 emission per person in 2002. The question may be asked whether it is ethical and socially responsible for the economically active countries to pollute more than others.

Another example is related to values and business practices. Figure 11.5 displays the study results of Transparency International, a Berlin-based organization. According to this worldwide study, the least ethically corrupted country is Finland, followed by Singapore and then Great Britain, Hong Kong, Germany, and the United States.

Yet another problem that has global ethics implications has to do with the law. In 1977, the U.S. Congress enacted the Foreign Corrupt Practices Act, which requires American parent companies and their foreign subsidiaries to abide by certain business standards (Baruch 1979). This act has three provisions:

A. **The Recordkeeping and Disclosure Provision.** This provision requires that companies shall " . . . make and keep books, records and accounts, which, in reasonable details, accurately and fairly reflect the transactions and disposition of the assets of the issuers." Not recording and disclosing payments (such as bribes) made to foreign recipients is an offense clearly prosecutable under this provision.

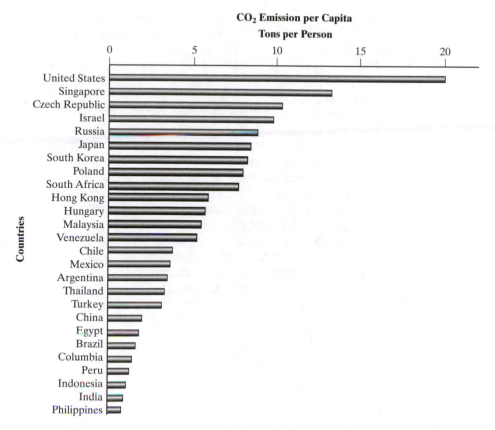

CO₂ Emission per Capita

Tons per Person

Figure 11.4   Per capita carbon dioxide emission by countries. *Source:* World Bank (2003).

B. **Internal Controls Provision.** This provision prescribes that companies have audit committees composed of independent, outside members of their boards of directors to provide independent financial audits. Not exercising proper managerial control in the use of funds (such as accepting bribes or other illegal payments) is a violation of this provision.

C. **Antibribery Provision.** This provision prohibits payments to foreign officials, foreign political candidates, or foreign political parties intended to corrupt those recipients who act in favor of the companies. Doing so is against the act even if it does not violate the laws of the respective countries involved. However, "grease payments" in the form of entertainment expenses and small gift items to minor officials are generally allowed for the purpose of facilitating transactions. The act is silent with respect to the actual dollar amount above which this provision is deemed violated, leaving such interpretations to the courts.

The act regards the integrity of management as a material factor. Engineers and engineering managers need to become familiar with and sensitized to these provisions

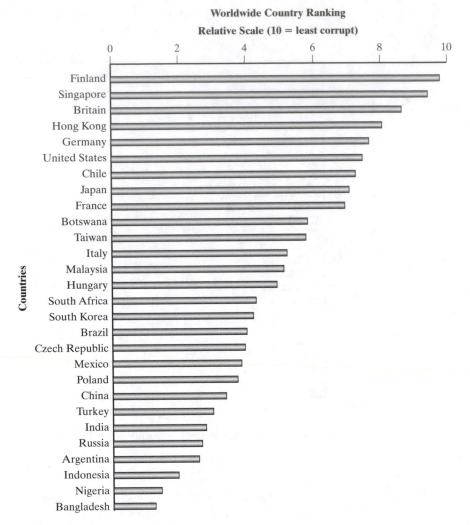

Figure 11.5 Corruption-based ranking of countries. *Source:* Transparency International, Berlin, Germany (2003).

when engaged in interactions with foreign personnel. When in foreign countries, the applicable local laws must be obeyed as well. Thus, it should be self-evident that any course of action that violates any laws, either of the home country or of the host country, must not be undertaken.

The issue related to the bribery of foreign public officials is a troublesome one. Some researchers, such as Cragg and Woof (2002), have questioned the positive impact of the Foreign Corrupt Practices Act on overall standards of international business conduct, particularly with respect to the bribery of foreign public officials.

Problems involving environmental discharge or child labor may cause engineering managers to be fearful of another law: The Alien Tort Claims Act has been used to prosecute U.S. companies for alleged human rights violations and environmental degradation abroad. Olsen (2002) reports three such industrial cases.

Even if all of the actions taken by engineering mangers are lawful, it is possible that not all lawful business decisions they make are ethical. It is usually quite complex to determine whether a given course of action taken in a host country is ethical or not, because the contrasts between the cultures, values, and customs of the host and home countries come into play. A number of examples cited by Donaldson (1996)* illustrate these differences:

1. Indonesians tolerate the bribery of public officers. Bribery is considered unethical and unlawful in the United States.

2. Belgians do not find insider trading morally repugnant. Insider trading is a criminal offense in the United States. For example, Martha Stewart was indicted and subsequently convicted for having committed conspiracy, making false statements, and obstruction related to her suspicious sales of InClone Systems stocks. She resigned as CEO of her company and was sentenced to serve time in prison.

3. In some countries, loyalty to a community, a family, an organization, or a society is the foundation of all ethical behavior. The Japanese people define business ethics in terms of loyalty to their companies, their business networks, and their nation. The Japanese are group oriented. In contrast, Americans place a higher value on liberty than loyalty and emphasize equality, fairness, and individual freedom over group achievements. Americans are focused on individualism.

4. The notion of a right or entitlement evolved with the rise of democracy in post-Renaissance Europe and the United States. This term does not exist in either Confucian or Buddhist traditions.

5. Low wages may be considered by workers in wealthy countries (e.g., the United States and Europe) as an exploitation of the workforce, but developing nations may be acting ethically when they accept low wages to induce foreign investment to improve their living standards.

6. Some Third World governments may want to use more fertilizer to enhance crop yields to combat food supply shortage problems, even though doing so would mean accepting a relatively high level of thermal water pollution.

7. A manager at a large U.S. specialty products company in China caught an employee stealing. She followed the company's practice and turned the employee over to the provincial authorities. The provincial authorities executed him.

8. In Japan, people who do business together exchange gifts, sometimes very expensive ones, as part of a long-standing tradition. Any foreign countries wanting to do business there will need to accept such practices as given.

9. Managers in Hong Kong have a higher tolerance for some forms of bribery than their Western counterparts, but they have a much lower tolerance for the failure to acknowledge the work of a subordinate. In some parts of the Far East, stealing credit from a subordinate for work or ideas is the most unethical activity.

*From T. Donaldson, "Values in Tension: Ethics Away from Home," *Harvard Business Review*, September–October 1996. Copyright © 1996; reprinted by permission of Harvard Business School Publishing Corporation, all rights reserved.

In the case of *The New York Times*, a staff writer stole credit by plagiarizing the writings of freelance writers and other sources. He did so for an extended period. In spite of multiple warnings, the responsible editors did nothing to stop this staff writer from publishing. Finally, the staff writer resigned and both editors were forced out.

10. Some Indian companies offer employees the opportunity for one of their children to get a job with the company once the child has completed a certain grade level in school. The company honors this commitment even when other applicants are more qualified. This perk would be regarded as nepotism in the United States, as it is against the principle of equal opportunity that jobs should go to the best-qualified applicants. On the other hand, some U.S. universities reserve certain admission quotas for children of alumni, major donors, and members of specific minorities.

11. The Swiss are known for their time sensitivity, whereas South Americans are known for their time laxity.

12. Forty percent of managers in the United States believe that the primary goal of a company is to make a profit, while only 33 percent in England, 35 percent in Australia, 11 percent in Singapore, and 8 percent in Japan share this belief.

There is a significant divergence in the perceived goals of a company from the manager's perspective. Once a company achieves its financial objectives, how should the money be distributed? Table 11.1 illustrates the emphasis of U.S. managers versus those of the general public in the United States. The self-centered approach of some U.S. managers could cause tensions and conflicts in values when dealing with foreign managers.

There is no international consensus on standards of business conduct (Williams 2000). Donaldson (1996) offers three basic principles as guidelines: (a) observing core human values, (b) showing respect for local traditions, and (c) focusing on the context when deciding what is right and wrong. Core human values determine the absolute moral threshold for all business activities. These values include the right to good health, economic advancement, and an improved standard of living. One must respect human dignity and recognize a person's value as a human being, and not treat others simply as tools. A good yardstick to use is the Golden Rule, which says "Do not do to others what they do not want done to themselves."

This principle requires that customers be treated well through the production of safe products, and employees through the maintenance of a safe workplace. Also, the local environment should be protected. Companies must avoid employing children and thereby preventing them from receiving a basic education. Local economic and education systems ought to be supported. Companies should forgo those business relationships which violate rights to health, education, and safety and which prevent the development of an adequate standard of living.

Donaldson (1996) classifies ethics conflicts into two types: conflicts due to relative economic development and conflicts due to cultural tradition. Ethical situations of the first type are related to wages, safety, and the environment; they arise from a low level of economic development in the host country in comparison with the developed country. To determine whether a given course of action is ethical, Donaldson suggests that the following question be raised: "Would the course of action under consideration be acceptable in my home country if our economic development were at the same stage as that of the host country?" If the answer is "yes," then the course of action is ethically acceptable. For example, if a specific developing country is currently at the stage comparable to the United States in the 1970s, then U.S. rules and regulations related to wages, safety, and the environment in practice during the 1970s, not those in the 2000s, should be used to assess any situation involving ethical conflicts in that developing country at the present time.

From time to time, courses of action must be taken even if they bring about conflicts due to cultural tradition if companies want to do business in a given host country. Generally, cultural tradition is to be respected and accepted. Saudi Arabia is known not to allow women to serve as corporate managers; most women there work in education and health care. Most foreigners who do business in Japan have now generally accepted the Japanese gift-giving tradition. Of course, compromises made in tradition-based conflicts must not violate core human values.

Some companies use a specific gift-giving strategy in order to be lawful, ethical, and compatible with local cultural tradition while promoting goodwill and fostering close working relationships. They elect to present two sets of gifts: a big and very expensive company-to-company gift and several small personal gifts, each being, for example, under $25. Past practices have shown that such an approach seems to work out well for the parties involved.

Again, when handling global problems of ethics, the following are the basic questions: "Is it legal with respect to laws in both the host and home countries?" "Does the planned course of action violate basic human values?" "Is it consistent with the local cultural norms?" "Is there a creative way to reconcile the ethics issues at hand?" Figure 11.6 presents the decision-tree diagram for global problems of ethics.

In handling challenging situations involving ethics in global settings, engineering managers need to uphold core human values, account for the relevant local traditions,

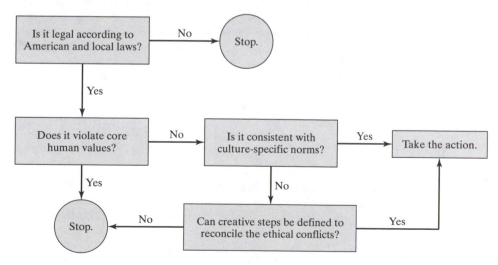

Figure 11.6   Decision tree for global ethics problems.

and be creative in problem solving to come up with a suitable and ethically acceptable course of action without violating laws (Tichy and McGill 2003).

Example 11.3.

In the past, Levi-Strauss engaged numerous contract manufacturers in Third World countries to produce athletic shoes. The company discovered in 1992 that two sewing subcontractors in Bangladesh employed children under the age of 14, in violation of company rules. The local economic condition was such that the children were contributing significantly to their family income. If they were discharged instantly, the lack of these income streams would cause economic hardship to the families involved. On the other hand, allowing the children to continue working clearly violated the stated company ethics policy prohibiting the employment of child labor.

What should Levi-Strauss have done?

Answer 11.3.

In fact, Levi-Strauss came up with a creative solution to this ethics dilemma. The children were sent back to school with tuition, books, and uniforms fully paid by the company. In addition, they continued to receive their full wages. They were promised jobs after they completed their education at the age of 14. This creative solution satisfied both Levi-Strauss and the children's families.

This is a practical example of the creative steps indicated in Figure 11.6.

Example 11.4.

Your company is among several U.S. suppliers actively competing for a major equipment project of a state-owned enterprise in Shanghai worth about $20 million. You are introduced to a Chinese "consultant" who offers to help. This consultant claims that he has contacts inside the enterprise's midlevel project evaluation teams and its top-level decision-making groups and that he can find out sensitive information for you, such as the equipment design features in the competitor's proposals. He can also relate back to you the enterprise's hints for improvements to your proposal and help move your project along in a multiple-round bidding process. If you do not win, you owe him nothing. However, if your project succeeds, you must pay him 5 percent of the final project value as a consulting fee. Your peers tell you that such an arrangement is quite normal in China and that a large part of the consulting fee goes directly to staff people in that enterprise and to the enterprise itself. Those who have rejected such help in the past have seen major equipment contracts awarded to less fussy competitors.

What should you do?

Answer 11.4.

Some Chinese state-owned enterprises use their own high-level employees to serve as such "consultants," with the ultimate purpose of driving down the final project cost. With the tacit support of the enterprise, each "consultant" forms a team consisting of employees situated in various departments and groups within the enterprise. In offering advice, they pass along information prepared to exert increased competitive pressure among the foreign suppliers. They also collect design intelligence so that the enterprise knows what critical questions to raise in project negotiations. It is one of those tools to help an unsophisticated technological buyer get the bid.

If your company needs the project to penetrate the Chinese market and views the winning of this project to be critically important for your company's future growth in the global market, then you should proceed to engage the consultant, as not doing so will render your company less competitive.

If your company regards this project to be useful, but not critically important to your future success, then stay on the high ground of morality and reject the services of the consultant.

Example 11.5.

You are a manager of a joint venture in Russia. One day you discover that your most senior officer in Russia has been "borrowing" equipment from the company and using it in his other businesses. When you confront him, the Russian partner defends his actions. After all, as a part owner of both companies, isn't he entitled to share in the equipment?

How do you propose to resolve this conflict?

(*Source*: S. Puffer and D. J. McCarthy, "Pinning the Common Ground in Russian and American Business Ethics." *California Management Review*, Vol. 37, No. 2, Winter 1995, pp. 29–46.)

Answer 11.5.

You should tell the Russian partner that the joint venture would be very happy to assist his other companies in any way. Then you should proceed to send him a bill for lease expenses of the equipment at a commonly accepted market rate. Tell him that, in the future, you would like to preapprove any such leases to make sure that the affected equipment is not needed in your own shop at the time.

11.7 CONCLUSION

Ethics in the workplace is a fundamental requirement for all members of a corporation—individual employees, managers of working groups, and executives. Everyone must diligently observe universal moral standards and basic human values, act legally, respect fully the local cultural traditions and practices, and extensively disclose the courses of action taken.

Most situations in the workplace involving ethics originate from a conflict between responsibilities and values. These situations must be carefully analyzed by the involved decision maker to arrive at a proper course of action. A number of models are included in this chapter to assist engineers, managers, and executives in carrying out such analyses.

Creating codes of ethics is helpful. Employees become more sensitive if they are exposed to more real-world cases with ethical implications. Knowing what is right and what is wrong is usually not sufficient in preventing unethical outcomes. Of critical importance to the success of a corporate ethics program is the disciplined enforcement of the code of ethics supported by top management. Good behavior should be recognized and rewarded promptly. Unacceptable behavior must be punished fully and in a timely manner.

The key to ethical behavior is honesty, fairness, and full disclosure. When selecting employees, engineering managers need to focus on their characters, as skills can more readily be imparted. Managers themselves need to be technically outstanding, managerially competent, and ethically sensitive.

11.8 REFERENCES

Adams, D. M. and E.W. Maine. 1998. *Business Ethics for the 21st Century*. Mountain View, CA: Mayfield Publishing Company.

Badaracco, J. Jr. 1997. *Defining Moment: When Managers Must Choose Between Right and Right*. Cambridge, MA: Harvard Business School Press.

Badaracco, J. L. Jr. 1998. "The Discipline of Building Character." *Harvard Business Review*, March–April.

Bagley, C. E. 2003. "The Ethical Leader's Decision Tree." *Harvard Business Review*, Reprint F0302C, February.

Baruch, H. 1979. "The Foreign Corrupt Practices Act." *Harvard Business Review*, January–February.

Boatright, J. R. 2000. *Ethics and the Conduct of Business*. 3d ed. Upper Saddle River, NJ: Prentice Hall.

Bowen, W. C., D. Bok, and G. Barkhart, 1999. "A Report Card on Diversity: Lessons for Business from Higher Education." *Harvard Business Review*, January–February.

Boylan, M. 2001. *Business Ethics*. Upper Saddle River, NJ: Prentice Hall.

Carver, K. A. and A. B. Livers. 2002. "Dear White Boss . . . " *Harvard Business Review*, November.

Cavanagh, G., D. Moberg, and M. Valasquez. 1995. "Making Business Ethics Practical." *Business Ethics Quarterly*, Vol. 5, No. 3, July.

Cragg W. and W. Woof. 2002. "The U.S. Foreign Corrupt Practices Act: A Study of Its Effectiveness." *Business and Society Review*, Spring.

Damian, G. and C. Stephen. 1996. *Business Ethics*. Oxford, UK: Oxford University Press.

Davis, M. 1998. *Thinking Like an Engineer*. New York: Oxford University Press.

Donaldson, T. 1984. *Case Studies in Business Ethics*. Englewood Cliffs, NJ: Prentice Hall.

Donaldson, T. 1996. "Values in Tension: Ethics Away from Home." *Harvard Business Review*, September–October.

Donaldson T. and T. W. Dunfee. 1999. "When Ethics Travel: The Promise and Peril of Global Business Ethics." *California Management Review*, Summer.

Elliot A. L. and R. J. Schroth. 2002. *How Companies Lie: Why Enron Is just the Tip of the Iceberg*. New York: Crown Business.

Garrett, T. M., R. C. Baumhart, T. V. Purcell, and P. Roets. 1968. *Cases in Business Ethics*. New York: Appleton-Century-Crofts.

Green, R. M. 1994. *The Ethical Managers*. New York: Macmillan College Publishing Company.

Harris, C. E. Jr., M. S. Pritchard, and M. J. Rabins. 2000. *Engineering Ethics: Concepts and Cases with CD-Rom*. 2d ed. Belmont, CA: Wadsworth.

Johnson, D. G. 1991. *Ethical Issues in Engineering*. Englewood Cliffs, NJ: Prentice Hall.

Johnson L. and B. Phillips. 2003. *Absolute Honesty: Building a Corporate Culture That Values Straight Talk and Rewards Integrity*. New York: AMACOM.

Kinni, T. 2003. "Words to Work By: Crafting Meaningful Corporate Ethics Statements." *Harvard Business Management Update*, No. C0301E.

Machan, T. R. (Editor). 1999. *Business Ethics in the Global Market*. Stanford, CA: Hoover Institution Press.

Martin M. W. and R. Schinzinger. 1996. *Ethics in Engineering*. 3d ed. New York: McGraw-Hill.

McHugh, F. 1998. *Business Ethics*. New York: Nichols Publishing Company.

Murphy, P. E. (Editor). 1998. *Eighty Exemplary Ethics Statements*. Notre Dame, IN: University of Notre Dame Press, 1998.

Nirvana, F. 1997. "Twelve Steps to Building a Best-Practice Ethics Program." *Workforce*, September.

Olsen, J. E. 2002. "Global Ethics and the Alien Tort Claims Act: A Summary of Three Cases within the Oil and Gas Industry." *Management Decision*, Vol. 40, No. 7, pp. 720–724.

Paine, L. S. 1997. *Cases in Leadership, Ethics, and Organizational Integrity*. Burr Ridge, IL: Richard L. Irwin.

Patterson J. and P. Kim. 1992. *The Day America Told the Truth*. New York: Dutton/Plume.

Pava, M. L. and P. Primeaux (Editors). 2002. *Re-imaging Business Ethics: Meaningful Solution for a Global Economy*. New York: JAI.

Pinkus, R. L. B., L. J. Shuman, N. P. Hummon, and H. Wolfe. 1997. *Engineering Ethics: Balancing Cost, Schedule and Risk, Lessons Learned from the Space Shuttle*. New York: Cambridge University Press.

Puffer, S and McCarthy, D. J. 1995. "Pinning the Common Ground in Russian and American Business Ethics." *California Management Review*, Vol. 57, No. 2, Winter, pp. 29–46.

Schminke, M. 1998. *Managerial Ethics: Moral Management of People and Processes*. Mahwah, NJ: Lawrence Erlbaum Associates.

Schwartz, M. S. 2002. "A Code of Ethics for Corporate Code of Ethics." *Journal of Business Ethics*, November–December, pp. 27–43.

Seebauer, E. G. and R. L. Barry. 2001. *Fundamentals of Ethics for Scientists and Engineers*. New York: Oxford University Press.

Seglin, J. L. 2000. "How to Make Tough Ethical Calls." *Harvard Management Update*, No. U0002D.

Thomas, R. R., Jr. 1990. "From Affirmative Actions to Affirming Diversity." *Harvard Business Review*, March–April.

Thomas, D. A. and R. J. Ely. 1996. "Making Differences Matter: A New Paradigm for Managing Diversity." *Harvard Business Review*, September–October.

Tichy, N. M. and A. R. McGill (Editors). 2003. *The Ethical Challenge: How to Lead with Unyielding Integrity*. San Francisco: Jossey-Bass.

Vaughan, D. 1996. *The Challenger Launch Decision: Risky Technology Culture and Deviance at NASA*. Chicago: University of Chicago Press.

Werhane, P. H. 1991. "Engineers and Management: The Challenge of the Challenger Incident." *Journal of Business Ethics*, Vol. 10, pp. 605–616.

Whitebeck, C. 1998. *Ethics in Engineering Practice and Research*. New York: Cambridge University Press.

Wilcox J. R. and L. Theodore. 1998. *Engineering and Environmental Ethics: A Case Study Approach*. New York: John Wiley.

Williams, O. F. (Editor). 2000. *Global Codes of Conduct: An Idea Whose Time Has Come*. Notre Dame, IN: University of Notre Dame Press.

World Bank. 2003. *The Little Green Book*. Washington, DC: The World Bank. (Also available on the Internet at *http://lnweb18.worldbank.org/ESSD/envext.nsf/44ByDocName/TheLittle-GreenDataBook2003FullDocument1MBPDF/$FILE/LittleGreenDataBook2003.pdf/*.)

11.9 QUESTIONS

11.1   Smith, an unemployed engineer who recently received certification as an engineer–intern from the State Board of Registration for Engineers and Land Surveyors, was seeking employment with a consulting firm. Engineer A, a principal with a large consulting firm, contacted Smith. After a long discussion concerning such matters as working conditions, salary, benefits, etc., Engineer A offered, and Smith accepted, a position with the firm. Thereafter, Smith canceled several other job interviews.

   Two days later, in a meeting with other principals of the firm, it was agreed by the firm's management (including Engineer A) that the vacancy should be filled by an engineering technician, not a graduate engineer. A week and a half later, Engineer A contacted Smith and rescinded the firm's offer.

Did the actions of Engineer A in his relations with Smith constitute unethical conduct? Why or why not?

(*Source*: Adopted from the files of National Society of Professional Engineers Board of Ethics Review).

11.2 As the business manager of your company, you are visiting several companies in Africa to promote new businesses. At the tail end of several successful rounds of negotiation, you are invited to attend a family banquet hosted by one of your potential business partners. This invitation represents a genuine sign of friendship and a commitment to good-faith business dealings in the future.

Would you be offended if the host wants you to pay for the food and drinks you enjoyed at the banquet when you depart? Explain your answer.

11.3 Cindy Jones, a chemical engineer with considerable experience in offset printing processes, was hired recently as an engineering supervisor by Company A. Before that, she had been working as a research chemist for a competing firm, Company B, where she invented a new formula and manufacturing process for press blankets. Jones's technique makes the blanket less prone to failure and produces better print quality. These press blankets are being marketed by Company B with great success.

When Jones was hired, there was no discussion about the new offset blanket during the interview. Jones was interested in moving into management; Company A had no openings available, whereas Company B was seeking to add managerial personnel with superior technical background.

One day soon after she had started her new job, Jones received an unexpected invitation to a staff meeting from the director of engineering. The meeting agenda focused on the formulas and manufacturing processes for offset blankets.

What should Cindy Jones do?*

11.4 Sara King is a member of the International Union of Operating Engineers. Through the union, she has secured a new job to operate a truck with an end loader at the XYZ Construction Company.

About two hours into her new job, the truck began to boil because of a leaky radiator. She stopped the truck and went to look for water.

About 100 feet ahead, Sara spotted a 5-gallon pail. On the way to get the pail, she happened to pass Joe Dow, an old union man, who was tending an air compressor. Joe Dow shouted, "Where are you going?" When Sara told him, Joe Dow replied, "I've got news for you. You are not going to get that pail. Understand? If you want to work on this job, you'd better start acting like a union worker, or I'll report you to the master mechanic. You'd better get back on the truck and wait for the foreman to get a couple of laborers to help you. Remember, if you stop your truck because of a boiling radiator and there's no pail within 40 feet of where you happen to stop, it's not your job to get a container."

Sara did not want any trouble. So she went back to the truck and waited for the foreman. It was two hours before the foreman came. In the meantime, seven other dump trucks and their drivers were idle. When the foreman finally did come, Sara explained the situation to him. The foreman said, "I'll get you a couple of laborers to draw some water." Sara explained further that she could easily have gotten the water herself earlier, but the

*Condensed and adopted from T. M. Garrett, R. C. Baumhart, T. V. Purcell, and P. Roets, *Cases in Business Ethics*, New York: Appleton-Century-Crofts, 1968. Copyright © 1968 Prentice-Hall, this material is used by permission of Prentice-Hall.

operator at the air compressor had told her to lay off. The foreman answered, "That's the way things are on this job. I don't want any trouble, so I do what the union people want."

Sara encountered other similar incidents as she continued on the job. The basic idea was always the same. Various craft unions decided on a lot of unreasonable restrictions that made a full day's work unproductive. The XYZ Construction Company had entered a cost-plus contract with the client, a steel company. So the more the employees loafed on the job and raised the cost, the more money XYZ Construction Company made. The steel company client was the one bearing the costs. In the long run, the consuming public ended up paying for this labor waste, which contributed to the increasing cost of steel.

Are there any ethics involved in this problem?*

11.5 Quick Meal is an international fast-food chain that operates in many countries. Company management wants to apply a uniform standard of business ethics, modeled after U.S. practices, to all of its stores worldwide.

When Quick Meal opened a new store in Country X, initially the local government cooperated fully. Then the government changed hands, and a corrupt group took over. Shortly thereafter, Quick Meal noticed that the general manager of the new store in Country X was providing free food and other concessions to governmental officials "under the table." The general manager was an American married to a local national. He was trying to get an "in" with the new government.

Store profits were still high, but Quick Meal decided to fire the general manager. The officials of the new government intervened and told Quick Meal to keep him or they would confiscate the local store. Quick Meal stuck to its decision and let the general manager go. The new government followed through with its threats and took away the local store.

A few years later, the government of Country X changed hands again. Although Quick Meal was promised some indemnity, there was still a considerable financial loss to the company. In spite of the fact that these losses were written off, some of the Quick Meal stockholders were unhappy with the company's decision regarding the general manager.

What should Quick Meal have done?*

11.6 Jane is a member of the board of directors of Power Company Z, which is considering the construction of a new power plant.

Coal-fired power plants emit sulfur dioxide into the atmosphere. Ambient air containing a high-concentration of sulfur dioxide is known to create acid rains, which damage crops and erode some metals (e.g., nickel and cooper). If number 2 oil is used as fuel instead, the sulfur dioxide emission of the power plant could be significantly reduced. However, replacing coal with oil will raise the fuel cost by about 20%.

Some directors believe that any increased costs would have to be reflected in higher prices. An increase in electricity price would create problems for the company. For example, the Public Utilities Commission may delay approving the proposal of rate increase. Consumers may react negatively to the price increase, which could hurt the company's public image.

Other directors are convinced that the company should not use methods which would increase expenses. They point out that diverse industries and motor vehicles are far more guilty of causing air pollution than the power industry. As one director put it, "Why

---

*Condensed and adopted from T. M. Garrett, R. C. Baumhart, T. V. Purcell, and P. Roets, *Cases in Business Ethics*, New York: Appleton-Century-Crofts, 1968. Copyright © 1968 Prentice-Hall, this material is used by permission of Prentice-Hall.

should we be leaders in this area when it is going to cost either stockholders or consumers a great deal of money?"

Jane knows that fuel represents only one-seventh of the total cost of generating and distributing electricity. She feels that the company has an obligation to protect public health as long as it can stay reasonably profitable. She believes further that the company should not allow purely business considerations to dominate its decisions in an area of such critical importance.

What do you think Jane should do?*

11.7 Company A recently bought a rock-crushing unit from Company B. This unit was expected to produce 750 tons of crushed rocks per hour, but has in practice been producing only 500 tons per hour.

Paul, president of Company A, complains to Gordon, sales manager of Company B, about the fact that he is now unable to fill contracts, which he secured based on the expected capacity of the machine. In some instances, he has been required to buy crushed rock—at retail prices—to satisfy his contract obligations. Furthermore, he is not able to repay the loan with the expected higher income from the increased production. Paul threatens to sue Company B, unless they return half of the purchase price of the equipment.

Frank, the foreman of Company A's new rock-crushing installation, and Elmer, the company's chief engineer, are not happy with the new equipment. However, they are not sure that company B is at fault. The contract for the new equipment specified that the unit should be able to crush 750 tons of properly graded limestone per hour. Company B had samples supplied by Company A and based its promise of performance on these tests. Paul had been using stone taken from several different company quarries. Both Frank and Elmer had objected to this since much of this stone was harder than that in the sample given to Company B. The equipment had not broken down, but it was not able to deliver the specified capacity.

Frank and Elmer discussed this matter and decided to present the problem to Paul. If Company B fought Paul's suit, Frank and Elmer would certainly be called on to testify. Moreover, they both felt that Company B has a right to know that Company A has been using a harder rock than that used in the tests.

Paul listened to Frank and Elmer, but was not convinced that he ought to inform Company B of the difference in rock hardness. Paul thought that the performance guarantee covered the crushing of rock for any and all of the company's quarries.

What course of action do you suggest for Company A?*

Web-Based Enablers for Engineering and Management

12.1 INTRODUCTION

The business environment in which all profit-seeking enterprises operate is changing rapidly due to advanced technologies and globalization. To remain competitive, companies need to adjust to trends developing in the 21st century economy.

Knowledge workers need to continuously update their skills on the job to remain marketable. The rapidly changing Web-enabled technologies, organizational design, and business models in the knowledge economy are important subjects for engineers to learn and master.

The specific objectives of Chapter 12 are manifold. The chapter discusses current examples of management and engineering functions that can be made more efficient by the use of Web-based technologies. It describes the need for various enterprises to become e-transformed. It also suggests specific areas where engineering managers are expected to add value in the e-transformation of engineering and technology enterprises.

In recent years, numerous progressive enterprises have been able to derive significant benefits from applying Web-based technologies. Table 12.1 summarizes some of these accomplishments.

Engineering managers are advised to keep abreast of new advancements and constantly add selected Web-based enablers to their personal files. Doing so will allow them to identify and use the pertinent tools to add value in a most efficient and timely manner.

TABLE 12.1 Benefits Derived from Using Web Technologies

Value	Specific Benefits	Company	Web-Based Applications
Cost Reduction	Cut $375 million annually from training budget and another $20 million in travel expenses	IBM	Collaborate and enhance skills on-line and use Web conferencing
	Save $25 million over 10 years to build and test new Stealth fighter plane	Lockheed Martin	Use Web to link 80 major suppliers
	Save $180 million via auctions in 2002 alone	GM	Electronic auctions to unload vehicles to auto dealers at the end of their leases
	Cut costs by $20 million annually over two years	GE	Automatically sift through data in commercial loans to weed out poor loan applicants
	Trim $100 million a year	Yellow	Analyze 60,000 orders daily to figure out the exact number of workers needed at 325 facilities
	Save $10 million annually in training costs and roll out new services more efficiently	Kinko's	Replace 51 employee training sites with e-learning in 2002
Customer Satisfaction	Promote customer satisfaction and increase sales by 15 percent	Gilbane	Manage large and complex construction projects over the Web
Productivity Gain	Increase sales productivity by 25 percent	Taylor-Made	Update sales and inventory figures via Web by 100 salespeople using handheld devices
Time Reduction	Decrease time to solve problems to months, instead of two to three years	Eli Lilly	Post problem on website and reward best ideas with cash prices
	Reduce gas station renovation time by 50 percent	BP America	Use Web to link builders, architects, and suppliers to revamp 10,000 BP America gas stations
Inventory Reduction	Cut inventory from 15 percent of sales to 12 percent	Whirlpool	Use Web to link factory with sales operation, suppliers, and key retail partners
	Scale down shoes order on speculation from 30 percent to 3 percent	Nike	Use Web to link Nike with manufacturing partners
	Cut inventory by 30 percent	Shiseido	Use Web to link sales outlets to sales staff and factory floor
Others	Provide more options, slash delivery time by one-third, and cut overstock.	BMW	Use Web to link dealership, suppliers, and factories and build to order most cars in Europe
	Whittle down steel cost by $17 per ton and the inventory and delivery times each by 50 percent	Posco	Use Web to plan steel production and track orders
	Increase output by 40 percent	Dell	Process orders from the Web and automate product assembly operations with robots

Source: Heather Green, "The Web Smart 50." *Business Week*, November 24, 2003.

12.2 WEB-BASED ENABLERS FOR ENTERPRISE MANAGEMENT

Customers, back-office workers, and supply-chain partners are the three major groups of stakeholders that form the backbone of a progressive enterprise. Their participation in the business activities of the enterprise makes it possible to generate and deliver customer value through integrated business applications. Enablers for Web-based management are software tools for managing these stakeholders. These tools are used for customer relationship management, enterprise integration and procurement, and supply-chain management.

Engineering managers need to become well versed in the use of such tools, which are becoming increasingly available in the marketplace.

12.2.1 Customer Relationship Management

Customer relationship management (CRM) focuses on the process of understanding and satisfying customers' present and future requirements. Engineers are usually engaged in some part of customer relationship management activities (Dyche 2002; Nykamp 2001; Burnett 2001). They are involved when the products of the company are of high technological content and require product customization, operations assistance, problem solving, application support, maintenance and repair assistance, spare parts, and other such services.

All customers are important, but some are more important than others, depending on the magnitude of their lifetime value (LTV). The LTV is the total revenue realized by the company from the purchases made by a customer over the period in which this customer continues to buy from the company. High-LTV customers should be managed differently from low-LTV customers.

It is known that a number of U.S. corporations have set up special corporate desks, each dedicated to specific major customers and manned by full-time employees who pay personal attention to all of the inquiries initiated by these high-LTV customers. At the other end of the spectrum, companies have set up extranets so that minor, low-LTV customers are encouraged to help themselves by accessing Web-based resources. Communications links with low-LTV customers may be via e-mail, text chat, Web-based callback request, and voice over the Internet. These extranets may contain product catalogs, installation and repair instructions, an FAQ (frequently asked questions) section, a news bulletin, discussion groups, chat rooms, users' groups, and other features. Surveys may be given to solicit customer feedback on product quality, desirable new features, service improvements, and other pertinent issues. Customer service centers are useful in coordinating the inquiries of minor- or medium-LTV customers who prefer human voices. An efficacius customer service center is expected to respond to customer inquiries within a very short period (say, roughly 20 minutes).

Customer relationship management has gained importance in recent years as a result of competition, globalization, the high cost of customer acquisition, and high customer turnover (Schmitt 2003). Surveys point out that U.S. corporations lose one-half of their customers every five years. The following statistics support the importance of this corporate function:

1. It costs five times more to sell to a new customer than to sell to an existing one.
2. A typical dissatisfied customer tells 8 to 10 people about the bad experience.

3. A company can boost its profit by 85 percent by increasing its annual customer retention by only 5 percent.

4. The odds of selling a product to a new customer are 15 percent; to an existing customer, 50 percent.

5. Seventy percent of complaining customers will do business with the company again if past service deficiencies are quickly resolved.

By managing customer relationships properly, an Internet-enhanced enterprise hopes to use existing relationships to increase revenue and apply integrated information to offer excellent customer service. The enterprise can also introduce more repeatable sales processes and procedures, conceive new value and instill loyalty, and implement a more proactive solution strategy. The key success factor for customer relationship management is the efficient integration of the following front-end activities:

1. **Sales.** Capture customer information, link sales and service functions, and cross-sell and up-sell other company products.

2. **Direct marketing.** Market new products and fulfill new orders.

3. **Customer service.** Access customer profiles in real time, create new value, instill loyalty, and deliver support functions.

4. **Call centers.** Respond with a human voice.

5. **Field service.** Deliver the needed maintenance or repair services and other special activities.

6. **Retention management.** Induce customer loyalty.

Charles Schwab has used a customer relationship management system developed by Siebel. Table 12.2 delineates the various dimensions of customer relationship management. Its specific functions are as follows:

1. Obtain detailed information about customers' behaviors, preferences, needs, and buying patterns.

2. Use that information to customize the relationship with the respective customer.

3. Apply that information to set prices, determine needs and desires of customers, and negotiate terms of purchase.

Fidelity Investments, Principal Financial Group, and several other mutual-fund companies have started using Web-based "e-401(k)" programs. The logic for doing so is compelling. When a customer places a phone call to a service center, it costs the company about $9 to respond. This cost is only $2 if the customer gets the required services from an automated voice-mail system. The cost is further lessened to 10 cents if the customer accesses a Web-based self-service system.

Autodesk Inc. is known to have applied an effective Web-enabled corporate help desk in providing high-tech help to customers. The company was able to whittle away the cost associated with internal phone-based support by 70 percent. Uniglobe.com uses a Web-enabled call center for travel services. A useful strategy is to build a self-service

TABLE 12.2   Dimensions of Customer Relationship Management

Dimension	Customer Relationship Management
Advertising	Provide information in response to specific customer inquiries
Targeting	Identify and respond to specific customer behaviors and preferences
Promotions and offering discounts	Tailored individually to the customer
Distribution channels	Direct or through intermediaries; customer's choice
Pricing of products and services	Negotiated with each customer
New product features	Created in response to customer demands
Measurements used to manage the customer relationship	Customer intention; total value of the individual customer relationship

knowledge base that allows customers to perform self-help tasks. Another strategy is to embed e-support capabilities into products. Hewlett Packard has done that.

iSky *(www.isky.com)* in Laurel, Maryland, provides an Internet-based real-time customer-care system that involves customer acquisition, retention, and optimization. It allows end customers to communicate with companies through any available media—phone, e-mail, and others. It offers self-help services by accessing the client's e-commerce site and using database and FAQ lookups, automated e-mail, and interactive voice response. It can move to human-assistance from an agent. The human-assisted customer care can be accessed through e-mail, text chat, voice-over Internet protocol, telephone callback, and the traditional teleservices support of phone or fax.

Waiting for customers to call for service is not what customer relationship management is all about. The company must also understand customers, anticipate their future needs, and take action to satisfy those needs. *Data mining* is a useful tool for profiling customers (by segmentation) and predicting their purchasing behaviors (Rud 2001; Kantardzic 2003; Pendharker 2003). Questions explored in data-mining analyses may include the following:

(a) Who are my best customers, and can I acquire more of them?

(b) How can I increase business with my best customers?

(c) Who are my worst customers, and should I salvage those relationships? If so, how?

(d) Why are some customers leaving?

(e) Are there other products or services that I can provide?

(f) How can I avoid acquiring unprofitable customers?

(g) What are the characteristics of prospective loyal customers?

(h) Which products and services are the prospective customers looking at or inquiring about?

(i) Which products has a single customer purchased most often?

(j) Are there customer complaints about product features, service quality, prices, or other issues?

Customer relationship management must determine which customers have the tendency to purchase what products and take proactive steps to promote these products to them (e.g., cross-selling and up-selling in order to derive increased business benefits).

A case in point is Caterpillar (Berry and Linoff 1999; Chopoorian, Witherell, Khalil, and Ahmed 2001). In 1997, this company set up a database of 100,000 client companies with known purchase records for fleets of truck engines from Caterpillar and its competitors. It then defined which client companies would have the best potential to buy from Caterpillar in the future by studying a predictive model created with data-mining techniques. Caterpillar initiated a direct-mail campaign to the resulting 10,000 to 20,000 client companies regarded as having high potential to purchase truck engines from Caterpillar. The resulting response rate to the direct mail campaign was 37 percent in 1999, compared with 16 percent in 1996 and the industrial average of only 3 to 5 percent for similar mailings. The conversion rate of non-Caterpillar clients was around 40 percent. As a consequence, Caterpillar's truck engine market share went up to 30.4 percent in 1999, from 23 percent in 1997.

In addition, the company built an intranet to post detailed information about each client company that included all past correspondences. This allowed its national and regional account managers to custom tailor communications packages on-line for individual client companies. All sales proposals for truck engine packages were Web based. Within 48 hours, a personalized package, complete with the salesperson's signature, could be sent. The Caterpillar case is a good example of how data mining can be applied to serve the business needs of the company.

Example 12.1.

In a recent issue of *Technology Review*, 10 emerging technologies that will affect the world profoundly were reviewed (Zacks, 2001). One of these 10 is data mining. Why is data mining so important?

Answer 12.1.

The 10 emerging technologies cited to have a profound impact on the world economy and on how we live and work are (1) brain–machine interface, (2) the flexible transistor, (3) data mining, (4) digital rights management, (5) biometrics, (6) natural-language processing, (7) microphotonics, (8) untangling code, (9) robot design, and (10) microfluidics.

Data mining refers to the analysis and nontrivial extraction of information from databases for the purpose of discovering new and valuable knowledge. The knowledge is in the form of patterns and rules, derived from relationships between data elements. Data mining emphasizes the use of computationally intensive machine learning tools in the analysis of large and complex databases. As processor speed is being increased constantly, difficult data-mining solutions are becoming attainable in a time frame that has become more and more practical to business decision makers (Berry and Linoff, 1997).

Implementing data mining usually involves the following stages: (1) setting specific goals for data-mining efforts, (2) data collection (data type), (3) data preparation (data segmentation, formatting, and quality control), (4) analysis and prediction (applying specific tools such as neural networks, decision trees, logistic expression, and visualization to build predictive models), and (5) measurement and feedback (applying models to bring forth results and taking corresponding business actions).

An important application of data mining is customer relationship management. Customer data are systematically analyzed to define customer preferences and characteristics so that products and services can be customized to grow better business results.

12.2.2 Enterprise Integration and Resource Planning

To achieve sustainable profitability, companies must plan, align, execute, and control all basic business processes. By streamlining all transactions and effecting data exchange between operations, companies can minimize inventories, shorten the cycle time to market, reduce costs, upgrade overall operational efficiency, and support customer service.

Enterprise resource-planning (ERP) software is designed to integrate into one information system all back-office operations, such as manufacturing, engineering, finance, distribution, procurement, decision support, knowledge management, marketing and sales, and other internal business functions (Meyerson 2002; Langenwalter 2000; O'Leary 2000). This building block is at the foundation of a progressive enterprise. The capability of enterprise resource planning and integration becomes particularly important for companies having numerous sites and multinational operations. Reviewed next are three specific components in enterprise integration systems.

A. **Procurement Resources Management.**   Procurement deals with the acquisition of typical operating resources, such as office supplies, services, travel, computer equipment and software, MRO (maintenance, repair, and operations) supplies, fuels, and training. Procurement activities involve identifying and evaluating vendors, selecting specific products, placing orders, and resolving problems and issues after purchase.

Traditional procurement follows a number of manually implemented and time-consuming steps that are subject to frequent human errors. These steps include the following:

1. Identify vendors who make engineered parts needed for the company's equipment and who satisfy a specific set of requirements defined by the company.
2. Exchange and review technical documents, such as drawings and product specifications.
3. Solicit bids and evaluate available offers.
4. Process the purchase orders within the company and obtain approvals.
5. Register orders and subsequent payment transactions into the company's back-office systems, involving accounts payable and the factory-receiving department.
6. Resolve human errors and transaction mistakes, such as wrong parts ordered, shipment dates entered incorrectly, and receiving locations improperly specified.

In contrast, Web-based procurement makes this process significantly more efficient, decreases administrative costs per item purchased, and is adaptable to the evolution of business models, because it can do the following:

1. Use on-line catalogs provided by suppliers.
2. Make use of an electronic data interchange (EDI) network to process documents, a network that currently has 400,000 members in active participation.

3. Utilize CAD software to communicate through drawings.
4. Monitor inventory levels and automate the procurement process.

On-line procurement is clearly the current trend. The objectives are to diminish order processing costs and cycle time, strengthen the corporate procurement capabilities, facilitate self-service, integrate with back-office systems, and raise the strategic importance of procurement.

In newer e-procurement models, reverse auction has been advanced as a tactic to procure well-defined items. Reverse auction works in favor of companies that enjoy a dominant buyer's position because of their large transaction volume. (See Section 9.6.2.) Companies engaged in reverse auction first prequalify selected vendors on the basis of product quality, financial strength, company reputation, and other factors. Then they publish the specifications of the goods needed and invite pre-qualified vendors to submit multiple rounds of bids before a predetermined date. The vendor with the lowest bid is awarded the supply contract. It is called "reverse auction" because the bid price is lowered in each consecutive round. The Internet greatly facilitates the process of reverse auction.

Microsoft is said to have spent millions of dollars developing MS Market, an on-line ordering system. It eliminates all paperwork; handles $1 billion in supplies; works in a distributed procurement environment; is particularly suitable for high-volume, low-dollar value transactions; processes orders linked to supply sources; and permits order status tracking. One important advantage of using such a procurement system is the elimination of rules and hidden procedures that greatly slow down the procurement process. MS Market scales down the purchase cycle from eight to three days and slices overhead costs by about 90 percent. It averages 100 orders and 6000 transactions per day.

Canadian Imperial Bank of Commerce (CIBC) has 1400 branch offices, 40,000 employees, and 14,000 suppliers. Using a software system marketed by Ariba, Inc. CIBC employees are able to buy on-line from preferred suppliers. CIBC decreased its procurement costs by 50 percent.

The following are additional examples of e-procurement software products available in the marketplace:

1. eProcurement (www.purchasingnet.com) is a Web-based procurement system produced by PurchasingNet (Lincroft, New Jersey). It accomplishes the following: (1) uses XML to support supplier catalogs, purchase orders, and invoices; (2) automates stock requisition by scanning inventory records and issuing requisition orders if preset inventory levels are met; (3) upgrades browser interface; and (4) integrates front and back office functionality. eProcurement operates on a Windows NT environment and is compatible with SQL and Oracle databases. Purchasing Net claims that the use of this system brought about savings of 5 to 15 percent in purchasing costs. A potentially useful application of the software is for the procurement of MRO (maintenance, repair and operations) parts, which are typically low-volume, high-variety, and noncore items.

2. ProcureIT 4.2.5 (www.procureit.com). Verian Technologies (Charlotte, North Carolina) markets this Web-based procurement system, which has a total of 14 modules. It automates purchasing with attachments (e.g., drawings and documents), processes orders, manages inventory, issues reports, performs audits, and

tracks the status of customer orders. The company claims that use of this software reduced its clients' purchase order costs by 80 percent (from $150 to $30 per order).

3. BuySite and MarketSite (www.commerceone.com) are marketed by Commerce One (Pleasanton, California). BuySite prepares requisition orders for office supplies, technical equipment, furniture, and other special items; gets approval; and issues purchase orders. MarketSite connects the company to qualified suppliers who post product catalogs on the Internet for easy selection. Schlumberger, an oil service company with $8.5 billion in sales and 60,000 employees, has used both the BuySite and MarketSite software products with success.

B. **Marketing and Sales Management.**   To service customers well, companies need to devise an easy ordering process, add value for the customers, increase sales force effectiveness, and coordinate team selling (Bishop 1998). The following processes are readily automated by the use of self-serve centers:

1. Sales configuration
2. Pricing
3. Quote and proposal generation
4. Commission and contract management
5. Order entry management
6. Product promotion

Whirlpool makes home appliances such as washers, dryers, dishwashers, dehumidifiers, microwave ovens, ranges, refrigerators, and air conditioners. It uses the Trilogy system (at *www.trilogy.com*) with success to integrate pricing management, product management, sales, commissions, promotions, contract management, and channel management. Other industrial users include Ford Motor Company, Xerox, Cooper Industries, Fisher Control International, Goodyear Tire & Rubber, Lucent Technologies and NCR.

C. **Decision Support and Knowledge Management.**   The functions of decision support and knowledge management focus on (1) business analysis, such as filtering, reporting, "what-if" analyses, forecasting, and risk analysis; (2) data capture and storage, including data warehousing, data mining, and query processing; (3) decision support, such as expert systems, case-based reasoning modules, and intelligent knowledge modules; and (4) data dissemination via proper means, including wireless or mobile front ends.

About 70 percent of Fortune 1000 companies have or will soon possess enterprise resources-planning (ERP) integration capabilities; among them are Coca-Cola, Cisco, Hershey Foods, Eli Lilly, Alcoa, and Compaq.

ERP vendors include Sap, Oracle, PeopleSoft, J. D. Edwards, and Baan.

Example 12.2.

What is a value chain? In what ways does the Internet affect the traditional value chain?

Answer 12.2.

A value chain designates the ways a firm organizes its business activities to sell products and services. The primary activities of a value chain include the following:

1. Identifying customers through market research and customer satisfaction surveys

2. Designing by concept research, product design, engineering, and test marketing

3. Purchasing materials and supplies by selecting vendors, ensuring quality, and keeping up timeliness of delivery

4. Manufacturing by fabrication, assembly, testing, and packaging

5. Marketing and selling through advertising, promoting, pricing, and monitoring sales and distribution channels

6. Delivering through warehousing, materials handling, and monitoring timeliness of delivery

7. Providing after-sales services and support through installation, testing, maintenance, repair, warranty replacement, replacement parts, and problem solving

The support activities of a value chain include the following:

A. Financing and administration—accounting, bill payment, borrowing, adherence to regulations, and compliance with laws

B. Human resource activities of recruiting, training, compensation, and benefits administration

C. Applying technology to streamline processes, facilitate communications, and eliminate wastes.

The Internet has exerted a strong influence on value chains. It caused disintermediation (e.g., the elimination of intermediate steps) and reintermediation (e.g., adding intermediate steps). (See Figure 12.1.)

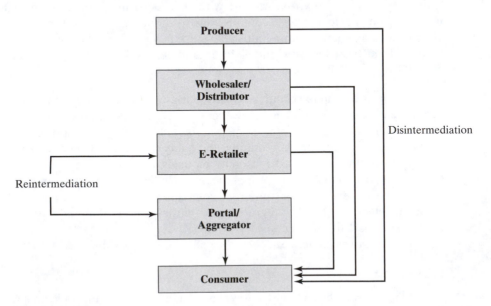

Figure 12.1 Disintermediation.

12.2.3 Supply-Chain Management

Supply-chain management involves the formation and maintenance of a complex network of relationships with business partners to the sources of, for example, raw materials; the manufacture and storage of intermediate products and finished goods; and the delivery of products. It focuses on the management among the partners of the flows of materials (tangible goods), information (demand forecast, order processing, and order status reporting), and finance (credit-card information, credit terms, payment schedule, and title ownership arrangements).

The goals of supply-chain management are to lessen the time to market, reduce the cost to distribute, and supply the right products at the right time (Franzelle 2002; Simchi-Levi, Kaminsky, and Simchi-Levi 2000; Ayers 2001). Achieving these challenging goals requires an integration of various operations, including market-demands forecasting; resources and capacity utilization subject to imposed constraints and real-time scheduling; and the optimization of multiple-objective functions involving cost, time, service, and quality. Web-based tools are available to assist in the management of such a complex supply chain. Examples of software products for supply-chain management include Advanced Planning and Optimization by *SAP*, and Rhythm by I2 Technologies.

The Visteon division of Ford makes chassis components and drive assemblies. It uses Rhythm by I2 Technologies to integrate intercompany processes. The benefits realized include a streamlined production schedule, the reduction of inventory by 15 percent, the removal of bottlenecks, and a shortened response time to orders.

There are a number of additional Web-based supply chain management enablers as illustrated in the next list.

A. EDI Linkage.   For automotive companies, the supply of engineered parts in a multiple-tiered vendor system requires a major managerial effort. Companies in other industries also recognize the efficiency gain and cost reduction that can be realized by streamlining the supply process.

Surveys indicate that over 70 percent of best-in-class companies have established electronic data interchange (EDI) systems with their suppliers to expedite communications, cut cycle time and error rates, standardize information transactions, and decrease purchasing expenditures. Users of EDI systems to manage their supply chains include General Motors, Ford, DaimlerChrysler, Sun Microsystems, Haggar Clothing, Wrangler, Xerox, and Honeywell.

B. mySAP ERP (*www.sap.com*).   mySAP ERP is a commercial, off-the-shelf enterprise resource-planning software package. It supports 25 specific functions related to accounting and finance, production planning and material management, human resources, and sales and distribution (Hernandez 1997; Williams 2000; Stengl and Ematinger 2001). It also supports order entry, facilitates supply-planning processes, and assists in customer service. Users of this system claim that the order entry process can be shortened from 18 days to 1 day and its financial close cycle from 8 days to 4 days.

Microsoft is said to have spent 10 months and $25 million to install mySAP ERP to manage its 25,000 employees and 50 subsidiaries. It claims to have realized an annual cost reduction of $18 million. Nestlé (maker of drinks, sweets, pharmaceuticals, and foods) uses the mySAP ERP application suite to manage its 498 factories, 210,000 employees, and operations in 69 countries.

Rhom & Haas company spent $300 million in 1999 to convert the company's 64 different information systems to SAS over a four-year period. By 2003, the company was able to trim 4000 workers and save $500 million in operating expenses. It projects an increase in the after-tax profit margin from 2 to 8 percent by 2005. Other companies that use mySAP ERP include Owens Corning, Colgate-Palmolive, and Warner-Lambert.

C. **Supply-Chain System** (*www.i2.com*). Toyota Motors buys over $1 billion a year in service parts, and its top 25 to 30 suppliers deliver about 80 percent of the total. It also plans to save $100 million in three years by increasing the level of ordering accuracy and inventory management. Toyota Motors uses the supply-chain system of i2, which consists of the optimization of fulfillment, logistics, production, revenue, profit, and spending, to build links with its dealers and suppliers. The specific goals are to (1) enable the suppliers to meet the company's annual demand forecast, which is updated quarterly for the needs of its 1500 dealers in the United States; (2) track delivery dates and lead time on-line and identify production bottlenecks; and (3) facilitate communications between the parties.

D. **SAS SRM** (*www.sas.com*). The minimum cost solution to a packaging, inventory, and distribution problem of United Sugars was found using the SAS SRM (supplier relationship management) solution. It involved 80 plants comprising packaging, storage, distribution, and transportation points; 250 sugar products; and 200 customers. The optimization model consists of 220,000 nodes, 1 million arcs, 3000 nonarc variables, and 26,000 linear side constraints. To generate solutions, the primal–dual predictor–corrector interior-point algorithm of the NETFLOW procedure is used, which performs the task in about 2.5 hours operating in a Windows NT environment.

The resulting optimal solution specifies the minimum cost schedule and defines the assets of packaging, distribution, and inventory that are needed to satisfy customer demand. The solution also allows United Sugars to reoptimize storage capacities, perform "what-if" analyses, improve inventory turns, handle customer sourcing requirements, and write suitable reports for users to view with Web browsers.

E. **Rail-ETA** (*www.rail-eta.com*). Kleinschmidt, Inc. (Deerfield, Illinois), offers this program as a real-time Web-based supply-chain solution for North American Railroad customers. It allows rail shippers and consignees to view rail shipment information and predict equipment arrivals. Specifically, its capabilities include the following:

1. Offering an in-transit visibility of inbound rail shipments via the Internet
2. Providing secure access
3. Conducting searches by purchase order or STCC (Standard Transportation Commodity Code)
4. Monitoring outbound shipments by plant and other criteria
5. Registering shipments from origin to destination
6. Estimating the time of arrival (ETA) based on historical data
7. Notifying customers by e-mail

Many companies apply enablers for Web-based management to derive efficiency and cost benefits. Other tools are being constantly utilized to augment the company's productivity and competitiveness. Engineering managers need to follow the literature closely so that they can offer the best advice in the selection and implementation of these advanced tools to their employers at any given time.

Example 12.3.

What is so special about Dell that caused this company to score increased sales in a world-wide-depressed PC market in recent years, while other major players like IBM, Hewlett Packard, and Gateway suffered significant declines in revenue?

Answer 12.3.

Dell should be examined from the standpoints of business models, operational excellence, and leadership qualities. At one point in time, Dell's stock had risen 29,600 percent to peak at $55 in 2000, although its price was around $35 in 2004 due to the general weakness of the U.S. economy in recent years.

Dell's success is due to several innovative business strategies (Magretta 1998; Sheridan 1999):

Direct Model. Dell invented the *direct model*, in which it sells *build-to-order* (BTO) computers directly to business or individual consumers. Bypassing so many intermediaries in the distribution channel significantly reduces costs. It also nurtures a close relationship with consumers, resulting in a better capture of the shifts in consumer needs, which are constantly being influenced by the rapid change in technologies.

Virtual Integration. Dell's success is also made possible by an excellent execution of its *virtual integration* strategy, whereby it coordinates all of its supply-chain partners to maximize value to its customers. Dell focuses on areas where value can be added, such as supply-chain management, product assembly, supplier selection for component changes and quality assurance, information flow, customer relationship management, segment gross margin, and cash-flow control.

In the rapidly-changing marketplace of computers, where technologies of processors, monitors, hard drives, other hardware components, and software all advance very swiftly, the old-fashioned business paradigm of "vertical integration" (i.e., we-have-to-develop-everything) is no longer applicable. Only buying, not building, and switching suppliers often will permit a company to introduce new products quickly enough to meet the customer's demands.

The competitive advantage of Dell lies in its capability of meeting customers' needs faster and more efficiently than any other company can.

Operational Excellence. Dell operates three assembly plants (Austin, Texas; Limerick, Ireland; and Penning, Malaysia) located near parts suppliers.

It forms partnerships with Sony for monitors, Airborne and UPS for logistics, unnamed third-party maintainers (TPMs) for after-sale service, plus others for additional computer components. Dell uses the in-depth expertise of TPMs to its advantage.

Dell focuses on the management of information flow between a few selected partners to increase the velocity of inventory movement by sharing its design database and methodologies and by maintaining an intranet (*www.premier.dell.com*). Suppliers are constantly informed of the changing marketplace demand. (See Figure 12.2.)

The company has established a Platinum Council to meet and talk with customers constantly. From the Dell Web page, customers can design computers by selecting various components. Once the order is submitted, Dell will ship out the completed product in three days. Ninety-seven percent of orders are shipped directly.

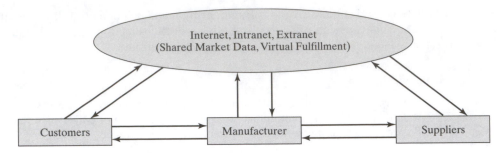

Figure 12.2 Dell model.

Dell realizes $10 million per day in sales from the Web page, although its major revenue comes from large corporations. Customers can track order status and receive technical support on-line by keying in a tag code for the purchased computer.

Dell's R&D is focused, not on creating new technologies, but on improving user experience, selecting relevant technologies, fostering ease of use, and facilitating cost reduction.

Dell achieves eight-day cash-flow conversion cycles, as customers pay by credit card and Dell pays suppliers via accounts payable.

In response to customer needs, Dell builds stability into its computers to make them "intergenerationally consistent" over numerous years.

In its plants, Dell uses one-person build cells to assemble computers and makes its own central-processing units. It maintains a six-day inventory for parts, but no finished-goods inventory.

Employees participate in profit sharing. Bonuses make up typically about 20 percent of a worker's total pay and are tied to certain key metrics, one of which is 15 percent year-to-year sales improvement.

Leadership Qualities. Dell believes that the most important leadership qualities in information economy are the following two:

1. An ability to process information and make decisions in real time to sense customer value shifts and to respond accordingly

2. Proficiency in managing a tightly controlled value chain formed by business partners contributing expertise. Integration of direct sales channels with the back-end supply chain is a key step.

12.3 WEB-BASED ENABLERS FOR ENGINEERING FUNCTIONS

Aside from the enablers for Web-based management already deliberated, there are Web-based enablers for engineering functions that may involve engineers and engineering managers more directly.

A large number of engineering and engineering-related functions will experience significant changes in the near future. In a descending order of engineering involvement, these functions include product design and development, project management, operations and manufacturing, engineering innovation, and maintenance. Engineering managers need to become familiar with all Web-based enablers related to these functions, as applying such knowledge will add value to their employers by achieving the goals of better operational efficiency and customer focus.

12.3.1 Web-Based Product Design and Development

Product design and development is a major function of engineering in companies for which products serve as the major source of revenue. Continuously offering new products is of vital importance to companies wishing to remain competitive in the long run. Engineers are the prime contributors to the design and development of technology-intensive products.

New products must satisfy one or more useful objectives (Otto and Wood 2001; Cross 2000; Ulrich and Eppinger 2000). They should solve a pressing customer problem. They should upgrade product features desired by customers, such as price, reliability, serviceability, availability, and usability. New products should incorporate functionality that adds value to customers. They should build on the company's core technologies and improve the individual consumer's quality of life or the future success of business customers. The principal goal of product design and development has always been to come up with products with suitable features that satisfy the needs of customers.

The marketplace has changed significantly in recent years. Customers have started to demand a faster response from producers. Competition has intensified due to the globalization of business activities. Market information has become widely available via the Internet to consumers and manufacturers alike; it demands a faster decision-making process by all companies involved in business. Moreover, the methodologies of product design and development have experienced major changes due to major advancements in computer hardware technologies, the communications capabilities of the Internet, and the sophistication of the modeling and analysis software technologies.

One survey suggests that product design may consume up to 80 percent of the total product cost. The Web-based engineering design tools reviewed in this section offer businesses the opportunity to significantly cut this cost.

In order to shorten time to market, out-innovate the competition, ensure product quality and reliability, and service customers properly, companies are tending to design and develop the core components of their products in-house. They outsource the remaining components to other business partners to utilize their special skills, technologies, and expertise in a mutually beneficial arrangement. Under these circumstances, product design and development may involve diverse team members of different technical disciplines, company orientations, geographical locations, engineering practices, and even cultural backgrounds.

For companies to succeed in such a challenging environment, attention must be given to several success factors, including communication, information sharing, collaboration, design verification, and team management.

A. Communication.   The first, most important step in product design and development is to understand the needs and wants of customers (Mello 2002). Missing or poorly defined customer requirements are the number-one reason for product failure. Proactive market research using such Internet-based tools as e-mail, discussion groups, chat rooms, and surveys via Web portals is essential to identify and define these requirements as key inputs for the conceptual design of a product. Once a preliminary product concept is defined, it can be tested via the Internet to gauge the response of potential customers. Doing so will allow further refinements to make the product more appealing before any significant investment is committed.

During the initial planning phase of product design and development, the Internet is particularly suitable as a research tool to invent concepts. Use may be made of semantic processing techniques with software such as Invention Machine (*www.invention-machine.com*) to generate concepts. The Internet can assist in checking for patentability of the product concept being considered. It can facilitate the location of licensable technologies for possible use in the product. Developers can use the Internet to review product standards (e.g., ASTM standards) available on the Web. Furthermore, the Internet can assist product developers in identifying vendor-supplied parts, including computer-aide design (CAD) models and specifications, for possible inclusion in the product. It is critically important for the team leader of the product design and development efforts to understand and manage the expectations of both upper management and all other parties involved.

While defining product concepts on the basis of customer feedback, the Internet can simultaneously be used to collect pricing data and major features of competitive products. Strategies of both target pricing (Section 9.6.3) and product positioning (Section 9.5.1) can then be applied to finalize the "bundle of benefits" to be offered.

Unimpeded communication between team members is critically important for the success of the product design and development efforts. Phones are inadequate for conveying the information encompassed by drawings and models. Faxes are often illegible. Face-to-face meetings are costly, and travel itself is time consuming. Also, a lack of visual focus allows the mind to wander. Web-enabled communications remove these hurdles. Web meetings (e.g., using NetMeeting software by Microsoft) may be held to facilitate audio and data transfer between team members via standard Web browsers. Delphi Automotive is said to have benefited from using NetMeeting, achieving the advantages of speed, travel cost savings, and a guaranteed "24/7" (24 hours per day, 7 days per week) approach to engineering. Besides e-mail, a Web-portal can be set up to facilitate interaction between team members and suppliers. Publishing the planning document for product design and development and the design guidelines on the Web will ensure access and promote understanding by all team members. Finally, communications tools via the Internet help overcome hurdles related to time zones.

B. Information Sharing.   Design data are usually proprietary in nature, and they must be managed with care. On the other hand, team members must have constant access to the data to iteratively refine a product design to meet the overall design goals of low cost, high quality, valuable features, and short time to market. Ready accessibility of design data also maximizes the use of available design expertise.

E-mail is convenient for short communications, but there are size limitations to sending attachments. The Intranet solves this concern because authorized team members can access detailed and restricted design data or other information regardless of the file size. A typical set of criteria used to select the right form of information sharing includes speed, format, security, ease of use, number of concurrent users, and compatibility with company's legacy systems.

Information of a graphics nature is usually needed for product design. A simple product design model can be transmitted by e-mail and reviewed over the phone. More complex, three-dimensional (3D) CAD models include geometrical data, material properties, surface finishes, grid data, and motion simulations of individual parts and complex assemblies. CAD models can be used to predict fit and performance, and to define optimal choices by performing "what–if" analyses. A single 3D CAD model can be several hundred megabytes in size. A model this large cannot be readily down-loaded to workstations because of limitations in transmission-line bandwidth and workstation disk space. The most common practice is to store the 3D CAD model in a server linked to a database. By using visualization software, a CAD file is then created on the server from the 3D CAD model. The visualization software renders a 3D CAD file with less complexity than the original model. It can work with file formats from a variety of CAD vendors (Brunet, Hoffmann, and Roller 2000).

C. **Collaboration.**   Product development is gradually changing from being a tra-ditional collocated process to a fast-paced, remote-global process. Requirements to shorten the product development cycle time, reduce product cost, and increase cus-tomer value tend to drive the product development process toward the "24/7" mode, while involving experts from geographical regions all around the world. To facilitate such a remote global product development process, it is necessary to have appropriate communications tools, the ability to exchange information accurately, the appropriate methods to coordinate projects, and the competency to attain productive collaboration between team members.

Many key product design activities are typically carried out concurrently and in-volve a number of team members with relevant expertise. For them to work collabora-tively, they need to view the same product model at the same time. To hold a collaborative product design session, team members use PC-based standard browsers to simultaneously access the server. Then they view, annotate, analyze, and update the displayed 3D CAD visualization file while communicating with each other via confer-ence calls or e-mails. The changes made to the design model are uploaded, and the revised product model is instantly available to all team members. It is this technology-enabled collaboration process that significantly shortens the product design time and upgrades the design quality (Creveling, Slutsky and Antis 2003) of new products.

Unigraphics (*www.unigraphics.com*) offers Product Vision, its visualization soft-ware, and the collaboration software of CoCreate (*www.cocreate.com*) is another tool in the same category.

As indicated, collaboration can clearly be enabled by currently available tech-nologies. However, the key success factor for truly useful product design collaboration resides in the participating team members themselves, in their willingness to contribute to the best of their abilities. Here lies one of the key roles of the team leader who man-ages the product design and development efforts. The team leader must apply his or her leadership skills to ensure congruence among the team members. The team leader needs to keep the team members motivated and focused on the goal, create mutual trust, and maintain a pleasant working environment so that everyone on the team wants to innovate for the overall success of the product development.

D. **Design Verification.**   The product design needs to be verified before the fin-ished product is made in large numbers. Traditionally, design verification is done by producing physical prototypes that are then tested for functionality, reliability, and serviceability. There are vendors who offer rapid prototyping services to fabricate physical product models of specific types directly from CAD files. An example of such a vendor is TDCi Solutions (*www.tdci.eu.com*), which offers a set of quality assurance (QA) solutions.

In recent years, large-scale analysis computer programs have become available, including finite element analysis (FEA) software such as DesignSpace from ANSYS, Inc. and computational fluid dynamics (CFD) software such as FIDAP from Fluent. In addition to these Internet-based applications service providers (ASPs) have started to offer "pay-as-you-use" services to verify CAD models with respect to operating stress, thermal, and flow conditions. These on-line analyses undoubtedly accelerate the product verification phase of the design and development effort.

E. **Configuration and Team Management.** Configuration management tools are available from CM Today (*www.cmtoday.com*) to assist the team leader in planning and controlling product configurations, managing product design changes, coordinating concurrent activities (such as procurement and parts manufacturing), modifying the supply-chain infrastructure in response to shifting customer requirements, solving problems, and achieving "right-first-time" product design and development execution.

The team leader must manage the efforts of the product design and development team in essentially the same manner as a project is managed. One of the key preparatory steps is to ready the team for those parts of the product design and development process executed by Web-based technology tools. Training everyone on the team to become familiar with the technical aspects of all of the tools is relatively easy. To get all of them to modify their old design practices, which may be individual centered, and to become comfortable in a group-oriented, collaborative endeavor may require strong prodding.

Conflict resolution is likely to be one of the key concerns for the team leader, as technically talented people can be inflexible in their views from time to time. Priority conflicts may arise, as team members from various business units, suppliers, and business partnerships may be pulled in different directions at the same time. Team members may have honest disagreements about design methodologies, material choices, and the availability of technology, depending on their prior experience and training. The diversity of team members may require the team leader to have a flexible leadership style. The team leader must exercise good managerial skills in planning, organizing, leading, and controlling to bring the product design and development process to a successful conclusion.

During the concurrent product design and development process, experts from marketing and sales, manufacturing, distribution, customer service, procurement, parts resources, financial control, and others may need to be brought in to contribute and innovate. All product data needs to be managed. This data may be in the various forms of CAD files, bills of material, parts numbers, process planning, technical publishing, word processing, tests, and product specifications, among others. The overall cost and schedule are tracked. In addition, design knowledge and lessons learned along the way must be preserved. At the conclusion of the product design and development process, a final report is issued.

The foregoing discussions have centered on what is needed to achieve success in product design and development. What follows is a group of Web-based engineering software tools that facilitate some of the tasks just described. Selected corporate users of these tools are also included.

Unigraphics (*www.unigraphics.com*).   The Product Vision software of Unigraphics, a division of EDS (Electronic Data Systems, Maryland Heights, MO), provides the visualization of 3D CAD models with browsers while the 3D module is stored on a Web-based server. General Motors has installed over 8600 licenses of Unigraphics, plus another 30,000 licenses of iMAN for vehicle development processes. The iMAN software automates the management of data and documents throughout the product life cycle. It focuses on information management, including product definition, variant configurations, changes, releases, workflow, and visualization; and it updates contents via iMAN portals. Currently, Northrop Grumman, BMW, Honda, Boeing, and Kodak are reported to also use this software product.

GS-Design (*www.gsdesign.com*).   Initially developed by Lockheed Martin for F-22 stealth fighters, GS-design is being marketed by Graphic Solutions (Milwaukee, WI) as a high-end, three-dimensional solid-modeling CAD system with a three-tier architecture (e.g., client/server/database). In this architecture, a thin client is downloaded to the user's PC. A centrally located Hewlett Packard application server hosts the GS design, and Hewlett Packard data servers host an Oracle database. This system permits collaborative design whereby geographically dispersed multiteam users can work concurrently. Each user may see the current state of the design in real time. The system provides secure central storage of CAD design data.

GS-Design is aerospace grade for ultralarge assemblies. It may be commercially subscribed on a "pay-as-you-use" basis over the Internet from CollabWare Corporation (Pittsburgh, Pennsylvania).

Alibre Design (*www.alibre.com*).   In 1995, Boeing, Ford, Kodak, MacNeal-Schwendler Corporation, and Structural Dynamics Research Corporation formed the Technology for Enterprise-Wide Engineering Consortium. In November 1999, this consortium released a Web-based, collaborative, mechanical product design tool which is being made available commercially, on a subscription basis, as Alibre Design.

The objective of this design software product is to facilitate ways to design, test, and market products by using the Internet as an enabling tool for real-time collaboration with suppliers and partners. These companies offer application service provider (ASP) service to facilitate cross-company product development efforts. It appears to be particularly useful to small or medium-size companies because there is no need for a major up-front investment for a computer-based design infrastructure.

CoCreate Software (*www.cocreate.com*).   This company, based in Fort Collins, Colorado, is a spin-off of Hewlett Packard. Its software OneSpace enables revisions of CAD models in real time over the Internet. Team members may see the master CAD model over a Web connection, make changes to the design geometry in real time, check design conflicts and modify the model on the spot over the Internet, and hold collaborative sessions with other members and outside suppliers. One Space supports all CAD systems offered by major vendors.

One Space is a client-server system in which displayed or viewed data are sent to each user's client, and the server processes the collaboration data. The server handles all functions of the model generation. The users can work on minimally equipped and minimally powered PCs, as only incremental model changes are sent over the Internet. It handles exact measurements of distance, length, diameter, angle, arc, area, volume, and other parameters from the viewed model. Native CAD data are imported and exported through CAD adapters, which are available for diverse CAD systems. Data can be displayed in drawings, DXF, DWG, PDF, Microsoft Office® documents, HPGL, TIFF and other formats. Data access, security, and data navigation capabilities are ensured. Finally, it has a Windows-like interface with pop-up menus and drag-and-drop capabilities. The software vendor claims that it has a high degree of flexibility.

By using OneSpace, Wavetek Wandel Goltermann has shortened evaluation cycles, ensured manufacturability by incorporating good ideas early in the product-design process, whittled down the number of needed prototypes by 50 percent, and increased customers' confidence in the company's ability to deliver products and updates on time. Other users of OneSpace include Hewlett Packard, Naval Undersea Warfare Center Division, and Aptec.

PTC software (*www.ptc.com*).   Parametric Technology Corporation (PTC) (Needham, Massachusetts) markets PTC software, which has been applied by Timberjack, a forest product company, to track 250,000 parts used in its forestry equipment in the past 50 years. PTC has also formulated an online parts manual and documentation to make product data more widely available to customers with browser-based interfaces.

The Pro/Engineer CAD package of PTC offers an electronic share space to present and refine ideas onscreen, allows access by standard Web browsers, and may be flexibly linked to various legacy database systems.

The software uses a standard Web browser and has 35 modules for all aspects of product development. Its users include DuPont, Royal Appliances, and Lockheed Martin Government Electronics Systems.

Navis Iengineer (*www.lucent.com*).   This software by Lucent supports several functions, including (1) dimensioning and measuring of edges, faces, arcs, points, objects, and assemblies; (2) checking for interference or collision of assemblies and parts; (3) marking up and annotating models; and (4) employing object simplification and real-time culling technologies to decrease the number of data points needed for the Web transfer to the user's rendering engines (i.e., PCs). The real-time culling technology determines which parts or objects to render to avoid submitting the unseen portions to users. Viewing the same model allows multiple team members to collaborate over the phone or via net conferencing.

The ipTeam (*www.nexprise.com*).   This *program manager* software, marketed by NexPrise, Inc. (Santa Clara, California), is a Web-based collaborative product development tool.

In a recent business article, Sikorsky Aircraft Corp, a United Technologies subsidiary, was reported to have applied ipTeam to design advanced helicopters. The design process involved 300 design team members, and this tool was able to track and record the chronology of the design process and offer version control and flexibility. Sikorsky claimed that the actual time for its preliminary design process was slashed by 30 percent because of ipTeam.

Webench (*www.national.com*).   With the motto "Going from concept to a working prototype in a few clicks," National Semiconductor (Santa Clara, California), originated the program Webench to design specific electronic products.

Webench chooses parts, composes and analyzes designs, provides virtual physical layouts with thermal modeling for populated board designs, conducts Web-based simulations, checks power supplies, evaluates microcontrollers, and builds prototypes to include bills of materials (BOM), parts numbers, and vendors with links for purchasing.

Miscellaneous.   Also available are tools that address other specific aspects of product design: (1) product simulation (*www.caciasl.com*), (2) decision analysis (*www.spotfire.com*), (3) risk analysis (*www.decisioneering.com*), and (4) 3D product engineering (*www.plm.3ds.com*), using a combination of four software modules (CATIA, DELMIA, ENOVIA, and SMARTEAM).

Application service providers (ASPs).   Some of the previously cited Web-based design tools are commercially available on a rental or lease basis from application service providers (ASPs). A Web portal is built to represent a single point of access to design information. Depending on the number of users and the length of the rental or lease contract, customization is usually possible. The rentable services include CAD model visualization, access to CAD modeling software, the use of the Web infrastructure—server, database, and thin clients—and communications enablers. It is particularly suitable for small to medium-size businesses to use ASPs to perform product design because of the low cost to apply high-end CAD and analysis packages. There is no need for a major up-front investment, or for maintenance and upgrade expenses. A suitable Web browser can access it from anywhere.

Besides the Web-based enablers, there are specific efforts that have been initiated by government and individual companies to advance Web-based product design capabilities. It is useful for engineers to stay informed of these developments. Examples of these development activities are illustrated next.

F. FIPER Program (*www.fiperproject.com*).   In 1999, a consortium led by Ohio Aerospace Institute initiated the development of open software called FIPER (Federated Intelligent Product EnviRonment Research). This program was funded for $21.5 million, 50 percent by U.S. National Institute of Standards and Technology (NIST) and 50 percent by six consortium members (e.g., Engineous Software, General Electric, Goodrich, Ohio Aerospace Institute, Parker Hannifin, and Stanford University).

The FIPER software, due to be completed in 2004, is for concurrent product design and manufacturing by team members in different locations and time zones. FIPER permits access to a centralized, real-time definition database that integrates design and analysis tools and automatically updates components geometry in response to cost, performance, and production issues. Through FIPER, manufacturers share knowledge to an extent not previously possible. Design expertise is collected and used to streamline the design process and minimize cycle time.

FIPER consists of the following building blocks:

1. **CAD tools**—Ansys, Nastran, and other mathematical modeling tools
2. **Design Drivers**—optimization, design of experiments, Monte Carlo Analysis, approximation methods, and DFSS tools
3. **Data Exchange**—table data, name–value, general text, Microsoft Excel, database, and PDM
4. **Cost Estimation**—cost models for materials and manufacturing processes
5. **CAD Tools**—models to depict the product design, provide interaction among team members, and ensure interoperability among various CAD systems
6. **KBE** (knowledge-based engineering)—source to preserve the best corporate practices and expertise related to product design
7. **Manufacturability**—model for manufacturing variation to reduce errors or damage during fabrication, assembly, or servicing
8. **Customer Tools**—to incorporate additional tools used by clients

It is anticipated that, when completed, this computer-based design and manufacturing system with Web-based capabilities will significantly advance the design and manufacturing technology and significantly cut the time and cost of product development, saving about $2.2 billion annually for U.S. manufacturers. This cost savings estimate is based on General Electric's calculation of engineering hours saved as a result of the FIPER capabilities for streamlining the design process and reducing testing requirements.

Engineering managers should keep themselves informed of the development status of this tool. FIPER is to be commercialized by Engineus Software (*http://www. engineus.com*). Its first version is expected to be issued in 2004.

G. **Ford Motor Company** (*www.ford.com*).   Ford spent $130 million to design C3P (a CAD-CAM-CAE product information management program) on an Extranet. Ford plans to add a browser-based tool to promote concurrent engineering on precision designs. The purpose is to eliminate product prototyping, permit instant design updates, and use the Collaborative Gateway of Proficiency, Inc., to convert proprietary CAD models into XML for subsequent transmission to browser-based clients.

Future advancements of Web-based tools for product design, with emphasis on speed, scope, and system compatibility, are anticipated. Computer hardware technologies are also expected to further raise processor speeds. Transmission bandwidth will continue to expand, causing data transfer between collaborating team members to become increasingly time efficient. Audiovisual capabilities will also improve, making long-distance human interactions more personal, as if participants were physically collocated. Collaborative design tools will become capable of broadly addressing all facets of the design process and facilitating refined interactions with CAD systems already in place.

Yet to be resolved are the following remaining hurdles: securing sensitive design data, ensuring network reliability and uptime, and resolving the design ownership issues for data and design brought about by extensive collaboration across supply chains and among companies.

12.3.2 Web-Based Project Management

Project management deals with the management of human and physical resources required to attain a well-defined project objective on time and within budget (Lewis 2002; Tinnirelle 2000). Companies may focus on civil construction, capital equipment, research and development, new product or process design, cost reduction, equipment retrofit, quality improvement, customer and market research, and other projects with high engineering content. Project team members may consist of company employees, contractors, suppliers, or customers.

Project management has become more difficult as business development scopes enlarge, the contents of projects expand beyond organizational boundaries, and project team members become geographically dispersed. Globalization and new business practices make project variables such as contractor interfaces and increased uncertainties for on-time and on-budget deliverables more complicated. A new collaborative work style demands a different project management paradigm—from "command-and-control" to quick collaboration—as well as an improved effectiveness that demands just-in-time development and Internet-speed deployment.

The best practices in project management focus on five key management issues: tasks and resources, cost, risk, communications, and knowledge.

A. Task and Resource Management.   The first and foremost step in managing a project is for the project manager to define the overall project objective and the associated budgetary, time, and other constraints, in consultation with company management. In the project-planning phase, tasks are formulated for implementation in parallel or in sequence, and their starting and completion dates are properly defined. Based on skills, experience, and availability, suitable people are assigned to these tasks, thereby establishing individual responsibilities and accountabilities. A Gantt chart or PERT (project evaluation and review technique) diagram is then used to graphically model the overall schedule and activities of the project. Tasks on the critical path are well defined, as these must be managed carefully. At that point, the anticipated completion date and budget for the project are iteratively finalized. The project manager initiates the project after having secured the approval of the project plan by company management and the commitment of all team members involved. The project manager also specifies the communications and data and document-transfer guidelines, as well as policies related to conflict resolution, problem solving, progress reporting, change management, and other issues. Training sessions are held to ensure that team members can apply all of the project management tools efficiently.

B. Cost Management.   Tools commonly used to manage costs include spreadsheet programs integrated with databases and automatic time entry systems. When incurred, actual expenses related to purchased items and capital assets are reported instantly. The total, cumulative cost incurred for the project is then determined and plotted graphically to compare with the planned budget. Progress reports are issued to keep company management informed of the project status. When necessary, remedial actions are initiated in a timely manner to minimize the impact of possible budgetary and other deviations.

C. **Risk Management.** The successful completion of diverse projects is inherently subject to uncertainties beyond the control of the project manager or team members. Examples of such uncertainties include weather conditions, labor strikes, priority shifts, delays in the delivery of materials, and the unavailability of key team members due to emergencies. Forward-thinking project managers should be aware of the relative impact of such uncertainties and be prepared to respond effectively.

Most project management software tools permit "what-if" analyses that determine the overall project outcome (e.g., schedule and budget) when one or more input variables (e.g., task duration, task expenditure, and task initiation date) deviate from the plan, while keeping all remaining input variables unchanged. The results of such analyses allow the project manager to understand the relative sensitivity of the overall project outcome to these deviations. By rank ordering potential plan deviations, the project manager is able to proactively devise suitable contingency plans that minimize adverse impacts on the overall project outcome. Past experience in handling similar projects is helpful in managing this type of project risk.

However, it should be mentioned that a "what-if" analysis has a fundamental deficiency: It varies only a small number of input variables while it keeps a very large number of remaining input variables constant. This deficiency is not consistent with how projects behave in the real world. Real-world projects have uncertainties associated with each and every project task, although these uncertainties are larger in some cases than in others. The overall project outcome model would be a truly realistic one if all input variables were allowed to vary within their respective ranges. The Monte Carlo simulation technique is capable of permitting all input variables of a risky project to vary simultaneously. (See Section 6.9.)

In recent years, companies in a large number of industries have started to apply the Monte Carlo simulation technique to manage risky projects on personal computers. All input variables may be modeled by a set of three values: the minimum, the most likely, and the maximum. The most likely value of an input variable corresponds to its value as originally planned in the conventional deterministic model, whereas the minimum and maximum values are the variable's possible lower and upper bounds, respectively. For each input variable, its lower and upper bound must be defined carefully to account for its range of potential variation. Each input variable is then modeled by a probability distribution function—Triangular, Gaussian, Beta, Poisson, or another—that portrays how the value of this input variable changes with probability and the highest probability being assigned to its most likely value. Upon initiation of the Monte Carlo simulation technique, all input variables are concurrently varied between their minimum and maximum values. The resulting project outcome as to schedule and budget is represented by a probability distribution function that has a set of minimum, most likely, and maximum values of its own.

By varying both the minimum and maximum values of specific input variables and the shapes of the distribution functions representing them, the project manager is able to ascertain their relative impacts on the overall project outcome. What the Monte Carlo simulation technique can deliver that the traditional "what–if" analysis cannot is two pieces of new information: (1) the absolute minimum and maximum values for the overall project outcome as to schedule and budget and (2) the probability (e.g., 80 percent) that the project will be completed within a specific budget and on or before a specific completion date. Both pieces of information are valuable to the project manager.

In fact, special cases are readily recovered by using the Monte Carlo simulation analysis. If all input variables are single valued (e.g., setting the minimum and maximum values of all input variables equal to their respective most likely values), then the overall project outcome will also be single valued. This result is equivalent to that produced by an ordinary deterministic project management tool. If one or more input variables are modeled by their three-valued sets while the majority of other input variables remain single valued, then the project outcome will correspond to that of a traditional "what-if" analysis.

Engineers and engineering managers need to familiarize themselves with advanced analysis techniques, such as the Monte Carlo simulation technique, which is also readily applicable to other engineering subjects involving uncertainties, such as system reliability, technology forecasting, and decision analysis.

D. Communication.   Communication between team members is a key factor for success in any project. As project numbers increase and project scope expands, team members may become diversified in composition and geographically dispersed. Team members communicate with one another through voice, text, video, data, and documents. Specific technologies must be actively applied to make communication between team members faster, easier, cheaper, of higher quality, and more productive. An open communication policy tends to encourage collaboration, which is essential for securing the success of any project.

E. Knowledge Management.   Managing knowledge within a project is a job of critical importance. At the end of a project, all team members usually gather to celebrate, give recognition to those who performed well, congratulate each other for the success achieved, and then immediately dash out to take on the next assignment. There may be debriefing meetings that summarize what went well and what did not go so well. In general, preserving the learning from each project is not always done systematically (Stokes 2001).

Oftentimes, there is a significant amount of knowledge worth preserving. Project management practices that yield positive results need to be documented. Mission-critical data are worth saving. Contingency plans that were successfully activated to eliminate unexpected risk reflect corporate expertise that needs to be recorded. Specific lessons learned from each project should be preserved.

Also important is the need to store useful knowledge in forms that make it readily reusable, such as searchable databases. Knowledge that is saved to add value, but not frequently applied, represents a waste of resources.

Software tools are available to facilitate various activities related to project management. What capabilities should one look for in a Web-based project management tool? Obviously, the project management tool must assist the project team, whose members may reside in distant locations and work in different time zones, in accomplishing the project objectives in time and within budget, while facilitating instant communications and data sharing. Friedlein (2001) offers one of the most comprehensive set of selection guidelines for Web-based project management tools, among those available in the literature. According to Friedlein, a good Web-based project management tool should be able to do the following tasks:

1. Offer easy access to the project status of assignments, costs, and the timeline in real time, via a browser-based client.
2. Have a portal for project-related information.
3. Import from and export to other project management tools.
4. Run programs on a Web server.
5. Manage contact information about every team member.
6. Map projects and assign tasks to team members (Gantt charts, histograms, and PERT charts).
7. Assign different views to allow each team member to see only the information and assignments pertinent to him or her.
8. Enable members to update tasks, add notes to tasks, attach documents to the list, and send e-mails to others.
9. Permit easy installation and use.
10. Offer sample templates for specific project topics that are readily modified to fit individual needs.
11. Possess the capability of tracking any number of tasks and milestones with associate links to Web sites, documents, and every other project.
12. Hold on-line, text-based chats for virtual meetings or threaded discussions for collaboration between team members.
13. Track resources allocation—load-leveling capabilities and membership rosters integrated with external directories.
14. Organize user interfaces.
15. Perform scenario analyses ("what–if").
16. Facilitate risk-adjusted modeling.
17. Document project outcome and learning.

Project management software tools have migrated from the infrastructure of centralized minicomputers, to PC desktops and local area networks (LAN), and, finally, to Web-based tools (Sommerhoff 2000). In the Web environment, project management tools involve

- E-mail
- Web pages to increase speed and availability of information
- GroupWare for project tracking and scheduling
- File transfer protocol (FTP) for transmission of documents and data
- Shared databases of multimedia formats
- Remote accessibility
- Integration with wireless and palm-top technologies

There are still two major constraints currently present in Web-based project management tools: bandwidth and infrastructure for multimedia transmission and data security.

Project management tools are advancing steadily. One website provides 120 questions to assist users by rank ordering commercially available project management tools (*www.checkspex.com*). Enumerated below are the major project management tools available to date:

A. **Primavera** (*www.primavera.com*). Primavera Systems, Inc. (Bala Cynwyd, Pennsylvania) has several modules useful for project management, including Project Planner to manage activities and resources, PrimeContract to manage costs, Evolve to manage the portfolio, and TeamPlay to promote the collaboration of participants.

   Toronto Transportation Commission (Toronto, Canada) successfully applied Project Planner by Primavera to manage 50 projects, each worth $2 million to $1 billion and having 100 to 600 activities, involving a total of 9000 people. A single master schedule is updated across a series of PCs linked to four servers on a LAN. Data are then converted to PERT charts and distributed to all involved. The software displays the links between each project and its associated activities. Communications are facilitated by e-mail, Web pages and other forms of Internet-based tools to keep everyone focused on the project goals. Resources to be deployed are categorized by discipline, workload, work hours available, and planned work to optimize resource loading for meeting project deadlines. Report generation may be in the form of paper, e-mail, and posting on TV or the Web. Data can be used for risk management purposes. The software allows project managers to keep in mind the big picture of all of their projects without getting lost in the details, but it also permits them to drill down for details if needed.

   Dick Corporation (Pittsburgh, Pennsylvania) utilized Primavera's Project Planner to manage the construction of a 115,000-square-foot addition to St. Francis Health System Hospital while the hospital was still operating. The software assisted in scheduling and coordinating all pertinent resources, including contractors, hospital personnel, and special equipment.

B. **Citadon CW** (San Francisco, California, *www.citadon.com*) provides an on-line collaboration workspace particularly suitable for managing construction projects in the building industry. Built on the Oracle database using Application Server 3.0 operating in a Windows NT environment, this software plans actions; tracks activities; coordinates the work of engineers, architects, and contractors; and automates various project components. Team members are connected to a password-protected Web site that contains project documentation, schedules, drawings, charts, and other project data files. Project managers have access to a messaging system and can monitor communications between team members.

   This software has been used by the San Francisco Public Works Department, Sun Microsystems, Stanford University, Fidelity Investment, and Charles Schwab.

C. **ProjectGrid** (Columbus, Ohio, *www.ProjectGrid.com*) provides websites on a subscription basis to host project-specific information. It functions as an on-line communications management tool for project management activities. It is capable of managing (1) financial information, (2) bid calendars, (3) project schedules, (4) work in progress, (5) site visit reports, (6) design drawings, (7) change orders, (8) contract invoices, (9) file downloads, (10) contract information, (11) administrative functions, (12) digital photo logs, and (13) other data. Team members access a password-secured website to receive updated project information.

This program may be particularly attractive to small and midsize companies.

D. **ProjecTrak** (*www.eden.com*) by *Eden Corporation* (Saratoga Springs, New York) is a leading Lotus Notes and Web-based project management tool composed of a number of customizable application modules, such as Project Manager, Bug Tracker, Help Desk, Asset Manager, Client Manager, and Software Designer. Since all modules are Web based, this tool allows users to access and upload data from anywhere. It focuses on creating competitive advantages for its users by efficiently and instantly sharing knowledge and information.

E. **ProjectTalk** (*www.projecttalk.com*) software by Meridian Systems (Folsom, California) has five modules to aid in project management: (1) document management, (2) cost control, (3) field management, (4) purchasing control, and (5) reports and queries. Through browsers, team members can access design drawings, job-site photos, project schedules, and over 400 reports. Prolog is linkable to "e-commerce" and "collaboration" modules of Projecttalk.com to facilitate Web-based communication and collaboration among team members. It is regarded as a standard in the AEC (architects, engineering, and construction) industry and is widely used by countless architects, engineers, general contractors, and public institutions.

F. **Worksolv** (*www.elite.com*). This project management tool by elite.com (Los Angles, California) connects team members with subcontractors and clients. It can be accessed via standard Web browsers and wireless Palm connections. It does full project budgeting, tracking, and reporting and claims to be easy to learn and convenient to use. It offers convenient access to subcontractors and clients and can be activated through subscription.

G. **Microsoft Project® 2002 and Project Central®** (*www.microsoft.com*). Project plans created on the desktop Microsoft Project 2002 program can be uploaded into Project Central, which resides on a Web server for storage in a linkable database (such as Microsoft SQL server, Oracle database, or others). Its capabilities include (1) creating Gantt charts for project and individual team members, (2) handling multiple project tasks and timelines, (3) producing summary reports, (4) updating project status and information by "time sheets," and (5) offering on-line help and tutorials (Chatfield and Johnson 2002).

Team members can access project information and view Gantt charts, multiple project tasks, timelines, and summary reports through a standard Web browser. Project status information can be added through time sheets that permit authorized team members to add tasks for themselves or others. Rules can be entered that, when triggered, will alert the project manager of unexpected deviations from the plan.

H. **Others.** Additional project management tools include (1) TeamCenter Project by Inovie Software (San Diego, California, *www.inovie.com*), and (2) Catalyst for Business Transformation by Cataligent Inc. (*www.cataligent.com*).

As evidenced by the preceding discussion, numerous Web-based project management tools are strong in some aspects and weak in others. There appears to be no single tool capable of satisfying all project management requirements. A growing trend in the field of project management is to lease advanced tools to manage specific projects. Some application service providers (ASPs) market Web-based project management tools. Leasing project management tools is more cost effective, meets the specific project needs better, and consumes less time for the company, compared with the option of either buying expensive software tools that have to be customized or building them in-house.

Application service providers are also capable of offering various additional services useful to the management of projects. These services include the following:

1. **Collaborative capabilities**—website hosting, e-mail, fax, Internet chat session, news group, message boards, scheduling on-line meetings, audio- and videoconferencing, electronic calendars, on-line paging, and access control (checking in and out)
2. **Work process management**—estimating and budgeting, task assignment and scheduling, job progress reports, purchase and procurement, accounting, and facilities management
3. **Work-flow capabilities**—document exchange (e.g., using PDF format), document management (review and editing of CAD models, photos, and drawings; control of file versions), data up- and download, approvals, change notification, transmittals, submittals, meeting minutes, and correspondence management

Example 12.4.

In the information age, every knowledge worker must know how to obtain useful information quickly from the Internet with a minimum amount of effort. As the amount of information continues to increase, information overload is a distinct possibility. Explain what one must know and do to surf the Internet effectively, find what one needs, and keep oneself current in selected topic-specific areas.

Answer 12.4.

The following are commonsense knowledge chunks related to extracting topic-specific knowledge from Internet-based information resources:

A. **Hardware and Software**

To access the Internet for an information search, one needs to have the use of a computer with the following capabilities:

(a) Modem (built-in or external) capable of dialing a local phone number
(b) Windows or other operating systems (Unix, etc.)
(c) Web browser software program (Microsoft Internet Explorer®, Netscape Navigator®) that enables the viewing of and searching for information
(d) Word-processing software
(e) Screen capture software that captures a photo, drawing, table, or graph contained in a document
(f) Internet connectivity

The user's computer must have a gateway to the Internet. A subscription-based Internet service provider (ISP) usually provides this gateway. An ISP-specific software program is then installed on the computer to check the user's identification, to automatically dial one of several local phone numbers, and to establish a gateway to the website maintained by the ISP. From this website, users may access other services (e.g., news reports, advertisements, etc.) offered by the ISP or may surf the Internet.

The speed of Internet access depends on the modem and the transmission line between the client (user's computer) and server (ISP's server computer). Examples of bandwidths include the following:

1. 14.4-kbps modem   14.4 kbps
2. 28.8-kbps modem   28.8 kbps
3. Digital (ISDN) phone line   128 kbps
4. T1 (dedicated connection)   1500 kbps
5. T3 (fiber optic backbone)   45,000 kbps

For students enrolled in a college-based degree program, computers with the aforementioned capabilities and built-in access to the Internet (including high-speed connectivity) are widely available.

B. **Access to the Internet**

Upon turning on an Internet-linked computer, the browser software icon is typically visible on the desktop. Double-click the browser software icon to start the browser program, type the URL of a database website or that of a search engine into the URL address bar, and click "go" to activate.

Digitized engineering, management, and scientific information (e.g., publications in journals, periodicals, newspapers, conference proceedings, and books) is typically searchable from databases. The outcome of a search from a database is a number of engineering, management, or scientific articles (either abstract or full text) related to the search topic. Examples of full-text databases include the following:

(a) ABI/Inform (business periodicals and academic journals of management)

(b) Business/Industry (900 business periodicals)

(c) Business Index ASAP (Infotrac—business, management, and trade publications, as well as local newspapers)

(d) Disclosure Global Access (U.S. corporate financial reports)

(e) Dow Jones Interactive (*The Wall Street Journal* and other business news)

(f) Gale Business Resources (448,000 U.S. and international companies)

(g) IEEE Xplore (100 IEEE journals and transactions, including those on engineering management)

One additional database worth mentioning is Compendex Plus (an engineering index), which offers abstracts, but no full-text documents.

Information of a commercial nature (e.g., companies or people offering expertise in specific domains, including products and services) is readily accessed through search engines. Search engines deliver a number of hyperlinked websites that offer additional information on products, services, and expertise related to the search topic.

C. **Conduct Search**

Upon reaching the home page of either a database or a search engine, the user enters a set of keywords to specify the search topic in the search box. Results produced are usually ranked on the basis of a combination of keyword matches, how often keywords appear in a document, and how often other sites link to a page. Some websites allows additional searches within the just-obtained search results to further refine the results, or "hits."

Experience has shown that it is advisable for the user to conduct topic-specific searches with several databases or search engines, as each one has access to only a limited number of resources.

The outcome of a search depends largely on how the keywords are entered. The use of Boolean operators (e.g., AND, OR, AND NOT, NEAR) and quotation marks around the key words will help narrow the search, as demonstrated in the example contained in Table 12.3.

TABLE 12.3   Internet-Based Search

Keywords	Number of Results	
	Alta Vista	Infotrac (ASAP)
Product Development Tools	129,693,425	123
Product Development	85,922,190	84,102
Product AND Development AND Tools	47,055	4,694
Product AND Development	439,369	92,559
"Product Development"	32,586	77,993
"Product Development Tools"	76	3

To attain optimal search results, users are advised to consult the search help files available in each database or search engine to understand how to perform advanced searches constrained by domain, time horizon, and other criteria.

Newer search techniques are being advanced that make use of natural language and navigation engines. The purpose is to produce faster and more relevant search results.

D. Evaluation and File Saving

Usually, most of the user's search results are at best only marginally related to the search topic. This initial list may contain thousands of documents. Additional manual editing is often required to narrow down the list further.

One way to do so is to scan the title of the document to gauge its relevance to the search topic and then open only those sites that appear to be relevant. Save the relevant information by selecting, copying, and pasting it into a word-processing file.

Some documents contain photos, tables graphs, or drawings that cannot be copied by the standard select copy and paste steps. However, these can easily be transferred into a word-processing file by using a screen capture software program.

E. Organization of Knowledge

The saved word-processing files should be organized according to search topic, database or search engine, and search dates. This should preserve source information and minimize the additional search efforts needed to remain current on the topics.

Read the saved documents to extract important knowledge chunks, and create summaries to review and update regularly as a step toward personal knowledge management.

12.3.3 Web-Based Manufacturing and Plant Operations

The operations of manufacturing plants, machine shops, chemical process plants, assembly plants, warehouses, distribution centers, customer contact centers, and others are important to any enterprise. Corporate and plant management closely monitors the relevant performance metrics of these operations, such as productivity, quality, efficiency, inventory turnover ratio, and asset utilization, because these metrics have a direct impact on the company's profitability (Janenko 2002; McClellan 2003; Gunasekaran 2001).

In recent years, Web-based tools have become increasingly available to assist in planning, benchmarking, analyzing, optimizing, controlling, monitoring, and correcting plant operations. The benefits realizable by using these tools include (1) instantly visualizing the plant status without time-zone or geographic restrictions, which makes operations transparent to relevant stakeholders; (2) reporting cost effectively; (3) planning and optimizing process and capacity efficiently; (4) managing inventory; (5) improving productivity; (6) enhancing enterprisewide integration; and (7) supporting decision making effectually. Some of these Web-based tools are as follows:

A. **CIMX** (*www.cimx.com*).   The CIMX software by CIMx Company (Milford, Ohio) manages data related to manufacturing. It links all plant floors of networked production facilities in a given supply chain and enables collaborative manufacturing without geographic constraints. CIMX devises, manages, and delivers work instructions, tooling information, quality standards, and best practices. It expedites planning and shares multimedia information (drawings, pictures, text, video, audio, and others) among participating groups.

Specifically, CIMX's CS/CAPP (client–server computer-aided process-planning) module supports the creation, management, and control of process plans, work instructions, and other manufacturing documents. Its CS/Tool (client–server tool and asset management) model assists in tool selection, tool ordering, crib functions, and gauge calibration. Its ShopBrowser model enables an electronic viewing of all manufacturing process instructions. Its QCP (quality control plan generation and management) module generates and manages control plans, using data from the QS-200 system and other data sources.

Users of CIMX's applications include GE Aircraft Engine, Boeing Space and Communications, Honeywell, and Ford. These users have reported the following benefits: (1) slashed planning time by 50 percent, (2) broadened the application of the company's best practices by 25 percent, (3) upgraded revision control by 10 to 40 percent, (4) reduced costs by 10 percent, (5) improved integration of control and process plans by 10 to 30 percent, and (6) sped up automatic creation of planning data and global sourcing by 35 percent.

B. **Agile PLM** (*www.agile.com*).   Agile Software Corp. (San Jose, California) markets a Web-based manufacturing system called Agile PLM that emphasizes communications by connecting original equipment manufacturers (OEMs) with suppliers and enables supply-chain partners to communicate over the Internet about new or changing product contents.

SCI (Huntsville, Alabama) uses its 2000 subscriptions to this software to manage 37 manufacturing facilities located in 37 countries.

C. **Web-based PLC** (*www.schneider-electric.com*).   Schneider Electric S.A. (North Andover, Massachusetts) was awarded a U.S. patent in May 1999 for a programmable logic control (PLC) linked to the Web by an Ethernet interface. The patent covers "transparent factory automation architecture," which establishes an open e-manufacturing environment with seamless communications between the manufacturer, the facility, and business systems.

The Web-based PLC couples an Ethernet interface module with a Web server on a PLC. The interface translates Ethernet, TCP/IP, and HTTP protocols used on the Internet into data recognizable by a PLC. The embedded Web server allows almost instantaneous transfer of data to and from factory floor equipment.

One very important application for this technology is the remote operation of process plants, such as air separation facilities, oil refineries, power plants, chemical-processing plants, and others.

D. **Embedded Systems.**   Embedded systems are very small, specialized computer systems stored on a single microprocessor. There are about 5 billion such smart and network-friendly microprocessors in use today. These embedded systems have widely known applications that, for example, (1) offer navigation assistance with the use of global positioning system (GPS) mechanisms, often integrated into airplanes and automobiles; (2) provide special services through personal digital assistants (PDA) to activate Web-enabled cellular phones; and (3) alert repair companies of product malfunctions.

It is conceivable that, over time, these embedded systems will find their way into the design of production equipment and systems, making remote operations increasingly more efficient, cost effective, and reliable.

E. **SAS** (*www.sas.com*).   SAS (Cary, North Carolina) produces a set of Web-based analysis programs that serve as decision support systems for various operations. Based on a network of Web servers, application servers, optimization servers, and data servers and built on multitier architecture, these programs have significant capabilities in data management, statistical analysis, data mining, and operations research.

For example, a user may use a Web browser to access a Web server to drive a specific application program hosted on an application server. The application server collects information to build a linear, dynamic, or mixed integer-optimization mathematical model to solve an operations research problem. A linked data server provides additional information needed for building this mathematical model. The application server initiates the optimization process performed by the relevant programs hosted on the optimization server. When it completes the optimization process, it sends an e-mail advising the user of the universal resource locator (URL) where solutions to the optimization problem are to be displayed. This multitier architecture allows the application server to have access to a number of other data servers that may host both the static data (such as product information, product specifications, product CAD models, and files) and dynamic data (such as the number of items in inventory, item prices, the quality of a manufactured product, and schedule deviations from the plan) related to a given application.

Sophisticated analysis programs, such as those marketed by SAS, offer valuable support to rapid decision-making processes performed by corporate management in relation to plant operations.

F. **RF Navigator EMMS** (*www.majuredata.com*). The RF Navigator EMMS software by Majure Data, Inc., (Rosewell, California) supports advanced inventory management capabilities (e.g., tracking, inspection hold/ release, shelf life, FIFO rotation, dock-shop concepts, and JIT concepts). Realizable benefits include (1) an increase in productivity by 20 to 40 percent by making the scheduling, planning, and distributing tasks more efficient; (2) the reduction of product recalls by verifying the components against the bill of materials; (3) the facilitation of managerial analysis, planning, resource allocation, and promotion of enterprisewide integration; (4) the refinement of customer service; (5) the management of inventory through accurate measurement, control, and analysis, and (6) the efficient reporting of production outputs.

Goodyear Tires and Rubber, Honeywell, Lucent Technologies, and Siemens Automotive use this tool.

G. **The "Build-to-Order" Concept.** A new manufacturing system concept is being tested in the U.S. automobile industry on the basis of the great success achieved by Dell in the PC industry. The Big Three automakers (GM, Ford, and DaimlerChrysler) are implementing a business plan to deliver customer-specifiable cars in 10 to 15 days. In a *build-to-order* (BTO) business model, cars are not made until they are ordered. It is anticipated that the supply-chain cost for BTO cars could be sliced by 14 percent (about $3,600 per car) when the new business model is fully implemented.

The current *build-to-stock* system suffers from high inventory and logistics costs and a lack of customer satisfaction. In addition, it does not allow automakers to gain insight into customers' future preferences and product demands because, in this current distribution mode, independent car dealers shield the carmakers from the customers. Furthermore, communication between automakers and Tier 3 and Tier 4 parts suppliers is poor, making it impossible for the automakers to change production plans in midseason even if new market information becomes available to them.

Several steps are needed for the Big Three to switch over to the BTO model: revamping core manufacturing, increasing supply-chain flexibility, and focusing on customer demand. Specifically, the auto manufacturers would need to

1. Communicate on-line with dealers who provide vehicle services to customers, and collect information such as customer preferences, buying trends, and customer's responses to marketing campaigns.
2. Use an internet portal such as Covisint, the e-marketplace, to process procurement, intensify collaboration between supply-chain members, and share business information such as demand forecasting, scenario planning, and others.
3. Apply enterprise resource planning systems to make inventory visible and production plans flexible.
4. Upgrade production systems from assembly lines (for mass assembly of cars) to assembly cells (to make individual BTO cars).
5. Simplify car design and combine components into subsystems that can be readily supplied by vendors.

A successful implementation of the build-to-order manufacturing strategy by the Big Three will have a profound impact on the manufacturing practices in all industries in the years to come (ReVelle 2002).

Example 12.5.

One of the newer strategies in manufacturing is *mass customization*, intended to market customized products by offering many choices to divergent customers and producing them at near mass-production costs. What does it take for a company to be successful in applying the mass-customization strategy?

Answer 12.5.

Mass customization has been cited as a new emphasis in manufacturing to strive for economies of scope (e.g., meeting customer's diversified needs by offering a large number of product features) with the economy of scale cost advantages of mass production. Two questions may be raised:

A. **Is Mass Customizing Worth Pursuing?** When dealing with mass customization, first the products must be identified that are projected to be profitable with sufficient market size and therefore worth mass customizing. It is a mistake to assume that every product will sell significantly better with more customized features. Certain products are more amenable to mass customization than others. Thus, the starting point is usually marketing. Specifically to be determined is which attributes add value to a large number of customers, among more choices offered to them?

   Research in literature suggests that good mass-market potential exists for product attributes for which there is a sharp difference in customer preferences. Examples are physical dimension, personal taste, and user requirements. The following examples describe some mass customization success stories:

1. Clothing—Levi Strauss offers 18 fabrics and five leg-opening designs and delivers customized jeans in eight days.
2. Home furnishings—Goodhome.com displays 45 decorated rooms and five distinctive lifestyles, with combinations of wallpaper, fabric, window treatments, and paint colors.
3. Footwear—Customatrix.com offers choices in materials, logos, colors, and other graphics to custom design athletic shoes.
4. Cosmetics and beauty care products—Reflect.com allows customers to choose from 300,000 customized product permutations and delivers the customized products in seven days.
5. Men's suits—Brooks Brothers uses a 3-D body-scanning technique to make custom suits in its made-to-measure program.
6. PC Computers—Dell pioneered its very own build-to-order program, delivering custom computers in three days.
7. Computer servers—Cisco Systems offers "Marketplace" for corporate customers to design combinations of routers, switches, and hubs.
8. Investments—Charles Schwab offers a mutual fund screener enabling customers to design personalized investment portfolios.
9. Toys—Mattel offers "My Design Barbie" for creating a toy friend for Barbie with specific hairstyle, hair color, complexion, and eye color.
10. Wireless phones—Point.com offers a "service plan locator" for designing wireless phones, service plans, and accessories.

   There are other products for which strong customer preferences could lead to mass market potential: sports equipment (golf clubs, bicycles, and skis), media (music, books, and photos), services (vacations, training, mortgages, and parties), and others (jewelry, accessories, gifts, and pet items).

Products with attributes that offer transitory values (e.g., novelty, curiosity, and surprise) may not be good candidates for mass customization. Examples of custom products in this category include (1) soap stamped with the user's name, (2) cookies glazed with the user's picture, and (3) bottled water with a custom taste.

Thus, determining whether a mass customization concept is worth doing from the business profitability standpoint is the first, very critical strategic planning step.

B. **Can Mass Customization Be Implemented Effectively?** Mass customization faces three critical challenges: (1) customer contact, (2) manufacturing technology, and (3) distribution logistics.

Companies implementing mass customization must be able to maintain contact with customers via the company's website, customer call centers, Extranets, or other means to elicit customer preferences. To assist customers in making choices in product attributes, a product configurator can be used to model the product and its attributes with parametric capabilities and modular options selection. An IT backbone system must be in place so that customer orders can be communicated to all partners in the supply chain to allow all of the required parts and subsystems to be made available.

The key questions related to manufacturing are (1) Which processes are needed to provide those product attributes that would add value to a large number of customers? (2) How flexibly can the current processes be modified to deliver the desirable product attributes? (3) Can the company redesign products to be more modular or configurable? and (4) How can new technologies be manipulated to bring about variety in attributes?

A flexible manufacturing process may be built on (1) standardization of parts and processes to eliminate duplication and to encourage design reuse and (2) modular design to assure interchangeability of parts and subsystems by creating precise and complete interface specifications, design checklists, and design rules. Supplier selection will become a key factor of success in mass customization. Collaboration across the supply chain is of vital importance. Upon receiving customer orders, parts or subsystems are manufactured from a base of preengineered modules and then are shipped out for assembling the final product. The "postponement" concept is practiced, in which the assembly of the final product is delayed as much as possible in the order-fulfillment cycle to retain the most flexibility in product customization. This is accomplished by positioning materials management and manufacturing activities of parts and subsystems further down the distribution channel. Assembly of the final product is typically done in locations close to customers so that rapid delivery of custom products is possible.

Distribution logistics refers to the physical means of delivering custom products to the right customers. Third-party logistics service providers are often involved in the global distribution and delivery of products. This step requires an extensive communication and coordination effort.

Mass customization requires a considerable amount of resources and technical talents for planning and implementation purposes. Suitable business partners must be engaged to form the required agile manufacturing system and supply chain to pursue the targeted mass market based on product customization.

Therefore, mass customization is an important new manufacturing strategy that can be profitably applied to some categories of products.

12.3.4 Web-Based Engineering Innovation

Continual innovation is the key to success for many enterprises. As a result of new products marketed and new patents issued, Microsoft, Cisco, IBM, DaimlerChrysler, General Electric, 3M, DuPont, Pfizer, DEKA corp., Rubbermaid, and other companies have been recognized as innovative.

Among the few best practices that foster corporate innovation are (a) recognizing innovation as a key to corporate survival in the long run, (b) committing sufficient resources to pursue innovative activities, (c) encouraging innovation from all organizational units in addition to R&D, and (d) selectively implementing innovative ideas to ensure business viability (Zairi 1999; Gulati, Sawhney, and Paoni 2003; Robert 1995).

Web-based tools are available to support the generation of innovative ideas. Support is available in the form of access to published information by using advanced search engines on the Internet. Support is achieved by exposing problems via Internet-based communications in order to engage more employees in finding solutions. Support is also obtained by using Web-based software tools to guide the generation of innovative ideas.

Invention Machine, Inc. (Boston, Massachusetts, *www.invention-machine.com*), offers solutions to problems related to intellectual assets management. To facilitate innovations, its software applies semantic processing technology that has been developed using the *theory of inventive problem solving* (TRIZ). The central principle of this theory is value analysis: To upgrade a product, place a value on each component, and then organize research according to those values. Invention Machine software searches deep into the Internet, analyzes large volumes of text, breaks down sentences, and reorganizes the contents into problems and solutions. Its Knowledgist program offers access to over 700 websites grouped by industry, reviews search results, sorts information according to user-defined parameters, and generates a structured index to allow an efficient review. Its Cobra program can "read" documents, pinpoint desired information, and extract only the needed sections for review.

Engineers using Invention Machine software are said to have significantly increased their innovative outputs, as measured by the number of original ideas advanced over a period of time. At least 500 companies are currently applying this software. Innovative products known to have been brought into being by the software include stronger parts for FormulaOne racing cars, better adhesives for Dow Corning, and a new type of filter for oil drilling equipment.

Assistance in generating innovation might be obtained from Dynamic Thinking (*www.dynamicthinking.com*), which specializes in providing a framework for understanding how ideas are formulated, promoted, and implemented.

The website yet2.com (*www.yet2.com*) offers a marketplace for licensable technologies. Performing patent searches on-line (*www.uspto.gov*) is already a common practice. Writing a provisional patent application can be assisted by using the software Patentwizard 2.0 (*www.patentwizard.com*). Filing a provisional patent application for a new innovation is usually advantageous, as it allows the inventor to test the market for the yet-to-be-patented idea or product for a period of one year before a formal patent application is submitted, thus preserving the inventor's original filing date.

These examples illustrates that some Web-based tools may augment the innovation process at any company. Indeed, the speed of access to information on the Internet will continue to increase. Search engines will become more powerful and intelligent. Knowledge extraction techniques will be further perfected. For the innovation process to be productive, however, these technological enablers are not enough. Employees have to be trained and effective methods must be incorporated into the workplace as a part of daily work life. Successful inventors must be properly recognized and adequately rewarded for innovation to flourish in a company.

12.3.5 Web-Based Maintenance

Once they become world-class manufacturing operations, firms realize the critical need to effectively maintain production facilities and systems. Operations productivity and product quality have their roots in maintenance functions, as poor maintenance raises downtime for manufacturing processes and malfunctioning equipment endangers product quality. It is thus vital for companies to integrate maintenance management as a part of the corporate strategy to ensure equipment availability and to control operations costs. To efficaciously executive a maintenance program requires a positive and uncompromising managerial attitude and commitment (Heizer and Render 1999).

Benchmarking with industrial companies via the Internet to define the best practices in maintenance is a good first step toward drafting a comprehensive plan. This plan may include activities such as maintenance targets, performance metrics, analysis and monitoring tools, and corrective measures. Establishing a *computerized maintenance management system* (CMMS) denotes another useful step. Internet-based communications tools allow personnel in companies with multiple plants and facilities to instantly update maintenance records and share the lessons they have learned. Also important are the following additional steps:

1. Establish teamwork to achieve understanding and cooperation among plant personnel when monitoring equipment use and tracking costs.
2. Conduct training to implement new maintenance concepts and practical methods to solve day-to-day problems.
3. Preserve and make available past corrective maintenance steps in order to facilitate their new applications.
4. Activate preventive maintenance to lessen the incidence of breakdown of critical equipment. For example, schedule maintenance at predetermined intervals—changing the engine oil every 3000 miles regardless of the actual chronological time elapsed.
5. Implement a predictive maintenance program based on past equipment mean-time-between-failure (MTBF) data and equipment reliability data available from industry sources, equipment vendors, and equipment or systems reliability analyses (e.g., Monte Carlo simulations).
6. Design and implement a proactive maintenance program that outlines steps to strengthen the preventive and predictive maintenance goals.

Such a Web-based maintenance system can be made accessible by Web browsers from anywhere. The system allows instant reporting of the maintenance status of all plants, facilitates quick decision making, implements means to correct deviations from the plan, minimizes operations cost increases, and preserves for future use maintenance lessons that have been learned.

12.4 E-TRANSFORMATION

E-transformation refers to the process by which old-economy companies apply e-business enablers to capture new value-creation opportunities in the rapidly changing marketplace.

Typically, old-economy companies are asset intensive. The companies respond slowly to changes in the marketplace and participate in a disconnected, nonnetworked economy. The knowledge-economy companies, on the other hand, are knowledge intensive, have short time-to-market capabilities, and compete in an extensively connected and fully networked economy (Tracy 2003).

The old-economy companies are production centered in that they first define what can be made using the resources at their disposal and then push the products they make onto the customers. They adopt an internal orientation. In contrast, the knowledge-economy companies produce what the customers need and want, where and when they want it. Knowledge-economy companies allow the customers to pull the products from their supply chains and adopt an external orientation.

A study offered by PricewaterhouseCoopers indicates that the types of capital emphasized by these two groups of companies are vastly different (as illustrated by Figure 12.3).

The harbingers of the knowledge economy are (1) new digital marketplaces, (2) the emerging role of alliances and hyperpartners, (3) new market indices, (4) the emergence of Internet protocol-based enterprise software providers, and (5) new cultural management and organizational expectations. It has been predicted that, by 2004–2005, the portion of the U.S. economy driven by electronic media (i.e., all B-to-B, B-to-C, and C-to-C transactions) will be greater than that driven by industrial companies.

What can an old-economy company gain by going through an e-transformation? Various old-economy companies want to become e-transformed so that they can service their customers faster, better, and cheaper. They also strive to attain additional business goals, such as (1) empowerment of customers, (2) promotion of trade, (3) improvement of business agility, (4) extension of enterprises in a virtual manner, (5) evolution and invention of products and services, and (6) development of new markets and customers.

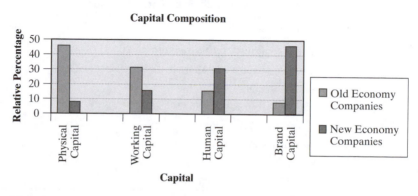

Figure 12.3 Capital composition of companies.

12.4.1 Strategies of E-Transformation

In order to serve customers faster, better, and cheaper, companies apply the following useful strategies:

1. Web-enable the customer interface by using websites and the Internet to facilitate order entry, distribute product information, offer frequently asked questions (FAQ's), configure products and processes, track orders, and deliver service and support.
2. Devise e-markets to promote procurement transactions.
3. Outsource all noncore operations that do not contribute to the company's competitive advantages. Noncore operations include human resources, distribution, customer support, manufacturing, financial accounting and management, procurement, and customer service.
4. Establish supply chains, manage inventory, and streamline communications channels (i.e., Intranets) to ensure information flow.
5. Enhance the operations efficiency of the company's back offices by using suitable enterprise integration software and hardware systems.
6. Realign the organizational structure to cut transaction costs and to speed up decision-making processes.

PricewaterhouseCoopers estimates that the benefits realizable by companies going through a successful e-transformation in the oil and gas industry are indeed significant (Paisie 2001). (See Table 12.4.)

It has been predicted that the market value of an old-economy company could more than double through an e-transformation.

The reduction of WACC in the example in Table 12.4 is the result of the company's ability to access cheaper capital in the equity markets because of the speed of information and streamlining of the cash conversion cycle. An e-transformed company is able to more quickly provide information to the public on strengthened

TABLE 12.4 Value Addition by E-Transformation

E-Transformation Steps	Value
Value of company yet to be transformed	1
Reduction of weighted-average cost of capital (WACC)	0.39
Decapitalization	0.34
Enterprise resources planning efficiency	0.12
Supply-chain optimization	0.12
Utilization of e-markets	0.12
Outsourcing	0.03
Value of transformed company	**2.12**

operations results and on the modifications to the traditional revenue streams and cost structures.

The decapitalization benefit in Table 12.4 refers to the reduction of capital utilization by divesting noncompetitive assets and operations. Doing this will allow the company to respond to changing market conditions with greater speed and flexibility and to capture future growth opportunities.

Improved enterprise resources planning tends to lower company overhead and minimize other internal transaction costs. Furthermore, the increased speed of information flow within the company allows "virtual daily close" and fast management decision making that leads to a reduction of risks to investors.

Supply-chain optimization improves assets utilization, increases operations margin, and scales down working capital requirements.

E-markets such as e-procurement and remote problem solving with MRO (maintenance, repair, and operation) websites can help whittle down life-cycle costs, increase EBIT (earning before interest and tax) margins, and reduce working capital.

Outsourcing decreases selling costs as well as general and administrative costs. Companies may typically outsource such operations as human resources, purchasing, financial operations, maintenance, and manufacturing to focus on core competencies.

As noted by Chang (2004), those value-addition activities studied by Paisie in Table 13-4 are likely to enhance primarily the operational excellence of a company. Competitors may readily apply similar tools to catch up. In order to build strategic differentiation of long-term and sustainable nature, new value-addition opportunities must be explored. Examples of such opportunities include creating novel business ventures based on corporate proprietary know-how and the development and deployment of global knowledge shops. It is estimated that the total market value of an old-economy company could easily triple as the result of capturing these new opportunities.

12.4.2 Best Practices of e-Transformation

Best practices of e-transformation are those known to have added value to companies that have already undertaken an e-transformation. According to a combination of several studies, the best practices of e-transformation include the following:

1. **Redesign the business model.** Companies should commit to innovation and be ready to compete against new entrants. Consider external resources such as alliances, partnerships, and outsourcing to achieve speed, flexibility, and cost goals.

2. **Get outside help.** Traditional business management mentalities may not be adequate for today's business. Consider bringing in new, competent leaders to head up the new initiatives. General Electric is known to have paired up its traditional business leaders with e-leaders who have demonstrated e-transformation capabilities.

3. **Form alliances and partnerships.** For activities such as sales support, marketing programs, and client deliveries, consider joint ventures or alliances.

4. **Initiate programs across business units.** Form cross-functional teams and focus on key broad issues to prevent e-transformation programs from becoming "siloed" or departmentalized.

5. **Formulate scorecards.** Establish clearly understandable metrics to measure progress on a regular basis. (See Section 7.6, "Balanced Scorecard.")

12.4.3 Critical Success Factors for e-Transformation

Critical success factors are those which have a profound impact on the achievement of success in e-transformation. The following summarizes a few such factors:

A. **Management commitment.** Without a firm commitment from the top of the company, no e-transformation program will succeed. Commitment means that people are dedicated, a budget is in place, and management is willing to change. Management demonstrates this by allocating a reasonable amount of time to experiment and showing its readiness and tolerance to allow failures. Risk taking must be accepted by all concerned.

B. **Champion.** A well-respected champion whose vision is clearly communicated and respected by all stakeholders must spearhead the transformation efforts. The champion must have a well-structured project plan approved and supported by top management.

C. **Customer focus.** The transformation efforts must be centered on creating better, faster, and cheaper products for customers. Inputs from customers are critically important. Consider offering an integrated experience (e.g., combinations of products and services) to satisfy customers' future needs.

D. **Collaboration with suppliers.** Use technology enablers to form partnerships and alliances. It is appropriate to collaborate with competitors to ensure supplies and then to compete against them for customers in order to attain profitability and a market-share position.

E. **Speed.** The implementation team must have a sense of urgency in carrying out project activities.

F. **Decision making.** Decisions need to be made swiftly and with employee empowerment.

G. **Focus on scalability: Think big.** This is to ready the company to capture future growth opportunities.

12.4.4 Specific e-Transformation Cases

In recent years, there have been a number of old-economy companies that have attempted e-transformations with considerable success.

A. **Cemex.**   Cemex S.A.DE C.V. (Monterrey, Mexico) is an asset-intensive, low-efficiency cement company. Cemex's business suffered from unpredictability of demand, which is rather common in the cement industry, as customers frequently change their orders just before shipment. Furthermore, there were 8000 grades of mixed concrete made by six plants. Cemex's communications problems were serious.

Cemex initiated a strategy of e-transformation to focus on managing information instead of assets, in order to resolve delivery and production difficulties that were impeding company profitability (Anonymous 2001). Specifically, the company took the following steps:

1. Linked delivery trucks to a global positioning satellite system so that dispatchers could locate all moving trucks and determine each truck's direction, speed, and cement contents. This made instant truck rerouting and last-minute delivery changes possible. It also minimized the average required lead time for delivering premixed and ready-to-pool cements to a work site from 3 hours to 20 minutes. The company used 35 percent fewer trucks and slashed costs in fuel, maintenance, and payroll.

2. Guaranteed the delivery of a perishable commodity product within minutes. In return, the company charged customers a premium price for this new value.

3. Used the Internet for customers, distributors, and suppliers to place and track orders and payment status. Time-consuming direct communications with customers were eliminated. The employees, so freed, were able to focus on improving services to build stronger customer relationships.

4. Accessed the company's operational details by management within 24 hours, compared with 30 days by the old system. Replaced investing in assets (such as trucks, ships, and staff) with building better managed information. The company's profits increased dramatically within a short period.

Recently, Cemex formed a separate business unit, Neoris, to actively market its e-transformation experience to other cement companies to generate additional corporate revenues.

B. GE Capital.   General Electric Capital offers equipment leases and financial services to customers. It initiated e-transformation activities in three specific areas: buying, making, and selling.

The company buys $5.5 billion worth of goods and services each year. A Web-enabled program was implemented to realize huge discounts and processing efficiency by centralizing procurement and narrowing specifications. The negotiation time was decreased by 50 percent. Staff travel time related to procurement was cut down significantly. The processing cost was also reduced from $50 to $5 per transaction.

In the production area, GE Capital created an on-line catalog to facilitate customer self-service. Customer call-center staff was drastically cut from 63 people to 13. Operation expenses were decreased by 20 to 40 percent in each business unit. Reducing process steps saved $30 million over two years.

In order to reach a wider circle of customers, Web portals were built. Customer feedback indicated that they preferred the speed and "24/7" convenience offered by the portals. Specifically, GE Plastics set up a website for all of its plastic products to keep information current, promote sales via the Internet, and reduce customer inquiries. Its product database is built on an Oracle system, which is accessible by Web browsers.

GE Capital plans to continue applying Web-enabled technologies to some of its internal functions, such as human resources, payroll, and related support activities.

C. **General Motors.**   General Motors is known to have taken a number of e-transformation steps to improve competitiveness.

One of the major obstacles GM had in product design was the variety of legacy systems and application and hardware platforms that impeded data sharing. After having standardized to a single CAD platform (e.g., Unigraphics and supercomputers), the GM vehicle design cycle time was slashed by more than 60 percent.

In the area of procurement, GM conceived an on-line Covisint procurement alliance with Commerce One, a software service provider.

For human resources management, GM applied PeopleSoft and realized a cost reduction of 50 percent. Employees can now conveniently access benefits and other human-resources information on-line.

Example 12.6.

Describe some well-known e-business models that have been proposed and pursued in recent years by the dot.com companies.

Answer 12.6.

Business models are methods of doing business by which an enterprise becomes and remains profitable. Typically, firms combine several models to pursue a Web strategy. E-business models are generally of the following specific types:

A. **Brokerage model.** A website is set up to (1) facilitate transactions (e.g., B2B, B2C, C2C), (2) charge a fee per transaction, (3) perform buy–sell fulfillment functions, (4) report results, and (5) provide access to information needed by the parties involved. (See Figure 12.4.)

Examples of such sites are the following:

1. E-Trade (financial securities fulfillment)
2. Expedia (travel services)
3. CarsDirect (autos)
4. MySimon (search assistance for the best price)
5. Priceline (demand collection, reverse auction, name your price)
6. ChemConnects (chemicals market exchange)
7. Mercata (aggregate buyers to benefit from volume-based purchasing)
8. DigitalMarket (connecting preferred distributors and buyers)

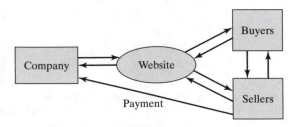

Figure 12.4   Brokerage Model.

(i) Yahoo's Store (set up merchants in a virtual mall for buyers to reach)

(ii) Amazon ZShops (transaction services, track orders, billing, and collection services)

B. **Informediary model.** The website offers information of value to individuals and businesses, provides free access to the site and free e-mails for personal communications, and collects users' profiles for sales. (See Figure 12.5.) The principal income is derived from the sales of consumer profiles to businesses.
Examples of such sites are the following:

1. NYTimes (uses registration to collect user information, free view of contents)
2. Emachines (free hardware in exchange for detailed information about Web surfing and purchasing habits)
3. Gomez (provides consumers with useful information about websites)
4. Netzero (provides free Internet access)

C. **Affiliate model.** The company sets up a network of affiliate merchants, each having specific products to sell. The website offers free information to consumers, who can point and click to reach a specific affiliate merchant. The site provides fulfillment services. The affiliate merchant pays a fee per concluded purchase. This is a pay-for-performance model. (See Figure 12.6.) Examples of such sites include (1) BeFree, (2) Affiliateworld, and (3) I-Revenue.net.

D. **Merchant/manufacturer model.** The company promotes goods (self-produced or vendor-supplied) made by partners for sale through its website, which offers information

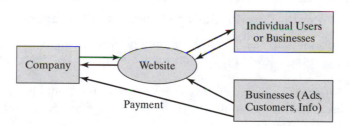

Figure 12.5   Informediary Model.

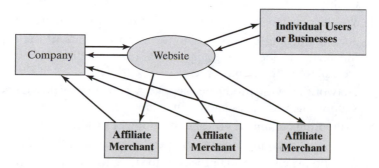

Figure 12.6   Affiliate Model.

access to customers and facilitates all on-line fulfillment functions. The company may also maintain brick-and-mortar storefronts to serve local customers ("surf and turf"). Examples of such sites include the following:

(a) Gap

(b) Lands' End

(c) Virtual Merchants (Amazon, EToys)

(d) Intel

(e) Apple

(f) Gateway

Under certain circumstances, channel conflicts may develop for businesses that maintain both on-line and physical storefronts.

E. **Advertising model.** The company establishes a website that offers contents free to consumers (e.g., news, search capabilities, chat rooms, stock quotes, forums, articles on specific topics, entertainment, etc.). The company's principal income is advertisements placed and paid for by merchants or businesses. In this model, the company works more or less like a radio or TV station; hence, it is also referred to as a broadcaster model. Examples of such sites include Google, AltaVista, and other search engines.

F. **Subscription model.** Users pay a subscription fee to access the website for premium news or specific reports, although some news is free. Examples of companies maintaining such sites include *The Wall Street Journal, Consumer Reports*, and AOL/Time-Warner.

G. **Others.** Engineering managers are advised to watch out for new models being constantly advanced. Diverse businesses may use more than one model in combinations to achieve Web success.

12.4.5 Potential Contributions by Engineering Managers

E-transformation is a critically important process for many enterprises whose ability to compete in the knowledge economy will depend on the degree of success achieved in this process.

Engineering managers have a lot to offer and should become active contributors to many transformation efforts. Examples of an engineering manager's contributions include the following:

1. Adopting new Web-based technology enablers to gain advantages in speed and quality

2. Web-enabling back-office engineering processes to increase productivity

3. Assembling and understanding the e-transformation best practices available in industry, and identifying applicable ways of effecting e-transformation in the workplace of their employers

4. Networking with producers, suppliers, and service providers to form potential supply-chain alliances and partnerships

5. Promoting the innovation and invention of products and services

6. Applying emerging technologies to confer competitive advantages

7. Simplifying engineering processes to strive for "doing it right the first time"

8. Identifying and applying best practices in engineering to foster value addition

9. Increasing production and manufacturing agility to slice costs and to shorten production cycle time

10. Implementing knowledge management and data mining to intensify the company's competitive strengths

11. Selecting pertinent performance metrics to monitor engineering activities

12. Optimizing distribution to realize logistic advantages

13. Managing concurrent multifunctional teams to serve customers better, cheaper, and faster

14. Offering innovative problem solving

15. Leading the company through changes

12.5 CONCLUSION

As the United States and world economies become increasingly digital, e-business, including e-commerce, is expected to play an increasingly important role in our society.

A digital economy places a strong demand on the speed with which transactions are fulfilled and on the variability of customers' needs. To be successful in the marketplace, businesses need to modify their organizational designs. They need to manage front-end customer relations and integrate back-end operations. They also need to ensure the required support of an efficient supply chain. Thus, smart enterprises combine supply- and demand-chain technologies with innovative business models designed to seamlessly integrate product development, manufacturing, distribution, marketing, and sales.

The flow of information is deemed to be the key management process of the future and must be accurately executed to maximize the value of products and services offered to customers.

Businesses in a digital economy tend to favor buying over building, in order to take advantage of the expertise of supply-chain partners. A key operational strategy is to integrate various supply-chain partners to maximize the value offered to customers.

In this chapter, we discussed Web-based technology related to product design and development, project management, plant operations, supply-chain management, engineering innovation, maintenance, procurement, and customer relationship management. These are only select examples of Web-based tools that are currently available in the marketplace.

Engineering managers need to constantly keep themselves current with the advancement of new technologies; assess the application potential of these new technologies, as well as new business models and innovative business practices; and selectively apply suitable ones to add value to their employers' businesses.

12.6 REFERENCES

Anonymous. 2001. "Business: The Cemex Way." *The Economist*, June 16.

Ayers, J. B. (Editor). 2001. *Handbook of Supply Chain Management*. Boca Raton, FL: St. Lucie Press.

Berry, M. J. A. and Gordon Linoff, 1997. *Data Mining Techniques*. New York: John Wiley.

Berry M. J. A and G. Linoff. 1999. *Mastering Data Mining: The Art and Science of Customer Relationship Management*. New York: John Wiley.

Bishop, B. 1998. *Strategic Marketing for the Digital Age*. Thousand Oaks, CA: Sage Publications.

Brunet, P., C. Hoffmann, and D. Roller (Editors). 2000, *CAD Tools and Algorithms for Product Design*. New York: Springer.

Burnett, K. 2001. *The Handbook of Key Customer Relationship Management: The Definitive Guide to Winning, Managing and Developing Key Account Business*. Upper Saddle River, NJ: Financial Times Prentice Hall.

Chang, C. M. 2004. "Opportunities to Add Value to Technology-Based Organizations." Proceedings of PICMET Portland International Conference on Management of Engineering and Technology, Seoul, Korea, (July 31–August 1).

Chatfield, D. S. and T. Johnson. 2002. *Microsoft Project 2002 Step by Step*. Richmond, WA: Microsoft Press.

Chopoorian, J. A., R. Witherell, O. E. M. Khalil, and M. Ahmed. 2001. "Mind Your Business by Mining Your Data." *SAM Advanced Management Journal*, Vol. 66, No. 2, Spring, pp. 45–51.

Creveling, C. M., J. D. Slutsky, and D. Antis, Jr. 2003. *Design for Six Sigma: In Technology and Product Development*. Upper Saddle River, NJ: Prentice Hall.

Coleman, P. and S. L. Nelson. 2000. *Effective Executive's Guide to the Internet: The Seven Core Skills Required to Turn the Internet into a Business Power Tool*. Redmond, WA: Redmond Technology Press.

Cross, N. 2000. *Engineering Design Methods: Strategies for Product Design*. 3rd ed. Chichester, NY: John Wiley.

Dyche, J. 2002. *The CRM Handbook: A Business Guide to Customer Relationship Management*. Boston: Addison Wesley.

Ficke, G. 2001. "E-Business Transformation." *Transmission and Distribution World*, Vol. 53, No. 8, July.

Franzelle, E. 2002. *Supply Chain Strategy: The Logistics of Supply Chain Management*. New York: McGraw-Hill (2002).

Friedlein, A. 2001. *Web Project Management: Delivering Successful Commercial Web Sites*. San Francisco: Morgan Kaufmann Publishers.

Green, H. 2003. "The Web Smart 50." *Business Week, November 24*.

Gulati, R., M. Sawhney, and A. Paoni (Editors). 2003. *Kellogg on Technology and Innovation*. Hoboken, NJ: John Wiley.

Gunasekaran, A. 2001. *Agile Manufacturing: The 21st Century Competitive Strategy*. Oxford, NY: Elsevier.

Haylock, C. F. and L. Muscarella, with R. Schultz. 1999. *Net Success: 24 Leaders in Web Commerce Show You How to Put the Internet to Work for Your Business*. Holbrook, MA: Adams Media Corporation.

Hernandez, J. A. 1997. *The SAP R/3 Handbook*. New York: McGraw-Hill.

Heizer, J. H. and B. Render. 1999. *Operations Management*. 5th ed. Upper Saddle River, NJ: Prentice Hall.

Janenko, P. M. 2002. *E-Operations Management: The Convergence of Production and E-Business*. New York: AMACOM.

Kalakota, R. and M. Robinson. 1999. *E-Business: Roadmap for Success*. Reading, MA: Addison Wesley.

Kantardzic, M. 2003. *Data Mining: Concepts, Models, Methods, and Algorithms*. Hoboken, NJ: Wiley-Interscience.

Langenwalter, G. A. 2000. *Enterprise Resources Planning and Beyond: Integrating Your Entire Organization*. Boca Raton, FL: St. Lucie Press.

Lewis, J. P. 2002. *Fundamentals of Project Management: Developing Core Competencies to Help Outperform the Competition*. New York: AMACOM.

Magretta, J. 1988. "The Power of Virtual Integration: An Interview with Dell Computer's Michael Dell." *Harvard Business Review*, March–April, p. 72.

McClellan, M. 2003. *Collaborative Manufacturing: Use Real-time Information to Support the Supply Chain*. Boca Raton, FL: St. Lucie Press.

Mello, S. 2002. *Customer-Centered Product Definition: The Key to Great Product Development*. New York: AMACOM.

Minoli, D and E. Minoli. 1998. *Web Commerce Technology Handbook*. New York: McGraw-Hill.

Meyerson, J. M. 2002. *Enterprise Systems Integration*. Boca Raton, FL: Auerbach.

Nykamp, M. 2001. *The Customer Differential: The Complete Guide to Implementing Customer Relationship Management*. New York: AMACOM.

O'Leary, D. E. 2000. *Enterprise Resource Planning Systems: Systems, Life Cycles, Electronic Commerce, and Risk*. New York: Cambridge University Press.

Otto, K. N. and K. L. Wood. 2001. *Product Design: Techniques in Reverse Engineering and New Product Development*. Upper Saddle River, NJ: Prentice Hall.

Paisie, J. E. 2001. "E-Transformation of Oil Companies Offers Unprecedented Value-Creation Opportunities." *Oil & Gas Journal*, January 18.

Pendharkar, P. C. 2003. *Managing Data Mining Technologies in Organization: Techniques and Applications*. Hershey, PA: Idea Group Publishing.

Piskurich, G. M. (Editor). 2003. *The AMA Handbook of E-Learning: Effective Design, Implementation, and Technology Solutions*. New York: AMACOM.

ReVelle, J. B. 2002. *Manufacturing Handbook of Best Practices: An Innovation, Productivity, and Quality Focus*. Boca Raton, FL: St. Lucie Press.

Robert, M. 1995. *Product Innovation Strategy Pure and Simple: How Winning Companies Outpace Their Competitors*. New York: McGraw-Hill.

Rosen, A. 1997. *Looking into Intranets and the Internet: Advice for Managers*. New York: AMACOM.

Rud, O. P. 2001. *Data Mining Cookbook: Modeling Data for Marketing, Risk and Customer Relationship Management*. New York: Wiley.

Schmitt, B. 2003. *Customer Experience Management: A Revolutionary Approach to Connecting with Your Customers*. New York: John Wiley.

Sheridan, J. H. 1999. "Focus on Flow." *Industry Week*, October 18, p. 46.

Silverstein, B. 1999. *Business-to-Business Internet Marketing: Five Proven Strategies for Increasing Profits through Internet Direct Marketing*. Gulf Breeze, FL: Maximum Press.

Simchi-Levi, D., P. Kaminsky, and E. Simchi-Levi. 2000. *Designing and Managing the Supply Chain: Concepts, Strategies, and Case Studies*. Boston: Irwin/McGraw-Hill.

Sommerhoff, E. W. 2000. "E-Commerce: Managing Design and Construction Online." *Facilities Design and Management*, October.

Stengl, B. and R. Ematinger. 2001. *SAP R/3 Plant Maintenance: Making It Work for Your Business*. New York: Addison Wesley.

Stokes, M. (Editor). 2001. *Managing Engineering Knowledge: MOKA—Methodology for Knowledge Based Engineering Applications*. New York: ASME Press.

Tinnirelle, P. C. (ed.) 2000. *Project Management*. Boca Raton, FL: Auerbach.

Tracy, B. 2003. *Turbostrategy: 21 Powerful Ways to Transfer Your Business and Boost Your Profit Quickly*. New York: AMACOM.

Ulrich, K. T. and S. D. Eppinger. 2000. *Product Design and Development*, 2d ed. Boston: Irwin/McGraw-Hill.

Williams, G. C. 2000. *Implementing SAP R/3 Sales and Distribution*. New York: McGraw-Hill.

World Bank. 2003. *The Little Green* Book. Washington, DC: The World Bank. (Also available on the Internet at *http://lnweb18.worldbank.org/ESSD/envext.nsf/44ByDocName/TheLittleGreen-DataBook2003FullDocument1MBPDF/$FILE/LittleGreenDataBook2003.pdf/*.)

Zacks, R. 2001. "Ten Emerging Technologies that Will Change the World." *Technology Review*, Vol. 104, January, p. 97.

Zairi, M. 1999. *Best Practice: Process Innovation Management*. Woburn, MA: Butterworth-Heinenman.

Zemke, R. and T. Connellan. 2001. *E-Service: 24 Ways to Keep Your Customers when the Competition Is Just a Click Away*. New York: AMACOM.

12.7 APPENDICES

APPENDIX 12.A. INTRODUCTION TO E-COMMERCE AND E-BUSINESS

E-business describes enterprises that electronically connect their key business systems directly to their critical constituencies—customers, employees, vendors, and strategic partners.

Several references are useful:

1. Anita Rosen. *Looking into Intranets and the Internet: Advice for Managers*. New York: AMACOM, 1997.

2. Pat Coleman and Stephen L. Nelson. *Effective Executive's Guide to the Internet: The Seven Core Skills Required to Turn the Internet into a Business Power Tool*. Redmond, WA: Redmond Technology Press, 2000.

3. George M. Piskurich (Editor). *The AMA Handbook of E-Learning: Effective Design, Implementation, and Technology Solutions*. New York: AMACOM, 2003.

4. Ron Zemke and Tom Connellan. *E-Service: 24 Ways to Keep Your Customers when the Competition Is Just a Click Away*. New York: AMACOM, 2001.

5. Ravi Kalakota and Marcia Robinson. *E-Business: Roadmap for Success*. Reading, MA: Addison Wesley, 1999.

The top six drivers of e-business are listed as follows:

1. **Speed.** Business solutions are available in days or weeks as opposed to months or years.

2. **Focus.** Today's executives know that anything that distracts their company from its subject matter expertise must be moved outside of the organization.

3. **Flexibility.** Creates a true "plug-and-play" approach to acquiring advanced business capabilities.

4. **Connectivity.** E-sourcing turns supply chains into fully integrated trading networks.

5. **Scalability.** The right solution can be put in place first and then easily grown as needed.

6. **Price.** Lower total cost of ownership and shorter time to benefit are fast tracks to success.

E-commerce, a subset of e-business, is broadly defined as the buying and selling of goods and services by using electronic transaction processing technologies. Haylock, Muscarella, with Schultz 1999) The roles of e-commerce are to (1) reduce transaction costs, (2) streamline information flow, (3) increase coordination of actions, (4) strengthen existing markets, and (5) create new markets.

The advantages of e-commerce include (1) an increase in sales by reaching narrow market segmentation in geographically dispersed locations (creating virtual communities) and (2) a decrease in costs by handling sales inquiries, providing price quotes, and determining product availability. On the other hand, the disadvantages of e-commerce are (1) the loss of ability to inspect products from remote locations, (2) the rapidly advancing pace of underlying technologies, (3) difficulty in calculating ROI, and (4) complex cultural and legal impediments.

E-technology encompasses all of the technologies and techniques necessary for doing business electronically. Some common terminology and definitions are listed as follows:

1. **Value-added networks (VANs).** These offer a number of advantages: (1) Users support only one communications protocol. (2) VANs record activity in an independent audit log. (3) VANs provide translation between different transaction sets. (4) VANs perform automatic compliance checks.

    There are also disadvantages to VANs: (1) There are numerous costs involved, including enrollment fees, maintenance fees, and transaction fees. (2) Business can become cumbersome for trading partners with different VANs. (3) Inter-VAN transfers do not always provide a clear audit trail.

    Figure 12.A1 illustrates how VANs work with buyers, sellers, and banks.

2. **Technology enablers.** The following two enablers are commonly engaged in e-commerce:

   A. **Legacy Data Systems**

      These systems contain historical company data, including engineering drawings.

   B. **EDI (Electronic Data Interchange)**

      EDI systems facilitate the electronic transfer of business and technical data (e.g., purchase orders, invoices, shipping notices) in a standard format between computers. They have four components: standards, software, hardware, and communications.

      ASC X12 Standards (Accredited Standards Committee X12) includes specifications for several hundred transactions sets (the names of the information for specific business data interchanges).

3. **Host.** A host is a computer directly connected to the Internet that stores Web pages.

4. **Plug-In.** A plug-in is a "helper" program, such as Adobe Acrobat Reader.

5. **Web browser.** A web browser is a software program permitting the viewing of programs specifically prepared in HTML/XTML formats and transmitted through the Internet. Currently, the two widely used browsers are Netscape and Microsoft Internet Explorer.

6. **Servers.** Servers are large computers that manage separate networks formed by clusters of individual computers. Information is transmitted between servers through (1) satellites, (2) microwaves, (3) Ethernet lines and fiber-optic cables, (4) television lines, and (5) phone lines.

7. **Messages.** All messages or information transmitted through the Internet are in the form of digitized text, graphics, still photos, animation, voice, or full-motion video.

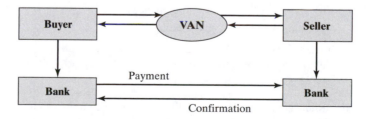

Figure 12.A1   Value-added networks.

8. **Packets.** Information is bundled into structures labeled with the network address of the recipient's electronic mailbox. Packet switching involves the following steps:

   A. Files and messages are broken down into packets that are electronically labeled with their origins and destinations.

   B. The destination computer collects the packets and reassembles the data from the pieces in each packet.

   C. Each computer the packet encounters decides the best route toward its destination.

9. **TCP/IP.** Packets of information are moved along Internet pathways by the Transmission Control Protocol/Internet Protocol (TCP/IP). TCP controls the assembly of a message into its smallest packets before transmission and reassembles them once they are received. IP sets the rules for routing packets from their sources to their destinations.

10. **Meta-Tags.** Machine-readable codes that are embedded in Web pages that Internet engines use to conduct a search.

11. **Framing.** Framing links the user from one site to another, whereupon the user accesses the link and the selected homepage appears surrounded by a "frame" containing the logo of the previous website and its advertisement. Framing may create the mistaken impression of affiliation and threaten relationships with advertisers.

12. **Deep linking.** This allows the user at one website to jump directly to a Web page deep within the interior of another website, bypassing the latter's home page and the advertisement.

13. **Spamming.** The practice of sending unsolicited junk mail, or spam, to e-mail accounts. A new anti-spam law, CAN-SPAM, was passed by the United States and went into effect on January 1, 2004. Under this law, marketers must (1) remove customers from their lists when requested, (2) provide automated opt-out methods which must remain workable for 30 days after mailing, (3) supply contact information (address and phone), (4) include truthful subject lines, (4) insert a notice that the messages is an ad, (5) display warnings if material is of sexual orientation, and (6) not engage in email harvesting.

14. **The World Wide Web.** An Internet-based application designed to provide a common protocol for identifying, storing, and retrieving data on computers around the world. Being a part of the Internet, it allows users to share information with an easy-to-use interface.

15. **Business process.** Processes that encompass all commercial and other business-related activities conducted by a firm. Examples include the following:

   (a) Transferring funds

   (b) Placing orders and sending invoices

   (c) Shipping goods to customers

   (d) Producing goods

   (e) Managing inventory

   (f) Managing distribution logistics

16. **Intelligent agents.** Programs that perform such functions as information gathering, information filtering, or mediation (running in the background) on behalf of a person or

entity. Examples include Auction Bot, Bargainfinder, Firefly, and Kasbach. Intelligent agents are used for

(a) Finding the best prices

(b) Defining specific product features

(c) Procurement (what, when, and how much to purchase)

(d) Stock alert (rules and conditions)

A number of software tools are available to construct Web pages. It is advisable to have objectives, establish links to secondary pages, build hyperlinks to pages of other websites, and permit any planned commercial transactions. The modes of access include (1) ordinary (dial-up) phone lines, (2) cable lines (cable modem), and (3) handheld equipment (e.g., the Palm Pilot) connected via satellite links.

Telnet allows users to log on and run a remote computer attached to the Internet (terminal emulation). Users can access Telnet through most Web browsers when client software is available. An example is the U.S. Library of Congress.

The Internet is widely accessed by private users for many activities, among them the following: (1) communicate by e-mail with family members, friends, including the transmission of documents, photos, diagrams, text files, graphics, video, and sound tracks; (2) chat on-line in newsgroups, mailing lists, and other such discussions; (3) read on-line news or magazines or conduct searches for data (in databases); (4) invest and trade securities (e-trade, etc.); (5) play games (entertainment); (6) make purchases (airline tickets, clothing, toys, books, CDs, candies, flowers, cars, computers, homes, mortgage loans, etc.); (7) get an education (e.g., the University of Arizona offers an on-line MBA degree program); (8) seek employment; and (9) look up the weather.

For business applications, the Internet is known for (1) B2B—business to business transactions: procurement, customer services, system and product design, etc.; (2) B2C—business-to-consumer transactions: e-retail, on-line order processing, merchandizing account, security; (3) C2C—consumers to consumers: auctions; (4) C2B—consumers to business; (5) B2G—business to government (e.g., FASA, the Federal Acquisition Streaming Act, specifies guidelines for registering, receiving, soliciting, and bidding for contracts); and (6) B2E—business to education; textbooks (Silverstein 1999; Minoli and Minoli 1998).

Online selling (e.g., B2C) is particularly good for the following types of goods: (1) hard-to-find specialty products and one-of-a-kind or regional items; (2) computer-related and high-tech products (from software to mouse pads to CD players); (3) information products such as reports, news, and data; (4) products with broad geographical or international appeal; (5) items that can be sold less expensively over the Internet than elsewhere; and (6) products with a high enough price tag to cover the cost of Internet selling in time, infrastructure, and shipping.

Successful B2C websites include the following: (1) Amazon.com offers books, CDs, and other items on-line, enters into contracts with producers for product supplies and contracts with logistics firms and warehouses for distribution and inventory management. Selling at below wholesale prices to secure customer traffic, Amazon.com makes money from on-line advertisement revenue. (2) Priceline.com holds auctions on-line for specific items. It is known for speed, multimedia presentation capabilities, breaking geographical and other constraints, and offering convenience to participating customers. (3) Cheaptickets.com offers airline tickets, with the price decreasing as the departure time approaches; (4) Landsend.com offers the new paradigm of shopping with a friend by pushing Web-based information to the friend's browser while talking on the phone.

Examples of additional websites are presented in Table 12.A1.

TABLE 12.A1 Examples of Web Sites

Name	Nature of Site
Amazon.com	Product sales
Blackboard.com	Education
Ebay.com	Customer value
Emusic.com	Distribution
Epinions.com	Customer empowerment
Hedgehog.com	Reverse auction
Mysimon.com	Comparison shopping
Nytimes.com	Information
Priceline.com	Services
Respond.com	Finding local providers
Travelocity.com	Travel services

APPENDIX 12.B. EXAMPLES OF NOTEWORTHY E-BUSINESSES

A. **Covisint.** On February 25, 2000, GM, Ford, DaimlerChrysler, and Renault-Nissan created an e-commerce venture, Covisint, to link with suppliers on-line to streamline their combined $240 billion in annual purchases. Convisint is headquartered in Southfield, Michigan, and operates in North America, Europe, Asia, and Latin America. It is estimated that 10 percent of the purchasing costs could be saved. In May 2001, PSA Peugeot Citroën joined this venture as an equity partner in Europe. Each company uses the system differently, and no company has access to information or activities of any other company. Besides cost reduction, timesavings in product development is a big benefit.

Covisint (*www.covisint.com*) is a massive dot-com startup, likely the world's largest Internet company and trade exchange. In addition to handling the $240-billion annual purchases of the Big Three, this venture handles another $500 billion of the Big Three suppliers each year. Oracle, Commerce One, and Sun Microsystems were the chosen technology partners in Covisint's venture. This new venture will likely set a global standard for online purchasing in the auto industry and perhaps in other industries.

The Covisint venture is a clear example of changes induced by the business-to-business use of the Internet in the way business is conducted. Goldman economists predicted that on-line transactions between companies, which at the time of this writing accounted for only 0.5 percent of all such businesses, would account for 10 percent by 2004, significantly reducing costs in various industries.

Other industrial initiatives are delineated in Table 12.A2.

In business-to-business exchange, the real value is created, not by the software technology suppliers (CommerceOne, Ariba, Oracle, etc.), but by the trading communities (auto, chemical, steel, etc.), whose members are in the driver's seat. However, since most businesses already have close ties with their key suppliers and already use current technologies (phones, fax, EDI, etc.), the additional value brought about by B2B may be only incremental.

B. **eBay.** A well-known dot com company, eBay facilitates trade between buyers and sellers in an auction format with no inventory, no special unique items, and no personal interactions.

TABLE 12.A2   Industrial Initiatives in E-Businesses

	Industry	Partners	Activities
Exostar.com	Aerospace	Boeing, Lockheed Martin, BAE Systems and Raytheon	Links 37,000 suppliers, airlines, and governments to represent a buying network of $17 billion in parts and supplies.
Novopoint.com	Food	Cargill and Ariba	Links food and beverage manufacturers and suppliers
ChemConnect.com	Chemical	Dupont, Bayer GE Equity and others	Chemical exchange for materials
e2open.com	Electronics	Some of the major electronics companies	Exchange for parts and supplies
Intercontinentalexchange.com	Energy	Duke Power, American Electric, Royal Dutch Shell, Elf, Cinergy, Consolidated Edison and others.	Largest exchange for natural gas and power in North America
XSAg.com	Agriculture	Various agriculture companies	Provides auction service on-line for chemicals and seeds

It maintains a current, up-to-date catalog that includes (1) item number, (2) item description, (3) price, (4) number of bids received, and (5) minutes left to bid. No fee is charged to buyers. Sellers pay insertion fees (25 cents to $2), feature fees ($2.00 to $49.90), and final-value fees (5 to 1.25 percent). Both buyers and sellers are required to enter binding on-line agreements to safeguard all transactions.

The unique feature of the eBay business model is that it links households all over the country and builds a global marketplace where none previously existed.

C. **Priceline.com**. U.S. patent number 5,794,207 was awarded to Priceline.com for its business model entitled "Bilateral Buyer-Driven Commerce." This was the first time a business model qualified for patent protection. Although many people doubted that such a patent could stand up to future court challenges, no one succeeded in invalidating it. Priceline.com continues to exist.

D. **Amazon.com.** On February 22, 2000, Amazon.com was awarded a U.S. patent for its "Affiliates Program," which allows owners of other websites to refer customers to Amazon in exchange for a fee. In other words, owners of websites can list items from an on-line retailer's (i.e., Amazon.com) catalog on their sites and can then link to the retailer to complete the transaction. The affiliate receives a percentage of the transaction as a referral fee.

Amazon.com was awarded a U.S. patent in September 1999 for its "1-Click" service that allows customers to shop without entering their shipping and billing information every time they buy. This is another example of business model innovation in the digital world.

E. **Global exchange services.** General Electric formed its new Internet service venture, Global eXchange services (*www.gxs.com*), as reported by *The New York Times* on March 6, 2000, to offer broad services, including auctions. The new venture involves 1500 people and is expected to bring in $1 billion in annual revenue within 4 to 5 years.

12.8 QUESTIONS

12.1   How are URL, domain name, and search engines defined? Use examples to explain the relationship between them. How can one make use of Web pages to promote business?

12.2   What are Internet, intranet, and extranet? How are they being used by numerous large and small companies today?

12.3   What are the standard markup languages used in the design of Web pages?

12.4   What are some of the legal issues related to the Internet and Web-based business transactions that remain unresolved at this time?

12.5   In implementing a computerized maintenance management system to reduce maintenance costs, what steps are taken?

12.6   What is Web mining, and how significant is it in generating useful results to support management decision making?

12.7   For the development of software products, the SCM (software configuration management) process is closely followed as a way to ensure performance and reliability while controlling costs. Explain what SCM can do and in what ways it is important that both developers and intended customers insist on SCM.

12.8   Although marketing and sales are not functions of engineering, they have a direct impact on product development and customer relationship management. Which Web-based tools are currently available to facilitate marketing and sales?

12.9   The business environment in the new millennium will continue to be fast paced, Internet enhanced, and globally oriented. Name a few factors that will affect the business successes of any companies in such a challenging environment.

Chapter 13

Globalization

13.1 INTRODUCTION

Globalization is defined by the International Monetary Fund as the growing economic interdependence of countries worldwide through the increasing volume and variety of cross-border exchanges in goods, services, international capital flows, and technologies.

Globalization is not a new phenomenon. International trade and commerce have a very long history. In recent years, the growth of the world economy and the migration of goods, services, capital, people, and technologies across borders have increased. The rapid expansion of the digital economy has also helped to accelerate the pace of globalization (Collier and Dollar 2002; Langhorne 2001). Some American companies are expanding to reach new global markets and foreign resources.

The world economy has become increasingly global, as current markets are more interconnected than ever before. Instead of only a few countries handling the trade of most currencies and goods, now many more countries play a part. American companies are actively pursuing markets in Asia, Europe, Latin America, and other regions. Some foreign-owned companies have achieved more sales revenues outside of their home countries than in their respective domestic markets (Govindarajan and Gupta 2001; Steger 2002). Table 13.1 lists several such global companies. Although the domestic auto market in Japan continues to fluctuate, Toyota is expanding aggressively in the United States, Europe, and China. As reported by Fortune Magazine recently, Toyota has now surpassed Ford as the number-two automaker in the world, just slightly behind General Motors in global unit sales.

John Zeglis, president of AT&T, said in 1999, "There are two kinds of companies in the future: those that go global and those that go bankrupt."

In this chapter, we will explore various management issues related to globalization (Rao 2001; Sullivan 2002). Our focal points are steps engineering managers may pursue to take advantage of the opportunities offered by globalization to add value.

TABLE 13.1   Percentage of Sales Coming from Outside of Home Market

	1993 (percent)	1998 (percent)
General Electric	16.5	30.1
Wal-Mart	0.0	13.8
McDonalds	46.9	61.5
Nokia	85.0	97.6
Toyota	44.6	49.6

13.2 GLOBAL TRADE AND COMMERCE

The world economy has become interdependent in recent decades. The worldwide integration of national economies—through the trade of goods, services, capital, and technologies—has become broad and deep. Figure 13.1 illustrates increased U.S. trade as a percentage of gross national product (GNP).

Another indicator of global trade activities is the steadily increasing number of strategic alliances formed across the 29 industrialized countries that are members of the Organization of Economic Cooperation and Development (OECD), as illustrated in Figure 13.2.

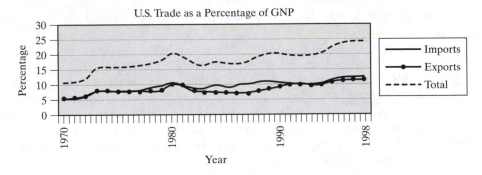

Figure 13.1   Growth of U.S. Trade. *Source*: STAT-USA.

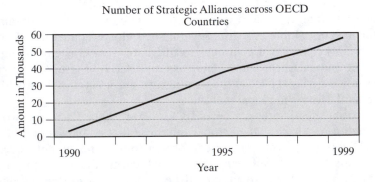

Figure 13.2   Increase in strategic alliances within OECD.

13.2.1 Multinational Enterprises

Multinational enterprises (MNEs) operate in more than one country. As of 2001, there were about 45,000 MNEs in the world. These enterprises play important roles in the global economy:

1. Holding 90 percent of all technology and product patents worldwide.
2. Conducting 70 percent of world trade, 30 percent of which is intracompany.
3. Pursuing diversified businesses, such as (1) mining; (2) refining and distributing oil, gasoline, diesel and jet fuel; (3) building energy plants; (4) extracting minerals; (5) making and selling autos, airplanes, communication satellites, computers, home electronics, chemicals, medicines, and biotechnology products; (6) harvesting wood and making paper; and (7) growing crops and processing and distributing food products.
4. Inducing governments to form treaties and trading blocks among the European Union, NAFTA, the WTO, the Multilateral Agreement on Investment, and the Uruguay round of GATT. These treaties tend to provide great power and authority for multinational enterprises to pursue globalization, thus increasingly undercutting the authority and power of national governments and local communities.

The 500 largest MNEs are responsible for 80 percent of all foreign direct investments. Of these MNEs, 443 are located in only three regions: the United States, the European Union, and Japan. (A detailed distribution of these major MNEs is presented in Figure 13.3.)

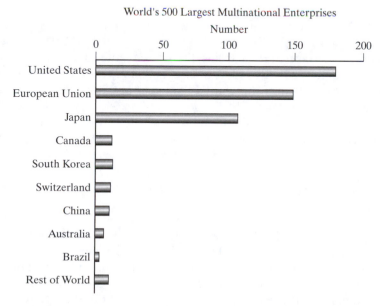

Figure 13.3   Current concentration of major MNES.

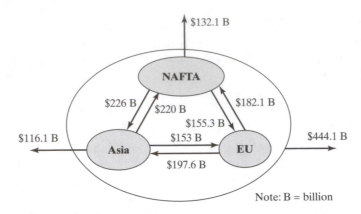

Figure 13.4   Trades within the triad regions for 1997.

Trade data from 1997 demonstrate the extent of dominance by the multinational enterprises operating in the triad regions (NAFTA, European Union, and Asia). The total export trade recorded for these three regions was $4,145.8 billion, composed of $2,092.3 billion from the EU, $1,010 billion from NAFTA, and $1,043.5 billion from Asia. The intercountry export within each region was 53.1 percent for Asia, 49.1 percent for NAFTA, and 60.6 percent for the EU. Thus, the net result was that the total export from these three regions to countries located outside of the regions was only 16.71 percent of the total. (See Figure 13.4.) As a result, some researchers charge that the operations of these multinational enterprises are de facto regional, not global (Rugman 2001).

If we take a longer-term view, we cannot afford to ignore the forecast made by the World Bank (1992). According to this forecast, by 2020 the largest economies in the world are projected to be China, the United States, Japan, India, and Indonesia.

Economic growth rates in emerging markets for the future are predicted to be 3 to 10 times that of the United States. About 50 percent of worldwide GDP (gross domestic product) would be generated in emerging markets. Consequently, we should expect that the situation just described (the MNEs having only regional operations for now and the home bases of the top 500 largest MNEs being concentrated in the triad regions) will surely change in the years to come. The roles played by the emerging countries in Asia, such as China, India, and Indonesia, could become substantial indeed. The extent of trade and commerce globalization is expected to further increase.

Goldman Sachs, an investment firm based in New York, studied the gross domestic products (GDP) of both the G6 and BRIC nations. BRIC is an abbreviation for Brazil, Russia, India and China. The G6 is composed of the United States, Japan, Germany, France, Britain, and Italy. Canada is normally a part of the G7 nations, but is excluded in this study because its GDP is only about 3 percent of the G7 total. Goldman Sachs predicts that by 2037, the total GDP of BRIC will match that of G6. (See Figure 13.5.)

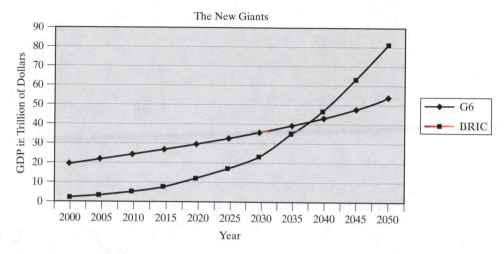

Figure 13.5   GDP forecasts for G6 and BRIC Nations. *Source:* Wilson and Purashothaman (2003).

Should this prediction hold true, it would mean that the next phase of globalization would be manifested in the expansion of global economic activities from being centered on the G6 nations at the present time to encompass the BRIC within the next 30-plus years, and that high economic growth rates would be found primarily in BRIC nations.

This type of forecast is, of course, valid only in the absence of any disruptive events, such as wars, global economic recessions, or natural disasters. Nevertheless, it does foretell the emergence of some developing economies and the increased degree of globalization in the years ahead.

13.2.2 Ownership of Global Companies

The five major stakeholders of any company are customers, employees, suppliers, investors, and the communities in which the company operates.

A large number of global companies manufacture products designed to reach global customers, employ workers from different countries, source materials and components from suppliers in global markets, interact with local communities at global locations, and have global shareholders. In recent years, countless countries have been setting up stock exchanges and security markets to attract foreign or domestic investments. Nowadays, it is easy for an investor to become a shareholder of any global company that is traded in one of many public stock exchanges.

As illustrated in Figure 13.6, the ownership of Nestlé, a well-known multinational enterprise, is quite global indeed. No single Nestlé shareholder owns more than 3 percent of the company stocks. This trend of global ownership is expected to continue as the capital markets become more accessible to investors residing in various countries. Over time, companies will diligently apply innovative global marketing strategies to sell products to global customers, in order to create value for a global ownership.

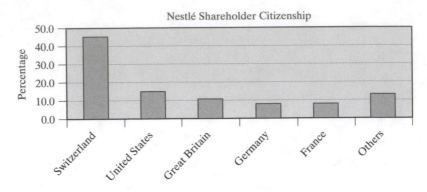

Figure 13.6   Ownership of Nestlé.

Example 13.1.

What are some practical reasons for a company to ever want to expand into the world of international business?

Answer 13.1.

The reasons for a company to want to expand into the global markets are plentiful. The following are the primary four:

A. **Desire to Expand Markets (Finding New Customers).** For companies whose products have been selling in a saturated domestic market at home, expanding into global markets represents an attractive opportunity to find new customers.

   Theodore Levitt (Harvard Business Review editor) proposes the idea that the characteristics of some products are converging, making them more and more universal, thus allowing companies to market efficiently to the global marketplace. Companies may be able to derive advantages based on the global economies of scale. However, in order to do well, companies need to understand the local customers, business practices, and cultures of global customers.

B. **Search for Natural Resources.** Companies pursue foreign investments to avail themselves of resources that may otherwise not be readily available. U.S. investments in Saudi Arabia and various U.S. offshore gas exploration projects are the prime examples.

C. **Proximity to Customers.** Companies expand into the global marketplace to be closer to their customers, for the sole purpose of understanding and serving them better, faster, and cheaper. Some products require customization in order to enrich the value offered to the customer. Customer satisfaction will increasingly become a key competitive focus. Companies in the position to understand their customers more thoroughly and which are able to customize their products will have a significant advantage in the marketplace.

D. **Labor Savings.** Today, certain developing countries offer skilled labor at a fraction of the cost needed to hire similar workers in the home countries of numerous major companies. Mexico is a prime example. Many U.S. companies have set up manufacturing shops in Mexico. The products made there are shipped back to the United States for distribution and marketing. Several other countries, such as China, India, the Philippines, and

Thailand, are also candidates for companies to realize labor savings in certain types of products or services.

Example 13.2.

When companies attempt to pursue global markets, which common entry strategies are deployed?

Answer 13.2.

There are several common entry strategies into global markets. In general, companies are well advised to first study the relative attractiveness of the target market (e.g., specific segments in different countries) by considering factors such as profitability, market size, and market growth. In addition, companies need to assess the degree of acceptance of their products in these targeted marketplaces (e.g., brand name, competitive position, market access, etc.). Once the most favorable product–market-segment pairs are selected, companies may pursue these global markets by

A. Exporting
B. Licensing or contracting manufacturing
C. Forming joint ventures with local partners
D. Creating a foreign branch of the company
E. Establishing a foreign subsidiary of the company

13.3 UNITED NATIONS STATISTICS AND GOALS

Statistics published by the World Trade Organization (WTO) indicate that world trade, defined as the total value of export, has increased about 500 percent over the last 30 years at a compound growth rate of 5.5 percent per year.

It is well known that countries open to global trade grow twice as fast as those which remain relatively closed to trade. At least two African nations, Nigeria and Tanzania, have chosen to rely on protectionism, foreign aid, and inefficient public policy. Today, they remain at the 1960s economic development levels of Malaysia, Thailand, and Indonesia. In recent years, Latin America has started to embrace market liberalization. It has abandoned its old policies of a dominant state presence in the economy, import substitution, and domestic industry protection. The results of these changes are encouraging, and more countries are expected to jump on the globalization bandwagon.

According to the World Bank (2001), the world output is projected to increase 33 percent from $30 trillion in 2001 to $40 trillion in 2010. The disposable income in regions like China, India, Southeast Asia, and Latin America will double over the same period. About 300 million people (roughly the size of the population of the United States) will join the thriving worldwide middle class in the next 10 years.

On the other hand, World Bank statistics also show that, from 1990 to 2000, only about 800 million people moved out of absolute poverty, which is defined as having less than one U.S. dollar per day of income. As of 2001, 50 percent of the world population lived on less than $2 per day. Eighty percent of the global population lives on less than 20 percent of the global income.

The UN has declared that one of its goals is to decrease the number of people in absolute poverty by 50 percent by 2015. Globalization is regarded as a key process in achieving this very meaningful goal (Lamberton 2002).

13.4 THE GREAT PHILOSOPHICAL DEBATE ABOUT GLOBALIZATION

Not everyone favors globalization. It creates winners and losers. Generally speaking, government leaders of both large and small countries are in favor of globalization because of its heightened opportunities for foreign direct investment, transfer of technology and best practices, and trade benefits. Business leaders are strong advocates of globalization for many reasons, including the following:

A. The greater flow of trade and investments stimulates economic growth.

B. Rising outputs bring about employment and income, which means higher living standards for consumers.

C. Higher living standards facilitate a greater social willingness to devote resources to the environment, education, health care, and other social goals.

D. Global competition keeps domestic business competitive and innovative, which leads to higher quality output and productivity.

E. Rapidly developing economies tend to generate a new middle class that is the bulwark of support for personal liberty and economic freedom.

Opponents of all stripes and creeds blame globalization for many of the world's ills (Sassen 1998). They are primarily from three major groups: labor union members, human rights activists, and environmentalists.

The labor union wants to protect local jobs in industrialized countries, as globalization will likely induce multinational enterprises to transfer manufacturing and other high-paying jobs to developing regions in search of cost competitiveness (Phillips 1998). They also raise the issues of child labor and forced labor in poor countries, citing past incidents of exploitation by multinational enterprises.

The human rights groups claim that local economic growth induced by globalization in some emerging countries may allow their respective dictatorships to stay in power longer, thus indirectly supporting continued suppression of the people. Globalization would become an inadvertent coalition partner in crime against humanity (Brysk 2002).

The environmentalists believe that, by relocating the manufacturing operations to developing countries with lower environmental control standards, the multinational enterprises are essentially exporting pollution and other environmentally unacceptable practices to the poor countries, thus causing irreversible damages (Asheghian and Ebrahimi 1990).

These three groups are united in their opposition to globalization. Their reasoning is further summarized as follows:

1. Globalization is a conspiracy of big companies exploiting small countries. It concentrates market power in the hands of a few large corporations, allowing them

to trample over smaller commercial rivals and flourish at the expense of small companies and consumers. Globalization is akin to companies without rules.

2. Globalization promotes the suppression of human rights in developing countries.
3. Globalization destroys the environment.
4. Globalization spreads terrorism, narcotics, disease, and money laundering (Condon 2002; Kugler and Frost 2001; Horowith 2001).
5. Globalization lowers labor standards and turns developing nations' workers into "slaves."
6. Information technology (IT) is a "tool for evil" in globalization.
7. Globalization takes away jobs from the United States.
8. Globalization undermines cultural diversity.
9. Globalization widens the gap between the rich and the poor.

Some of the arguments in opposition to the antiglobalization views are enumerated next:

1. **Representation.** Numerous antiglobalization demands reflect the values of young, middle-class U.S. and EU consumers. They may not be the true representative voice for the Third World countries they claim to speak for.

2. **Dominance.** Globalization does not mean the triumph of giant companies over small ones. A case in point is Nokia versus Motorola. Nokia is small in size and dynamic in marketing strategy; this attests to the fact that corporate size is not a requirement for global success. Globalization does shift the balance of advantages from local incumbents (big or small) to foreign challengers. Such protection barriers as the high cost of capital and the difficulty of acquiring new technology, are gradually removed over time.

3. **Environment.** It is partially correct that globalization may indeed affect the environmental conditions in some developing countries. Their competition to attract foreign investment could accelerate the import of production plants that generate carbon dioxide, toxic wastes, and other environmentally unacceptable discharges.

   It is true that any pollution discharged into the ambient air by a production plant is bad for the environment. In the United States, various environmental regulations for reducing harmful emissions were enacted only after many tough struggles between big business and government. The key issue is how to balance the value created by greenness with that produced by wealth and economic progress. While the United States can afford to go green at the present, its current environmental standards may not be appropriate to impose on other countries that are in situations comparable to the United States in the 1960s.

   Thus, the acceptable degree of greenness for a given country is not to be decided by rich countries' environmentalists. For India and other countries, wealth generation may be more pressing in the short run than environmental greenness. This is why the governments of various developing countries welcome globalization and do not share the views of environmentalists.

   However, some local governments and global companies have different views. Levi Strauss established ethical manufacturing standards for its overseas operations. Home Depot adopted an ecofriendly lumber supply program. Starbucks buys coffee from farmers who preserve forests.

4. **Labor Standards.** The claim that globalization diminishes labor standards is a questionable one. The key issues involved are wages, work conditions, and child labor.

   Most foreign direct investment (FDI) is value driven and does not primarily chase after low wages. For example, the United States has a positive FDI, meaning that the total amount of foreign investment in the United States is larger than the U.S. investment abroad. Clearly, this surplus FDI is not driving down U.S. labor standards.

   The governments of developing countries oppose the imposition of the current U.S. and EU labor standards onto their regions, as doing so will cause them to lose the wage advantages they currently enjoy. Imposing an external wage standard that is not locally sustainable can be harmful, as evidenced in Germany. After reunification, West Germany imposed its high wage standards on East Germany. The result was an economic disaster: There was zero growth and high unemployment in the East. The governments of developing countries argue that the Asian "Tigers" (e.g., Taiwan, South Korea, Hong Kong, and Singapore) have convincingly shown the road to prosperity for developing economies. Each of these countries started out with low wages and cheap exports and then allowed wages and income to rise gradually in concert with the increased value of products and the improvement of their workers' skills. Local wages must be sustainable in local economies. As expected, all developing countries insist on speaking for themselves and want more investment, freer trade, and better enforcement of local laws, not the imposition of foreign wage standards.

   In general, global companies do provide higher wages to their workers than their local rivals. Some global companies have started paying attention to workplace conditions as well. Gap and Nike are said to have adopted codes of conduct for their overseas plants.

   Child labor is commonly accepted on American farms today and was legal during the long period when the United States was a developing country. Imposing 21st century labor standards on today's developing countries thus runs the risk of appearing hypocritical. For many families in developing countries, child labor may be a major source of income, just as it is on American farms today and was for others many years ago.

5. **Human Rights.** The argument that globalization supports human rights suppression is a questionable one. According to Maslow's hierarchy of needs, once a person's physiological needs (clothing, shelter, transportation, and other subsistence needs) are met, the next higher levels of needs (social acceptance, peer recognition, and self-actualization) become activated in search of continued personal satisfaction. Accordingly, the local population will most likely seek more freedom of speech, rights of assembly, and respect for human rights over time, but only after their basic subsistence needs are met. An example is Taiwan, which transformed itself peacefully from a dictatorship to a democracy through its rapid advancement of global trade and economy.

6. **Side Effects.** Globalization promotes free trade and the exchange of goods, services, information, money, and technologies across national borders. Indeed, there are no effective solutions for minimizing the detrimental side effects of increased flows of terrorism, narcotics, disease, and money laundering, unless the governments involved are committed to forcefully combating them. In addition, an aggressive implementation of some of the following programs may help:

   A. Enhance educational training for the poor (postsecondary, vocational).
   B. Make social services more widely available.
   C. Adopt policies to strengthen the productive capabilities of all, including the low-income groups.
   D. Set up safety nets (e.g., social security, unemployment insurance, etc.) for those who are in need.

   It is true that what one believes depends on where one stands. Winners and losers have different views on globalization. In the United States, the steel industry is in deep trouble, due mainly to cheap imports. The U.S. textile and farm industries need governmental subsidies to survive. On the other hand, U.S. high-technology industries (e.g., electronics, computers, airplanes, appliances, consumer goods, telecommunications equipment, banking services, and others) are benefiting tremendously from an open global market.

   Globalization is an inevitable and unstoppable trend that unfortunately causes dislocations. A prudent approach should be to go forward with globalization while initiating steps to minimize its detrimental side effects. Examples of such steps include asking MNEs to support education and job-skills retraining in emerging countries, encouraging MNEs to adopt responsible environmental and labor practices when producing products in developing countries, and promoting democracy and respect for human rights.

13.5 IMPACT OF CATASTROPHIC EVENTS ON GLOBALIZATION

In recent years, several major events have had a profound impact on the world economy, political stability, and peace. These events include the terrorist attack against the Twin Towers of the World Trade Center in New York City on September 11, 2001; the Iraqi war in 2003; and the Severe Acute Respiratory Syndrome (SARS) outbreak in South China in 2003 that spread rapidly to Hong Kong, Singapore, Toronto, Taiwan, and other locations. The immediate consequences of each of these events have been an increase in the cost of doing international business and changes in business relationships between the United States and other countries. A number of projected factors may exert a cooling effect on globalization:

1. Since the war against terrorism may be protracted, insurance premiums may be raised because of heightened security concerns.
2. Increasing security risks reduce the willingness of business people to travel internationally and may lead to a reduction in team performance, collaboration, information sharing, and knowledge management.

3. A higher return may be demanded to compensate for increased investment uncertainties.

4. Heightened border inspections may slow cargo movements and force companies to stock more inventories (such as spare parts).

5. Tighter U.S. immigration policies could curtail the inflow of skilled and blue-collar workers (e.g., from Mexico and Canada to the United States).

6. Time horizons for international projects may be shortened when companies make new foreign direct investment.

7. The availability of global equities may drop because a smaller number of investors are willing to take the added risks involved. Foreign direct investment to specific countries regarded as posing a high risk (e.g., India, Pakistan, the Philippines, parts of South America and Southeast Asia, most of the Arab world, and Russia) may be cut.

8. Because of disagreements with U.S. foreign policy, some businesspeople from Third World countries may become reluctant to make deals with American businesses.

There are countless specific examples of the rising costs and risks of doing international business. U.S. expatriates are leaving Indonesia because of the radical Islamic unrest against U.S. and British interests there. Cargo-laden trucks are taking seven hours to cross the Laredo, Texas, border crossing, compared with only two hours before the September 11, 2001, terrorist attacks. Delphi Automotive Systems, which operates 56 plants in Mexico, scheduled 200 trucks per day to bring products into the United States before the terrorist attacks. Now, the company ships parts in smaller lots more frequently so that it can redirect shipments to planes, boats, or helicopters if the transportation situation so requires.

It may take several years for these effects to dissipate and for the world economy to resume a normal growth pattern.

13.6 NEW OPPORTUNITIES OFFERED BY GLOBALIZATION

Globalization offers unique opportunities for the creation of value. These globalization-specific opportunities are not open to those businesses which do not globalize. It would appear to be obvious that businesses in pursuit of globalization should explore these opportunities to the fullest.

According to Gupta and Govindarajan (2001), there are five globalization-specific opportunities: adapt to local market differences, exploit economies of global scale, exploit economies of global scope, tap optimal locations for activities and resources, and maximize knowledge transfer across locations.

13.6.1 Adapting to Local Market Differences

Among local markets, there are major differences in language, culture, income levels, customer preferences, distribution systems, business practices, and marketing environment. Companies need to adapt their products, services, and processes accordingly.

Business Week is known to have North American, Asian, and EU editions. Baskin Robbins introduced ice cream flavored with green tea in Japan. Coke markets Asian tea (sokenbicha), English tea (kochakaden), coffee drinks not offered by local competitors, and fermented milk drinks in Japan.

Anheuser-Busch Inc. successfully marketed its premium and regular beer products in aluminum cans in diverse countries except China. After a careful market study, the company adapted to local conditions by switching to glass bottles and marketing its premium-grade beer in large bottles to restaurants and regular-grade beer in small bottles to average consumers who buy from the local supermarkets. Budweiser scored a huge sales success in China, because this marketing strategy added value to all three parties involved. Chinese customers are known to like to show off by ordering premium beer in large bottles when inviting friends to eat in restaurants, typically ordering one big bottle for each friend at the table. When they buy beer to consume at home, they want to save money, as no one else is around for them to impress. Furthermore, using glass bottles that could be readily sourced locally and recycled pleased the Chinese government. These practices raised the local labor involved in bottling the contents of the products and eliminated solid-waste disposal problems brought into being by aluminum cans. For the company, this strategy whittled down the beer product cost by doing away with the need to import expensive aluminum cans from the United States.

Whirlpool markets the White Magic washing machine. The company runs a global factory network that makes basic models with 70 percent common parts; the remaining parts are readily modifiable to suit local needs. For the Indian market, it conceived a TV-based advertisement program to associate Indian housewives' belief that white means hygiene and purity with Whirlpool washing machines designed to be capable specifically of washing white fabrics in local water. The company offered incentives for local retailers to stock washing machines and hired contractors conversant in 18 local languages to deliver products and collect cash payments. Annual sales of Whirlpool washing machines went up from $110 million in 1996 to $200 million in 2001—an impressive 80 percent gain.

Kodak has had tough competition in analog film sales and photo processing services from Fuji in Japan, Agfa in Germany, and other global players. In China, Kodak markets its franchise business, the chain of Kodak Express photo supply and development shops, to small entrepreneurs by (1) supporting the franchisee by offering Kodak equipment as collateral to secure local bank loans, and (2) supplying monthly training services to transfer know-how. Kodak was able to establish about 10,000 Kodak Express shops in China by the end of 2001. Its Chinese market share increased from 30 percent in 1995 to 60 percent in 2001.

Not adapting to local conditions could lead to business failure. Walsin-CarTech, a joint venture of CarTech with Walsin-Lihwa in Taiwan, planned to build a steel mill in South Taiwan to produce 200,000 tons per year of stainless steel and carbon bar, rod, and wire products for the world markets. Unexpectedly, the local farmers around the intended plant site delayed the installation of electric power lines until they were financially compensated. The Taiwan government also complicated the plant's permitting process. Meanwhile, competitors added their production capacity for stainless steel. The plant needed more investment capital to build than originally expected.

The two-year delay in the plant startup caused the joint venture to miss the window of opportunity. Subsequently, CarTech abandoned the joint venture in 1998 and moved on to form a steel joint venture in India.

Adapting to local markets will likely allow companies to increase their market share, augment their gross margin due to enhanced value to customers, and neutralize local competition. However, the cost increase associated with local adaptation must be commensurate with the value added to customers, inducing them to pay for the higher price charged. TGI Friday's incorporated many local dishes (e.g., kimchi) into its menu when it entered the Korean market. This strategy backfired because Korean customers wanted to visit TGI Friday's to taste American, not Korean, food.

The degree of local adaptation may shift over time as the result of the global media, international travel variables, and a steady reduction in income disparity. Companies must constantly adjust their local adaptation strategies.

13.6.2 Economies of Global Scale

Companies may realize economies of global scale by taking a number of steps, such as (1) spreading fixed costs—R&D, operations, and advertising; (2) reducing capital and operating costs per unit when capacity is increased; (3) pooling purchase power—volumetric discounts and lower unit transactions costs by sourcing from a few large suppliers; and (4) creating a critical mass of talent—centers of excellence for specific products and technologies.

Autobytel refined a global baseline architecture that consisted of software modules that can be snapped together in various combinations, depending on the local needs. There are hooks for adding customer software when required. New features invented for a specific country may be incorporated back into the baseline if it seems likely that they will be used elsewhere.

There are a number of counterbalancing factors to consider. Too much centralization in product manufacturing can mandate higher costs of distribution. Concentrated production can also isolate the company from the targeted marketplace. Procurement from a few suppliers generates dependency and constraint, insofar as supply disruptions related to labor unrest, access to world-class technologies, and utilization of existing competencies are concerned.

13.6.3 Economies of Global Scope

Globalization allows products and services that do not require local adaptation to be marketed to multiple regions and countries. Companies can benefit by

1. Providing coordinated marketing approaches for standard products (e.g., PCs, software products, ketchup used in McDonald's) to achieve greater consistency in quality, faster or smoother coordination, and lower unit transaction costs
2. Leveraging market power and customer-specific insights, as a global supplier understands a global customer's value chain better and hence is better prepared to serve. For example, FedEx, as a multilocation logistics service provider, better understands the needs of Laura Ashley, a multilocation global customer.

3. Specifying the same hardware platform design for all global locations. GM uses Unigraphics as its common computer-aided design and manufacturing tool and design environment, making it easy for global engineers to collaborate and do design work 24 hours a day.

In 2003, IBM entered an eight-year contract worth $1.2 billion to take over the North American and European information technology operations of the French tire company Michelin.

However, there is a challenge facing the management of centrally coordinated marketing programs: How should businesses reconcile the tension between the needs of headquarters and those of the regional units in the actual delivery of products and services?

13.6.4 Location-Based Optimization

This is another globalization-specific opportunity for companies to add value. Certainly, the intercountry differences in location-based cost structure and services must be considered. By optimally selecting the location for each activity in the value chain (e.g., R&D, procurement, component manufacturing, product assembly, marketing, sales, distribution, and service), global companies can secure advantages in several areas.

A. **Performance Enhancement.**   To build and sustain world-class excellence conferred by talents, speed of learning, and the quality of external and internal coordination, Fiat chose Brazil, not Italy, as the place to design and launch its "World Car," the Palio. Microsoft established a corporate research laboratory in Cambridge, the United Kingdom, rather than in the United States.

B. **Cost Reduction.**   Cost is, of course, a major concern to any company. Cost considerations relate to factors such as local manpower and other resources, transportation and logistics, government incentives, and local tax structures. For example, Texas Instruments set up a software development unit in India, and Nike sources the manufacture of athletic shoes from Asian countries (China, Vietnam, Indonesia, and others).

C. **Risk Reduction.**   Beside economic and political risks, there are also currency risks associated with devaluation. A company might need to spread the manufacturing operations across a few locations to minimize such risks.

For instance, Texas Instruments has been designing integrated circuits in India since 1986, Sun Microsystems has hired Russian scientists for software and microprocessor research, and CrossComm Corp has its communications software written by Poles at the University of Gdansk.

To capture location-based opportunities to create value, companies need to have the right management skills with the flexibility and the ability to foster coordination.

Ford relocated some manufacturing operations to Mexico to become more selective in hiring, to achieve a reduction in turnover, and to realize better productivity by training. Ford was able to achieve lower wage rates as well as higher productivity than it would have been able to do in the United States.

Location-specific conditions do evolve with time. Companies must be flexible in shifting production should the location-based conditions no longer justify a continuation of production at a given site.

Coordination is of critical importance for companies to maximize the value generated by location-based opportunities. Texas Instruments conceived the product concept of its TCM9055 (high-speed telecommunications chip) in collaboration with engineers in Sweden. It developed the product in France with the use of software tools advanced in Houston, manufactured the product in Japan and Dallas, and tested the product in Taiwan.

13.6.5 Knowledge Transfer across Locations

The global company may add value by actively transferring knowledge across locations. Knowledge about product or process innovations and about risk management options are of particular value.

A. **Product and Service Innovations.**   Sharing new ideas among subsidiaries eliminates the "reinvention of the wheel" and speeds up product and process innovation.

Proctor & Gamble used ideas conceived at different centers to develop Liquid Tide in 1980: They built upon technologies developed in Cincinnati (resulting in a new ingredient to help suspend dirt in wash water), Japan (cleaning agents), and Brussels (ingredients that fight the mineral salts present in hard water).

Proctor & Gamble applied an efficient stocklist-based distribution system from India to Indonesia and China and thus significantly minimized its cost of innovation.

In 1997, ABB, a $23 billion industrial-product company headquartered in Zurich, Switzerland, shifted 1000-plus manufacturing jobs from Western Europe to emerging economies over a five-year period for the purposes of increasing efficiency, exploiting lower wages, and becoming more responsive to customers in growth markets. ABB set up a system that propels local ideas for new products and projects around the world in just three weeks. On the basis of key words contained in the proposal, principal global players comment and sign off within an allocated period; this minimizes the time from idea to approval.

B. **Reduced Risks of Competitive Preemption.**   By rapidly transferring new innovations to all global locations, the global company can lessen the danger of losing ideas to competitors for replication in other markets.

Generally speaking, there are two types of knowledge that are important to a company. Codified knowledge is typically embodied in chemical formulas and engineering blueprints and is documented in operations manuals. Such knowledge is readily transferable. On the other hand, the tacit knowledge embedded in people's minds, in behavior patterns, and in the skills of individuals or teams may be difficult to transfer. Examples of such tacit knowledge include the vision of a road map of new technologies or competency in managing global customer accounts. Managers in global companies need to find effective ways to transfer tacit knowledge across subsidiaries.

It is a natural tendency for people to want to preserve specific competencies (e.g., manufacturing superiority) for survival and competitive reasons. Global companies need to systematically recognize unique know-how that is worth transferring and encourage knowledge sharing across locations. All subsidiaries must also be encouraged to learn from peer units instead of being handicapped by the "not-invented-here" syndrome that some locations develop.

Global companies are blessed with location-based, value-addition opportunities not readily available to companies that are domestically focused. Global leadership is needed to take advantage of these unique opportunities to confer competitive advantages.

13.7 PREPARATION FOR GLOBALIZATION

Pursuing globalization successfully requires that global companies understand the success factors gleaned from the experiences of others and that they are properly prepared. Preparation addresses the issues associated with global virtual teams, management styles, globalization pathways, international perspectives, and personal readiness.

13.7.1 Success Factors for Globalization

In order to attain long-term profitability, global companies must (a) build customer relationships supported by superior, worldwide, uniform service; (b) possess wide and deep knowledge about customers; (c) have strong and easily recognizable brands; (d) hire and retain talented people; and (e) organize global virtual teams to effectually implement global strategies.

Other key traits of companies that have successfully entered the global market include

A. Home market strength, which provides a solid revenue basis for global expansion
B. A global business model that is easily replicated and scaled up to multiple markets
C. A powerful vision that motivates employees and communicates core values
D. Strong leaders who can articulate and carry the message globally

General Electric (GE) has been recognized as the master of globalization. It has moved a large number of plants to countries of lower labor cost. Specifically, GE locates and relocates manufacturing plants to locations where the GE quality standards can be met at the lowest cost. GE pursues the business tenet of "continuous mobility." As the emerging economies continuously improve their production skills and standards, GE relocates plants from one location to another to keep costs under control. General Electric Medical Systems, a division of GE, is known to have moved production plants from Paris to Budapest, from Milwaukee to Mexico City, and from Japan to Shanghai and Bangalore. According to Jack Welch, "Ideally, every plant you own would be on a barge." In Figure 13.7, quality standards are plotted against countries. The x-axis is also indicative of the relative cost for achieving a specific quality standard. It is generally expected that outputs of higher quality demand larger efforts and incur higher costs. Some developing countries are unable to produce quality beyond a certain level due to limitations in technology and skills. However,

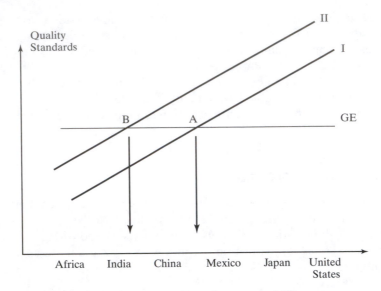

Figure 13.7   The concept of continuous mobility.

as the developing countries steadily upgrade their engineering and production skills, the quality-cost curve shifts to the left and moves from I to II. Most global companies will want to maintain high quality standards for their products, while keeping costs competitive. For example, if GE operates at point A at a given time, then it is only logical that GE seeks to operate at point B, whenever it is feasible to do so, in order to reduce its cost base. Thus, moving its production plants periodically to lower cost countires, while maintaining predetermined quality standards, is the basis of GE's "continuous mobility" strategy.

Of the 150,000 GE foreign workers, 15,000 are in Mexico, 10,000 in China, and 15,000 in India. (See Figure 13.8.) GE also pushed its suppliers to relocate abroad (e.g., Mexico) to cut costs and increase profits.

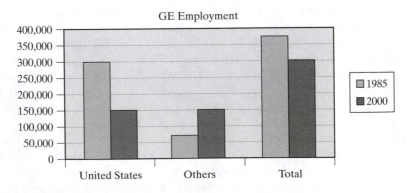

Figure 13.8   General Electric Employment.

13.7.2 Global Virtual Team

To achieve global success, global virtual teams must function well, as the pace of change has increasingly forced global organizations to be more outward looking, market oriented, customer focused, and knowledge driven (Lipnack and Stamps 2000). Global virtual teams would typically be composed of members who are geographically dispersed, each with specific technical or business competencies, cultural and language backgrounds, working habits, and variable comfort levels with technology accessibility and utilization. Global teams may be ineffective because of cultural values, cultural and language differences, and other factors, such as the leader's approach, lack of organizational support, or individual rewards overshadowing the team's success. (Dreo, Kunkel, and Mitchell 2002; Hoefling 2001).

The cultural barriers just mentioned could come from

1. Function—due to differences in reasoning styles, reactions, and getting motivated by people in various professions, such as engineers and marketing personnel
2. Organization—due to different value perceptions and behaviors
3. Nationality—due to different styles of human interaction because of national origin (e.g., in the United States the emphasis is on the individual, whereas in East Asia the emphasis is on the group and on reaching a consensus)

Cultural issues need to be overcome. Metrics should also be set for goals and performance, and these should be focused on deliverables so that all team members fully understand their respective accountabilities. Managers also should convince all team members of the value of change made possible by the team activities.

Selecting the right team members is a crucial first step for the leader of any global virtual team. Preference should be given to people who can act on their own when needed. One model of membership selection is to find regional alliance partners who have the necessary core competencies to execute a global, centralized strategy. The best kind of alliance partners are those who think ahead, demonstrate commitment with investment, understand the company's requirements, share the same vision, and operate on behalf of the company.

When building global virtual teams, leaders need to pay attention to factors known to have a direct impact on team success. These factors include clearly articulating common goals, being aware of overlapping competencies and skills, acknowledging each other's contributions and needs, formulating clear procedures and ground rules for working together, and establishing common rules and technologies for sharing information and data.

In order to operate global virtual teams effectively, proactive attention and preventive maintenance are needed. Members may need to be trained properly. A constant monitoring of the team progress is advisable. Roles and responsibilities must be clearly communicated and emphasized. To get team feedback, frequent communication is needed.

Communication is the key to keeping teams together and on track. Poor communication can bring forth stumbling blocks affecting team success. Some of these stumbling blocks are (1) commitment of team members may be misdirected from global activities to local priorities (the "out-of-sight, out-of-mind" syndrome); (2) trust between global team members may not be strong enough; (3) time-zone differences, resulting in work-time overlaps, can discourage frequent communications; (4) language barriers, as English is not universally spoken, can make verbal and written communication in other countries uneasy and time consuming for some; (5) culture differences produce different work habits and value perceptions, which can cause communication to be less free and open. Team leaders need to proactively schedule conference calls, in addition to e-mail, and have quarterly face-to-face meetings if budgets and schedules so allow.

Example 13.3.

Global teams are often deployed to handle important tasks, such as devising specific implementation plans, proposing market entry strategies, and designing global products.
How do you make global teams efficacious?

Answer 13.3.

The following generalized steps could help make a global team efficacious in achieving its intended objectives:

A. Select a team leader who is well recognized for leadership quality, interpersonal skills, and managerial capabilities. Above all, the team leader must enjoy the strong support of the company's top management. The president or another suitable top executive of the company should announce the appointment of the team leader to demonstrate company commitment.

B. The team's objectives must be specified, including standards to measure progress and the expected outcomes. The potential impact of the team efforts on the company's profitability is clearly understood.

C. The team leader specifies the qualifications of the team members on the basis of the expertise and level of experience needed to achieve the team objectives.

D. The team leader solicits suggestions from various regional management centers and receives assurance that all required local support (time off, secretarial work, analysis, use of local facilities and engineering, marketing, and other resources) will be offered to team members.

E. The team leader interviews specific candidates and selects the team members with the concurrence of local management. Team members with different skills and expertise may be needed as the team progresses through various stages, resulting in a constant flow of people moving into and out of the team. By organizing the team properly, the team leader ensures that a good working atmosphere prevails at all times.

F. With the support of local management, the team leader should compile a comprehensive roster of corporate talents (who is specialized in doing what, for how long, with what accomplishments, etc.). This roster is constantly maintained and could be used by all team members.

G. The team leader makes sure that all members receive proper teamwork training regarding

   1. Working in teams
   2. Communications, including the use of associated equipment and tools
   3. Problem solving
   4. Available support functions and resources that can be tapped
   5. Group goals and expectations

H. The team leader creates guidelines for fostering communications between team members (e.g., Intranet, videoconferencing).

I. The team leader enhances personal interactions and cooperation, and builds trust and confidence among team members by inviting all to be physically present at the first team meeting.

J. The team leader plans out and implements subsequent team meetings to focus on achieving the team objectives. He or she encourages all to communicate, prevents the dominance of meetings by a few, and takes into account the diverse cultural backgrounds of the respective members.

K. The team leader assigns members specific actions and steps to carry out (e.g., analysis, focus group inputs, activity-based costing, etc.).

L. The team leader conducts field trips, on-site visits, and other activities needed to collect data and to exercise judgment.

M. The team leader engages outside consultants and other resources to provide benchmarks, suggest alternatives, or overcome bottlenecks if needed.

N. The team leader solves problems, resolves conflicts, and secures support functions needed for all team members.

O. The team leader provides regular reports to all regional managers concerning the teams progress and team member performance.

P. The team leader strives to achieve a consensus on major issues. He or she makes sure that the outcome serves as a valid solution to the problem under consideration.

Q. The team leader reports the final outcome of the team to the company president or to other upper management. He or she gathers the whole team to make a formal presentation of the results and to celebrate the successful completion of the team effort.

R. The team leader documents the experience and preserves the learning gained by the team efforts, by collecting inputs from all team members.

13.7.3 Management Style

Management style plays a critically important role in globalization, because globalization makes communication and personal interaction necessary between people of different cultural, business, and personal backgrounds. For example, American managers may grow up in hierarchical and command and control systems. They typically perform thorough competitive analyses, using strong analytical tools and strategic audits. They focus on short-term profit objectives. They value being goal and achievement oriented. A great number of them are competitive, aggressive, ambitious, and intolerant of poor performance. Their management style is also influenced by American education, politics, and internal and external reward systems. In contrast, the Japanese management style is characterized by teamwork, market-share objectives, commitment to quality, and a philosophy that says, "The nail that stands up gets pounded back down."

It is thus important for managers of global companies to recognize and accept extreme differences in management styles practiced by people in different geographical regions. In fact, there is no style—American, Chinese, Japanese, German, French, or British—that must be rejected. To be successful in a global economy, a manager must accommodate the priorities of other cultures by stressing shared goals and a common outlook; remaining open minded; adapting to the local culture, business practices, and value systems; and avoiding both cultural and intellectual arrogance (Schneider and Barsoux 2003; Brett 2001).

Companies involved in global businesses typically change their management styles over time. Initially, some global companies direct worldwide activities by home-country standards, adopting central decision-making and control paradigms. As the global companies increase their foreign investments, home-country standards become a reference basis for managing worldwide operations by pursuing a model of decentralized and autonomous global operations. Gradually, the global companies build a global network and follow a transnational strategy that is integrated and interdependent.

Example 13.4.

In pursuing global business, one commonly practiced strategy is to elicit the maximum possible collaboration with the right business partners in the host country. This is because their local knowledge is of tremendous value in facilitating the adaptation of foreign-made products or services to local market needs and for problem solving. Creating trust among the partners will naturally be a critical first step towards implementing such a strategy.

What are some American management practices that can be counterproductive in winning the collaboration of foreign partners?

TABLE 13.2   American and Japanese Management Practices

	United States	Japan
Employment	Short term	Lifetime
Decision making	Individual	Group consensus
Responsibility	Individual	Collective
Evaluation	Rapid	Slow
Control	Explicit formal	Implicit informal
Career path	Specialized	Nonspecialized
Concern	Segmented	Holistic

Answer 13.4.

As background information, Table 13.2 contrasts the typical American management practices with those of the Japanese.

The following list describes some of the American practices that could be counterproductive in winning the collaboration of foreign business partners:

A. Exhibiting a highly competitive and arrogant personal demeanor and an ignorance of local culture, customs, and other differences in business practice, alienates the foreign partners. Here are two examples of well-known language blunders: (1) An "escrow account" in English means a "gyp account" in French. (2) When an issue is "tabled" in America it means that it is not to be brought up again; but in England, it means the exact opposite—that it will be brought up.

B. Being highly impatient and pushing aggressively for instant decisions fails to allow time for the foreign partners to create group consensus for the decisions at hand.

C. Emphasizing short-term profitability is a barrier to recognizing the business goals of the foreign partners of seeking long-term, broad-based collaboration.

D. Being proud of risk taking, and exhibiting decisiveness, a command-and-control rationale, and an excessive profit motive, fails to recognize that the values favored by the foreign partners might be different.

E. Adopting the "ugly American syndrome" makes American managers insist that foreign partners do exactly what Americans do.

13.7.4 Strategic Pathways to Globalization

Companies pursue globalization along any or all of geography-based, product-based, customer-focused, and Internet-based pathways.

The *geography-based pathway* is a pathway in which companies pursue globalization in geographical areas that have common cultural and linguistic ties—Canada and England for U.S. companies, China for Taiwanese companies, Southeast Asian countries for Chinese companies, and African countries for French companies.

Following the *product-based pathway*, companies conceive and perfect specific products that do not require local customization. The companies then distribute the products globally wherever there is a demand for them.

According to the *customer-focused pathway*, global companies follow their major clients to foreign markets with a basket of products to serve the needs of the local customers of these clients more efficiently. Examples of such baskets of products include the combination of insurance, banking and securities offered by Citicorp, and that of logistics and inventory management used by FedEx.

The *Internet-based pathway* prescribes that companies devise a Web presence and leapfrog over other competitors to reach end users in numerous global markets.

13.7.5 International Perspective

"Thinking globally and acting locally" is regarded as a best practice for a global company to keep things in perspective while achieving practical results. Without an international perspective, global managers have a disadvantage in the global economy of the 21st century (Garten 1999).

13.7.6 Globalization Mistakes

There are a number of mistakes commonly made by companies pursuing globalization. Among them are a *lack of company commitment*, when companies do not make a firm and sufficient corporate commitment to people, capital, and time; and *low management attention*, when senior managers get involved only when there is a crisis affecting earnings. Oftentimes, companies assign *low priority*, viewing the international businesses as "incremental," and do *not engage foreign partners decisively* (take a minority position when entering a joint venture with local partners). Should the business relationship turn adversarial, these companies can get blocked out of the target markets; thus, one should always attempt to keep 50–50 ownership to stay even.

A *lack of cultural sensitivity and understanding* has been known to be a major source of frustration for global managers. An American in Japan is described in the following example (Glover, Freidman, and Jones 2002):

A young American manager was sent to Japan to work with the Fuji villagers on a forest project. During his first week in Fuji, he requested a local village chief to send "three men to do an eight-hour job clearing a field." Each of the three men was to be paid an hourly wage.

The next morning, 40 able-bodied men from the village showed up to do the work. The American manager asked the group to select three men, reasoning that, as he did not need all 40 of them, he would send the remaining 37 back to the village. The chief responded that, if all 40 of them cleared the field, they would complete the work in one to two hours and then could go back to the village to do other work. Furthermore, the chief requested that the men not be paid individually. He would take the money and put it in to the village fund, a traditional communal means for distributing money equally.

The American manager sent the chief and all of his 40 men away and paid higher wages to three Fujian Indian contract workers he got from a nearby city a week later. He complained that the Fujian villagers were not motivated to be productive and they did not seem to have any individual initiative.

This was clearly a case of a cultural clash between the occidental approach to productivity based on "scientific management" and the oriental approach of getting the work done in a speedy manner by a collective work group. Their culturally conditioned views of productivity were different. The American manager was trapped in the "one best way" he believed in, namely, "three people to do an eight-hour job." He was unable to see possibilities of adapting to the concern of the chief, who needed to secure external funds to augment the overall village operations, while delivering work at a faster rate by using all of his able-bodied men. The chief thought it ought to be the same to the American manager, as the total cost remained the same, whether the work was done by 3 or 40 people, as long as it would be at the same hourly wage rate. Thus, the chief was equally frustrated by this exchange and viewed any future interchanges with this inflexible American manager with suspicion from that point on.

Example 13.5.

In the 1980s, a lot of multinational companies were eager to conquer foreign markets. In a hurry, they committed a large number of culturally insensitive blunders, contributing to major marketing and business failures at the time. Name a few such embarrassing examples.

Answer 13.5.

The examples of culturally induced mistakes are plenty. Reviewing them from time to time is useful only for the purpose of preserving the learning opportunities they offer:

1. Chevrolet introduced "Nova" in Puerto Rico and found out only later that *No va* means "doesn't go" in Spanish.

2. Ford introduced a low-cost truck, the Fiera, into some developing countries without success. It turned out that *fiera* means "ugly old woman" in Spanish.

3. Esso, the oil company, went to Japan. The phonetic pronunciation of *Esso* in Japanese means "stalled car," which was not helpful in promoting the sales of gasoline there.

4. Cadbury Schweppes, an English food company, introduced its Rondo soft drinks into the United States. It failed badly, although it was a success in England. Later, they found out the people in the United States thought Rondo was a dog food.

5. Rolls Royce, before Mercedes acquired it, marketed a car called "Silver Mist." When that name was translated into German, the "mist" became "excrement." It forced Rolls Royce to change the name.

6. McDonald's promoted its food products in Japan using white-faced clowns at one time. White face in Japan is a death symbol.

Source: David A. Ricks, *Big Business Blunders: Mistakes in Multinational Marketing*, Homewood, IL: Richard D. Irwin, 1983.

13.8 GLOBALIZATION DRIVERS

There are a number of driving forces present in the global economy. Each company may be driven to globalization by a different set of drivers, such as market reach, cost, competition, and government.

Market drivers include worldwide increases in per capita income that result in greater purchasing power and an increasing demand for goods worldwide.

Another market driver is the convergence in lifestyles, tastes, aspirations, and expectations of consumers. An additional market driver is increased global travel, which bring about a new class of global consumers. A further market driver is the creation of larger future markets in emerging countries. Over 90 percent of the world's population is outside of the United States. Companies need to pursue globalization to reach these extended markets (Johansson 2000; Cundiff and Hilger 1988).

To be located close to customers is an important corporate marketing strategy. There are about 17 automobile assembly plants built by foreign carmakers on American soils. As recently as June 2003, Nissan motors announced an investment of $250 million to relocate its production of the Pathfinder SUV (2005 model) from Japan to Smyrna, Tennessee. This relocation program would create 1500 new jobs in the United States. Also well known are the investment examples of General Motors, Ford and Volkswagen in China, and Renault in Japan.

Cost drivers include lower manufacturing and production costs (due to lower labor costs), economies of scale, accelerating technological innovations, and upgraded transportation and logistics. Some companies seek a cost advantage as the primary motive to go global.

Competitive drivers include (1) global competitors with speed and flexibility; (2) the increased formation of global strategic alliances, resulting in a proliferation of partnership relations with suppliers, customers, and competitors (Calpan 2002); and (3) more countries becoming attractive battlegrounds. Creating competitive advantages is the principal goal for some companies to pursue globalization.

Government drivers of globalization include the emergence of trading blocks (EU, NAFTA), a large scale of privatization (Brazil, China, etc.), and (3) a reduction of trade barriers (WTO). Companies go global to take advantage of the benefits made possible by these official or semiofficial government bodies.

Ernst & Young conducted a survey of more than 300 CEOs in 1993. The top 10 drivers in the global race were recognized as follows:

1. Increased speed of delivery to customers
2. Enriched ties with strategic partners abroad
3. Enhanced support of domestic customers' international operations
4. Meeting of cultural needs of foreign customers
5. Access to new technologies
6. Avoidance of overseas protectionism
7. Reach for lower taxes and government benefits
8. Access to foreign technical and management talent
9. Utilization of low-cost labor
10. Avoidance of domestic regulatory constraints

Indeed, numerous forces of significant magnitude are driving companies toward globalization.

13.9 IMPLEMENTATION ISSUES RELATED TO GLOBALIZATION

When pursuing globalization, the management of global companies can benefit from the international business experience gained by other companies. This section addresses emerging issues and the ways some global companies conduct global businesses.

Global companies face two emerging issues. The first is *fairness*. Traditionally, in the market economy, major multinational profit-seeking companies have pursued globalization. Home-country governments of these companies tend to set the macro-economic policy and rules of the game, and they do not always have the interests of developing countries in mind. Calls have been issued by some developing countries to seek global governance, with the participation of all developing countries in order to ensure fairness to all involved.

The second issue is *conflicts of interest*. In industrialized countries, workers in certain "old-economy" sectors (e.g., mining, textiles, agricultural and other manufacturing enterprises) face unemployment when jobs are transferred to developing countries that offer lower labor rates and are more competitive. Workers in the "knowledge economy" sectors—electronics, computers, and high-tech export businesses—are gaining. Surveys indicate that people with low incomes are generally opposed to globalization, while those with high incomes favor it.

Protests staged against the WTO by the joint forces of labor (stumping for work rules to protect U.S. jobs), environmental (promoting the reduction of pollution), and human rights groups (protesting for the elimination of political and religious suppression) at various international places in recent years have indicated clearly that not everyone is in favor of globalization.

Even inside developing countries (e.g., China), globalization is not welcome by all. Increased privatization, foreign investment, market opening (telecommunications, banking, financial services, and others), and increased foreign trade can facilitate the destruction of countless existing state enterprises and thus cause massive unemployment in the state sector. Hence, globalization brings gains to some sectors and losses to others, as discussed previously. Global companies need to devise long-range programs to address these issues in order to sustain the benefits they realize from globalization.

Global companies engaged in international business may be classified into one of the four groups listed next, according to their corporate behavior in conducting global business.

13.9.1 The Defender

These companies are internally focused. They have no global orientation and no international element in their business strategies. They are focused on domestic markets and make no effort to understand other markets and cultures. They have limited skills and knowledge to pursue foreign markets. They look for governments to provide protection against foreign intrusion (e.g., through trade barriers, quota, duties, laws, and special agreements). Their view is, "What is different is dangerous." Examples include (1) the U.S. steel industry, which sought quotas from 1960 to 1980 to restrict Japanese steel imports to the United States; (2) the U.S. footwear industry, which attempted in vain in 1980 to get import protection; and (3) the U.S. textile and machine tools industries, which got government relief against imports in 1975. Today, many of these industries are under the protection of bankruptcy laws or import tariffs. Some of them have barely survived under governmental subsidization programs.

13.9.2 The Explorer

These companies are largely inwardly oriented, with dominance in the domestic markets. They are aware that opportunities may exist abroad. They move into foreign markets very cautiously after closely studying the opportunities available. They have some knowledge about the markets abroad and possess a restricted set of skills to pursue them. Overall, they have small international business revenues. The home-based headquarters controls their businesses. They may pursue some export and franchising activities, but with rather limited investment commitment. Companies in this category include Seiko and Lotus (which has been acquired by Microsoft).

13.9.3 The Controller

These companies are more externally oriented than the explorers. They want to control the market abroad. They have sufficient knowledge and skills to pursue foreign markets, but have a limited global mind-set. They generate a significant amount of overseas sales revenues with major investment commitment. They impose their home culture and practices on overseas operations, although they do tailor some strategic decisions to suit the local cultures or to optimize the interests of their home office and the local markets. They maintain financial and strategic control at the home office, while allowing some independence in overseas activities. Examples of companies in this category include Coke, McDonald's, and Pizza Hut.

13.9.4 The Integrator

These companies have a global perspective based on heightened awareness (knowledge) and strengthen abilities (skills). They form a worldwide web of relationships, partnerships, and alliances with suppliers, developers, designers, distributors, competitors, and customers. They reconfigure these relationships over time as new threats and opportunities arise. They coordinate, rather than control, these networks of business partners. They focus on overall organizational effectiveness in delivering products of value to customers. They understand, bridge, and resolve differences between people, companies, values, and cultures. Their core strategy is to win in the marketplace by leveraging, sharing, and nurturing complementary capabilities. In this group of companies, General Electric, Toyota, and Dell are known to have formed networks with primary, secondary, and tertiary suppliers and subcontractors. To be globally successful, companies need to walk, talk, and act like integrators.

13.10 QUALITY OF GLOBAL LEADERSHIP

Global management is demanding, indeed. According to Lamberton (2002), global managers must possess certain characteristics and savvy to be successful in a global environment.

13.10.1 Inquisitive Mind

Global business is highly complex and uncertain, due to variations in cultural, linguistic, political, social, and economic conditions. Global managers must constantly learn in order to succeed. Constant learning requires an inquisitive mind. Successful global leaders are adventuresome, curious, and open minded.

Inquisitiveness strengthens personal character growth, characterized by emotional connection to people and uncompromising integrity; and duality, the capacity to handle uncertainty and the ability to balance tensions. Personal integrity inspires staff trust and commitment, which in turn affect the implementation results of any global strategy.

13.10.2 Global Mind-set

To be effective in global business, managers need to have a global mind-set, which will enable them to do the following (Jeannet 2000):

1. Extend concepts and modes from one-to-one relationships to holding multiple realities and relationships in one's mind simultaneously. Then act skillfully on this more complex reality (global think).
2. Change management orientation from taking individual initiatives to adopting team and group initiatives.
3. Focus simultaneously on hard issues (low-cost producers, bottom lines, budgets, manufacturing, marketing, distribution, head count, and finances) and soft issues (value, culture, vision, leadership style, innovative behavior, and risk taking).
4. Balance the pressures of global integration (product standardization) and local responsiveness (adjusting to the needs of local markets). Recognize the interdependence of the global economy and view the world from a broad perspective. Seek trends that affect company business, balance contradictory forces, rethink boundaries, and build and maintain organizational networks at the global level.

5. Serve as a catalyst within the company, being sensitive to, and capable of, managing cultural diversity. Become more tolerant of other people and cultures. Consider culture diversity an asset. Connect emotionally with people and the worldwide organization. It is worth noting that European managers are said to be more accustomed to exposure to cultural diversity than American managers.

6. Recognize complex patterns in the global environment and thrive on ambiguity. Become proficient at managing uncertainty and dealing with conditions that change constantly and are inherently complex.

7. Preserve a unique time and space perspective. Take a long-term view, extending personal space in geography and relationships.

8. Exhibit business and organizational savvy. Recognize opportunities, grow in knowledge of available resources, and be capable of mobilizing them to take advantage of opportunities.

13.10.3 Knowledge and Skills

Global managers must possess specific knowledge and capabilities to succeed. They should have a mastery over technology (information systems, telecommunications, and operations) and use it effectually. They need to be aware of the social and political features of different countries. They should be familiar with the specific culture and cross-cultural issues that affect management. Of great importance is their understanding of the global competitive practices in manufacturing and communications, such as total quality management, just-in-time delivery, factory automation, employee involvement, and outsourcing. Also helpful are some general knowledge about business and industry and the skills required to put knowledge into action, to become acculturated, to lead, and to motivate a diversified workforce.

13.10.4 Global Business Savvy

Of critical importance to global companies is the business savvy of their global leaders to size up business opportunities and to have a vision of doing business worldwide.

A. Recognize Global Market Opportunities.   The ability to recognize new opportunities is a key leadership quality of global managers. They need to be able to

(a) Assess the cost and quality differences in production outputs and inputs, and exploit cost differentials for land, energy, labor, raw materials, and people talents.

(b) Identify market needs for goods and services from a deep and broad knowledge base, having mastered finance, accounting, marketing, human resources, operations, international relations, economics, industry conditions, and strategy disciplines.

(c) Size up opportunities for efficiency gains by (1) eliminating redundancies to wring out costs, (2) using economies of scale in procurement, (3) pursuing standardized outputs, and (4) selling to multiple markets.

(d) Create competitive advantages by forming supply-chain networks involving strategically selected, local and global partners with complementary resources and expertise.

B. **Envision Doing Business Worldwide to Ultimately Make Money.**   The manager should also have a good overall perspective of what the company's business has to offer—what the core is, why the core is what it is, and what drives the core. This understanding is combined with the fundamental good business goal of making money for the company.

13.10.5 Demonstrate Global Organizational Savvy

Global managers are required to demonstrate organizational capabilities. These capabilities are built upon specific qualities:

A. **Know Your Company.**   Global managers must have an intimate knowledge of their own companies with respect to subsidiaries' product lines, cost structures, and overall competitiveness. They should know the location and quality of technological resources available, including physical assets and managerial and employee talents. Global managers must be known to the company's key decision makers by having served on key committees, participated in task forces, and attended critical meetings.

B. **Mobilize Resources.**   Global managers need to be able to mobilize resources to take advantage of global opportunities. Establishing trust with top management and key decision makers will assure their favorable response to these mobilization efforts.

C. **Develop Insight.**   Global managers must be able to identify critical knowledge and capabilities beyond merely understanding policies and programs.

D. **Keep Current.**   Global managers must keep themselves constantly informed of what is presently going on at the headquarters.

13.10.6 Personal Preparation

Global businesses require managers to be patient with, tolerant of, and open-minded toward divergent cultures, customs, and business practices. They should be dedicated to the mission at hand and assume a flexible negotiation style to win. They should possess stamina to endure personal hardships, the personality to effectively handle uncertainties and ambiguity, and the conviction that what is different is not necessarily dangerous (Dalton, Ernst, Deal, and Leslie 2002; Marguardt and Berger 2000; McCall and Hollenbeck 2001).

The manager needs to recognize that, for global businesses to succeed, the new model of responsiveness, partnership, teamwork, and decentralization must replace the early management model of efficiency, hierarchy, control, and centralization.

Not everyone has the desire to become a global leader. Those who want to be global leaders need to become proactive in seeking opportunities for leadership development. Personal preparation can assist in refining these desirable traits.

13.11 PRODUCTION ENGINEERING IN A GLOBAL ECONOMY

Global companies needs to scale down the capital investment for production, shorten the time to market, change product features flexibly to meet the requirements of the changing local markets, ensure product quality, and offer after-sales support and services to maintain long-term profitability. As a result, production engineering will experience significant changes in the global economy (Philips 1998).

To compete effectively on a global basis, companies need to aggressively engage partnerships, form supply alliances, and create networks with domestic and international suppliers of raw material, subassemblies, semifinished goods, and others to make finished products flexibly and speedily (Calpan 2002). The production engineering issues involved in global environments will be significantly different from those in traditional old-economy companies.

In old-economy companies, where vertical production has been the norm, production engineering deals with the manufacturability of products and the application of resources (technologies, labor, materials, plan layout, utilities, etc.) to optimize production.

In the knowledge economy, in which global companies pursue production via networked partners, each having its own production capabilities and engineering, production engineering will need to deal with issues in different ways. Some of these issues and methods are as follows:

A. **Product Design and Specification.** Communicate and enforce interface specifications of all parts produced by the partners. Standardize interfaces to promote the exchangeability of parts in order to offer different product features. Effect infrastructure, hardware and software systems to facilitate supply-chain management. Integrate middleware, EDI, systems technologies, regulations, production practices, and others to improve efficiency.

B. **Manufacturing.** Selectively apply Web-based enablers in manufacturing and operations to gain advantages. Produce core parts, and assemble the finished products with vendor-supplied parts. Refine an assembly procedure to safeguard product integrity. Conduct statistical process control (SPC) to ensure system quality of the assembled, finished products. Pursue ISO and other quality process certifications of networked partners.

   With GE's example of "continuous mobility" as a guide, companies should explore ways to have quality products made at locations where the loss is lowest. To implement such a manufacturing strategy, production engineering at headquarters must be prepared to transfer manufacturing and technologies processes to new recipient groups whenever needed.

C. **Management.** Share information and foster collaboration between the networked global partners. Preserve knowledge and apply global experience and learning. Strengthen a customer relationship management system by tapping into the expertise, innovation, and knowledge base of the networked partnership. Assist in setting up an enterprise resource-planning system to optimize the utilization of corporate assets. Make the production process transparent to, and traceable by, customers.

D. **Logistics.** Ensure just-in-time (JIT) delivery and optimize transportation logistics.

E. **Inventory Control.** Balance loads between global sites to adjust to changing supply and demand and marketplace conditions.

F. **Risk Management.** Respond to labor, politics, currency, and unexpected changes and conflicts.

Production engineering students are typically involved in facilitating the transformation of old-economy companies into knowledge-economy companies engaged in globalization.

13.12 JOB MIGRATION INDUCED BY GLOBALIZATION

Globalization has a profound impact on jobs, which are migrating to and from the United States as the push for globalization intensifies. For example, when foreign products (such as textiles, shoes, toys, and cheap electronics) overwhelmed U.S. markets in the 1980s, some American workers suffered. When Japanese automakers set up shop in the southern part of the United States in the 1990s to compete against Detroit carmakers, their investments created new jobs for American workers. When Nokia expanded its cell phone business into the United States, American workers benefited.

To maintain the competitive edge, global companies are constantly looking for ways to offer products and services better, cheaper, and faster. One way of achieving cost competitiveness is to practice the "continuous mobility" strategy spearheaded by General Electric, seeking and finding the least-cost production sites that achieve acceptable product quality. GE moves its production plants to new locations whenever doing so will result in cost advantages without incurring penalties in quality. Factors in favor of outsourcing include reduced telecommunications costs, increasingly more powerful computers, and rising levels of skills abroad.

Besides reducing the cost of doing businesses, many companies outsource work to conserve investment capital; shorten time to market; build a variable cost structure; secure needed skills, technologies, and expertise; and realize a round-the-clock operation.

Over the years, a large number of American-based multinational companies have implemented strategies that cause engineering and other jobs to migrate to Third World countries. Figure 13.9 exibits the reduction in manufacturing jobs in the United States during the last several decades.

It is clear that engineering and blue-collar jobs related to manufacturing have been disappearing at an alarming rate. However, as presented in Figure 13.10, American manufacturing productivity has risen constantly over the last decade due to automation and a shift to higher value work.

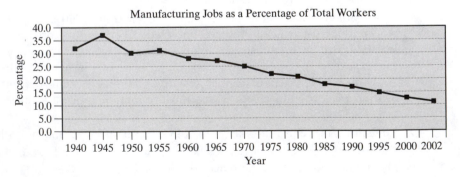

Figure 13.9   U.S. Manufacturing jobs as a percentage of the total workforce. *Source:* U.S. Department of Labor.

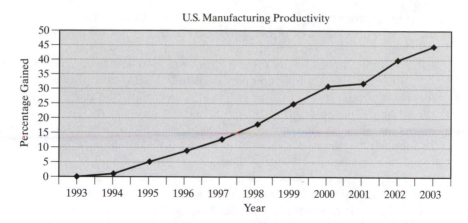

Figure 13.10   U.S. Manufacturing productivity. *Source:* U.S. Bureau of Labor Statistics; Federal Research Board.

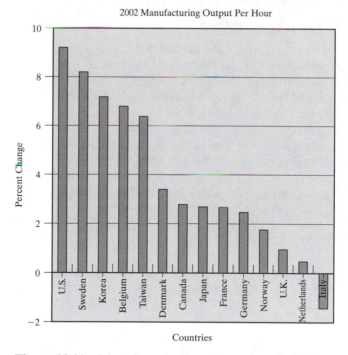

Figure 13.11   Manufacturing output per hour in 2002
Source: U.S. Bureau of Labor Statistics 2004 A.

For the year 2002, U.S. productivity reached an annual growth rate of 9.2 percent, the highest worldwide. (See Figure 13.11.) U.S. Bureau of Labor Statistics (2004 B) has determined that outsourcing, to both overseas and domestic partners, has contributed about 1.5 percent toward this 9.2 percent growth.

Innumerable U.S. firms are involved in outsourcing engineering and science related work to developing countries. Engardio (2003) offers a comprehensive list of American companies engaged in outsourcing engineering and science jobs, including General Electric, Flour, Intel, Oracle, Texas Instruments, Hewlett Packard, Boeing and others.

Other big technology companies, including IBM and Microsoft, are bringing in foreign workers to America on L-1 visas, which allow the companies to pay workers their home-country wage rates for as long as seven years. The federal government places no limits on the number of L-1 visas issued to a company (Bridger 2003). Ford announced its plan to purchase auto parts worth $1 billion from China in 2004. With the stroke of a pen, Ford shifted thousands of jobs to China.

A large number of companies are also engaged in outsourcing various nonengineering jobs overseas (e.g., customer service, microbiology research, tax return preparation, back-office support, IT applications, and CT scan interpretation work). In fact, McCarthy (2002) predicts that 3.3 million U.S. white-collar jobs valued at $136 billion in wages will migrate abroad by the year 2015. (See Figure 13.12.)

The white-collar job migration is a serious problem that must not be overlooked. It is a moving train viewed by many to be unstoppable.

13.12.1 Global Pie Concept

Is white-collar job migration all that bad? It is certain that some white-collar workers will be adversely affected. Some analysts believe, however, that the overall impact of offshoring may actually be positive for both the United States and the world at large. Agrawal and Farrell (2003) estimated that the benefit of about $1.45 may be derived from $1 spent by America in offshoring, suggesting that the global pie will be made bigger, not smaller, by the offshore investments. The key point is, of course, that the offshoring investment is not predicted as a zero-sum game, wherein the gain of one party is at the expense of the other. Rather, both parties are expected to benefit in the process.

13.12.2 How to Survive the White-Collar Migration

Agrawal and Farrell (2003) assume in their studies that the value of U.S. labor reemployed is $0.45 to $0.47 for each U.S. dollar invested in offshoring. This means that the

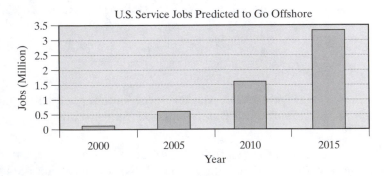

Figure 13.12 U.S. white-collar jobs predicted to migrate abroad. *Source:* McCarthy (2002).

investment funds freed by going offshore can be effectually deployed within the United States (e.g., hiring employees to do higher value work and initiating leading-edge R&D).

Engineers whose jobs may be outsourced unexpectedly need to ask the question "What will it take for them to become part of these 'reemployed resources' that companies will want to use the freed investment to engage?" There are a number of steps engineers may take to augment their relative competitiveness in this situation.

A. **Get Advanced Training.**   Currently, the white-collar jobs migrate primarily from the United States to four lower wage countries: China, India, the Philippines, and Mexico. These recipient countries bring forth an overwhelmingly large number of engineering and science graduates at the B.S. level every year. In Figure 13.13, the data points for the years 1989 and 1999 originated from the National Science Foundation, whereas those for 2009 are predictions. In other words, the number of U.S. graduates at the B.S. level are predicted to become a smaller percentage of the five-country total— only 15 percent—by 2009. Each U.S. engineer or scientist graduating with a B.S. degree will have six other engineers or scientists as competitors in the job market. Similarly, the M.S. and Ph.D. graduates in the United States will make up about one-third of the five-country total by 2009. Each U.S. engineer or scientist with a M.S. or a Ph.D. degree will have only two others from the four low-wage countries to compete against.

These numbers suggest that it would be useful for American engineers to get advanced degrees as one of several steps to decrease the adverse effects of white-collar migration. Having an advanced degree is likely to be useful, but is still not a guarantee that an engineer will survive the threat imposed by the white-collar job migration.

B. **Practice the "Steady-Ascent" Strategy.** Engineers should constantly seek to upgrade the ways they perform their jobs, so that their employers can provide better, faster, and cheaper products and services to their customers. Innovations are desirable at the interface between engineering, design, and manufacturing to minimize labor, cut waste, enhance quality, and speed up time to market. The added value concept is quite

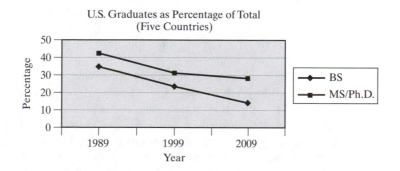

Figure 13.13   U.S. engineering and science graduates as percentage of the five-country total. *Source:* National Science Foundation (2004).

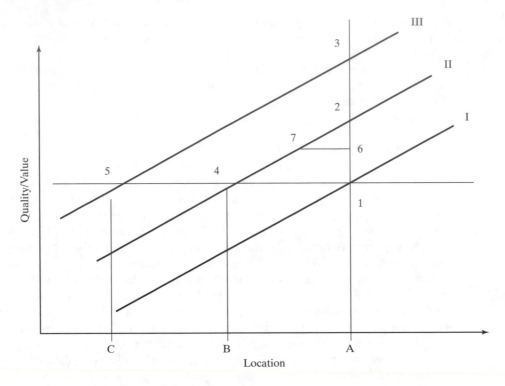

Figure 13.14 Value addition in time.

ordinary in itself. However, it becomes critical in times of white-collar job migration. (See Figure 13.14.)

Quality and value versus location is plotted in Figure 13.14. The horizontal axis denotes production sites, from low cost on the left to high cost on the right. Production cost at A is higher than at B, which in turn is higher than at C. The straight lines I, II, and III symbolize the linear relationship assumed to exist between the quality–value combination and cost. It is reasonable to assume that the higher the product or service quality is, the higher its production cost will be. In the course of time, such a straight line shifts from I to II to III, reflecting the constant strides made by developing countries in their capabilities of manufacturing products and supplying services at high quality, while keeping their wages at relatively low levels.

Let us assume that Point 1 is the current point of operation, with a quality–cost relationship depicted by the straight line I and its location selected at A (e.g., the United States). As the developing countries upgrade their engineering and production skills and become increasingly capable of supplying work of acceptable quality, the quality–cost curve available to U.S. firms moves from I to II. Under these circumstances, U.S. firms will logically shift production and other work from point 1 to 4 and relocate the associated operations from A to B (e.g., to India or China), while keeping the quality at an acceptable level. Should another country offer even lower wages in the future, the U.S. firms will not hesitate to move their facilities again from B to C to operate at point 5. This is consistent with the concept of "continuous mobility" practiced by General Electric and other firms. (See Figure 13.7.) The relocation of operations from A to B and from B to C will ensure that the firms remain cost competitive over a long period, without sacrificing the quality of their products or services.

For individual engineers to ensure that they do not get displaced by white-collar migration, Figure 13.14 offers an obvious strategy. It is a given fact that the developing countries will constantly upgrade their engineering and production skills, as illustrated by the shifting of the straight line from I to II to III. Therefore, in order for engineers to survive the white-collar job migration, they need to periodically add enough value to ascend from point 1 to 2 and from 2 to 3 while employed in the home country (A in this example). In fact, should the individual reach point 3 after the quality–cost line II is reached, his or her job may be outsourced. Specifically, if the value contributed by the engineer reaches Point 6 rather than Point 3, then employers will seek to outsource the work to operate at Point 7, the lowest-cost operating point for the same value/quality.

The key message here is that engineers must pay attention to the rapid progress made by developing countries and use these external benchmarks as the correct yardsticks to measure their own performance so that the incremental value they add is always on par with or higher than that which can be readily gained by their employers through offshoring. Furthermore, this value must be added in time for them to remain competitive from the value–cost standpoint. Besides the magnitude of the value contributed, timing is critical. Engineers need to avoid being caught off guard in this dynamic environment. It is like rowing a boat upstream; resting means falling back.

Thus, the traditional mode of continuing education (e.g., taking a course or two from time to time) will probably not be sufficient to ensure job security. What is needed is fast learning guided by constant external benchmarking and adding value before equivalent value can be attained at lower cost (via going offshoring).

As global employers pursue continuous mobility, global engineers need to practice the steady-ascent strategy in a timely manner.

C. **Be Selective.**   Engineers are advised to be selective in choosing activities and functions that add recognizable value to their employers. Adhering to the following guidelines will make it more likely that engineers will be marketable at any given time:

(a) Stay close to the employer's core competencies, which are usually preserved and nurtured in-house.

(b) Learn to absorb complex information quickly and solve technical, business, and people problems creatively. Focus your own work on maximizing its value added.

(c) Become versed in interacting with workforce members of diversified backgrounds. Baby boomers who are managing the current operations of many companies are expected to retire in the next two decades. The gap left behind by them will need to be filled by capable people who can direct teams of diversified cultural and business backgrounds (Kaihla 2003).

(d) Acquire the capabilities of designing and advancing the next generation of products and technologies and pursuing work that has high innovative content. Apply new technologies to wring out product costs. Dell uses robots to assemble computers quickly and reliably in the United States, negating some of the cost advantages of moving plants offshore.

(e) Cultivate visions with originality and a global orientation, and demonstrate leadership in strategic planning.

(f) Maintain a broad business network that can be tapped into to add value for challenging situations.

(g) Avoid attaching yourself for too long to functions or activities that are readily outsourced. Study your industry. Stay away from labor-intensive "grunt" work—tasks that can be performed by following well-specified procedures and are more or less mechanical or operational in nature. Such work is easily learned and performed by other engineering graduates in low-wage countries. Examples include low-level engineering design, assembly operation, customer service, procurement, project management, troubleshooting equipment, engineering analysis using canned software programs, and laboratory tests.

13.13 CONCLUSION

Globalization continues to be an inevitable business trend in the 21st-century economy. Three of the top five economies in the world are likely to shift to Asia by 2020.

Globalization creates ample value-added opportunities to those engineering managers who are properly prepared and equipped with the required global mind-set, knowledge, and savvy.

Adjustments are needed for American managers to become effective in the global arena. Their traditional American business attitude may need to be modified. They need to become more tolerant of, and adaptive to, that which is foreign and different. Past experience can serve as a useful basis for guiding future progress.

Engineering managers are encouraged to continuously follow the globalization process, to become sensitized to all issues involved in globalization, to make useful contributions to high-value work, and to prepare to lead in the globalization of the economy.

13.14 REFERENCES

Agrawel, V. and D. Farrell. 2003. "Who Wins in Offshoring?" *The McKinsey Quarterly,* No. 4 (Global Directions).

Asheghian P. and B. Ebrahimi. 1990. *International Business: Economics, Environment and Strategies.* New York: Harper & Row.

Brett, J. M. 2001. *Negotiating Globally: How to Negotiate Deals, Resolve Disputes, and Make Decisions Across Cultural Boundaries.* San Francisco: Jossey-Bass.

Bridger, C. 2003. "Some Firms Are Sending Their White-Collar Work Overseas." *The Buffalo News*, August 30.

Brysk, A. (Editor). 2002. *Globalization and Human Rights*. Berkeley, CA: University of California Press.

Calpan, R. 2002. *Global Business Alliances: Theory and Practices*. Westport, CT: Quorum Books.

Collier P. and D. Dollar. 2002. *Globalization, Growth and Poverty: Building an Inclusive World Economy*. Washington, DC: The World Bank.

Condon, B. J. 2002. *NAFTA, WTO and Global Business Strategy: How AIDS, Trade and Terrorism Affect our Economic Future*. Westport, CT: Quorum Books.

Cundiff, E. W. and M. T. Hilger. 1988. *Marketing in the International Environment*. Englewood Cliffs, NJ: Prentice-Hall.

Dalton, M., C. Ernst, J. Deal, and J. Leslie. 2002. *Success for New Global Managers: What You Need to Know to Work across Distances. Countries and Cultures*. San Francisco: Jossey-Bass.

Dreo, H., P. Kunkel, and T. Mitchell. 2002. *Virtual Teams Guidebooks for Managers*. Milwaukee: ASQ Quality Press.

Engardio, P. et. al. 2003. "The New Global Job Shift." *Business Week*, February 3.

Garten, J. E. 1999. *World View: Global Strategies for the New Economy*. Boston: Harvard Business School Press.

Glover, J., H. Friedman, and G. Jones. 2002. "Adaptive Leadership: When Change Is Not Enough." *Organization Development Journal*, Part 1, Summer.

Govindarajan, V. and A. K. Gupta. 2001. *The Quest for Global Dominance: Transferring Global Presence into Global Competitive Advantage*. San Francisco: Jossey-Bass.

Gupta, A. and V. Govindarajan. 2001. "Converting Global Presence into Global Competitive Advantage." *The Academy of Management Executive*, May.

Hoefling, T. 2001. *Working Virtually: Managing People for Successful Virtual Teams and Organizations*. Sterling, VA: Stylus Publishing.

Horowitz, L. G. 2001. *Death in the Air: Globalism, Terrorism & Toxic Warfare*. Sandpoint, ID: Tetrahedron.

Howard, T. J. 1995. *Global Expansion in the Information Age*. New York: Van Norstrand Reinhold.

Jeannet, J.-P. 2000. *Managing with a Global Mindset*. Upper Saddle River, NJ: Financial Times/Prentice Hall.

Johansson, J. K. 2000. *Global Marketing: Foreign Entry, Local Marketing and Global Management*. 2d ed. Boston: Irwin/McGraw Hill.

Kaihla, P. 2003. "The Coming Job Boom." *Business 2.0 Magazine*, September.

Kugler, R. L. and E. L. Frost (Editors). 2001. *The Global Century: Globalization and National Security*. Washington, DC: National Defense University Press.

Lamberton, D. 2002. *Managing the Global: Globalization, Employment and Quality of Life*. New York: St. Martin's Press.

Langhorne, R. 2001. The Coming of Globalization: Its Evolution and Contemporary Consequences. New York: Palgrave.

Lipnack, J. and J. Stamps. 2000. *Virtual Teams: People Working Across Boundaries with Technology*. 2d ed. New York: John Wiley.

McCarthy, J. 2002. "3.3 Million US Services Job to Go Offshore." *Forrester Research* (TechStrategy Brief), November.

Marguardt, M. J. and N. O. Berger. 2000. *Global Leaders for the Twenty First Century*. Albany, NY: State University of New York Press.

McCall, M. W., Jr. and G. P. Hollenbeck. 2001. *Developing Global Executives: The Lessons of International Experience*. Boston: Harvard Business School Press.

National Science Foundation. 2004. *Science and Engineering Indicators 2004*. Washington, D. C.

Phillips, B. 1998. *Global Production and Domestic Decay: Plant Closings in the US*, New York: Garland Publishing.

Rao, C. P. 2001. *Globalization and Its Managerial Implications*. Westport, CT: Quorum Books.

Ricks, D. A. 1983. *Big Business Blunders: Mistakes in Multinational Marketing*. Homewood, IL: Richard D. Irwin.

Rugman, A. 2001. *The End of Globalization*. New York: AMACOM.

Sassen, S. 1998. *Globalization and Its Discontents*. New York: New Press.

Schneider, S. C. and J.-L. Barsoux. 2003. *Managing Across Cultures*. 2d ed. Upper Saddle River, NJ: Financial Times/Prentice Hall.

Steger, M. B. 2002. *Globalism: The New Market Ideology*. Lanham, MD: Rowman & Littlefield Publishers.

Sullivan, J. J. 2002. *The Future of Corporate Globalization: From the Extended Order to the Global Village*. Westport, CT: Quorum Books.

Thurow, L. C. 1997. "New Rules: The American Economy in the Next Century." *Harvard International Review*, Winter.

U.S. Bureau of Labor Statistics. 2004A. U.S. Had largest manufacturing productivity increase in 2002. Washington, D. C. (March 29)

U.S. Bureau of Labor Statistics. 2004B. The Effect of Outsourcing and Offshoring on BLS Productivity Measures, Washington, DC (March 26).

Wilson, D. and R Purushothaman. 2003. *Dreaming with BRICs: The Path to 2050*. Global Economics Paper #99, Goldman Sachs, New York (October 1).
<www.gs.com/insight/research/ reports/99.pdf>.

World Bank. 1992, *Global Economic Prospects and the Developing Countries*. Washington, DC: World Bank.

World Bank. 2001. *Global Economic Prospects and the Developing Countries*. Washington, DC: World Bank.

13.15 QUESTIONS

13.1 For products intended for global markets, customers' wants and needs are different from one market to another. How can a centralized global team build up a product to serve as a "platform" for the global market?

13.2 Japanese companies face challenges similar to those faced by U.S. companies in that low-cost manufacturing capabilities are readily available in such countries as China, India, the Philippines, and Mexico. How can the Japanese companies plan to deal with these challenges?

13.3 It can be argued that democracy and capitalism are concepts that are fundamentally incompatible with each other. Democracy is built on the principle of equality—one person, one vote—regardless of the individual's intelligence, wealth, work ethic, or any other features that may distinguish one individual from another. Capitalism, on the other hand, fosters inequality. It uses incentive structures to encourage hard work and wise investment to realize differences in economic returns. Because future income from investments (in human or physical assets) depends on current income, wealth tends to generate wealth, and poverty tends to constrain the individual's economic growth. The cycle is self-reinforcing: success breeds success, and failure compounds failure. "The economically fit are expected to drive the economically unfit out of existence. Thus, there are no equalizing feedback mechanisms in capitalism" (Thurow 1997).

What are some of the remedies capitalistic countries have introduced to mitigate such inequality? Would globalization compound this condition in a capitalistic and democratic country? Why, or why not?

13.4 Globalization, which causes the countries involved to become more interconnected, clearly has tremendous social and political implications. It also has a cultural dimension to it, due to worldwide communications that facilitate the global connections. Cultural globalization may lead to a more civic global society with a greater consensus on civic values. It may also diminish the rich diversity of human civilization, as the Asian, Islamic, South American, and other non-Western values become increasingly generic. For many, the preservation of distinct cultural traditions is a very serious matter.

Is globalization a form of Western imperialism that may homogenize non-Western values? Why or why not? Can homogenization be avoided or mitgated?

13.5 During the new century, increased flows of products, services, technologies, capital, and workers across national borders will affect the economical, social, and political life of everyone involved. The United Nations is expected to play a critical role in this increasingly dynamic environment.

In your opinion, what should be the major missions of the United Nations in addressing these issues?

Chapter 14

Engineering Management in the New Millennium

14.1 INTRODUCTION

Tomorrow's corporate America, due to advancements in technology and the expansion of the global market, will be quite different from that of the recent past (Corta 2000; Anonymous 2000).

Advancements in communications technologies have made various business transactions cheaper, faster, and better at meeting customer demands than before. Web-based tools are available to foster information sharing and knowledge management, leading to improved supply-chain management, mass customization, benchmarking, customer relationship management, and e-transformation, on top of adding value to the traditional engineering functions (e.g., product design, project management, plant operations, and manufacturing) (Deetz, Tracy, and Simpson 2000).

Rapid changes in production, service, and distribution technologies will continue to result in extensive organizational reform. Globalization will cause engineering organizations to be managed differently than in the past. It is important for today's engineers and future engineering managers to proactively anticipate the future, to recognize the significant changes lying ahead, and to prepare to seize new opportunities associated with these major changes.

In this chapter, we first contrast past engineering management in corporate America against that of the new millennium (Schmidt 1999). Specifically, we address the issues related to new trends emerging in our economy. The differences between companies in the old and new economies are reviewed. The characteristics of knowledge-economy companies and their likely strategies are examined next. We then discuss the new types of workforce and the new responsibilities of managers in progressive companies. Finally, we outline the significant contributions expected of engineering managers in these companies.

Business development is critical to the success of 21st-century companies (De-Palma 2002; Johansson 2000). The extension of market reach is due primarily to globalization. (See Chapter 13.) Capable enterprises are constantly looking for new customers in emerging markets to expand sales, seek resources and location-based synergies to enhance their competitive advantages, and exploit economies of global scope to mitigate business risks. Reaching out to global markets will be a major driving force affecting innumerable engineering enterprises in the new millennium.

In this chapter, we explore various management issues applicable to the marketplace and emphasize new challenges that engineering managers are expected to face in the new millennium.

14.2 FUTURE TRENDS

Before we talk about specific changes expected in corporate America, a brief review of some of the new trends in our economy is in order. Several major trends already noticeable in the knowledge economy are induced by the growth of Web-based enablers and by the increasing demands of customers for better, faster, and cheaper products and services. The common underlying threads among these trends appear to be *effectiveness* (customer and environment), *efficiency* (internal structure and operation activities), and *integration* (one-stop consolidation). Discussed below are specific trends related to customer focus, enterprise resource planning and application integration, supply strategy, knowledge management, changes in organizational settings, and population diversity.

14.2.1 Customer Focus

In the knowledge economy, customer service will be driven by several distinct characteristics (Hiebeler, Kelly and Ketterman 1998; Schmitt 2003).

A. Speed of Customer Service.   It is preferable to reduce the processing time between searching, selecting, and entering and fulfilling orders. There will be no more excessive handoffs. Companies are moving toward a seamless integration of steps to accept orders, trigger receivables, send orders to production, route requisitions to warehouses, activate shipments by logistics partners (e.g., UPS, FedEx), replenish inventory, update accounting, replenish stock with suppliers, and track delivery status to ensure on-time delivery (Buss 1999).

B. Customer Self-Service.   Companies involved in real estate, insurance, travel, car buying, auction, parts sourcing, and retailing are increasingly moving toward empowering customers to serve themselves by creating 24/7 (i.e., 24 hours a day, 7 days a week) systems and cutting out intermediaries. The following three companies are successful examples of the trend to encourage and empower customer involvement in self-service:

- Gateway. Customers define their own needs, configure systems, place orders, pay for new computers, and get limited support.
- E-Trade. Customers trade securities without broker involvement, using a 24/7 website.
- Microsoft Expedia. Customers make reservations on-line to book flights and receive confirmations at a lower cost, while enjoying faster service.

C. **Integrated Solutions for Customers.**   Customers have moved away from best-of-breed individual solutions toward integrated systems. A specific example is Microsoft Office Suites with integrated functionality.

Customers are motivated by the desire to spend less time shopping, shop at one-stop stores, make fewer shopping trips, and face fewer choices that may be time consuming and difficult. The following are examples of organizations responding to these desires:

- Gap. A one-stop clothing and accessory provider that encourages convenience-based "package" purchases. Mannequins are outfitted with shirts, blue jeans, belts, baseball caps, sunglasses, socks, shoes, gloves, and a knapsack, marketing a hip image.
- Citicorp. In 2002, this company started to offer the combined on-line services of banking, credit cards, automobile insurance, brokerage and investing, mortgage and loans, and e-mail cash for money abroad.
- Automatic Teller Machines (ATMs). These machines are being expanded by countless financial institutions to include Web-based services such as e-mails, on-line purchases, and transactions.

D. **Customized Service and Sales.**   Statistics indicate that diverse companies lose 50 percent of their customers every five years. Generally, it costs 5 to 10 times more to obtain a new customer than to retain one. One approach taken by some companies is to train service people to cross-sell and up-sell.

For example, Home Depot emphasizes service to do-it-yourself customers and provides customers with easy access to information. The company starts offering service before sales and continues its customer service after sales.

E. **Consistent and Reliable Service.**   Customers prefer to have single points of contact, which requires the company's service calls to be coordinated with supply-chain business partners (i.e., those members of the extended enterprise family under outsourcing, alliance, or partnership agreements).

F. **Flexible Fulfillment and Convenient Service.**   Gevalia Kaffe imports coffee and makes home deliveries. It performs 200,000 transactions per week and has built an e-business infrastructure to take orders, find the lowest-cost routing distance between the customer and the nearest warehouse, check inventories, issue shipping orders, and activate shipping by the networked partners who deliver coffee to home addresses.

G. **Transparent Sales Process.**   Customers typically want to know the order status, product information, pricing, and availability (Larson and Lundberg, 1998). UPS is known to have set up a 24/7 sophisticated information system that performs the following functions:

- Tracks air and ground parcels at any time and from anywhere
- Achieves flawless delivery, which is now becoming the norm rather than the exception.

Another example is Solectron, which makes circuit boards and electric assemblies for such customers as IBM, HP, and Intel and has plants in California, Washington,

Malaysia, France, and Scotland. The company has refined a shop-floor tracking and recording system (STARS) that uses bar codes to enable customers to track the status of their orders. The enhanced process transparency helps create new demand while retaining satisfied customers.

H. **Continuous Improvement in Customer Service.**   Constantly learning new ways to enhance customer service is one of the keys for companies to succeed in today's marketplace. For example, Nordstrom bends over backwards to please customers with gold-plated service and a no-questions-asked return policy. This company found that 90 percent of its business is from a loyal 10 percent of its shoppers. (This is an example of the Pareto principle, described in Chapter 5.) The methods Nordstrom applies to achieve a high degree of customer retention and sustainable innovation are outlined as follows:

- Motivate employees by paying them very high rates of commission.
- Use undercover shoppers to evaluate service and give cash rewards to employees who achieve a perfect score.
- Delegate authority downward to allow autonomy and local decision making.

Nordstrom's strategy is compatible with the *Service Profit Chain Model* (Heskett, Jones, Loveman, Sasser, and Schlesinger, 1994). This model is unique in that it originates from company leadership (e.g., vision, values, energy, concern, and discipline). Superior leadership creates a good place for employees to work (e.g., through workplace design, job description and latitude, selection and development, communication and information, and tools for serving customers), which in turn produces satisfied employees. Happy and productive employees serve customers better. Satisfied customers create increased profitability to the company.

I. **Technology-Enabled Services.**   The trend is toward an integration of various means of access available to customers, such as the Web, direct dial-up, interactive voice response, and kiosks.

For example, customers are projected to be in the driver's seat in influencing the development of IT technologies (Moschella 2003). The waves of IT technology move from a technology-centered paradigm to a customer-centered practice.

The first, system-centric, wave focused on developing proprietary systems. The second, PC-centric, wave centered on working out the hardware and software standards. The third, network-centric, wave dealt with Internet standards. The fourth, customer-centric, wave addresses the information content and transaction standards. This last wave aligns more closely the values of IT technologies with the needs of customers.

14.2.2 Enterprise Resource Planning and Application Integration

In the past, companies concentrated on achieving optimum performance in various individual functional departments. Each of these departments had the tendency to pursue optimal operations within its own boundaries. Each department, similar to a silo, acted like a tightly controlled organizational unit with limited communications capabilities to the outside world across its boundaries, except through its top. In practice, the deficiencies created by poor, uncoordinated transactions between the functional groups more than offset the benefits of the local optimization they achieved. Thus, over time, it has become evident that corporate competitive advantages can be achieved only through a proper integration of the individual functional departments of, for example, engineering, manufacturing, design, customer service, procurement, finance, and accounting. Today's trend is toward enterprise application integration (Myerson 2002).

An integration of disparate departmental functions permits a greater access to information, while it links employees, business partners, and customers more effectively. There are many enterprise application software products on the market that facilitate such integration. The obvious benefits created by this type of integration are speed, accuracy, and cost reduction, because of reduced manual handoffs (e.g., manpower), which allows business decisions to be made faster and at higher quality and lower cost.

A. Integrated Communications Systems. The trend is toward an integration of networks composed of telephones, cable TV, wireless, and computer data. The "last mile" bandwidth problem (from telephone switching in the office to the home) will likely be solved using fiber-optic systems. AT&T, Sprint, and MCI/Worldcom are said to be working on the integration of voice and data services.

Browsers and modems are used as customer home contact points today. WebTV could very well replace these in the future.

B. Wireless Applications. The use of airwaves for other services in addition to phone calls is expected to increase. Data transfer to mobile units allows managers and leaders to make important decisions from anywhere at any time. For example, Palm Pilot (3Com) is a two-way personal communication tool in text format that displays real-time flight schedules, news headlines, and on-line transactions such as movie ticket purchases, stock trading, etc.). The Nokia 9000 Communicator is a phone, Web browser, and personal-messaging and data-organizing system.

C. Leveraged Legacy Systems. Middleware consists of connectivity products that link the existing legacy systems on mainframe computers with client and servers and the Internet. Middleware provides the important function of making existing data widely available to employees and customers.

14.2.3 Supply Strategy

Traditional enterprises create vertically integrated organizational structures and amass heavy physical assets to achieve competitive advantages in the marketplace. Knowledge-economy companies form flexible supply-chain partnerships to increase the speed to market and vary the product features to better satisfy the ever-changing needs of customers. Use is made of the technological and marketing expertise of the networked partners. Competitive advantages are thus derived from their capability to form such knowledge-intensive business networks, rather than from the value of capital assets piled up on the ground.

Outsourcing is a major trend that favors the formation of virtual enterprises. A single company working alone is no longer viable, as diverse competencies are needed to compete in today's marketplace. In these virtual enterprises, certain noncore business processes will be strategically outsourced to achieve higher earnings and more pronounced competitive advantages. Examples include

- Niagara Mohawk, which outsources human resources and purchasing functions, and
- United Technologies and American Express, which outsource their procurement functions to IBM.

Other companies have started outsourcing many of their noncore operations, concentrating instead on doing what they do best. For example, Sun Microsystems has focused on design, electing to contract out or purchase all workstation components. This strategy permits Sun to (1) introduce new products rapidly, (2) achieve better quality, (3) ensure dependability, (4) shorten speed to market, (5) gain flexibility, and (6) realize cost advantages.

Another example is Sara Lee, which focuses on building new products, managing brands, and building market shares, while outsourcing the production of Leggs hosiery, frozen desserts, Wonderbras, Coach briefcases, and Kiwi shoe polish.

In essence, this strategy allows the company to jump on and slide down someone else's experience curve (e.g., Boston Consulting Group's 85-percent curve) to maximize benefits while maintaining flexibility in switching partners.

The current push for better asset utilization (Return on Assets [ROA]) helps move companies toward becoming knowledge intensive (through supply-chain and marketing), rather than capital-asset intensive (through in-house production). Better asset utilization can be achieved by creating contract partnerships, setting up global production networks, keeping overhead costs low, changing products frequently, and innovating through technology.

U.S. companies that market products to end users have also started to outsource such corporate functions as production, back-office work, logistics, after-sales service, procurement, inventory management, and new-product design. As this outsourcing trend continues, more demands are created for such service functions. The economy of scale dictates that new, vertically integrated parts and service providers are likely to be created for the following purposes:

A. Vertically integrated parts suppliers are becoming huge multinationals in their own right, and they are tightly integrated to create efficiency (e.g., Delphi Automotive). Their factories are designed to be quickly rearranged in order for the same shop to make different products for different client companies. The focus is to manufacture products and to operate at capacity almost all the time. They get cheap components by buying in quantity. Parts may come from various regions of the world. Their gross margins are relatively small (6 to 8 percent on sales), but they generate good return on equity (for example, 20 percent).

B. Integrated service organizations take orders from banks, automakers, and pharmaceutical companies to handle financial advisement, accounting, and other services. Services offered to different industries are now bundled together. Certain design service providers may have teams of industrial, mechanical, and chip engineers scattered around the world.

Engineers in these product and service provider companies will need to work in an interdisciplinary environment, as dictated by the companies' mission to serve clients in different industries. Since these product and service providers conduct no R&D, perform no marketing, and develop no products, product innovations will have to come from their client companies' own R&D departments or from third-party startups.

Managers in the client companies will need to learn how to supervise the interactions between diversified product and service providers, and to ensure that their breakthrough idea—now entrusted to an outside provider to practice—will not benefit the competition. This is an important issue related to knowledge management.

14.2.4 Knowledge Management

Knowledge management refers to activities related to the preservation and enterprisewide application of corporate expertise and know-how to create competitive advantages.

Business success depends on innovation and the expertise of competent knowledge workers, whose insights need to be properly preserved by documentation, knowledge sharing, recruitment, and retention. The importance of knowledge management is becoming increasingly evident in the knowledge economy, wherein knowledge-intensive companies forcefully strive, as a corporate strategy, to attract talented and more mobile knowledge workers.

14.2.5 Changes in Organizational Settings

There will be two broad types of corporations in the near future. The first type generates and markets products and services to consumers (the end users). These companies retain core competencies and outsource everything else to selected service providers. A network of contract partners may produce the products or services, each being particularly efficient and superior in supplying specific components or elements. Outsourcing will succeed for these companies, as it has for Dell Computers, with products that are more than adequate for the customers and that are composed of current technologies available from the company's networked partners. Outsourcing requires a perfect understanding of what customers need and how the products are specified. Under these conditions, companies compete on the basis of speed, flexibility, and cost. Going virtual is thus becoming useful. This is a major change in organizational settings, moving toward the virtual and away from vertical integration.

The other type of corporation will supply specialized parts, designs, and services to client companies under contracts. Acquiring various manufacturing facilities and organizing them into vertically integrated enterprises are the functions of these companies. Their manufacturing plants are laid out in a flexible manner in order to respond to divergent needs of global clients. This is a second change in organizational settings moving toward additional vertical integration.

Some virtually organized companies are swinging back toward vertically integrated operations to perform their creative and developmental tasks. The reason for this is that, to achieve breakthroughs in products and technologies, innovative performance is needed. Usually, virtual companies do not have sufficient information or

TABLE 14.1   Population Diversity

	1995 (percent)	2003 (percent)	2050 (percent)
Whites (non-Hispanics)	73.60	69.00	50.10
Blacks (non-Hispanics)	12.00	12.60	14.52
Hispanics	10.20	12.53	24.40
Asians	3.30	3.86	7.86
Others	0.90	2.01	3.12

Source: United States Census Bureau (2004).

resources to enable their networked partners to make parts for innovative products. On the other hand, a vertically integrated company can readily amass all of the needed resources under one roof and thus can be effective in developing new technologies and products. Cisco systems has integrated its operations to develop new optical networks. So the pendulum swings back a little, from virtual to vertical integration, for those companies which intend to come up with innovative products.

Not all companies should become virtual; neither should all functional elements of a company be vertical. The organizational form may have to be frequently adjusted according to the company's objectives, strategies, and changing environments.

14.2.6 Population Diversity

The U.S. Census Bureau predicts that there will be significant shifts in the American population by 2050. (See Table 14.1.)

For example, the projected increase in Hispanics is due primarily to immigration. As a consequence, the U.S. workforce is expected to become more diverse (Livers and Caver 2003). Managers may need more diversity training, as more women and minorities are expected to attain key positions.

The trends just described have become noticeable in the marketplace since the turn of the millennium. All of these trends have important effects on how business priorities are set and how companies are run. In the sections that follow, some of the anticipated changes will be elucidated, along with the challenges facing engineering managers in the new millennium.

14.3 OLD-ECONOMY AND KNOWLEDGE-ECONOMY COMPANIES

Engineering enterprises in the old economy differ from those in the knowledge economy in many ways. These differences, which were partially discussed in previous chapters, are reviewed next.

14.3.1 Old-Economy Companies

Typically, old-economy companies are of a pyramidal structure. From the top, the CEO directs the company president, vice presidents, directors, and managers to lead, plan, organize, and control the workers at the bottom. The hierarchies are rigid and the company boundaries are clearly marked.

These companies are capital intensive. They build such physical assets as facilities and equipment to create competitive advantages. They produce tangible goods such as cars, machines, refrigerators, and foods. Their business models are created around fixed assets, working capital, and economies of scale. They derive strengths from mass production, offering the same products at lower and lower prices. Examples include Ford, Kodak, USX, and other "brick and mortar" companies.

Shareholders own the company's equity, including the physical assets. All physical assets are readily fenced in and chained down so that company management can exert full control over these important assets that serve as the basis of the company's competitiveness in the marketplace.

Companies deploy in-house capabilities. Major functions are vertically integrated, starting from raw materials, work-in-progress, finished goods, distribution, marketing, and sales, and ending with service to customers. The emphasis is on self-sufficiency, stability, and incremental growth.

These companies create intensive rivalries with competitors to motivate employees and to ensure success in the marketplace. Companies are run as "going concerns," intended to last forever.

In old-economy companies, innovative ideas typically float up the chain of command to be screened and reshaped for further evaluation before finally being presented for decision making and possible implementation sometime in the future. Decision making is resource centered. Resource centering is the optimum use of limited resources available to the company. Daniel Carp, CEO of Eastman Kodak, has said, "By the time the idea went in to the CEO, it had a yellow ribbon tied around it."

These companies maintain a staff to process order entry, invoicing, accounts receivable, accounts payable, procurement, shipping, delivery, and other functions. Such processes take time and cost money. Also, internal financial performance reports are typically prepared on a monthly basis.

14.3.2 Knowledge-Economy Companies

Knowledge-economy companies center on a small number of core operations. They have a permanent staff to advance technologies, but they assume a predominantly weblike organizational design that links business partners, employees, external contractors, suppliers, and customers in various collaborative arrangements. All participants in the networked arrangements grow interdependently. The networks remain flexible, and the business and activity boundaries of the companies are vague and fluid. Knowledge-economy companies aim at achieving future growth by way of networked partnerships.

These companies are idea intensive, as ideas are the key assets in an information-based economy. The assets are in employees' heads, not physically constructed on the ground. They provide digitizable goods (e.g., data, software, books, news, music, movies, games, financial services, and advertising). Knowledge-economy companies also e-transform themselves to produce commodity-type goods (e.g., jeans, cosmetics, computers, clothing, and sneakers). Their business strategies emphasize cost reduction by applying technologies secured through networks, and they bring breakthrough ideas to their markets first.

They derive strength by mass customization, wherein customers are given tools to design individually preferred products. Excellent examples include the following:

1. Dell offers custom-designed computer systems from a group of well-defined system components (memory size, hard-drive capabilities, modem type, monitor, etc.) with price and delivery options.

2. Levi's presents custom-designed Spin jeans for special fit, style, and color, with 1.7 million variations to choose from.

3. Proctor & Gamble offers custom-designed cosmetics and perfumes with up to 50,000 different choices.

4. Nike customizes sneakers by features, styles, and colors.

5. American Quantum Cycles custom-designs motorcycles that use different seats, handlebars, and paint colors.

6. Charles Schwab's mutual funds screener allows customers to design their own investment portfolios.

7. Mattel's My Design Barbie lets customers build a friend for Barbie by choosing hairstyle, color, complexion, and eye color for a doll.

8. Cisco's Marketplace allows business customers to create routers, switches, and hubs to build specific computer hardware systems.

9. Point.com's Service Plan Locator allows customers to choose wireless phones, service plans, and accessories.

Such choice boards allow companies to first sell and then produce, instead of the former way of spending money to create products and services, hoping to eventually sell some or all of them.

These companies have high price-to-book (P/B) value ratios and high per employee market values. For example, Microsoft employed 31,000 people at one time and has a market value of $600 billion. Yahoo has a P/B ratio of 40.

Shareholders of these companies do not have true ownership of the intellectual property assets, as ideas reside inside employees' heads and cannot be controlled. Because of workforce mobility, this is a significant risk factor for knowledge-economy companies.

Typically, knowledge-economy companies build their core competencies, outsource the rest, and are virtually integrated. They nurture an array of formal and informal networks of partnerships and alliances consisting of free agents, outside contractors, designers, prototype producers, manufacturers and distributors, and others. As partnerships and alliances shift, these networks change accordingly. They focus on achieving advantages in technology and in time to market in order to realize revolutionary growths.

By outsourcing noncore activities (manufacturing, product design, accounting, procurement, customer services, janitorial services, and others), companies save time and management attention, if not money, to go after big opportunities of the e-future. Well-known examples include these companies:

- Cisco Systems, the most networked organization, with all administrative functions conducted over the Internet, owns only 2 of 34 plants that make its products, and 90 percent of the company's orders are not handled by human hands.
- Dell manages the most efficient supply-chain network, maintaining an inventory for 6 days only.
- General Motors created a "sensing and responding" supply chain for delivering customized vehicles in 10 to 12 days in 2003–2004.

Sometimes companies collaborate even with competitors to achieve win–win solutions. For example, Covisint is an electronic trading exchange created collaboratively by several automakers, namely, GM, Ford, DaimlerChrysler and Renault-Nissan, to streamline parts procurement and generate revenue.

Excellent ideas are instantly transmitted through the weblike organizational design. Empowered employees make decisions to center on creating competitive advantages through shortened time to market and better customer satisfaction, which will bring about an increase in top line.

Companies achieve "virtual financial close" as the processes of order entry, invoicing, shipping, accounts receivable, accounts payable, procurement, and others are digitized to function seamlessly, making all financial performance data only "one click away."

Massive digitalization allows companies to remove people from various routine assignments and thus achieve advantages in cost, speed, and quality of information.

In the United States, interaction costs, which are the expenses incurred to get different people and companies to work together to create and sell products, account for about 50 percent of all labor costs. Because these interactions are automated in knowledge-economy companies, their initial productivity gains are predicted to reach 20 to 40 percent per year. These companies embrace digitalization through the deployment of suitable enterprise integration software systems. Specific examples of unit cost reduction strategies include the following:

1. **Bank transactions.** A transaction by a bank teller costs $1.25, compared with 54 cents by phone, 24 cents by ATM, and 2 cents by the Internet.
2. **Procurement.** This cost is $140 for people procuring parts and supplies, compared with $11.70 by an Internet-based catalog system (Corning, for example).
3. **Handling job applicants and resumes** costs $128 by a people-based procedure, but only 6 cents when accomplished by digitizing the process and eliminating manual labor.
4. **Customer inquiry processing** costs $50 over the phone, compared with 50 cents by the Internet (General Electric, for example).

The key is to make Internet use a strategic priority. Spotting opportunities early and exploiting them fast will work. Knowledge-economy companies regard human

TABLE 14.2   Contrasting Views of the Corporation

Characteristics	Old-Economy	Knowledge-Economy
Organization	Pyramid	Web/network
Focus	Internal	External
Style	Structured	Flexible
Source of strength	Stability	Change
Structure	Self-sufficiency	Interdependencies
Resources	Physical assets	Information/ideas
Operations	Vertical integration	Virtual integration
Products	Mass production	Mass customization
Financials	Quarterly	Real time
Inventories	Months	Hours
Strategy	Top down	Bottom up
Leadership	Dogmatic	Inspirational
Workers	Employees	Employees and free agents
Job expectations	Security	Personal growth
Motivation	To complete	To build
Improvements	Incremental	Revolutionary
Quality	Affordable best	No compromise

Source: Business Week, August 28, 2000, p. 87.

skills, expertise, and relationships as the most precious resources in an organization, whereas the 20th century "brick and mortar" companies emphasize physical assets of production. Table 14.2 provides a very concise, yet comprehensive, summary of the differences between old-economy and knowledge-economy companies.

14.4 CHARACTERISTICS OF PROGRESSIVE COMPANIES

Progressive companies are those knowledge-economy companies which make it a strategic priority to utilize Internet-based technologies, invent innovative business models, create supply-chain networks, empower a diversified workforce, and pursue customer satisfaction in order to achieve business success in the marketplace.

The major trends discussed in the previous section tend to shape the progressive companies of the new millennium. These companies look for real-time customer feedback (e.g., via on-line chat rooms) to get customers to talk about performance and problems. They focus on a market segment of one, delivering mass-customized products and services. They induce loyalty with the personalization of customer relationships. They view the enduring relationship with employees as an enormous asset because these employees connect the company to its partners. Their total employees consist of permanent staff, temporary service people, contract workers, free agents, and consultants. The companies are further linked to the employees of their networked partners. It is projected that, by 2010, 41 percent of the workforce will be working on a contract basis. Like actors and athletes, talented businesspeople will have agents to represent them.

A case in point is Nokia, a cell phone giant in Europe. The company entered the U.S. market successfully with only five permanent employees by outsourcing sales, marketing, logistics, and technical support. Nokia employees work with a mix of contract teammates from around the globe, many of whom they never meet face to face. Every project calls for a new team composed of people with special talents and skills. In such a dynamic setting, information that is more than one hour old will be viewed with skepticism. Everyone's performance is carefully and constantly evaluated. If there is no contribution, there will be no continued membership.

Progressive companies streamline most staff jobs by employing certain types of enterprise-planning systems. They devote special attention to supply-chain management, enterprise integration, customer relations, knowledge management, Web-enabled transactions, and globalization.

Companies in the new millennium must manage all stakeholders well in order to survive and flourish. There are five categories of stakeholders: (1) *Investors* are needed for capital to grow. (2) *Customers* provide the sources of sales revenue growth and earning stability. (3) *Employees* provide productivity and client satisfaction. (4) *Suppliers* provide competitive advantages in technology, innovation, and supply-chain management. (5) *Communities* influence the company's reputation, as well as local regulations, and legislation that the company must manage. Some of the unique characteristics of these companies are described next.

14.4.1 Complex Organizational Design

The organizational form of progressive companies may be large and complex. They have less mass, as various in-house functions are performed by alliances, joint ventures, and partners in networks, and they tend to constantly shift portfolios of businesses and assets. The new trends discussed previously reflect the competitive needs for operational efficiency and customer focus. The drive for efficiency and a customer-centered corporate strategy is enabled by available technologies.

A traditional value chain is typically driven by core competencies (as depicted by the flowchart in Figure 14.1), starting from what the companies can do best and then defining products and services that the customers may want.

A reverse-value chain begins with customer needs (defined by extensive customer study and market research) and works backward along the fulfillment chain into company capabilities and functions. (See Figure 14.2.)

Enterprises organized according to the reverse-value chain tend to be more capable of satisfying the needs of customers, a key requirement for business success in the 21st century (Hines 2000).

Figure 14.1 Traditional value chain.

Figure 14.2   Reverse-value chain.

Another guideline useful for organizational design is to choose a narrow focus. Few organizations can do many things well; none shine in every dimension of business, such as cost, quality, price, convenience, and ease of use.

Examples of a narrow focus in business include service, operation, and innovation. *Service excellence* focuses on strong customer relationship management that anticipates customers' needs, enables self-service, and offers value. *Operational excellence* stresses the optimal leverage of assets, efficient transactions, customized solutions (sales intelligence, management of customers' expectations), the use of measurement systems, outsourcing noncore processes, end-to-end process effectiveness, and others. *Continuous innovation excellence* exhibits product leadership to delight customers, grows by merger and acquisitions, proactively educates customers about how to use and benefit from new products, and encourages innovation.

A progressive enterprise must link three major groups of stakeholders together: (a) customers, (b) back-office workers, and (c) supply-chain partners. After these groups of stakeholders are properly linked together, the design of a progressive enterprise organization may contain major units as illustrated in Figure 14.3.

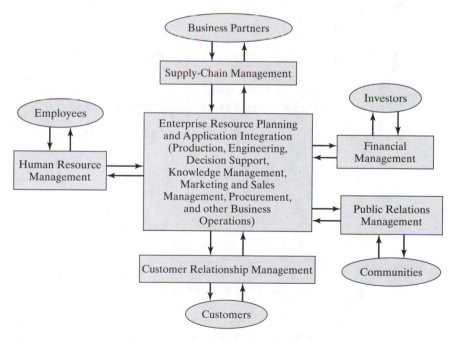

Figure 14.3   Organizational design of a progressive enterprise.

14.4.2 Global Reach

Progressive companies serve global markets by increasing product specifications and otherwise offering products that differ from those in the domestic markets, derive more revenues from overseas markets than domestic markets, and have future growth opportunities weighted toward emerging markets. (See Chapter 13.)

14.4.3 Partnerships

Progressive companies are dynamic. They constantly create external networks to access product innovation, process technologies, manage working capital, maintain market access, and cultivate knowledge. These critical linkages are based on mutual benefit, trust, empathy, and good communication—and they are vigorously protected (Segil, Goldsmith, and Belasco 2003).

14.4.4 New Composition of Employees

The full-time regular employees in progressive companies are typically young and technically literate; they embody the core competencies of these companies. Some of these employees may reside abroad, have modern attitudes toward the employer–employee relationship, and change jobs frequently, preferring opportunities for interesting and challenging work that stretches their abilities.

Supporting these core employees are various part-timers, independent contractors, agency temporaries, and employees of vendors and consultants who will also be engaged. Therefore, the company may have a smaller share of support staff provided by outside agencies.

Example 14.1.

What are some new characteristics of workers in the 21st century?

Answer 14.1.

Workers in the new century may have these characteristics:

A. Free agents can now sell their skills around the world via the Internet; this was impossible to do not too long ago.

B. Professional groups are likely to offer the senses of identity and community, health insurance, and other benefits needed by free agents who float from one company to another.

C. Each employee may have as many as 20 different jobs throughout a career of 45 years (an average of 2.5 years per job). They tend to constantly bargain for better deals within their organizations (e.g., stock options, a sign-up bonus, new projects, Thursdays off, an August sabbatical, etc.).

D. Workers seek to acquire a broad set of marketable skills, as companies will continue to outsource white-collar jobs and spread centers of excellence around the world to seek advantages in cost, speed, and expertise.

E. Managers and professional workers need to be flexible and adaptable to organizational changes and become cosmopolitan, equally at ease both at home and abroad.

14.4.5 Management Reporting Layers

Progressive companies will have fewer reporting layers between business units, divisions, and executive management. Business units will increasingly operate autonomously and be measured by specific performance metrics (e.g., a balanced scorecard; see Section 7.6).

14.4.6 Customer Sophistication and Demand

In the new millennium, customers will be better informed, more sophisticated, and increasingly demanding. They will use the Internet and access information, demand product customization, require price transparency, and conduct comparison shopping. Customer satisfaction will hinge on product quality, after-sales support and services, and value for the money. More data on customer satisfaction will be published on the Web, exposing the market leaders and followers in every industry. Progressive companies should be well prepared to serve their customers.

14.4.7 Public Image

Because a good public image is regarded as a competitive advantage, progressive companies nurture a positive reputation of being reliable and ethical. Public image is expected to become increasingly important to the stakeholders of all companies.

Communities and governments will demand more assurances from companies to protect public interests from ill-conceived mergers and acquisitions, dangerous operating practices, antisocial business decisions, and oligopolistic behaviors. Examples of such governmental intervention include (1) Alaska's opposition to BP Amoco's acquisition of ARCO to take a 70-percent share of Alaska's oil and gas reserves; (2) Italy's accusing Coke of distorting competition rules; (3) The U.S. Department of Justice (DOJ) accusing Microsoft of monopolizing the Windows operating software market; and (4) The U.S. DOJ approving the merger of AOL and Time Warner only after having studied it for more than one year.

14.4.8 Stock Market Valuation

The stock valuation of progressive companies will be based on returns on their intangible assets instead of their tangible assets. The intangible assets of a company include (a) brand capital (customers and the community) and (b) knowledge capital (strategic suppliers and employees). Progressive companies are capable of optimizing the return on these intangible assets that form the primary basis for the companies' competitiveness in the marketplace.

14.5 TRANSITION TO THE KNOWLEDGE ECONOMY

Numerous old-economy companies need to modify their business practices in order to become successful in the knowledge economy (Weiss 2000).

For companies in the new millennium, there are several critical corporate values. *Speed* is important, as time may be more valuable than money. Companies looking for speed may favor the strategy of buying over building. *Talents* are crucial, as companies compete largely on the capabilities of attracting the best people, empowering them to innovate, and creating an environment that makes the best people want to stay. *Market dominance* refers to the company's emphasis on long-term market position rather than quarter-to-quarter increments in performance. *Customer orientation* signals the value placed by the company on tracking the needs of customers and creating supply chains to satisfy them. *Efficiency* is a critical corporate value: Companies streamline operations using advanced software technologies such as enterprise integration systems to adapt to a changing marketplace. Finally, *outsourcing* represents a preference for the company to farm out noncore activities in order to conserve resources, time, and management attention.

(For a discussion on achieving speed and customer orientation, see the specific example of Cemex in Section 12.4.)

Outsourcing cuts down a company's investment, shortens its time to market, and enables it to take advantage of the unique skills and expertise of the supply partners in the production of products and services. For example, Cisco Systems is known to have dropped in-house R&D in favor of buying start-up companies for its technological needs. Companies need to aggressively engage partnerships, form supply alliances, and create networks with domestic and international suppliers of raw materials, subassemblies, semi-finished goods, and others to make finished products flexibly and speedily. Outsourcing facilitates mass customization by which product features are flexibly modified to meet customers' needs. The major issues involved in outsourcing and in the management of supply-chain networks are addressed next (McGrath and Hoole 1992; Venkatesan 1992).

14.5.1 Product Design and Specification

Managers need to communicate and enforce interface specifications of all parts manufactured by the partners. Standardizing interfaces to promote exchangeability of parts and create varying product features is essential. Effecting infrastructure hardware and software systems facilitates supply-chain management. Companies need to integrate middleware, EDI, systems technologies, regulations, production practices, and other elements of their business.

14.5.2 Manufacturing

Managers should take the following actions: (1) Selectively utilize Web-based manufacturing and operations enablers to attain business advantages; (2) make critical parts in house and combine with vendor-supplied parts to assemble the finished products; (3) specify assembly procedures to ensure product integrity; (4) conduct a statistical process control (SPC) to ensure system quality of the assembled finished products; and (5) pursue ISO and other quality process certifications of networked partners.

14.5.3 Management

Companies need to (1) share information and collaborate with networked global partners; (2) preserve corporate knowledge and ensure global application and learning; (3) set up a customer relationship management system by tapping into the expertise, innovation, and knowledge base of the networked partnership; (4) assist in implementing an enterprise resource-planning system to optimize the utilization of corporate assets; and (5) make the production process transparent to be traceable by customers.

Engineering managers may be particularly qualified to facilitate the transitions of old-economy companies into knowledge-economy companies. Companies of the 21st century will form more external relationships with networked suppliers, key customers groups, or client companies for which products or services are made on contract. The uniqueness of these relationships can be attributed to their characteristics:

- **Transience.** A contractual relationship may last for a short period of time, until the market conditions call for change.
- **Diversity.** Partners may be global, with varying degrees of differences in culture, language, working norms, and value systems.
- **Coordination.** Strong coordination capabilities are needed, along with project management skills, to handle human relations, risk management, conflict resolution, tolerance for uncertainty, problem solving, motivation capabilities, and communication.

Managers also need to spend increasing amounts of time nurturing external relationships.

14.5.4 Logistics

Managers must assure just-in-time (JIT) delivery and optimize transportation logistics.

14.5.5 Inventory Control

Balancing loads between global sites will enable the company to adjust to changing supply and demand and marketplace conditions.

14.5.6 Risk Management

The company must proactively respond to labor, political, currency, and other unexpected changes and conflicts.

14.5.7 Component Design

Reducing the number of global component suppliers may be accomplished by minimizing the number of components in products. Designing each component innovatively by the use of common components minimizes reengineering to meet local needs. The company should strive to exceed local quality requirements to create a competitive advantage.

14.5.8 Component Manufacturing

Linking design with manufacturing allows production costs to move down the learning curve. Successful companies produce high-labor-content parts in low-labor-cost areas and use skilled labor for other high-technology components.

14.5.9 Product Assembly

Assembly of finished products in plants close to the local markets assures response to the local content laws, import duties, final tests, and quality control.

14.5.10 Load Balance

Transfering components to assembly plants and shifting like components from plant to plant balances the load among regions.

14.5.11 Procurement

Procuring almost all parts centrally, except low-cost and low-volume items, helps the company to realize economies of scale. Vendors are to be centrally qualified.

14.5.12 Marketing and Sales

The company needs to coordinate marketing and sales forecasts centrally for global markets, on the basis of local inputs.

Example 14.2.

> Companies in the future will rely more and more on temporary workers, specialized vendors, and consultants to flexibly satisfy unique needs and contingencies. Employer–employee relations will become peer-to-peer relations rather than hierarchical ones. Engineers will find that their careers are less stable than in previous generations.
>
> In your opinion, what is the most important preparation for engineers under this volatile, uncertain, and dynamic scenario?

Answer 14.2.

> Engineers need to proactively manage their own careers more closely than ever before, while keeping their skills, knowledge, and experience marketable. A good way to focus is to pursue lifetime learning about the changes in technologies, tools, industry, and business. Engineers who do not take care of their own careers will rapidly become redundant.
>
> The employer's position is well stated by Bahrami (1992): "You own your own career, we provide you with opportunity!"

14.6 PERSONAL STRATEGIES FOR THE FUTURE

Engineering managers need to personally ready themselves to meet challenges in the future related to the application of Internet-based technologies and the extension of market reach and globalization (Stettner 2000; Kamp 1999). To acquire the skills to be successful in the future, engineers should

1. Get postgraduate education, which will be increasingly essential to achieving personal prosperity in the future. Learn how to learn quickly. Constantly hone skills and acquire broader perspectives.
2. Become efficient in digesting vast amounts of information, measuring trade-offs, and making the right choices. Be computer and Internet literate, develop

conceptualization skills, and acquire fast reading and knowledge-processing capabilities.

3. Acquire a good understanding of new products and markets in transition, which are often defined by ever-changing technologies.

4. Be a capable communicator, compromiser, and leader of change.

5. Foster external relationships and networks of partners to succeed. The number and complexity of partnerships are expected to increase.

6. Lead with suitable style. Visualize yourself as the director of a symphony orchestra or ballet company. The leader sets the vision, but allows the individual "artists," all leaders in their own right, to fashion their own roles in the performance. The ability to be assertive and ultimately emerge with a common direction to which all commit is the key.

7. Supervise temporary employees, outside contractors, and other service providers. The number of permanent employees will decline and of transient workers will increase.

8. Become familiar with the capital market, as leaders will spend more and more time attracting venture capital. Skills to communicate with analysts and investors are important.

9. Prepare for intensive enterprise integration due to globalization. Exercise "pattern recognition" to spot trends that may affect the company's future.

10. Acquire the following minimum survival skills for 21st century office workers:

    (a) Be good at something the world values. Keep up marketable skills to offer value above and beyond what other engineers and managers elsewhere can offer. These skills include

    - Leadership—technology planning, management of diversified employees, and perspectives with global orientation are essential

    - Technologies—include core, emerging, and innovative. There must be a love of technology present. Since technology changes everything, embracing it is essential.

    (b) Stay away from service technologies, which are readily outsourced. These include software programming, procedure or manual preparation, basic design assignments, customer support, product testing, experiments, statistical studies and analyses, and other rudimentary engineering jobs.

    (c) Also, stay away from those company functions which are being taken over by enterprise integration software systems (e.g., Oracle, Ariba, I2, Baan). Understand clearly what these software systems can do, and then focus your efforts on acquiring skills for the high-level work that such software systems cannot perform.

11. Create networks. Whom do you know? What do others know that you can use? It is extremely important to link with future project mates and peers who appreciate your talents and capabilities. Determine whom you can call upon to further your current and future projects and goals.

12. Nurture an entrepreneurial instinct. Seize opportunities for new projects and act on them.

13. Do marketing. "Early to bed, early to rise, work like hell, advertise!" Tell your story to let people know more about you; have a personal website; attend trade show presentations.

14. Grow better constantly. Manifest a passion for renewal.

14.7 CONTRIBUTIONS IN THE NEW MILLENNIUM

Significant changes are anticipated as a result of the transformation of old-economy (vertical) companies to knowledge-economy (virtual) companies.

Knowledge-economy companies rely more on brand, customers, and knowledge to compete and less on hard assets, such as the plants, equipment, real estate, and traditional distribution channels of the old-economy companies. Engineering managers will play important roles in this transition.

14.7.1 Technologies

Applying technological intuition to lead in business planning involves the deployment of emerging technologies (Spencer and Johnston 2003; Anonymous 2003).

Example 14.3.

Continuous improvement is a critical requirement for all companies, domestic or global. How can the process of continuous improvement be implemented in a global company?

Answer 14.3.

Continuous improvement helps companies to remain competitive in the marketplace. To manage the continuous improvement process, management should do the following:

A. Specify objectives in consultation with top management. Assign the responsibility of managing the global continuous improvement process to someone with visibility. Commit resources and establish a central office to coordinate the global continuous improvement efforts. The mission of this office must be communicated to all global employees.

B. Define specific goals in a number of areas, on the basis of inputs from various divisions, using standards derived by gleaning the available best practices in the industry.

C. Create a number of task forces by interviewing and selecting capable and devoted people who have expertise in diversified disciplines and who are from various operational units in different regions. Each task force should be empowered to pursue improvement ideas in a specified domain.

D. Hold teamwork training sessions for members of diversified cultural and technical backgrounds. Visit team members in various locations to establish contact, assure understanding, and build trust.

E. Set up a communications system (e.g., an intranet, videoconferences, regional meetings, and phone, fax, multimedia technologies) to enable members to interact constantly. Apply Web-based tools to foster close collaboration.

F. Create "suggestion box systems" at each site for members to solicit and obtain inputs from knowledgeable employees. All employees are encouraged to contribute new ideas for improvement.

G. Empower the task force teams to implement improvement ideas deemed useful and to apply resources made available from the central continuous improvement office.

H. Reward employees who suggested those creative ideas which produced positive results after implementation. Present awards in well-advertised meetings to promote the continuous improvement effort. Publish awards and the positive results in company newsletters and other suitable media to practice positive reinforcement.

I. Encourage the cross-pollination of ideas from global employees at various locations to take advantage of their diversified experience and viewpoints.

J. Summarize and publicize results on a regular basis.

14.7.2 Innovations

Applying innovations to shorten product development cycles, reduce time to market, cut costs, and add features will create competitive advantages (Cook 1998; Teague 1999; Zairi 1999). Companies that create new businesses follow a typical six-step process:

1. Define and develop core technologies to solidify company strengths (best practices, continuous improvement). Core technologies represent the foundation for corporate partnerships and alliances.

2. Create products and services built around core technologies to differentiate from others and to succeed in the marketplace.

3. Validate market acceptance of proposed product and service concepts.

4. Conduct technical feasibility tests to produce and deliver the proposed product or service. Focus on networked partner supply chains, distribution logistics, sales, and services.

5. Acquire money from inside the company or from venture funding.

6. Implement the approved business plan by systematically using improved best practices.

It is noteworthy that four of the preceding six steps need significant innovative input from engineers and engineering managers.

14.7.3 Value Addition to E-Transitions

Managers need to apply Web-based technologies to various corporate functions to improve efficiency; develop mass customization ("build-to-order") systems—using design, production, logistics, and service—to strengthen corporate competitiveness; and transform other corporate functions to add business values.

14.7.4 Customer and Knowledge

The management of customers, knowledge, and connections is of critical importance.

    A. Relationship Management. Customer relationship management involves analyzing data generated by call centers and other sources and devising programs to foster relationships with customers.

B. **Connectivity Creation.**   Connectivity development, maintenance between business partners (e.g., suppliers, alliance partners, employees, and customers), and contents management (e.g., software selection, hardware specification, system design, interface design, quality assurance, control of proprietary information, and others) are essential.

C. **Knowledge Management.**   "It is not who you know. It is what who you know knows." How can your knowledge and theirs be combined to create new win–win opportunities?

14.7.5 Social Responsibility and Leadership

Engineering managers should become visible leaders in society by speaking out more forcefully on broad issues of social interest on behalf of their corporations and professional associations. Examples of such broad issues include

1. Ethics and integrity of business and leadership
2. Environmentally friendly corporate policies to combat global warming
3. Globalization and the environment
4. National energy policies that focus on conservation or exploration
5. The question of taxation on the Internet
6. The patient's bill of rights in dealing with HMOs
7. Investment of social security benefits by individuals

Example 14.4.

In the new millennium, engineering managers will need to be prepared to lead and manage technology-intensive companies and industries. What specific technology management activities should engineering managers be concerned with?

Answer 14.4.

Generally speaking, there are eight activities engineering managers should pay attention to (National Research Council 1987):

1. Integrating technology into the overall strategic objectives of the firm (strategic planning)
2. Getting into and out of technologies effectually (gatekeeping)
3. Assessing and evaluating technology more effectively (making decisions and choices)
4. Conceiving better methods for transferring and assimilating new technology (managing technical knowledge)
5. Reducing new-product development time (creating supply chains and applying innovations)
6. Managing large, complex, and interdisciplinary or interorganizational projects, programs, and systems (leading cross-functional or global teams)
7. Managing the organization's internal use of technology (leading teams and managing projects)
8. Leveraging the effectiveness of technical professionals (managing knowledge workers)

Engineering managers focus not only on R&D management or management of technical professionals, but also on managing manufacturing and process technology, new-product development, and other technology-intensive functions in an organization. In addition, they strive to implement projects and programs correctly the first time and dedicate their efforts to continuous improvement. They are capable of commercializing technology products because they possess the required background and training in activity-based costing (cost accounting), NPV analysis (financial management), and customer relations management (marketing management). The critical roles of engineering managers are well recognized, as technology is of strategic importance to the national economy.

14.8 THE CHALLENGES AHEAD

This book is aimed at assisting engineering graduates in assuming leadership positions in enterprises of the 21st century. Many of these enterprises are affected by rapid changes in technology and the fast-paced advancement of globalization.

The first part of the book consists of the basic functions of engineering management, such as planning (Chapter 2), organizing (Chapter 3), leading (Chapter 4), and controlling (Chapter 5). These functions provide engineers and engineering managers with foundation skills to manage themselves, staff, teams, projects, technologies, and global issues of importance.

Best practices are emphasized as the pertinent standards for goal setting and performance measurement. Engineering managers solve problems and minimize conflicts to achieve the company's objectives. They use the Kepner–Tregoe method to make rational decisions and take lawful and ethical actions. They engage Monte Carlo methods to assess projects involving risks and uncertainties. They engage emerging technologies, motivate a professional workforce of diversified backgrounds, develop new generations of products and services in a timely manner, and constantly surpass the best practices in industry.

The roles of engineering managers in strategic planning, employee selection, team building, delegating, decision making, and managing creativity and innovation are explained. The augmentation of managerial competencies is emphasized.

The second part of the book covers the fundamentals of business, including cost accounting (Chapter 6), financial accounting and analysis (Chapter 7), managerial finance (chapter 8), and marketing management (Chapter 9). This part is written to enable engineers and engineering managers to acquire a broadened perspective of the company business and its stakeholders and to facilitate their interactions with peer groups and units.

These chapters also prepare engineering managers to make decisions related to cost, finance, products, service, and capital budgets. Discount cash flow and internal rate of return analyses are reviewed. These discussions are of critical importance, as decisions made during the product design phase typically determine up to 85 percent of the final costs of products. Additional demonstrations are presented of activity based costing (ABC), to define indirect costs related to products and services, and of economic value added (EVA), to determine the real profitability of an enterprise above and beyond the cost of capital deployed.

Also presented is capital formation through equity and debt financing. Resource allocation concepts based on adjusted present value (APV) for assets in place and option pricing for capital investment opportunities are addressed as well. By understanding the project evaluation criteria and the tools of financial analyses, engineers and engineering managers will be in a better position to secure project approvals. A critical step to refining technological projects is the acquisition and incorporation of customer feedback. For leaders and managers to lead, they must meet the major challenges that arise from the initiation, development, and implementation of major technological projects that contribute to the long-term profitability of the company.

The important roles and responsibilities of marketing in any profit-seeking enterprise are then introduced, along with the supporting contributions expected of engineering managers. Various progressive enterprises are increasingly concentrating on customer relationship management to grow their businesses. This customer orientation is expected to continue to serve as a key driving force for product design, project management, plant operation, manufacturing, customer service, and countless other engineering-centered activities.

The third part of the book addresses five major topics: engineers as managers and leaders (Chapter 10), ethics in engineering and business management (Chapter 11), Web-based enablers for engineering and management (Chapter 12), globalization (Chapter 13), and engineering management in the new millennium (Chapter 14). These discussions provide additional building blocks to enhance the preparation that engineers and engineering managers must undertake to assume technology leadership positions and to meet the challenges in the new millennium.

Engineers are known to possess strong skill sets that enable them to do extraordinarily well in certain types of managerial work. However, some of them may also exhibit weaknesses that prevent them from becoming effective leaders in engineering organizations or even from being able to survive as engineers in the industry. The expected norms of effective leaders are described. Steps enabling engineering managers to augment their leadership qualities and attune themselves to the value-centered business acumen are elaborated. Certain outlined steps should be of great value to those engineering managers who want to become better prepared to create new products or services based on technology, integrate technology into their organizations, and lead technology-based organizations.

A number of tried and true rules are included to serve as suitable guidelines for engineering managers to follow in becoming excellent leaders. Above all, engineering managers are expected to lead with a vision of how to apply company core competencies to create value, insights into how to seize opportunities offered by emerging technologies, and innovations in making products and services better, faster, and cheaper, so that they constantly improve customer satisfaction. The concepts of value addition, customer focus, time to market, mass customization, supply chains, enterprise resources integration, and others are also elucidated.

Although engineers are already known to be ranked high in trustworthiness and integrity (ahead of businessmen, bankers, certified public accountants, lawyers, and others), it is important for all engineers and engineering managers to remain vigilant in observing a code of ethics. Other topics related to ethics are discussed, with examples of several difficult ethical dilemmas.

The changes wrought by the Internet are transforming most aspects of company business, including information dissemination, product distribution, and customer service.

As processor design, software development, and transmission hardware technologies continue to advance, their roles in business will surely grow steadily and affect various functions of engineering management in the future. Progressive engineering managers need to know which Web-based enablers for engineering and management are currently available. They must identify which ones can be applied effectually to promote product customization, expedite new products to market, align supply chains, optimize inventory, foster team creativity and innovation, and enhance customer service. Presented in considerable length is a comprehensive set of Web-based tools related to product design, manufacturing, project management, procurement, plant operations, knowledge management, and supply-chain management.

Globalization expands the perspectives of engineers and engineering managers further with respect to divergence in culture, business practices, and value. Globalization is a major business trend that will affect innumerable enterprises in the coming decades. Engineers and engineering managers must become sensitized to the issues involved and prepare themselves to contribute to enterprises wishing to cash in on the new business opportunities offered in the emerging global markets. They need to be aware of the potential effects of job migration caused by globalization and take steps to prepare themselves to meet such challenges. The major hurdle for engineers and engineering managers to overcome is failing to create global technical alliances to take advantage of new technological and business opportunities.

Engineering management will face external challenges in the new millennium. What these specific challenges are, how engineering managers need to prepare to meet these challenges, and how to optimally make use of location-specific opportunities to create competitive advantages have been examined. Progressive companies change organizational structures, set up supply chains, expedite e-transformation, and implement advanced tools to serve customers better, cheaper, and faster.

Globalization is also expected to constantly evolve. The United Nations has predicted that, by the year 2020, three of the five biggest national economies will be located in Asia. There will certainly be winners and losers as businesses become more and more global. It is important for future engineering managers to explore prudent corporate strategies for engineering enterprises in the pursuit of globalization, while minimizing any detrimental impact on the environment and maintaining above-standard human rights and labor conditions.

As depicted in Figure 14.4, this book is intended to assist engineers and engineering managers in creating individual leadership pyramids that, when constantly nurtured and reinforced brick by brick, allow them to be recognized and remain visible as leaders for a long time to come.

How can engineering managers exercise their leadership roles to add value to the stakeholders of the company? The model depicted in Figure 14.5 illustrates a possibility. From left to right, engineering managers apply resources (displayed in the second column from the left), initiate specific processes (lined up in the third column from the left), create values (registered in the fourth column from the left), and eventually benefit specific stakeholders (enumerated in the right-hand column).

There are three specific resources engineering managers may activate: product design, ERP and other Web-based tools, emerging technologies, and vision and innovations.

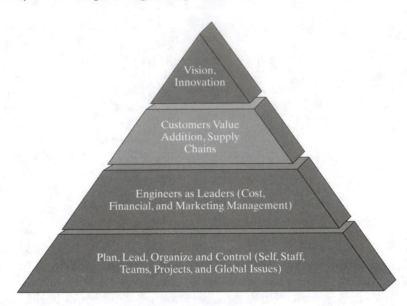

Figure 14.4 Leadership pyramid.

Engineering managers utilize product design, ERP, and other Web-based tools to manage global supply chains, strengthen products, effectuate the company's e-transformation, achieve optimization of global resources, and supervise global projects and teams. They use emerging technologies to develop products, promote the e-transformation of the company, and achieve best practices with respect to the environment. They lead with vision and innovations to spearhead product design and development, initiate e-transformation, optimize the use of global resources, manage global projects and teams, and create best practices with respect to the environment, ethics, and other community values.

The successful management of global supply chains leads to better, cheaper, and faster products and services and more effective collaboration with suppliers. The process of developing products and services with the help of Web-based tools or other emerging technological aids leads to better, cheaper, and faster products and services, more effective collaboration with suppliers, and a stronger global competitive stance for the company. E-transformation of the company brings about to swifter decision making and a more robust global competitiveness. Optimization of global resources leads to stronger global competitiveness, improved employee satisfaction, and surpassing innovations. Moreover, as the company follows best practices with respect to ethics and the environment, it projects a superior image of corporate citizenship to the public.

The creation of better, cheaper, and faster products and services adds value to customers. Improved collaboration with suppliers makes suppliers happy. Better and faster company decision making pleases the shareholders. Stronger global competitiveness adds value to shareholders. Improved employee satisfaction and cutting-edge innovations add value to both shareholders and employees. Excellent corporate citizenship favors the communities as well as the employees.

To confront the management challenges of the 21st century, engineering managers need to manage from the inside as well as from the outside, to lead from the present to

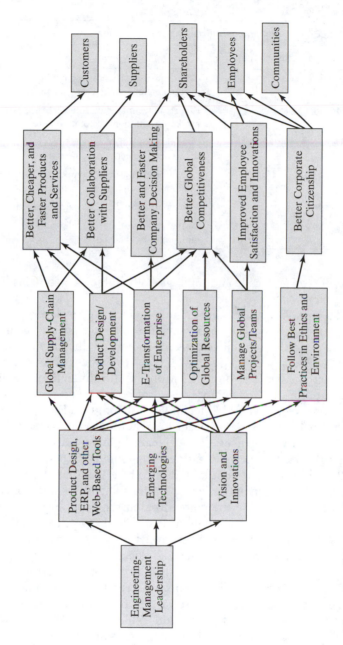

Figure 14.5 Value-addition model for engineering managers.

TABLE 14.3   Management Challenges for Engineers

Manage/Lead/Act/Think	Focuses
Inside	Core competencies, cost and quality control, production and engineering functions
Outside	Emerging technologies, supply chains, market orientation, customer relationship
Today	Organize, control, plan, lead teams and projects; do things right
Tomorrow	New projects, new core competencies, new products, new markets; do the right things
Local	Implementation details, local adjustments
Global	Global resources, global scale and scope, global mindset and savvy

the future, and to act locally and think globally. Table 14.3 contains descriptions of this alternative viewpoint.

On the *inside*, engineers and engineering managers plan, organize, lead, and control to implement projects and programs. They manage people, technologies, and other resources to add value to their employers. They strengthen the company's core competencies and come up with products with features that customers want. They effectively define (by activity based costing and Monte Carlo simulations), monitor, and control costs. They appraise the company's financial position and seize the right moments to initiate major projects with high technological contents. These projects are supported by rigorous financial analyses in order to meet tough corporate evaluation criteria.

On the *outside*, engineers and engineering managers keep abreast of emerging technologies and screen new technologies that might affect the company's products or services. They proactively identify and introduce Web-based tools related to product design, project management, plant operations, facility maintenance, and knowledge management to streamline the company's current operations. They define the best practices in the industry, emulate them as standards by which to evaluate their own in-house practices, and relentlessly strive to surpass these best practices. They look for potential supply-chain partners whose alliances could create competitive advantages for their employers in production, distribution, product customization, or after-sale service. They are sensitive to their constant need to improve the management of customer relationships. Through the use of web portals and other current technologies, they meet this need. They strive to add value to all stakeholders—customers, employees, suppliers, investors, and the communities in which the company operates.

Example 14.5.

In the new millennium, innovative ideas, rather than physical assets, will enable companies to compete efficaciously in a global-knowledge economy. Usually, innovative ideas come from knowledge workers who are typically inventive, independent, and mobile. No single company is capable of "chaining" down these workers, as they are happy to be there, but ready to move on at any time.

It is likely to be a major challenge for engineering or technology managers to foster innovation on a continued basis in such an environment. What might be a good strategy for engineering and technology managers to adopt in order to secure a constant flow of innovations potent enough to sustain the relative competitiveness of their employers?

Answer 14.5.

During the last century, a number of well-known companies (IBM, AT&T, Dupont, GE, Merck and others) achieved remarkable business success by emphasizing R&D in-house and innovation on the inside. They proudly advertised the number of U.S. patents they received per year as an indication of their inventive power. They kept a large number of experts on their payroll to foster innovations. Many of these giants have since left their historical mission of inventive discovery. Some have also abandoned the past practice of not sharing with others those inventions which did not fit their respective corporate strategies at the time.

Companies in the knowledge economy have implemented a flexible technology strategy with great success. Known examples include Microsoft, Cisco, Dell, and Pfizer. Because skilled workers are mobile, companies can no longer count on in-depth development of innovations on the inside. In order to secure a constant inflow of creative ideas, they pursue outside innovations (e.g., by acquisition, joint venture, or contract research) deemed useful to foster their corporate objectives. The emphasis has been shifted from in-depth innovation within a discipline to innovation with breadth and integration across disciplines.

Procter & Gamble (P&G) is known to be an aggressive acquirer of creative ideas from the outside. In 2001, 10 percent of P&G products came from outside sources, and this percentage is expected to rise to 50 percent by 2006. P&G has also decided to make a patented technology available to outsiders, including competitors, if it is not used by at least one internal business unit within three years (Chesbrough 2003).

To meet the new challenge of creating a constant flow of creative ideas, engineering and technology managers must scan promising innovations on the outside (universities, start-ups, competitors, and others) and integrate them for profitable internal applications. Any inside innovations that do not conform to the corporate objectives are to be aggressively marketed to outside companies to generate licensing revenues.

For the *present*, engineers and engineering managers focus on keeping the company smoothly operating by "doing things right." They pay attention to details. They introduce a balanced scorecard to make sure that both financial and nonfinancial metrics are selected and deployed to monitor and evaluate the company's performance. They attempt to continuously augment current company operations. They take care of assignments (e.g., cost control, waste elimination, etc.) that should be accomplished for the company to achieve business success in the short term.

For the *future*, engineers and engineering managers seek e-transformation opportunities to create corporate competencies in the long term. These are profit-producing opportunities that may be created by significantly enhancing the value of the company's products to customers through, for example, distribution, price, service, features, and ordering processes. Managers develop and introduce new products in a timely manner to ensure sustainable profitability for the company in the future. They envisage a vision for the future, contribute to new company strategies related to technology, and assist company management in defining "the right things to do."

Several U.S. physicists determined in 2003 that the strength of the earth's magnetic field has decreased 10 percent over the past 150 years. They estimated that the field could vanish altogether 1500 years from now. It is amazing how far forward they can cast scientific predictions. Yet, engineering managers need to be more practical and concentrate on the next 1.5 to 15 years.

Example 14.6.

Forecasting future market conditions and technologies is a difficult, but necessary, skill for companies striving to sustain business success. Looking out for emerging technologies should be the primary role of engineering and technology managers.

What might be a good strategy for engineering and technology managers to become sensitized to forecasting technology and scanning emerging technologies so that they fulfill their important role of serving as technology "gatekeepers"?

Answer 14.6.

Different engineering and technology managers will have different preferences in fulfilling this important role of forecasting. One possibility is to adopt the following logical sequence of steps:

1. Compose a "wish list" of technologies that would make the company's current products cheaper, faster, and better. Define desirable new product features based on customer inputs and the technologies required for their development. Define new product concepts and the requisite technologies that might be compatible with the current product lines marketed by the company.
2. Understand some of the emerging technologies noted in the literature (for example, see Table 2.2).
3. Determine the useful technologies that might be available during the next 5 to 10 years to support the current products, product enrichments, or new product concepts.
4. Assess the development activities associated with these useful technologies in universities, start-ups, technology incubator firms, contract research companies, or other organizations, both domestic and global, to gauge their quality and readiness for commercialization.
5. Make specific recommendations in a timely manner to secure the supply of such new technologies for enhancing the commercial success of the company's products.

At the *local* level, engineers and engineering managers seek to best utilize the resources available (for example, people, technology, and business relationships) to achieve the company's objectives. They adjust to local conditions and take lawful, ethical, and proper action to discharge their daily responsibilities. They maintain their local networks of professional talents and business relationships to heighten the company's productivity. They communicate their experience and preserve lessons learned so that others at different sites within the company may benefit.

At the *global* level, engineering managers are sensitive in their pursuit of the optimal use of location-based resources to realize global economies of scale and scope and to derive both cost and technology advantages for their employers. They create global networks of professional talents and business relationships and exploit innovative business opportunities. As companies pursue globalization over time, these professionals acquire a global mindset, become globally business savvy, and ready themselves to exercise leadership roles in international settings.

Example 14.7.

"Think globally and act locally" has been the general guideline offered to managers at the headquarters of global companies that seek to achieve success in the global markets. The logic is rather compelling.

For those managers of global companies that operate in local regions or markets, perhaps the guideline should be "Think locally and act globally."

What is your opinion on these guidelines?

Answer 14.7.

The principal objective of requiring headquarters managers of global companies to "think globally and act locally" is to make sure that the company's products and services are sufficiently adjusted to the needs of local markets, while enjoying the economies of scale advantages of being global.

Numerous local managers of global companies have created innovative strategies and achieved remarkable success, because of their understanding of the culture and customs in local markets. Oftentimes, the same insight and innovative strategies need only be applied to other regions, with only minor modifications, for the headquarters managers of the global companies to realize the economies-of-scope advantages on a global basis.

Gurcharan Das (1993) spoke in favor of the concept of "thinking locally and acting globally," as a result of his personal experience as a local manager of Procter & Gamble in India.

Figure 14.6 elucidates the interrelationship between these six-dimensional challenges.

Figure 14.6   Six-dimensional management challenges for engineers and engineering managers.

Because many companies are affected by the rapid advancement of technology and the fast-paced advancement of globalization, the new millennium both creates ample opportunities for, and poses new challenges to, engineers and engineering managers (Kouzes and Poser 2002; Drucker 1999; Deep and Sussman 2000; Palus and Horth 2002; Jacobson with Setterholm and Vollum 2002). Those engineers and engineering managers who capture these new opportunities and meet these new challenges will be rewarded.

14.9 REFERENCES

Anonymous. 2000. "The 21st Century Corporation." *Business Week*, August 28, pp. 75–212.

Anonymous. 2003. *Best Practices: Ideas and Insights from the World's Foremost Business Thinkers*. Cambridge, MA: Perseus Publishing.

Bahrami, H. 1992. "The Emerging Flexible Organization: Perspectives from Silicon Valley." *California Management Review*, Summer, pp. 32–52.

Buss, D. D. 1999. "Embracing Speed." *Nation's Business*, Vol. 89, June, p. 12.

Chesbrough, H. 2003. "The New Business Logic of Open Innovation." *Strategy & Innovation*, July 1.

Cook, P. 1998. *Best Practice Creativity*. Brookfield, VT: Gower Publishing.

Corey, E. R. 1999. "A Note on Case Learning." *Harvard Business School Note*, No. 9-899-105, April 27, 1999.

Corta, J. W. 2000. *Twenty-first Century Business*. Upper Saddle River, NJ: Prentice Hall.

Cruickshank, H. 2003. "The Changing Role of Engineers." *Engineering Management*, February.

Das, G. 1993. "Local Memoirs of a Global Manager." *Harvard Business Review*, March–April.

Deep, S. D. and L. Sussman. 2000. *Act on It: Solving 101 of the Toughest Management Challenges*. Cambridge, MA: Perseus Publishing.

Deetz, S., S. J. Tracy, and J. L. Simpson. 2000. *Leading Organizations through Transition: Communications and Cultural Change*. Thousand Oaks, CA: Sage Publications.

DePalma, D. A. 2002. *Business without Borders: A Strategy Guide to Global Marketing*. New York: John Wiley.

Drucker, P. F. 1999. *Management Challenges for the 21st Century*. New York: Harper Collins Books.

Heskett, J. L., T. O. Jones, G. Loveman, W. E. Sasser, Jr., and L. A. Schlesinger. 1994. "Putting the Service-Profit Chain to Work." *Harvard Business Review*, March.

Hiebeler, R., T. B. Kelly, and C. Ketterman. 1998. *Best Practices: Building Your Business with Customer-Focused Solutions*. New York: Simon & Schuster.

Hines, P. 2000. *Value Stream Management: Strategy and Excellence in the Supply Chain*. Upper Saddle River, NJ: Financial Times/Prentice Hall.

Jacobson, R. with K. Setterholm and J. Vollum. 2000. *Leading for a Change: How to Master Five Challenges Faced by Every Leader*. Boston: Butterworth-Heinemann.

Johansson, J. K. 2000. *Global Marketing: Foreign Entry, Local Marketing, and Global Management*. 2d ed. Boston: Irwin/McGraw Hill.

Kamp, D. 1999. *The 21st Century Manager: Future-Focused Skills for the Next Millennium*. London, UK: Kogan Page, 2d ed.

Kouzes J. M. and B. Z. Poser. 2002. *The Leadership Challenge*. 3d ed. San Francisco: Jossey-Bass.

Larson, M. and D. Lundberg. 1998. *The Transparent Market: Management Challenges in the Electronic Age*. New York: St. Martin's Press.

Livers, A. B. and K. A. Caver. 2003. *Leading in Black and White:* Working Across the Racial Divide in Corporate America, San Francisco: Jossey-Bass.

McGrath, M. E. and R. W. Hoole. 1992. "Manufacturing's New Economies of Scale." *Harvard Business Review*, May–June, pp. 94–102.

Moschella, D. 2003. *Customer Driven IT: How Users Are Shaping Technology Industry Growth.* Boston: Harvard Business School Press.

Myerson, J. M. (Editor). 2002. *Enterprise Systems Integration.* Boca Raton, FL: Auerbach.

National Research Council. 1987. *Management of Technology: The Hidden Advantage.* Washington, DC: National Academy Press.

Palus, C. J. and D. M. Horth. 2002. *The Leader's Edge: Six Creative Competencies for Navigating Complex Challenges.* San Francisco: Jossey-Bass.

Segil L., M. Goldsmith, and J. Belasco. 2003. *Partnering: The New Face of Leadership.* New York: AMACOM.

Spencer, R. H. and R. P. Johnston. 2003. *Technology Best Practices.* Hoboken, NJ: John Wiley.

Stettner, M. 2000. *Skills for New Managers.* New York: McGraw-Hill.

Schmidt, J. A. 1999. "Corporate Excellence in the New Millennium." *Journal of Business Strategy*, November, p. 39.

Schmitt, B. 2003. *Customer Experience Management: A Revolutionary Approach to Connecting with Your Customers.* New York: John Wiley.

Teague, P. E. 1999. "Surprise, Innovation Is Key to Success." *Design News*, April 5, p. 14.

United Nations. 2002. *World Summit on Sustainable Development.* Johannesberg, South Africa: Heinrich Böll Foundation.

Venkatesan, R. 1992. "Strategic Sourcing: To Make or Not to Make." *Harvard Business Review*, November–December, pp. 98–107.

Weiss, A. 2000. *Good Enough Isn't Enough: Nine Challenges for Companies That Choose to be Great.* New York: AMACOM.

Zairi, M. 1999. *Best Practices: Process Innovation Management.* Boston: Butterworth-Heinemann.

14.10 QUESTIONS

14.1 Sustainable development refers to work that simultaneously satisfies economical, social, and environmental requirements (United Nations, 2002). It is self-evident that work must be economically viable so that customers are willing to pay for the work supplied. Work must also be safe and otherwise socially compatible. Furthermore, work needs to be environmentally acceptable in that harmful discharges are minimized, wastes are decreased, material and energy resources are conserved, and any other detrimental impact on the environment is minimized.

Some academicians suggest that it is the engineer's responsibility to attain the ideals of sustainable development. They view it as the major challenge facing engineers in the future (Cruickshank 2003). Do you agree with this notion? Why or why not?

14.2 Leading technological innovation will be a major challenge for engineering managers in the new millennium. What are some of the success factors for technological innovation?

14.3 The new millennium is expected to see continued changes in communications technologies, business practices, worker diversity, customer empowerment, and marketplace conditions. Name a few leadership qualities that are deemed essential for engineering managers to achieve success in the new millennium.

14.4 U.S. productivity has improved noticeably in recent years, averaging 4 to 5 percent per year, while the U.S. economy grew by only 3.5 percent. The gain in productivity was due, in large part, to the use of technology, in addition to longer working hours by those who are lucky enough to have jobs. According to the Economic Policy Institute, the average U.S. worker has added 199 hours to a year since 1973. The United States achieved the per-hour productivity of $32, compared with $38 for Norway, and $34 for Belgium. In other words, U.S. workers are simply working longer, not necessarily better or smarter. They take less annual vacation time (only 10.2 days, on average), compared with 30 days in France and in Germany.

At the same time, a large number of U.S. companies are aggressively outsourcing work to low-wage countries, such as China, India, the Philippines, and Mexico. A 2003 study released by the University of California at Berkeley indicates that as many as 14 million U.S. service jobs are in danger of being shipped overseas.

Who is responsible for this peculiar position that U.S. workers are being forced into? How can U.S. companies meet the new challenge of improving the quality of life for their workers, without sacrificing the companies' relative competitiveness in the marketplace?

14.5 In the new millennium, speed to market will be a major harbinger of competitiveness. "The early bird gets the worm," as the saying goes. How can engineering managers meet this challenge by leading their companies to be the first movers?

Appendix

Selected Cases of Engineering and Business Management

The case method is well established and widely practiced in management schools. Engineers and engineering managers are advised to consult a recent publication regarding the advantages of learning from cases (*Source*: E. Raymond Corey. "A Note on Case Learning." *Harvard Business School Note*, No. 9-899-105, April 27, 1999.)

Samples of engineering and business management cases are selected to assist engineers and engineering managers in gaining management and business perspectives. These cases, illustrating the complex engineering and management issues involved, may be studied to augment the topics covered in this book. The author has used all of them over a period of time with success. Engineers and engineering managers are strongly encouraged to study these and other cases to derive useful lessons from them.

Class surveys indicate that engineering students generally like case discussions, as they enhance critical thinking and stimulate active participation. They like the fact that well-organized case studies are the next-best avenues, besides working in the right industrial settings, for them to acquire useful experience in dealing with real-world management issues. They understand also that the principle of "more in and more out" applies here, in that the more effort they put into preparing themselves for the class discussions, the more valuable insights they will get out of the cases discussed.

It is generally advisable to specify a list of questions for each case ahead of time for students to focus their preparation on. It is equally important to ensure that the discussions are not dominated by a few students in the class, that most students are encouraged to speak up, and that the discussions are driven by thought-provoking questions. Each case takes, on average, the class time of about 60 to 90 minutes. Usually, students like the instructor to give them a written case summary that captures the salient points of the case discussion. Past experience indicates that asking students to summarize in writing what lessons they have learned from each case, without reiterating the case details, helps them to internalize the "take-away" messages.

Engineers and engineering managers may purchase case materials from the publisher, Harvard Business School Publishing Company (*http://www.hbsp.harvard.edu*). Any academic educator may register at the same website of Harvard Business Online for Educators free of charge. Once registered, the educator may download a free copy of any of these cases for evaluation purposes. To use these cases in class, students need to purchase the original copies, possibly at an academic discount price. For information regarding order processing, contact 1–800–545–7685 or *http://www.custserv@hbsp.harvard.edu*.

TABLE A.1   Classification of Engineering and Business Management Cases

Case Reference	Subjects Covered	Primary Discipline	Products Involved
Case No. 1	Implementation of e-commerce, internet, information technologies, computer networks	Management of information system	Industrial gases (oxygen, nitrogen, argon)
Case No. 2	Marketing of new products, innovations, manufacturing product development, product life cycle	Operations management	Mass spectrometers
Case No. 3	Product positioning based on exterior styling, computer-aided technologies, organizational and process changes	Operations management	Automobiles
Case No. 4	Product development to overcome negative publicity, pricing strategy, business strategy, product management	Marketing	Oilfield pumping motors
Case No. 5	Marketing fundamentals of industrial products (pricing, product development, public relations, sales, target markets)	Marketing	Equipment to dispense adhesives
Case No. 6	Manufacturing cost accounting (budgeting, accounting, financial management, operations management)	Accounting and control	Manufactured goods
Case No. 7	Strategic planning for the next two years based on CEO's current performance (competitive strategy, general management, industrial analysis, business policy)	General management	Machinery products
Case No. 8	Distribution channels and conflicts (e-commerce, Internet, marketing, pricing, product management, retailing)	Marketing	Printers
Case No. 9	Operations management (considering both technical and human issues in a changing environment)	Operations management	Automotive components

(*Continued*)

TABLE A.1   (Continued)

Case Reference	Subjects Covered	Primary Discipline	Products Involved
Case No. 10	Organizational design at various stages of company development (entrepreneurial management, divisional structure)	Human resources management	Technology services
Case No. 11	Management of teams (group behavior, innovations, product development, project management, collaboration)	Organizational behavior and leadership	Various technical products
Case No. 12	Product development and business teams (integration of team capabilities, marketing, manufacturing, operations management)	Operations management	Winchester disk drive
Case No. 13	Strategic decision making involving business or production expansion (trade-offs of options, competitive strategies, organizational behavior)	Competitive strategy	Automobiles
Case No. 14	Problem solving related to supply-chain difficulties (flexibility, just-in-time delivery, production planning and control, operations management).	Operations management	Commercial motors
Case No. 15	Activity based Costing (cost allocation, pricing, profitability analysis, cost analysis)	Accounting and control	Brass valves, pumps, and flow controllers
Case No. 16	Capital budgeting decision related to an enterprise resource planning system (cash flows, working capital, NPV analysis, financial analysis)	Finance	Consumer appliances
Case No. 17	Product and brand management (brand image, product positioning, international competition, marketing strategy)	Marketing	Power tools
Case No. 18	Niche marketing (channel conflicts, customer segmentation, product and service offering, competitive strategy)	Operations management	Computers
Case No. 19	Global challenges (organizational behavior, global leadership, management of change, manufacturing operations)	General management leadership, strategic planning	Automobiles
Case No. 20	Start-up challenges (global strategy, alliances, international businesses)	Competitive strategy	Semiconductors
Case No. 21	Cooperative strategy, distribution channels, high-technology products, strategic marketing plan	Entrepreneurship	Software products
Case No. 22	Resource allocation decisions and interactions among groups	Team operations and corporate governance	Plastics products
Case No. 23	Supply strategies for global markets and decision making	General management	Medical products

Twenty-three cases are included here. (See Table A.1.) Three of these cases (No. 10, No. 15, and No. 17) are Harvard Business School's best-selling cases. Table A.1 groups the cases according to the main subjects covered in this book.

The set of cases listed in Table A.1 covers technology products (both industrial and consumer) of large and small firms engaged in domestic and international businesses. The cases outlined in Table A.2 address various engineering management issues discussed in this book. Table A.2 relates these cases to the chapters involved.

TABLE A.2 Chapter Coverage Corresponding to Cases

Chapter Number	Chapter Coverage	Applicable Cases
1	General Introduction	
2	Planning	Case No. 7 (Strategic planning)
		Case No. 19 (Planning for company revival)
3	Organizing	Case No. 10 (Organizational design)
4	Leading	Cases No. 11 and No. 12 (Teams)
		Case No. 13 (Decision making using Kepner–Tregoe method)
		Case No. 19 (Leadership)
		Case No. 22 (Decision making in groups)
5	Controlling	Case No. 9 (Operations management)
		Case No. 14 (Problem solving)
6	Cost Accounting	Case No. 6 (Manufacturing, accounting)
		Case No. 15 (Activity-Based Costing)
7	Financial Analysis	Case No. 16 (Income statement and balance sheet)
8	Financial Management	Case No. 16 (Capital budgeting)
9	Marketing Management	Case No. 17 (Product/brand management)
		Case No. 3 (Product positioning)
		Case No. 2 (Marketing of new products)
		Case No. 5 (Marketing management)
		Case No. 18 (Niche marketing)
		Case No. 8 (Distribution channels)
		Case No. 21 (Strategic marketing)
10	Engineers as Managers and Leaders	Case No. 1 (e-transformation)
11	Ethics	
12	Web-Based Enablers	Case No. 4 (Product development)
13	Globalization	Case No. 20 (Global business)
		Case No. 23 (Global supply strategy)
14	Engineering Management in the New Millennium	Case No. 19 (New challenges)

LIST OF CASES

Case No. 1   F. Warren McFarian and Melissa Dally. "Electronic Commerce at Air Products." *Harvard Business School Case*, No. 9-399-035, August 19, 1989.

Case No. 2   Steven C. Wheelwright and Sandeep Dugal. "Associated Instruments Corporation: Analytical Instruments Division." *Harvard Business School Case*, No. 9-689-052, March 16, 1989.

Case No. 3   Stefan Thomke and Ashok Nimgade. "BMW AG: The Digital Auto Project (A)." *Harvard Business School Case*, No. 9-699-044, November 1, 2001.

Case No. 4   E. Raymond Corey. "Dominion Motors and Controls Limited." *Harvard Business School Case*, No. 9-589-115, September 23, 1992.

Case No. 5   John A. Quelch. "Loctite Corporation: Industrial Products Group." *Harvard Business School Case*, No. 9-581-066, Rev. July 15, 1991.

Case No. 6   William J. Bruns, Jr. "Monterrey Manufacturing Company." *Harvard Business School Case*, No. 9-197-023, Rev. December 18, 1996.

Case No. 7   Richard G. Hamermesh and Evelyn T. Christiansen. "International Harvester (B-1)." *Harvard Business School Case*, No. 9-381-053, June 30, 1986.

Case No. 8   Rajiv Lal, Edith D. Prescott, and Kirthi Kalynam. "HP Consumer Products Organization: Distributing Printers via the Internet." *Harvard Business School Case*, No. 9-500-021, Rev. March 22, 2000.

Case No. 9   Amy C. Edmondson and Mikelle F. Eastley. "GM Powertrain." *Harvard Business School Case*, No. 9-698-008, April 26, 2000.

Case No. 10   Nitin Nohria and Julie Gladstone, "Appex Corp." *Harvard Business School Case*, No. 9-491-082, February 10, 1992.

Case No. 11   Anne Donnellon and Joshua D. Margolis. "Mod IV Product Development Team." *Harvard Business School Case*, No. 9-491-030, Rev. March 5, 1991.

Case No. 12   Steven C. Wheelwright and Clayton Christensen. "Quantum Corp.—Business and Product Teams." *Harvard Business School Case*, No. 9-692-023, February 17, 1992.

Case No. 13   Anita McGahan and Greg Keller. "Saturn Corp.: Module II Decision." *Harvard Business School Case*, No. 9-795-011, August 18, 1994.

Case No. 14   David Upton and Andrew Matherson. "EG&G Rotron Division." *Harvard Business School Case*, No. 9-695-037, April 14, 1997.

Case No. 15   William J. Bruns, Jr. "Destin Brass Products Co." *Harvard Business School Case*, No. 9-190-089, April 24, 1997.

Case No. 16   Richard S. Rubak, Sudhaker Balachadran, and Aldo Sesia. "Whirlpool Europe." *Harvard Business School Case*, No. 9-202-017, November 26, 2002.

Case No. 17 Robert J. Dolan. "Black and Decker Corporation (A): Power Tools Division." *Harvard Business School Case*, No. 9-595-057, March 30, 2001.

Case No. 18 Frances X. Fei, Youngme Moon, and Hanna Rodriguez-Farrar. "Gateway: Moving Beyond the Box." *Harvard Business School Case*, No. 9-601-038, May 9, 2002.

Case No. 19 Michael Y. Yoshino and Masako Egawa. "Nissan Motors Co. Ltd.—2002." *Harvard Business School Case*, No. 9-303-042, December 9, 2002.

Case No. 20 George Foster, Christopher S. Flanagan, Paul L. Wattis, and Phillis Wattis. "NetLogic Microsystems." *Stanford University Case*, No. E94, June 28, 2001.

Case No. 21 Andrian B. Ryans. "FastLane Technologies." *Richard Ivey School of Business, University of Western Ontario Case*, No. 98A006, November 10, 1999.

Case No. 22 Anonymous. "Peterson Industries." *Harvard Business School Case*, No. 9-396-182, February 12, 1996.

Case No. 23 Tarun Khanna. "General Electric Medical Systems." *Harvard Business School Case*, No. 9-702-428, February 26, 2003.

Index